D0204200

Wan[illegible]
91 [illegible]
White Plains,
NY 10606

Sept. 1990

Springer Texts in Statistics

Advisors:
Stephen Fienberg   Ingram Olkin

Springer Texts in Statistics

Alfred	Elements of Statistics for the Life and Social Sciences
Blom	Probability and Statistics: Theory and Applications
Chow and Teicher	Probability Theory: Independence, Interchangeability, Martingales. Second Edition
Christensen	Plane Answers to Complex Questions: The Theory of Linear Models
du Toit, Steyn and Strumpf	Graphical Exploratory Data Analysis
Kalbfleisch	Probability and Statistical Inference: Volume 1: Probability. Second Edition
Kalbfleisch	Probability and Statistical Inference: Volume 2: Statistical Inference. Second Edition
Keyfitz	Applied Mathematical Demography. Second Edition
Kiefer	Introduction to Statistical Inference
Kokoska and Nevison	Statistical Tables and Formulae
Madansky	Prescriptions for Working Statisticians
McPherson	Statistics in Scientific Investigation: Basis, Application, and Interpretation
Nguyen and Rogers	Fundamentals of Mathematical Statistics: Volume I: Probability for Statistics
Nguyen and Rogers	Fundamentals of Mathematical Statistics: Volume II: Statistical Inference
Peters	Counting for Something: Statistical Principles and Personalities

(continued after index)

Glen McPherson

Statistics in Scientific Investigation

Its Basis, Application, and Interpretation

With 58 Illustrations

Springer-Verlag
New York Berlin Heidelberg
London Paris Tokyo Hong Kong

Glen McPherson
Department of Mathematics
The University of Tasmania
Hobart, Tasmania 7001
Australia

E-Mail Address: mcphers@hilbert.maths.utas.oz.au

Editorial Board

Stephen Fienberg
Department of Statistics
Carnegie-Mellon University
Pittsburgh, PA 15213
USA

Ingram Olkin
Department of Statistics
Stanford University
Stanford, CA 94305
USA

Mathematics Subject Classification (1980): 62-01, 62-07, 62P10, 62P15, 62P25

McPherson, Glen.
Statistics in scientific investigation : its basis, application, and interpretation / Glen McPherson.
p. cm. — (Springer texts in statistics)
1. Research—Statistical methods. 2. Science—Statistical methods. 3. Statistics. I. Title. II. Series.
Q180.55.S7M36 1990
507.2—dc20 89-21711

Printed on acid-free paper.

© 1990 Springer-Verlag New York Inc.
All rights reserved. This work may not be translated or copied in whole or in part without the written permission of the publisher (Springer-Verlag New York, Inc., 175 Fifth Avenue, New York, NY 10010, USA), except for brief excerpts in connection with reviews or scholarly analysis. Use in connection with any form of information storage and retrieval, electronic adaptation, computer software, or by similar or dissimilar methodology now known or hereafter developed is forbidden.
The use of general descriptive names, trade names, trademarks, etc., in this publication, even if the former are not especially identified, is not to be taken as a sign that such names, as understood by the Trade Marks and Merchandise Marks Act, may accordingly be used freely by anyone.

Camera-ready copy prepared using LaTEX.
Printed and bound by R.R. Donnelley & Sons, Harrisonburg, Virginia.
Printed in the United States of America.

9 8 7 6 5 4 3 2 1

ISBN 0-387-97137-8 Springer-Verlag New York Berlin Heidelberg
ISBN 3-540-97137-8 Springer-Verlag Berlin Heidelberg New York

Affectionately dedicated to my family,
Wendy, Ewen, and Meegan

Contents

Preface

In this book I have taken on the challenge of providing an insight into Statistics and a blueprint for statistical application for a wide audience. For students in the sciences and related professional areas and for researchers who may need to apply Statistics in the course of scientific experimentation, the development emphasizes the manner in which Statistics fits into the framework of the scientific method. Mathematics students will find a unified, but non-mathematical structure for Statistics which can provide the motivation for the theoretical development found in standard texts on theoretical Statistics. For statisticians and students of Statistics, the ideas contained in the book and their manner of development may aid in the development of better communications between scientists and statisticians.

The demands made of readers are twofold:

- a minimal mathematical prerequisite which is simply an ability to comprehend formulae containing mathematical variables, such as those derived from a high school course in algebra or the equivalent;
- a grasp of the process of scientific modeling which comes with either experience in scientific experimentation or practice with solving mathematical problems.

The base on which this book is developed differs from that found in the myriad of standard introductory books on "Applied Statistics." The common approach takes what might be termed the statistician's view of Statistics. The reader is presented with the tools of Statistics and applications of those tools based on a methodological taxonomy. There is a danger with this approach that the essential structure and function of Statistics will not be grasped, leading to misuse of Statistics. In books divided on methodological grounds, the opportunity for comparing alternative methodologies is diminished, lessening the guidance in method selection.

The opening chapters of this book provide a users' view of Statistics in the form of a unified and cohesive framework for Statistics on which to hang the theory and applications that follow. The role of probability, examples of statistical models, and the general theoretical constructs are fitted within the basic structure. When considering applications of Statistics, grouping is based on experimental questions rather than methodological considerations.

The material in this book is a portion of a comprehensive structure of Statistics which I have constructed during my twenty years of academic life as a teacher and consultant. The remaining material is to be published in two more books that are in preparation. The structure has been guided by two basic principles. They are:

- a need for a simple but comprehensive view of Statistics; and
- the requirement that any description should be understandable by readers in their own *language*.

The *comprehensive* aspect is difficult to satisfy because there are a number of schools of Statistics. The contents of this book naturally reflect my leanings, but I believe that a blend of different approaches is optimal and I have attempted to develop a view of Statistics which presents the several approaches as variations on the one theme rather than as antagonists.

Since my aim is to explain Statistics in the language of the experimenter, I make many references to real life experiments. With few exceptions these are based on data from my own consulting work, although in some cases I have used a little 'statistical' license because the experimenter did not wish to have the material published; because I was critical of the experimental practice or the statistical analysis employed; or because a modification of question or data suited my needs. I have been forced to simplify or vary the scientific explanation of many of the examples in order that the statistical point is not submerged in the description. I apologize in advance for the scientific incompleteness of the descriptions, and urge that the conclusions and interpretations be considered only in the context of examples of statistical applications. The use of real life examples is necessary to establish the importance of experience in any application of Statistics and to emphasize that real life applications of Statistics involve more than a routine process of putting numbers into formulae and mechanically drawing conclusions.

Deciding which topics to include and the depth to which chosen topics should be covered has been one of the most difficult aspects in the preparation of the book. I am conscious of the fact that others would have chosen differently and I would appreciate comments from readers as to areas which they believe should be included, expanded, or perhaps excluded. While I have made every endeavor to make the book error free, I am only too aware that in a book of this size and breadth there will be typing and possibly statistical errors. While I acknowledge assistance of others in the preparation of the book, I take sole responsibility for these errors.

Computational Considerations

Computers are now an integral part of statistical analysis. The presentation of methods in this book is based on the assumption that readers have access to at least one general purpose statistical package. In general, methods have been restricted to those that are widely available. Only where the simplicity of methods makes the use of electronic calculators feasible, or where computational facilities may not be readily available, are computational details described.

For the various methods and examples in the book, the relevant sets of commands for programming in MINITAB and SAS are available on a set of disks together with the data from all tables, examples, and problems in the book. These may be obtained from the author at the address provided on the copyright page.

Exercises

A selection of exercises is provided to accompany most chapters. The exercises are primarily concerned with the application of Statistics in scientific experimentation. There are a limited number of theoretical exercises and some simulation exercises. Many of the exercises require the use of a statistical computing package. Solutions to exercises including the relevant programs written in MINITAB and SAS are available from the author at the address given on the copyright page.

Acknowledgements

Over the many years in which the structure underlying this book has been in the process of development, many people have offered valuable advice and engaged in useful exchanges concerning the nature of Statistics and the role of Statistics in scientific experimentation. I express my gratitude to all of those people. One person who stands alone in having freely offered his extensive wisdom and knowledge throughout the project is my long time colleague and friend, David Ratkowsky. Our many thought-provoking discussions (and arguments) have greatly enriched the content of the book. Jane Watson, who worked with me in teaching courses based on the book, offered many valuable suggestions. Scientists within the University of Tasmania and elsewhere generously allowed me to use material from consulting projects. Many variations and refinements have come from the use of earlier versions of the book by my students and scientific colleagues. The Institute of Actuaries kindly permitted the use of the data in Table 1.2.2 from an article by R.D. Clarke.

Typing of earlier versions of the book was undertaken by several secretaries in the mathematics department at the University of Tasmania, most notably Lorraine Hickey. However, my deepest gratitude goes to the developers of the scientific word processing package T3, who have provided me with a simple means of preparing the final draft myself. Meegan assisted with the production of the figures in the book, Ewen assisted with data checking, and Wendy spent many hours proofreading and checking the final draft.

Tasmania, Australia Glen McPherson

Plan of the Book

The development of the book traces the manner in which students and scientists might naturally wish to seek knowledge about Statistics according to the following scheme.

Scientific Investigations

Scientific investigations provide the motivation for Statistics. The idea of *scientific* investigation suggests an objectivity in the examination of the information collected in the course of the investigation. The research worker may use the study to investigate personal beliefs or hypotheses. The conclusions reached must, however, be free of any bias on the part of the experimenter. Statistics has a very important role to play in helping to maintain objectivity. This is established in **Chapter 1** which introduces and illustrates both the descriptive and analytic roles of Statistics in scientific investigation.

The Analytic Role of Statistics

While the involvement of Statistics in scientific studies is varied both in respect of areas of application and forms of statistical analysis employed, there is a general structure which provides a unified view of Statistics as a means of comparing *data* with a *statistical model.* **Chapters 2, 3**, and **4** describe this structure.

Probability and Statistics

It becomes apparent in developing a means of measuring the extent of agreement between the data and the statistical model that probabilistic ideas are involved. **Chapter 5** provides a link between the role of Probability in statistical analysis and the formal basis and rules of Probability as a discipline in its own right.

Statistical Models

There prove to be a small number of statistical models which cover a vast majority of scientific applications of Statistics. These models are introduced in **Chapter 6** and the basic information required in the comparison of these models with data is introduced in **Chapter 7**.

Mathematics and a Theory of Statistics

While this book is developed from the perspective of the user of Statistics, it is important even for non-mathematical users to appreciate the need for a mathematical structure for Statistics. **Chapter 8** provides a description of the basic approaches and tools used in statistical analysis. There is a consideration of the task faced by statisticians in defining the *best* statistical method for a given situation.

Examples of Statistical Applications

Chapter 9 is the first of a number of chapters which provide an illustration of statistical analysis. The primary purpose of such chapters is to establish a simple, but generally applicable, approach to the employment of statistical methods. Such examples are designed to give the reader the insight to select and apply statistical methodology.

Model and Data Checking

A statistical model generally contains assumptions which are distinct from the question of interest to the experimenter. These assumptions are statistical expressions of the structure of the experimental set-up or population which is under investigation. If there are errors in these assumptions or in the data which provides the factual information, then erroneous conclusions may arise from the analysis. **Chapter 10** contains an explanation of the nature of the statistical assumptions which commonly appear in statistical models, and presents methods for detecting errors in both these assumptions and in the data which are employed in the statistical analysis.

Defining Experimental Aims

A valuable role of Statistics is to be found in its requirement for a precise statement of experimental aims. **Chapters 11** and **12** examine this aspect using two areas of application. In Chapter 11, interest is in the average value of a quantity under study. In Chapter 12 the problem is to decide if one group is different from another or one treatment is better than another.

Types of Investigation

In **Chapter 13** a distinction is made between *observational studies* in which the investigator merely acts as an observer and *designed experiments* in which the experimenter plays a more active role. In the latter type of study, the experimenter assigns subjects or experimental units to treatments for the purpose of comparing the effects of the treatments. A special type of observational study known as a *survey* is also introduced.

The special methods employed in designed experiments for the selection of sample members and their allocation between treatments are introduced.

Comparisons Involving More Than Two Groups or Treatments

The methods introduced in Chapter 12 for the comparison of two groups or treatments are extended in **Chapters 14, 15,** and **16** to cover situations in which there are more than two groups or treatments. **Chapter 15** presents an important class of experimental designs and associated statistical methods which permit the separation of treatment effects from the effects of factors which are not of interest to the experimenter.

Relationships

In cases where more than one quantity is under investigation, the aim may be to examine if changes in one quantity are associated with changes in other quantities. **Chapter 17** introduces models and methods for this purpose. Where a relationship exists, it may be exploited to predict values in a *response* variable from values of *explanatory* variables. The means by which this is done is described in **Chapter 18**.

Variability

There are circumstances in which the measurement of variation or equivalently, measurement of precision, are of interest. **Chapter 19** is concerned with the definition and measurement of these quantities.

Cause and Effect

One of the common areas of interest in scientific investigation is the exploration of a cause-effect structure for a collection of variables. The extent to which Statistics may be used in this process is the topic of **Chapter 20**.

Time Related Changes

The nature of statistical models in studies of changes in response over time, and how such models are employed is the topic of **Chapter 21**.

Computers and Statistics

Chapter 22 provides a general discussion of various aspects of the role of the computer in statistical analysis and in data storage and analysis.

Using the Book

The book is designed to act primarily as a text book or a book to learn about Statistics and subsequently as a reference book.

As a *text book*, the core material is contained in Chapters 1 to 11 and the minimal introduction is provided by studying the following topics.

1. *Statistics in scientific experimentation*

   Sections 1.1 and 1.2.

   (Section 1.3 on descriptive Statistics should also be studied by students with no previous background in Statistics.)

2. *The structure and function of Statistics*

   Section 1.4 and Chapters 2, 3, and 4.

3. *The role of probability in Statistics*

   Sections 5.1 and 5.2 provide an informal introduction. The more mathematically inclined students should also work through Sections 5.3 and 5.4.

4. *Statistical models*

   Sections 6.1 and 6.5 introduce two important statistical models and Sections 7.1 and 7.3 provide basic information for the application of these models.

5. *Theoretical constructs*

   Chapter 8 should be studied in its entirety by mathematically inclined readers. Those with limited mathematical background should read Sections 8.1, 8.2.3, 8.2.5, 8.3.1, 8.3.3, and 8.3.5.

6. *A simple application of Statistics—proportions and probabilities*

   Section 9.1 and Sections 9.2.1 to 9.2.4.

7. *Model and data checking*

   Chapter 10.

8. *A simple application of Statistics—averages*

   Section 11.1 introduces an experimental problem and considers its translation into a statistical problem.

Section 11.2.1 discusses the choice of statistical methodology.

Section 11.2.2 read in conjunction with Sections 7.7.2 and 7.7.3 provides a method for studying medians based on a distribution-free model.

Section 11.2.3 provides a method for studying means based on a distribution-free model. Section 11.2.4 read in conjunction with Section 7.4 provides a method for studying means based on the assumption of a Normal distribution.

9. *The role of computers in Statistics*

   Chapter 22.

Beyond this basic introduction, the course may be extended with a consideration of the following topics.

1. Comparing two groups or treatments

   Chapter 12

2. Design of experiments

   Chapter 13

3. Comparing three or more groups or treatments—an informal introduction

   Sections 14.1.1 to 14.1.4

4. Linear and multiplicative models and methodology

   Chapter 14

5. Experimental and treatment designs and their applications

   Chapter 15

6. Comparing frequency tables

   Chapter 16

7. Studying relations between variables

   Chapter 17

8. Use of supplementary information—regression analysis

   Chapter 18

9. Variability and variance components

   Chapter 19

10. Cause and effect

    Chapter 20

11. Studies involving time or time-related variables
    Chapter 21

Commonly Used Symbols

Mathematical Symbols

exp The exponential function, e.g., $\exp(y)$

log The natural logarithm function, e.g., $\log(y)$

$|\ |$ Two bars enclosing a symbol, a number, or an expression indicate that the number or the value of the symbol should be treated as a positive value irrespective of the actual sign, e.g., $|+2|$ and $|-2|$ are both equal to 2

$\simeq$ approximately equal to

$n!$ pronounced *n factorial*—defined in Appendix 1, Chapter 5

$\binom{n}{x}$ pronounced $N\ C\ X$—defined in Appendix 1, Chapter 5

$\min(x, y)$ the smaller of the values contained in parentheses

Statistical Variables

x, y, etc. lowercase symbols in italics are generally used to represent response variables

y_i, y_1, etc. a subscript is employed where it is necessary to distinguish different variables

y_{ij} multiple subscripts are introduced where there is need to identify variables associated with different groups (Section 10.3.1)

y_0, x_0, etc. the subscript 0 is an abbreviation for the word *observed* and indicates the numerical value observed in the data for the response variable which it subscripts

$\bar{y}$ the statistic denoting the arithmetic mean of the response variable y in the sample (Section 1.3.9)

$\tilde{y}$ the statistic denoting the median of the response variable y in the sample (Section 1.3.8)

$\hat{y}$ a hat appearing over a variable indicates an estimator of the variable it covers

Parameters

A, B, etc. uppercase letters in italics generally denote parameters of probability distributions or component equations

$\hat{A}, \hat{B}$, etc. a hat appearing over a parameter indicates an estimator (i.e., statistic) of the parameter it covers

A, a a lowercase italic letter when used in conjunction with its uppercase form denotes an estimate of the value of the parameter

Sampling Distributions and Probability Distributions

$\pi(\ )$ the probability distribution of a random variable (Section 5.4.1) or the sampling distribution of a response variable (Section 5.2.2), e.g., $\pi(y)$

$\pi_S(\ )$ the joint probability distribution of random variables (Section 5.4.4) or the sampling distribution of a sample (Section 5.2.3)

$\pi_{\mathbf{st}}$ the sampling distribution of a statistic (Section 5.2.4)

Special Distributions

$N(M, \sigma^2)$ Normal distribution with mean M and variance σ^2 (Sections 6.5 and 7.3)

$N(0,1)$ *Standard* Normal distribution (Section 7.3.2)

z Widely used to denote a variable which has Standard Normal distribution (Section 7.3.2)

$\mathbf{z}_\alpha$ Value of a Standard Normal variable which satisfies the probability statement $\Pr(z > \mathrm{z}_\alpha) = \alpha$ (Section 7.3.2)

$t(\nu)$ t statistic with ν degrees of freedom (Section 7.4)

$\mathbf{t}_\alpha(\nu)$ Value of a $t(\nu)$ statistic which satisfies probability statement $\Pr\{|t(\nu)| \geq |\mathrm{t}_\alpha(\nu)|\} = \alpha$ (Section 7.4.2)

$\chi^2(\nu)$ Chi-squared statistic with ν degrees of freedom (Section 7.5)

$\mathbf{X}^2_\alpha(\nu)$ Value of a $\chi^2(\nu)$ statistic which satisfies the probability statement $\Pr\{\chi^2(\nu) \geq \mathrm{X}^2_\alpha(\nu)\} = \alpha$ (Section 7.5.2)

$F(\nu_1, \nu_2)$ F statistic with ν_1 and ν_2 degrees of freedom (Section 7.6)

$\mathbf{F}_\alpha(\nu_1, \nu_2)$ Value of an $F(\nu_1, \nu_2)$ statistic which satisfies the probability statement $\Pr\{F(\nu_1, \nu_2) \geq \mathrm{F}_\alpha(\nu_1, \nu_2)\} = \alpha$ (Section 7.6.2)

1

The Role of Statistics in Scientific Investigations

1.1 Scientific Investigation

Statistics means different things to different people. The everyday view is of the study of sets of numbers. Mathematicians seek a mathematical structure, statistical consultants might include the design of experiments and interpretation of experimental findings, and experimenters often regard computing as part of Statistics.

This book presents a particular view of Statistics which is useful to scientists, technologists and science-based professionals—agricultural scientists, engineers, economists, medical researchers—anyone who had need of Statistics as a tool in *scientific investigations.* An understanding of Statistics—the way it works, when and how it is used, how to interpret its findings—is provided within the framework of scientific investigations.

Before beginning the task of providing a structure for Statistics, it is important that we are in agreement as to what constitutes scientific investigation. This is particularly so for students who have not had the experience of undertaking investigations or experiments.

The popular image of a *scientific* investigation is of white-coated persons in a laboratory, usually under the control of an eccentric genius who is about to make a world-shattering discovery. This last aspect is strengthened by the popular media which delights in publicizing *important breakthroughs* in science, particularly in the medical area.

The fact is that revolutionary science with this romantic appeal is only a small fraction of scientific research and development. Typically, scientific investigation proceeds in a slow and painstaking way, with accumulated factual information from prior investigations being linked with the theories of the investigator to develop the next stage of the investigation. Rarely is an experiment which is done in isolation of great value. Rarely are experiments sudden urges. Indeed, to ensure they are providing answers to precisely the questions being posed by the investigators, they must be carefully considered and planned. Here is often found the distinction between the experienced research worker and the raw recruit. The former is typically more skilled in the design of experiments which have well-defined objectives and the capacity to meet these objectives. The raw recruit is likely to be overambitious in the objectives and produce a design which leads to ambiguity in its findings.

Most importantly, *breakthroughs* generally come not from someone shouting "Eureka," but from reasoned argument based on sound fact and scientific principles. The genius is the person who can see something in the facts which others have not seen; who can devise an experiment which will allow a specific question to be answered free of extraneous factors; who can see the flaws in the arguments of others.

Some experiments are designed with the anticipation of finding massive changes or differences which will offer clear cut *yes* or *no* answers. Statistics has little to offer in this type of experiment. The growing importance of Statistics in scientific research and development lies in the increasing number of investigations which are searching for evidence of much more subtle or complex changes. In these cases there are likely to be many factors which are involved and it is the way in which these factors interact that must be understood. This is so whether it is cancer which is being studied by medical researchers, the effect of man on the environment being studied by ecologists, or the effect of financial management being studied by economists.

The increasingly complex and varied nature of investigations, the greater expense of research and development, and the more technical nature of the conclusions places greater pressure on scientists to use more efficient designs, methods of analysis and to more effectively present findings. This in turn demands a better understanding of tools such as Statistics.

Twenty years ago, the typical introduction to Statistics for scientists was a collection of statistical method in a 'recipe book.' The scientists simply selected, from the small number provided, the method which was listed as being applicable in a given situation, applied the formulae given and drew a conclusion according to a clear cut rule. The methods were simple and the interpretation simplistic in most cases.

Young scientists who attempt to adopt this approach today are likely to find themselves without funds for research or perhaps without a job. *Understanding* Statistics and its place in scientific investigation is now an essential part of scientific training. The remainder of this section and the following three chapters are devoted to an explanation of the basic structure of Statistics within the framework of scientific investigations, with the aim of providing a basic understanding of Statistics.

1.2 Statistics in the Experimental Process

In seeking to understand Statistics within the context of scientific investigations, you should set as a goal, answers to the following three questions:

Why use Statistics is scientific investigations?
When is Statistics used in scientific investigations?
How is Statistics used in scientific investigations?

As information, ideas, and explanations are presented to you in this book, keep these questions in mind.

Those scientific investigations in which Statistics might be employed are characterized by the fact that (i) they generate *data*, and (ii) there is an *experimental set-up* which is the source of the data. Scientists construct *models* of the experimental set-up and use the data to accept or modify those models.

What are *data*? How can you identify the *experimental set-up*? What is the nature of the *model* of the experimental set-up as is employed in Statistics? These questions are explored in detail in Chapters 2 to 4. For the present, the aim is to give you a simple overview.

1.2.1 Experimental Set-Ups

For statistical purposes there are two descriptions of experimental set-ups which have common application.

1. *Populations.* The first, which is most useful in areas of biological and social sciences, views the object of investigation as a *population* or collection of objects from which a *sample* of population members has been drawn for examination in the investigation. This is the view which is useful, for example, in surveys which seek information on the collective opinion of people in a society.
2. *Physical processes.* The second useful description of an experimental set-up is of a physical process which successively generates outcomes. A favorite academic example is the tossing of a coin where the outcome is the appearance of a head or a tail showing uppermost on the coin. In economic terms, the daily rates of exchange of a currency may be viewed as the outcomes from the complex process which is determined by a country's economic and political management.

Whatever the view taken of an experimental set-up, the essential feature is the existence of well-defined characteristics of the population or process about which information can be obtained from a *sample* of members or outcomes.

1.2.2 Data

The information collected from the sample of members or outcomes provides the factual information or *data* used in studying the population or process. You may have the idea that data are necessarily sets of numbers. In fact, data may involve more than numbers as the information collected in the study reported in Example 1.1 illustrates.

Example 1.1 (Physiotherapy Training Before Childbirth). *Helen Lawrence is a physiotherapist who provides pre-natal training to assist* *women prepare for childbirth. As part of a study undertaken to examine the factors which are associated with pain experienced during childbirth, she collected the information recorded in Table* 1.2.1.

Table 1.2.1. A Study of Factors Affecting Pain Experienced During Childbirth (Reference: Helen Lawrence, physiotherapist, Hobart, Tasmania)

Physiotherapy Trained Group

Subject	Response PRI	Response REAC	Time in labour (hours)	Marital status	Age	Attitude	Analgesic
1	4.7	2	4.5	M	22	NS	4
2	4.7	3	8.0	M	24	C	5
3	6.0	2	6.5	M	24	C	4
4	4.6	2	5.0	M	29	C	3
5	7.1	3	7.0	M	23	C	4
6	3.6	4	6.0	M	34	C	3
7	7.5	3	4.5	M	21	C	3
8	4.6	3	11.0	M	26	C	3
9	8.4	4	9.0	S	24	NS	4
10	3.3	3	6.0	M	24	NS	4
11	5.7	5	6.0	M	29	NS	4
12	3.8	2	3.0	M	21	NS	3
13	9.2	4	18.0	M	24	NS	6
14	9.3	5	11.0	M	25	C	6
15	8.0	3	24.0	M	24	C	6
16	7.5	5	12.0	S	22	W	6
17	7.7	4	6.5	M	30	C	6
18	8.6	4	14.0	M	24	C	4
19	9.5	5	13.0	S	27	C	4
20	9.3	4	15.0	M	20	NS	6
21	8.1	2	12.0	M	24	NS	6
22	10.0	5	9.0	M	28	C	7
23	8.0	3	7.0	M	28	C	7
24	8.4	5	7.7	M	25	W	6
25	8.2	5	8.5	M	29	C	1
26	7.5	4	15.0	M	24	NS	6
27	6.3	2	8.0	M	21	C	1
28	7.9	3	6.5	M	33	C	4
29	6.5	4	17.0	M	22	NS	6
30	7.8	4	9.5	M	31	W	4
31	6.9	3	9.0	S	24	NS	6
32	4.8	4	7.5	M	27	NS	6

Table 1.2.1 (*cont.*)

Subject	Response PRI	Response REAC	Time in labour (hours)	Marital status	Age	Attitude	Analgesic
33	7.0	2	12.5	M	22	C	4
34	7.0	4	8.5	M	30	NS	6
35	8.3	5	10.5	M	19	C	6
36	7.8	4	24.0	S	27	C	7
37	7.0	1	3.5	M	25	W	1
38	7.3	3	7.0	M	25	NS	3
39	6.9	4	7.0	M	27	C	5
40	8.9	4	7.5	S	18	NS	6
41	9.3	4	7.5	M	25	W	6
42	8.0	4	15.0	M	30	NS	6
43	6.6	4	13.0	M	33	NS	4
44	3.4	2	5.5	M	22	W	3
45	7.2	2	22.0	M	27	C	7
46	6.5	5	12.0	M	29	C	3
47	7.5	5	12.0	M	26	C	6
48	9.0	5	13.0	S	26	NS	6
49	5.8	2	5.5	M	25	NS	4
50	9.5	5	6.5	M	29	NS	6
51	3.0	1	7.0	M	26	NS	3

Others Group

Subject	Response PRI	Response REAC	Time in labour (hours)	Marital status	Age	Attitude	Analgesic
1	4.5	3	6.5	M	23	NS	4
2	4.5	3	10.0	S	18	NS	4
3	8.0	3	7.5	M	30	NS	4
4	5.2	3	3.5	M	20	NS	4
5	7.0	5	12.5	M	21	NS	5
6	4.6	5	10.5	M	25	C	4
7	6.2	1	2.0	S	20	C	2
8	4.3	5	21.0	S	18	NS	6
9	4.6	3	7.0	S	21	NS	4
10	5.9	4	3.5	S	20	C	4

Table 1.2.1 (*cont.*)

Subject	Response PRI	Response REAC	Time in labour (hours)	Marital status	Age	Attitude	Analgesic
11	5.6	4	11.0	M	30	NS	4
12	5.3	2	2.5	M	24	NS	4
13	7.9	4	10.0	M	23	NS	4
14	8.6	3	11.5	M	30	NS	2
15	4.0	3	5.0	S	18	W	4
16	4.6	3	2.0	M	29	C	2
17	6.6	4	7.5	M	32	C	4
18	7.7	4	7.5	M	24	C	4
19	6.6	3	2.0	M	27	NS	4
20	9.6	4	7.0	M	23	NS	4
21	9.3	5	6.5	M	21	C	6
22	9.7	5	17.0	S	23	W	7
23	6.5	4	9.5	M	26	NS	5
24	2.9	3	10.0	S	18	NS	3
25	6.5	3	9.5	M	23	W	3
26	6.5	4	9.5	M	26	NS	5
27	3.2	4	4.5	S	25	C	3
28	7.3	4	7.0	S	18	W	3
29	9.3	4	2.5	S	20	NS	4
30	5.3	4	7.0	S	18	W	4
31	5.2	5	5.5	S	17	NS	4
32	9.1	5	9.5	M	19	NS	7
33	8.7	5	16.5	M	23	W	5
34	6.7	3	4.0	M	28	NS	2
35	7.9	4	10.5	S	20	W	6
36	4.9	4	8.0	M	23	C	3
37	5.9	5	6.0	M	37	W	7
38	6.3	5	9.5	M	21	NS	5

Key to symbols used in the table:

PRI: Pain rating index — a rating based on the woman's assessment in which higher values indicate greater level of pain.

REAC: Pain reaction — a rating based on the assessment of the nurse in which higher categories mean greater level of pain.

M or S: married or single.

Attitude: C = confident; NS = not sure; W = worried.

Analgesia: 1 = nothing; 2 = NO_2; 3 = <100mgs pethidine; 4 = >100mgs pethidine; 5 = morphine; 6 = epidural; 7 = general anaesthetic.

The information in Table 1.2.1 would all be classified by statisticians as *data.* Note that some information is in the form of categorical information—*married* or *single* in the case of the variable *Marital status.* Note also that some of the categorical information is in a disguised form. Whereas *Marital status* is identified by entries in a column in the table, there is another variable which might be called *Training type* with categories *Physiotherapy trained* and *Others.* The classification for this variable is achieved by grouping sample members from the same category together.

Chapter 2 is devoted to a consideration of the different types of data and the manner in which data are collected.

1.2.3 Statistical Models

The fact that an experimental set-up is being studied suggests that not everything about it is known. The reason for undertaking a scientific investigation is to learn more about unknown features of the experimental set-up by comparing expectations of outcomes under the model with what has actually been observed in the data. A famous example is reported in Example 1.2.

Example 1.2 (Studying the Pattern of Bomb Hits in London in World War II). *Towards the end of World War II, the Germans launched the so-called flying bombs in London. There was an obvious strategic interest in England to know the accuracy with which the bombs could be guided to a target. If the bombs could be accurately directed, it would be wise to diversify administrative and engineering sites. If, alternatively, the bombs were landing randomly over the city, it would be more efficient for production to maintain closely connected units.*

A statistical model was proposed which assumes each bomb was equally likely to land anywhere in the city and that the bombs were directed independently of one another. Based on this model, statistical theory provides a means of assessing whether the observed outcome is likely to have arisen under the proposed model. This is done by comparing what has been observed with what is expected under the proposed model in the absence of chance variation. The statistical analysis establishes if the difference between observation and expectation can reasonably be explained as chance variation. If not, the adequacy of the model is called into question.

For the statistical analysis, the city was divided into 576 *regions of equal area and the number of hits recorded in each region was determined. The data collected are summarized in Table* 1.2.2 *with a listing of the number of regions suffering* $0, 1, 2, \ldots$ *hits. The expected number of regions to suffer* $0, 1, 2, \ldots$ *hits under the proposed model is available for comparison. Formal statistical analysis was used to provide evidence on the acceptability of the proposed model. (The methodology and an analysis of these data are contained in Chapter* 16.)

Table 1.2.2. Distributions of Regions by Bomb Hits (Source: Clarke [1946])

Number of hits	0	1	2	3	≥4
Observed frequency	229	211	93	35	8
Expected frequency under model	227	211	98	31	9

Example 1.2 contains some of the ingredients which provide the reason for statistical involvement in scientific research. The expected distribution of hits is not in exact agreement with the observed distribution. Does this mean the model is incorrect? Not necessarily, since the data provide incomplete information on the mechanism which generated them. Allowance must be made for the vagaries of chance. While the contribution of chance to the location of a single bomb hit cannot be stated, the statistical model can characterize the effect of chance on the general pattern of hits.

The nature of models and the type of information and conjecture which is found in them is the topic of Chapter 3.

In the case of Example 1.2, the formal processes of Statistics were employed to make a comparison between the model and the data. In this role, Statistics plays an active part in the decision making process. There are many situations in which Statistics plays only a passive role in that it is used merely to condense and summarize information in the data. This aspect of Statistics is referred to as *Descriptive Statistics.* It is nonetheless an extremely important area of Statistics and is the subject of Section 1.3.

1.2.4 Statistical Advisors

The influence of Statistics on scientific experiments extends beyond the analysis of data. Since the data provide the factual information about the experimental set-up, the manner in which the sample providing the data is obtained necessarily has an important bearing on the interpretation and value of the data. A well-publicized example was provided by a presidential election in the U.S.A. where a telephone survey pointed to a landslide win for the Republican candidate in days when telephones were predominantly a possession of the rich. In fact the Republican lost the election. The reason is rather obvious—a majority of telephone subscribers may have supported the Republican candidate, but they were only a fraction of the population who voted.

To avoid serious misuse of Statistics and, on a more positive note, to improve the effectiveness of the role of Statistics in scientific investigation, it is common for scientists to seek assistance from statistical advisors or consultants. The manner in which they can contribute to the experiment is summarized in Figure 1.2.1. If you have access to an experienced statistical consultant and if you have any doubts as to the best way to conduct the experiment, the appropriate time to seek advice **is before the experiment begins,** that is, at the planning stage. Statisticians have specialized training in the development of experimental designs and in the methods of sample selection. This knowledge is of little value after the experiment has been conducted.

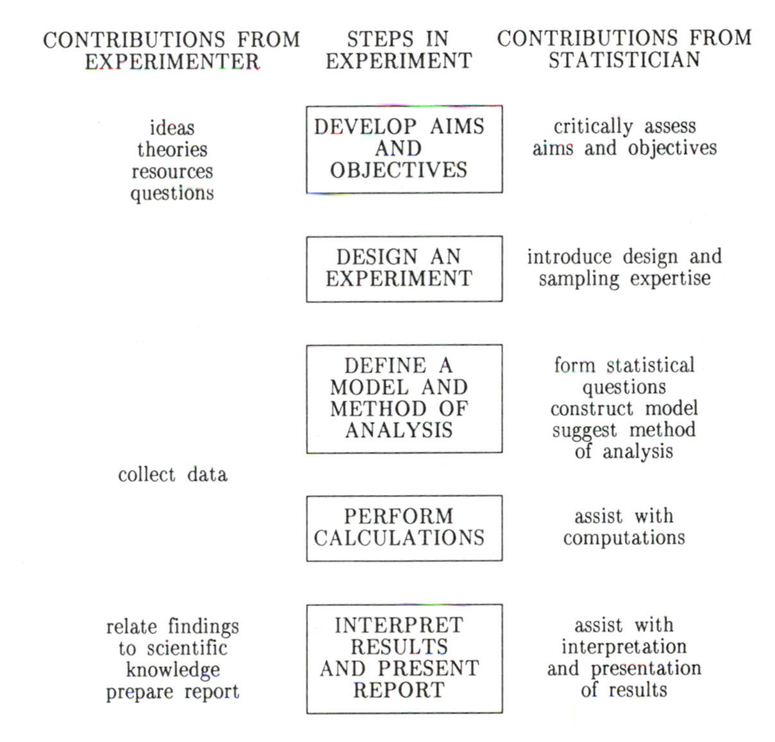

Figure 1.2.1. The role of statisticians in scientific experimentation.

Young or inexperienced research workers are the most likely to gain from this preliminary discussion with the statistician. It is usual for such workers to attempt a much more complex experiment than is feasible with the available resources. Also, because of their lack of experience, they commonly have not thought of all possible criticisms of their experimental design. An experienced statistical consultant can help these scientists avoid serious mistakes by simply questioning them about the relevance of the proposed experiments to the hypotheses or theories they plan to test.

Coupled with their statistical expertise, statistical consultants usually have extensive experience in constructively criticizing experimental aims, applications for research grants, reports of experimental findings, etc. These skills can be of great benefit to scientists in an era when scientific journals have become much more demanding of the use of correct statistical methods, companies have become much more demanding of clarity and simplicity in presentation of reports of scientific work, and research funding is typically becoming more difficult to obtain.

1.3 Descriptive Statistics

A simple definition of Descriptive Statistics is *the summarization and condensation of data.* The summarization may be in tabular form or in graphical form. The condensation involves the reduction of information in the data to a single number of a set of numbers.

A rule which reduces data to a single number is referred to by statisticians as a *statistic.* (To avoid confusion, in this book the discipline of *Statistics* is distinguished from these rules called *statistics* by commencing the former with an upper case S.)

There are a multitude of forms of Descriptive Statistics. Apart from some which are in common use, there are forms which are specific to particular scientific disciplines and there is scope for individual scientists to define their own forms of tables, graphs or statistics to improve the presentation of results of an investigation. In this section, examples of some of the commonly used forms are illustrated. Look to publications within your own area of interest for specialized forms of data summarization. Recommended reading when there is need for graphical presentation of results is Cleveland [1985].

While there is no underlying philosophical or mathematical structure to Descriptive Statistics, there are numerous rules which should be observed in the application of this area of Statistics. These will be noted in the appropriate places.

Before the application of computers for analysis of data became routine, there were constraints on the production of Descriptive Statistics and the analyst would typically be selective in the number and nature of the methods employed. Today, most scientists have the capacity to apply a vast array of procedures for summarizing and displaying information in a set of data. There is a need to employ common sense in making a selection of methods. Appreciate also, that complete automation of the operation removes from the investigator the *feel* for the data which can be very important in detecting features which lie outside the scope of the computer-based processes. After all, these processes will only display the features which were selected as important by the analyst and/or the person who devised the computer program.

1.3.1 Non-Graphical Data Presentation: Ordered Responses

When examining a set of measured responses, it is valuable to order the numbers from smallest to largest.

Ordered sets of numbers not only permit a simple visual check on the range of values, they also are a tool in the area of Statistics known as *Data Analysis* which, as one of its functions, seeks to identify errors in the data. Consider, for example, the data provided in Table 1.3.1.

Table 1.3.1. Physical Activity and Levels of Blood Chemicals (The data below were obtained from blood samples taken from runners in a 'fun run' in Hobart. The runners are divided into two groups described as Trained and Untrained.) Levels in Blood Before ("Pre") and After ("Post") the Race

(a) Untrained runners

Place	Glucose		Sodium		Potassium		Cholesterol	
	Pre	Post	Pre	Post	Pre	Post	Pre	Post
871	67	100	141	144	4.3	4.8	233	277
942	46	73	141	143	3.9	4.3	238	244
292	67	89	141	142	4.4	4.3	219	224
972	66	95	142	143	4.0	4.5	156	160
861	70	76	142	138	4.2	4.5	252	247
314	60	108	146	145	4.1	4.1	234	246
547	72	77	146	142	3.8	4.2	159	162
831	76	88	143	141	4.6	4.7	237	231
522	72	104	142	140	4.5	4.7	183	180
620	53	108	144	141	5.1	4.9	194	194
283	66	84	142	144	4.0	4.9	168	150

(b) Trained runners

Place	Glucose		Sodium		Potassium		Cholesterol	
	Pre	Post	Pre	Post	Pre	Post	Pre	Post
134	72	224	140	134	4.2	4.4	211	196
17	79	87	147	138	4.2	4.5	211	205
7	95	305	140	139	4.0	3.5	206	185
1	84	307	137	136	4.3	3.5	241	228
18	87	185	137	138	4.4	4.4	208	212
48	85	150	137	140	4.4	4.5	207	237
51	77	216	138	140	4.4	4.5	210	231
84	35	141	141	140	4.3	4.4	225	222
5	79	192	137	140	3.9	4.2	176	208
35	86	230	136	138	4.1	4.2	218	228
22	69	241	140	141	4.2	4.2	192	204

The *pre-glucose* reading for the *Trained* runners contains one value which is clearly detected as being different from the others and this is highlighted by ordering the values:

35 69 72 77 79 79 84 85 86 87 95

This display alone does not suggest there is an error in the reading of 35. If the analyst has knowledge which indicates that a reading of 35 for blood glucose level is surprisingly low or if there is a reason for expecting the readings from different individuals to cluster together, then the value might be regarded as unusual and in need of further examination.

1.3.2 Stem and Leaf Charts

The data which are presented in Table 1.3.2 comprise 137 readings. Even though the readings are presented in order of magnitude, the points of interest tend to be camouflaged by the mass of figures. There is a simple technique for displaying the information which serves both to display pattern in the collection and as an aid in later more detailed investigation. The technique produces what is termed a *stem and leaf chart.* The process is more simply described through example than definition. A stem and leaf chart for the data of Table 1.3.2 is produced in Table 1.3.3.

Table 1.3.2. Blood Cholesterol Levels (mgm%) from 137 Workers in a Zinc Processing Plant (Data supplied by courtesy of Prof. Polya, Chem. Dept., Univ. of Tasmania.)

145	145	146	158	158	163	168	168	168	172
174	175	175	175	175	175	175	178	180	180
180	181	181	187	187	192	194	194	195	196
196	198	198	198	198	201	201	201	204	204
204	205	205	205	205	205	206	206	206	207
208	209	211	211	211	212	212	212	214	214
214	216	217	217	217	217	218	218	218	218
221	221	221	221	223	224	225	225	227	227
227	227	228	230	235	235	235	236	237	238
238	239	241	242	242	243	243	345	245	245
247	248	252	253	253	253	254	254	255	256
256	257	257	261	262	266	267	267	268	268
268	268	273	274	278	283	284	285	289	292
294	296	300	301	302	314	331			

Table 1.3.3. Stem and Leaf Chart Based on Data in Table 1.3.2

Cumulative Frequency	Frequency	Stem	Leaves
3	3	14	556
5	2	15	88
9	4	16	3888
18	9	17	245555558
25	7	18	0001177
35	10	19	2445668888
52	17	20	11144455555666789
70	18	21	111222444677778888
83	13	22	1111345577778
92	9	23	055567889
102	10	24	1223355578
113	11	25	23334456677
122	9	26	126778888
125	3	27	348
129	4	28	3459
132	3	29	246
135	3	30	012
136	1	31	4
136	0	32	
137	1	33	1

The first three lines are reproduced below:

Column	1	2	3	4
	3	3	14	556
	5	2	15	88
	9	4	16	3888

The first two digits in column three constitute the *stem* of the numbers and the single digits which collectively form the number in column four are the *leaves.* The set

Column	3	4
	14	556

is an abbreviation for the numbers 145 145 146.

The number in column two records the number of readings with the corresponding stem and the number in column 1 provides a cumulative count.

The only practical difficulty in producing a stem and leaf chart is in the choice of a stem. It is not always desirable to make the number of classes, i.e., the number of stems, equal to the number of possibilities formed by deleting the last digit of each number. The only essential requirement is that the interval between successive stems should be a multiple of 1, 2 or 5 times a power of ten. In choosing classes, the two conflicting aims are

(i) to minimize the number of classes in the chart; and

(ii) to ensure that the number of leaves for any one stem is not too large (a convenient upper limit is 30).

Table 1.3.4 contains an example in which the interval has been set at 0.5 rather than 1.0 to meet the second condition.

Table 1.3.5 contains data which pose a different problem. The numbers range from 91 to 2351 for the readings on girls. To ensure that the number of classes is not excessive, the simplest approach is to ignore the units digit in every number. By dividing each number by ten and only retaining the whole number portion, 91 becomes 9, 316 becomes 31 and 2351 becomes 235. Note that in this case, it is not possible to reproduce the original data. There has been some *condensation* of the data, although it is unlikely to be of any significance.

1.3.3 Frequency Tables for Measured Responses

There are both data sets and circumstances for which stem and leaf charts do not provide the best condensed view of the data. Table 1.3.6 illustrates such a situation. A summary table of two sets of data is presented in which the observations were obtained from whole numbers in the range 0 to 8. With this limited range of whole number entries, it is simpler to present a table of counts of the number of occurrences of each value. Thus at Site 1, there were 7 occasions on which no cars were observed, 26 occasions on which one car was observed, etc. This form of table is termed a *Frequency table*. In this case the table contains all the information which could be presented in a stem and leaf chart.

Where the response shows continuous variation, the Frequency table will generally contain less information than the corresponding stem and leaf chart, but has the advantage of a more condensed form.

As with the stem and leaf chart there is a subjective element in forming the classes of a Frequency table. As a guide, aim to have, on average, one class for every 15–25 responses. Use 10 classes if there are about 150–250 responses and 20 classes if there are 300–500 responses. The most appropriate number of classes depends to some extent on a suitable class width. In the table illustrated in Table 1.3.7, a class width of one unit seemed to be a reasonable choice. This led to an average of 25 responses per class.

Table 1.3.4. Stem and Leaf Chart for 250 Readings in a Quality Control Study (The following data were obtained from a quality control check on limit thermostats. The thermostats were nominally set to switch off when the temperature reached 80° C. The figures recorded are the temperatures at which the thermostats actually switched off.)

Cumulative Frequency	Frequency	Stem	Leaves
1	1	73.	1
1	0	73.	
2	1	74.	4
3	1	74.	8
4	1	75.	4
4	0	75.	
12	8	76.	00222344
20	8	76.	55688889
32	12	77.	012222233344
48	16	77.	5566666778889999
66	18	78.	000111111223333344
80	14	78.	56666677789999
104	24	79.	000000111233333334444444
132	28	79.	5555566677777777888888999999
156	24	80.	000111222222222233344444
173	17	80.	56667777777889999
191	18	81.	000000111233333444
206	15	81.	555555678888899
216	10	82.	0000111244
231	15	82.	566666777788889
240	9	83.	112222344
246	6	83.	677799
246	0	84.	
247	1	84.	6
247	0	85.	
248	1	85.	7
249	1	86.	0
250	0	86.	8

Table 1.3.5. Energy Intakes (Megajoules) Over a 24-Hour Period for a Sample of 15-Year-Old School Children (Source: Dr. D. Woodward, Dept. of Biochemistry, Univ. of Tasmania.)

(a) Girls (sample size 110)

91	316	362	431	469	489	499
504	505	509	519	538	548	553
557	575	580	583	583	586	601
609	609	613	623	648	669	674
693	697	697	704	715	721	725
727	729	736	738	752	755	760
764	769	772	776	782	784	790
803	809	815	816	826	844	847
857	877	880	887	901	911	912
929	929	929	947	952	953	967
970	972	973	981	984	998	1031
1043	1055	1056	1077	1079	1086	1089
1089	1097	1107	1121	1125	1146	1166
1195	1245	1280	1310	1312	1338	1383
1384	1404	1431	1469	1617	1626	1675
1713	1829	1868	2269	2351		

(b) Boys (sample size 108)

457	484	487	489	539	543	639
645	649	651	656	710	715	719
736	743	753	761	764	774	779
790	796	846	850	877	883	884
899	921	926	930	947	959	980
991	1025	1057	1059	1079	1081	1087
1104	1109	1110	1118	1118	1132	1153
1154	1155	1164	1168	1181	1182	1187
1203	1212	1231	1234	1235	1281	1297
1320	1322	1326	1353	1382	1414	1415
1417	1425	1445	1459	1479	1484	1511
1569	1577	1597	1601	1641	1660	1670
1685	1695	1715	1769	1781	1786	1831
1843	1928	1986	2039	2070	2107	2223
2296	2313	2364	2419	2480	2575	2697
2710	2925	3362				

Table 1.3.6. Data from a Study of Traffic Flow (The data presented below come from a survey by third-year Statistics students of traffic flow at two sites along a road.)

Table of Frequencies — Number of cars passing per minute

No. of cars	0	1	2	3	4	5	6	7	8
Site 1	7	26	32	33	9	7	3	3	0
Site 2	21	21	24	20	14	11	6	1	2

Table 1.3.7. Frequency Table Based on Data in Table 1.3.4

	Class Limits[2]		Class[3]		Cum[5]	Rel[6]
Class[1]	lower	upper	Midpoint	Freq[4]	Freq	Cum Freq
75.9 or less		75.95		4	4	0.016
76.0 – 76.9	75.95	76.95	76.45	16	20	0.08
77.0 – 77.9	76.95	77.95	77.45	28	48	0.19
78.0 – 78.9	77.95	78.95	78.45	32	80	0.32
79.0 – 79.9	78.95	79.95	79.45	52	132	0.53
80.0 – 80.9	79.95	80.95	80.45	41	173	0.69
81.0 – 81.9	80.95	81.95	81.45	33	206	0.82
82.0 – 82.9	81.95	82.95	82.45	25	231	0.92
83.0 – 83.9	82.95	83.95	83.45	15	246	0.984
84.0 or more	83.95			4	250	1.000

Explanation:

1. defines the range of recorded values for each grouping or class
2. based on the fact that the underlying variable (temperature) is continuous, the class limits represent boundaries between the groupings on a continuous scale
3. given the lower limit is l and the upper limit is u, the midpoint for the class is $\frac{1}{2}(l+u)$
4. frequency — the number of responses in the data set falling into a class
5. cumulative frequency — the number of responses in the data set falling into that class or a lower class
6. relative cumulative frequency — the proportion of responses in the data set falling into that class or a lower class

1.3.4 Frequency Tables for Categorical Responses

When the responses are categorical, the Frequency table is the typical form of condensation of the data. For the quantity *Marital status* in Table 1.2.1, the Frequency table would have the form:

Table of Frequencies

Marital Status		
single	married	total
21	68	89

Alternatively, a table of relative frequencies may be preferred, e.g.,

Table of Relative Frequencies

Marital Status	
single	married
$\frac{21}{89} = 0.24$	$\frac{68}{89} = 0.76$

As a variation, it is often preferable to present the information in the table as percentages rather than proportions.

Frequency tables may also be produced for combinations of variables. Again using Table 1.2.1 to provide the example, a table of frequencies is provided for the quantities *Marital status* and *Physiotherapy training:*

Table of Frequencies

	Marital Status		
	single	married	total
trained	7	44	51
untrained	14	24	38
total	21	68	89

In this case, the table of relative frequencies may take three different forms. If there is interest in the relative numbers of single and married women between the trained and untrained group, figures in the body of the table should be divided by corresponding row totals; if there is interest in the relative numbers of trained and untrained women between the single and married groups, the figures in the body of the table should be divided by corresponding column totals; if there is interest in the relative numbers in each of the four category combinations, the figures in the body of the table should be divided by the grand total. As for the table based on one variable, proportions may be converted as percentages if these are judged to be easier for the reader to comprehend.

Careful thought should be given to the most effective form of presentation.

1.3.5 Line Diagrams

A simple graphical expression of the information in an ordered set of numbers is illustrated in Figure 1.3.1. This diagram comprises the plot of each response on a scaled line with repetitions being shown stacked on top of one another and is called a *Line diagram.*

The Line diagram for the Trained group in Figure 1.3.1 clearly displays the one low value.

Visual comparison of Line diagrams based on data from the same response variable, but selected from different groups, displays both differences in the average responses and differences in the patterns of response as illustrated by Figure 1.3.1.

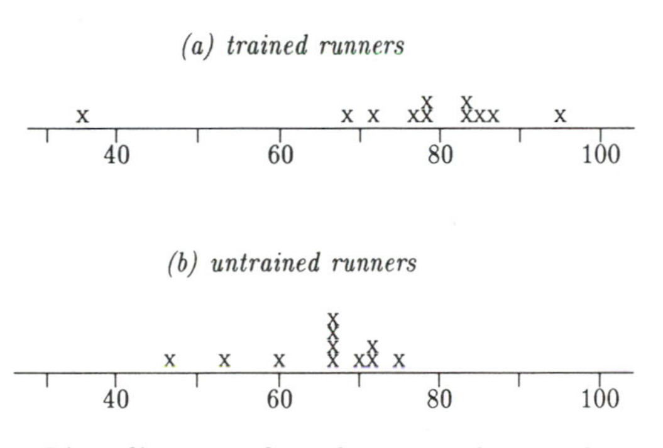

Figure 1.3.1. Line diagrams based on pre-glucose data in Table 1.3.1.

Line diagrams are most appropriate when sample sizes are small and are recommended as a standard presentation when the sample size is less than thirty. For larger sample sizes, some form of condensation is desirable and *Box-plots* (Section 1.3.10) are recommended.

1.3.6 Histograms

A graphical version of the data in a frequency table is provided by constructing a rectangle over the interval corresponding to every class such that the areas of the rectangles are proportional to the numbers of sample members in the class. Such a figure is called a *Histogram* and is illustrated in Figure 1.3.2.

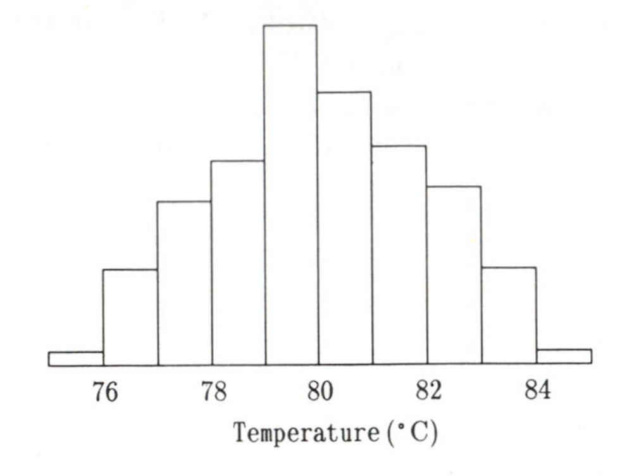

Figure 1.3.2. Histogram based on data in Table 1.3.4.

1.3.7 Scatter Plots

The methods presented so far for measured variables have been for data from a single response variable. If the investigator is interested in examining the relation between two variables there may be value in plotting the paired readings on cartesian coordinates. Such a plot is referred to as a *Scatter plot.* An example is given in Figure 1.3.3 based on pre- and post-readings of glucose for Trained runners with the data taken from Table 1.3.1. The points plotted are (72, 224), (79, 87), ..., (69, 241). The Scatter plot not only permits an assessment of the form of the relation between the variables but also has the potential to highlight any odd pairs of readings.

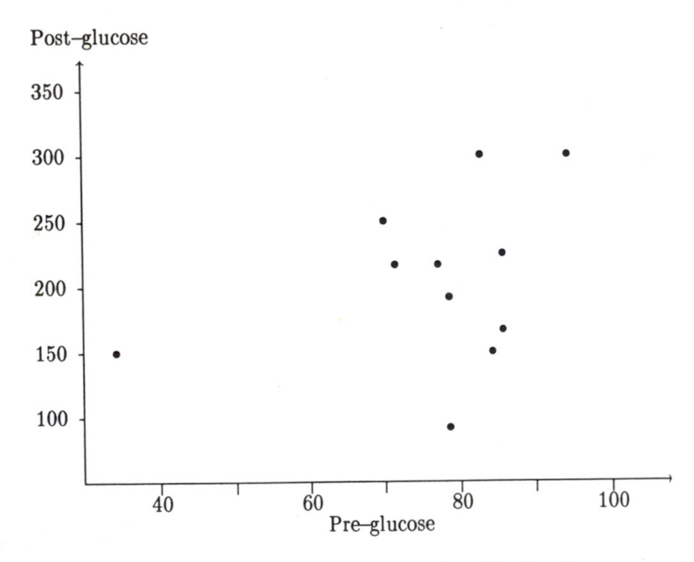

Figure 1.3.3. Scatter plot based on data for Trained runners in Table 1.3.1.

Where the aim is to examine the manner in which values of one variable change in relation to changes in another variable, the scatter plot gives information on the nature of the trend line and the extent of scatter of the points about the trend line.

1.3.8 Percentiles and Quantiles

These are statistics which have application with a set of measured values for a variable. As the name suggests, *percentiles* are based on division into 100 units, which statisticians refer to as the *first percentile, second percentile,* etc.

The Pth *percentile* should ideally divide the data such that there is a percentage P of the numbers below the Pth percentile and a percentage $100-P$ above the Pth percentile. Thus, the 5th percentile should have 5% of values below it and 95% of values above it.

If there are exactly 100 numbers all of different value, this condition can always be satisfied. For example, the 10th percentile would lie between the 10th and 11th numbers if the observations were ordered from smallest to largest.

Using the convention that the numbers comprising the data, when ordered from smallest to largest, are represented as $y_{(1)}, y_{(2)}, \cdots, y_{(100)}$, the 10th percentile would assume a value between $y_{(10)}$ and $y_{(11)}$. Clearly this does not uniquely define the 10th percentile since it may take any value between $y_{(10)}$ and $y_{(11)}$ and still satisfy the requirements stated above.

There is not universal agreement as to which number should be assigned to a percentile when it lies between two different numbers, although in this case, the value which is midway between the 10th and 11th ordered observations would have wide acceptance. Thus, in the case where there are 100 observations, the Pth percentile could be determined from the formula $\tilde{y}_P = \frac{1}{2}[y_{(P)} + y_{(P+1)}]$.

To determine $\tilde{y}_P$ when the sample size is different from 100, the above formula is inapplicable. If the sample size is n and $p = P/100$, then, by analogy with the above formula, $\tilde{y}_P = \frac{1}{2}[y_{(np)} + y_{(np+1)}]$. In general np will not be a whole number. For example if the sample size is eleven and the 25th percentile is required, np is $11 \times 0.25 = 2.75$.

There is not universal agreement as to the definition to employ in this case. Different statistical packages will possibly yield different values because they rely on different definitions. Equation (1.3.1) offers a weighted average which yields the above formula for $\tilde{y}_P$ when np is a whole number.

$$P\text{th percentile} = \tilde{y}_P = (1-a)y_{(b)} + ay_{(b+1)} \tag{1.3.1}$$

where n is the number of observations; $y_{(1)}, y_{(2)}, \cdots, y_{(n)}$ is the set of observations ordered from smallest to largest; $p = P/100$; b = largest whole number less than $np + \frac{1}{2}$; $a = np + \frac{1}{2} - b$. The construction of the formulae based on (1.3.1) are illustrated below for sample sizes of 11 and 12 and for percentiles 49, 50, and 51.

	p	$np + \frac{1}{2}$	b	a	percentile
1. $n = 11$					
	0.49	5.89	5	0.89	$0.11y_{(5)} + 0.89y_{(6)}$
	0.50	6.00	5	1.00	$0.00y_{(5)} + 1.00y_{(6)}$
	0.51	6.11	6	0.11	$0.89y_{(6)} + 0.11y_{(7)}$
2. n=12					
	0.49	6.38	6	0.38	$0.62y_{(6)} + 0.38y_{(7)}$
	0.50	6.50	6	0.50	$0.50y_{(6)} + 0.50y_{(7)}$
	0.51	6.62	6	0.62	$0.38y_{(6)} + 0.62y_{(7)}$

Median. The 50th percentile has a special name, the *median,* and it is used widely to identify the central or average value of the set of numbers. Equation (1.3.1) simplifies to $\tilde{y}_{50} = \frac{1}{2}[y(\frac{1}{2}n) + y_{(\frac{1}{2}n+1)}]$ if n is even or $\tilde{y}_{50} = y_{(\frac{1}{2}n+\frac{1}{2})}$ if n is odd.

Interquartile range. The difference between the 75th percentile and the 25th percentile is called the *interquartile range* and is used as a measure of the variability in the set of numbers.

Illustration. Based on the 137 cholesterol readings in Table 1.3.2, the 25th, 50th and 75th percentiles are respectively 198.0, 218.0 and 252.25. The interquartile range is 54.25.

Quantiles. Percentiles have the special property that they derive from division into 100 units. The more general division is provided by *quantiles.* If there are K units then the possible quantiles are the first, second, third, ..., Kth quantiles. The definition of the Qth quantile when there is division into K units which has the same basis as that given above for the Pth percentile is defined by (1.3.2).

$$Q\text{th quantile given } K \text{ units} = (1 - a)y_{(b)} + ay_{(b+1)} \tag{1.3.2}$$

where all terms have the same meaning as in (1.3.1) except for p which is equal to Q/K.

Where K equals four, the quantiles are referred to as *quartiles* and there are the relationships:

$$\begin{aligned}\text{1st quartile} &= \text{25th percentile}\\ \text{2nd quartile} &= \text{50th percentile} = \text{median}\\ \text{3rd quartile} &= \text{75th percentile.}\end{aligned}$$

1.3.9 Moments

Another important group of statistics, which can be applied to a set of numbers measured on a scale, are collectively called *moments.* To provide formulae which define moments, consider the set of numbers to be represented as $y_1, y_2, \ldots, y_n$. Then the kth *moment*, m'_k, is calculated from (1.3.3).

$$m'_k = (y_1^k + y_2^k + \cdots + y_n^k)/n \quad \text{for} \quad k = 1, 2, 3, \ldots . \tag{1.3.3}$$

Arithmetic mean. The most widely employed moment is the first moment which is known as the arithmetic mean (often simply abbreviated to the *mean*) which is computed from (1.3.4).

$$\bar{y} = (y_1 + y_2 + \cdots + y_n)/n = \sum_{i=1}^{n} y_i/n. \tag{1.3.4}$$

The arithmetic mean is commonly used to measure the average value of a set of numbers.

A second set of moments, *the corrected moments,* uses the arithmetic mean in its formulation. The kth *corrected moment*, m_k, $(k = 2, 3, \ldots)$ is calculated from (1.3.5).

$$m_k = [(y_1 - \bar{y})^k + (y_2 - \bar{y})^k + \cdots + (y_n - \bar{y})^k]/(n-1). \tag{1.3.5}$$

Sample variance. The most widely used of the corrected moments is the second corrected moment which is called the sample variance and is computed from (1.3.6).

$$\begin{aligned} s^2 &= [(y_1 - \bar{y})^2 + (y_2 - \bar{y})^2 + \cdots + (y_n - \bar{y})^2]/(n-1) \\ &= \sum_{i=1}^{n} (y_i - \bar{y})^2/(n-1). \end{aligned} \tag{1.3.6}$$

This provides a measure of variability in the set of numbers.

Sample standard deviation. The square root of the sample variance is termed the sample standard deviation and is denoted by the symbol s. Neither the (sample) variance nor the (sample) standard deviation have a simple scientific interpretation except when used in conjunction with a special statistical model which will be introduced in Section 6.5.

Illustration. From the data in Table 1.3.1 for the pre-race glucose readings of untrained runners, the arithmetic mean and the standard deviation are computed to be 65.00 and 8.85 respectively.

1.3.10 Characterizing the Pattern of Variation

Individual statistics such as the mean and the median each provide information on a specific property of a set of data. There are occasions on which a general picture of the pattern of variation in the data is desired. This was the case in the study which produced the data in Table 1.3.5. There the investigators sought to characterize the patterns of energy intakes for 15-year-old girls and boys based on these data.

The most obvious way of characterizing the pattern of variation in a collection of measured numbers is to present a pictorial display. Line diagrams and histograms provide presentations which are easily understood by non-statisticians.

For a non-graphical presentation, the use of percentiles is strongly recommended. Not only do percentiles have a simple interpretation individually, but they can be considered in combination to give further useful information.

The other set of statistics which is available is the set of moments. As noted earlier, beyond the first moment, i.e., the arithmetic mean, there generally is not a simple interpretation of the numerical values of moments. Nevertheless, it has been common practice for the mean and standard deviation to be required in many areas of science. There are two possible misinterpretations which can arise from this practice. If the mean is presented with the standard deviation following in parenthesis, i.e., $\overline{y}(s)$, there is the possibility that s will be interpreted as an absolute error. That is, it might be assumed that the population mean is certain to lie no further from $\overline{y}$ than a distance s. The second potential misinterpretation results from a rule which states that there is a 95% probability that the interval $\overline{y}-2s$ to $\overline{y}+2s$ will include the value of the population mean. This is correct only for a special model and the non-statistician is unlikely to possess the expertise to ensure that the correct interpretation is applied.

Given that percentiles are to be used for the characterization, the choice of which percentiles to present depends both on sample size and purpose of application. The following comments may be of assistance.

1. The median ($\tilde{y}_{50}$) and the interquartile range ($\tilde{y}_{75}-\tilde{y}_{25}$) provide basic information on the location and spread of the values while the more extreme percentiles indicate the extent to which the responses spread in either direction.

2. By comparing the statistics $\tilde{y}_{50} - \tilde{y}_{25}$ and $\tilde{y}_{75} - \tilde{y}_{50}$ it is possible to judge whether the pattern is symmetric or skewed. (Symmetry is implied by the two differences being of similar magnitude.)

3. It may be useful to regard a pair of statistics such as $\tilde{y}_5$ and $\tilde{y}_{95}$ as defining the range of typical values for a population. For example, the typical range of weights for boys of a given age might be defined

in this way. Take care to emphasize, however, that values outside the range are not necessarily abnormal.

4. Depending on sample size, the extreme percentiles may be very sensitive to variations in individual responses. As a guide, it is suggested that percentiles are not presented which are outside the limits defined in Table 1.3.8.

Table 1.3.8. Guidelines for the Smallest and Largest Percentiles to Present as a Function of Sample Size

Sample Size	Percentiles	
	smallest	largest
less than 5	$\tilde{y}_{50}$	$\tilde{y}_{50}$
5 – 49	$\tilde{y}_{25}$	$\tilde{y}_{75}$
50 – 99	$\tilde{y}_{10}$	$\tilde{y}_{90}$
100 – 199	$\tilde{y}_{5}$	$\tilde{y}_{95}$
200 – 399	$\tilde{y}_{2.5}$	$\tilde{y}_{97.5}$

Illustration. For the energy intake data for girls in Table 1.3.5, the following information was obtained:

	value	5th	25th	50th	75th	95th
Percentile	value	489	674	846	1086	1675

Based on these figures, it may be concluded, among other things, that:

(a) the middle fifty percent of intakes lies between 674 and 1086 megajoules and the middle ninety percent of intakes lies between 489 and 1675. (Perhaps this latter range might be regarded as the *typical* range, although with care begin taken to point out that an intake outside this range does not imply the intake is abnormal—that is a decision for the medical profession to make.);

(b) the distribution of energy intakes is skewed towards the upper end since $\tilde{y}_{95} - \tilde{y}_{50} = 829$ is much larger than $\tilde{y}_{50} - \tilde{y}_{5} = 357$.

Box-plots. These provide, in the manner illustrated in Figure 1.3.4, the following information:

(a) the position of the median (*);

(b) the 25th and 75th percentiles (the extents of the box);

(c) points equal to one interquartile range below the 25th percentile and above the 75th percentile; and

(d) all individual values which lie outside these limits.

In Figure 1.3.4, the median value is 79 which is represented by the ∗, the 25th percentile is 73.25 which is the position of the left upright of the box, the 75th percentile is 85.75 which is the right upright of the box. The vertical bar to the left of the box is one interquartile range (12.50 units) from the left edge of the box, the vertical bar to the right of the box is one interquartile range to the right of the right edge of the box, and the reading 35 is indicated by a **x** to indicate that it lies outside the vertical bars.

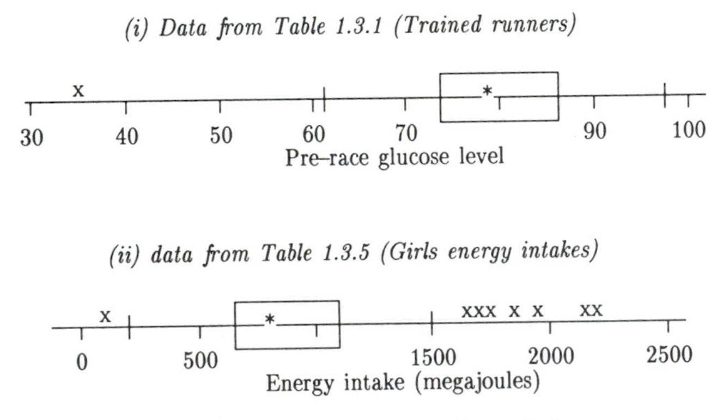

Figure 1.3.4. Box-plots for selected data sets.

1.3.11 Describing the Average Value of a Set of Numbers

Technically, the arithmetic mean is regarded by statisticians as the average value.

Where the information is to be presented to a readership with a limited mathematical background or where it is being used in the sense of typical or normal, the median may sometimes prove more meaningful since there is no mathematics required in explaining what it represents.

A disadvantage of the median where results are to be combined from different data sets is the necessity of obtaining the complete data sets to compute the median for the combined sample. The arithmetic mean has the property that the means from the individual data sets may be averaged to obtain a value for the combined set.

Illustration. The energy data for girls in Table 1.3.5 yield values of 918 and 846 for the mean and median respectively. The larger value for the mean indicates the great influence of a small proportion of very large energy intakes on the statistic.

1.3.12 Measuring Variability in Data

In samples of less than ten observations, the difference between the largest and smallest observation, known as the *range*, provides a simple measure of variability. In larger samples, the *interquartile range* is generally to be preferred as it utilizes more information in the data. The *variance*, or equivalently, the *standard deviation*, is widely employed but, as will be explained more fully in Section 6.5, it has meaning only in a restricted situation.

1.4 Analytic Statistics

Descriptive Statistics provides scientists with factual information by summarizing or highlighting features in the data. It does not offer a means of utilizing that information to pass judgment on the acceptability of hypotheses which the research worker may have proposed or to provide an indication of the precision with which numerical quantities are estimated. There is a branch of Statistics which may be employed for this purpose. In this book it is termed *Analytic Statistics*.

The manner of passing judgment may be subjective, relying on unstated rules laid down by the individual analyst, or it may be objective, depending on a formal structure which is generally agreed upon by a wide body of people. In this book, subjective methods are grouped together under the title of *Data Analysis* and objective methods, where there is a probabilistic basis, are grouped together under the title *Inferential Statistics*.

1.4.1 Data Analysis

The techniques which are collected together under the title of Data analysis serve the following three objectives:

- preliminary examination of data to detect errors in the data or model;
- provision of simple statistical methods to avoid the need for more formal or complex procedures of Inferential Statistics; and
- secondary examination of data for evidence of special features which could have scientific importance even though they do not come within the list of questions or conjectures stated by the investigator.

The procedures which come within the umbrella of Data analysis are many and varied. Some techniques are visual, others involve extensive computations.

It is essential to distinguish Data analysis from examination of data without reference to an underlying model. Consider the following set of numbers which constitute data collected from a sample of seven persons:

$$220 \quad 220 \quad 220 \quad 245 \quad 245 \quad 260 \quad 480.$$

Clearly, the last number is much larger than the first six numbers. Could this mean the last number is an error, perhaps wrongly recorded or the result of a faulty measuring instrument?

To answer that question we must have some information on (a) the nature of the experimental set-up; and (b) the manner in which the sample was selected. In statistical terms, we must have a model which describes the pattern of variation expected.

In fact, these data come from a sample of seven employees in a large industrial organization. The numbers are weekly salaries of the seven employees. The first six salaries are for unskilled workers and the last for the manager of the chemical laboratory. There is neither an error nor anything odd about the last number.

Experienced researchers practice this form of *preliminary* Data analysis as a routine part of performing the scientific investigation. Many would be surprised that it is considered part of Statistics—"common sense" might be their description.

There are three important reasons for regarding this type of investigation as part of Data analysis and hence part of Statistics. The first is to illustrate that Statistics is not separated from other components of scientific investigation by a large gulf—there is a continuum from non-statistical to statistical analysis. The second is to emphasize that the statistical model contains an expression of the experimental set-up and the most effective application of Statistics can only come about if all relevant information is included in the model. The third is to emphasize to young and inexperienced investigators the importance of critically examining data at the preliminary stage of analysis.

Exploratory data analysis is the name given when data are examined in ways which are not directly suggested by the questions posed by the investigator. While it is difficult to conceive of a sensible scientific investigation in which the investigator does not have questions or hypotheses which have prompted the study, there is often scope for examining the data in ways which are not specifically aimed at providing answers to those questions. The aim is to detect characteristics of the data which are not consistent with the presumed model. In this way there is the possibility of (i) detecting something of scientific importance which can then become the reason for a further study; or (ii) establishing an error in the model, which, if left unaltered, could seriously distort the findings about the primary questions asked by the investigator.

1.4.2 Inferential Statistics

The basic function of the methods of Inferential Statistics is to provide an objective measure of closeness of a statistical model and a data set. The measure always has a probabilistic base. The model is presumed to provide a description of the experimental set-up which generated the data. Thus, if the model and data are *close*, the model is presumed to be a reasonable description of the experimental set-up which generated the data.

Given this information, there is scope for

(i) comparing different models in order to decide which is the best supported model; or

(ii) deciding if an individual model might reasonably have generated the data.

The first approach might be of use to a botanist who is attempting to classify a plant or a technologist who is required to decide between different production methods.

The second approach is more widely used in science. It has application not only when there is only one model proposed, but also when there are many possible models and interest lies in deciding which of the possible models are consistent with the data, rather than in choosing between them.

At this introductory stage, it is not possible to provide a deeper insight into Inferential Statistics. There is a need to better understand the scope of *data* (Chapter 2) and *statistical models* (Chapter 3) before an insight can be given into the basis of the process of Inferential Statistics (Chapter 4).

Problems

1.1 *Error checking*

For each *pre-race* variable in Table 1.3.1, (i) order the data for the *trained* runners from smallest to largest; and (ii) present the 11 numbers in a line diagram (Section 1.3.5). Note any readings which you might consider unusual.

1.2 *Graphical presentations — scaled responses*

(a) Construct separate stem-and-leaf charts (Section 1.3.2) for boys and girls from the data in Table 1.3.5 but use the same categories for both sexes to allow a comparison of patterns of energy intake. Use this information to construct histograms for the two sexes.

(b) Construct separate Box-plots (Section 1.3.10) for girls and boys from the data in Table 1.3.5.

(c) Briefly describe the differences in energy intake patterns for boys and girls in the study.

1.3 *Graphical presentation — scatter plots*

Using the data for the *trained* runners given in Table 1.3.1, plot *pre-race* versus *post-race* readings for each of the variables. Is there evidence of a relation between pre- and -post race levels of any of the variables? If there is, explain the nature of the relationship.

1.4 *Frequency tables — categorical variables*

(a) Based on the data in Table 1.2.1, prepare a frequency table with the following structure.

		Pain Reaction				
		1	2	3	4	5
Attitude	Confident					
	Not sure					
	Worried					

(b) Alter the table to present the information as percentages which sum to 100 across each row. Based on the information in this table, does the level of *pain reaction* appear to be related to *attitude*?

1.5 *Percentiles*

(a) Based on the data in Table 1.3.5, compute 5th, 25th, 50th, 75th and 95th percentiles for (i) girls' energy intakes and (ii) boys' energy intakes (Section 1.3.8).

(b) Compute the interquartile range and difference between the 95th and 5th percentiles for each sex.

(c) Briefly describe the differences in characteristics of energy intake between the samples of boys and girls which is contained in the information obtained in parts (a) and (b).

1.6 *Basic statistical information*

The following numbers are extracted from Example 11.3 and record percentage monthly increases in property values for 28 properties in a city.

0.05	0.99	0.13	1.45	0.74	5.21	1.04	1.51	1.68
0.93	3.01	0.88	1.72	0.73	0.69	1.33	1.01	0.83
1.44	1.81	0.80	0.91	0.95	1.14	0.26	0.97	1.30
0.74								

(a) Order the data from smallest to largest value and construct a *line diagram.*

(b) Present the maximum and minimum values and compute the arithmetic mean (Section 1.3.9) and the median (Section 1.3.8).

(c) Compute the difference between the largest and smallest values (the *range*), the interquartile range (Section 1.3.8), the variance and the standard deviation (Section 1.3.9).

1.7 *Statistics applied to frequency data*

The data in Table 1.3.6 comprises two sets of 120 readings summarized in frequency tables. To compute the mean from the data in either table, the application of (1.3.4) would require the 120 readings to be considered individually. The following formula simplifies the calculation.

$$\overline{y} = \sum_{i=1}^{k} f_i c_i \Big/ \sum_{i=1}^{k} f_i$$

where k is the number of categories (9 in Table 1.3.6); f_i is the frequency in the ith category; c_i is the response associated with the ith category.

(a) Use the above formula for $\overline{y}$ to establish the mean number of cars passing both sites are identical.

The simplification in formulae may be extended to the variance and leads to the following alternative expression for the sample variance to that given in (1.3.6).

$$s^2 = \sum_{i=1}^{k} f_i (c_i - \overline{y})^2 / (n-1)$$

where $n = f_1 + f_2 + \cdots + f_k$ is the sample size.

(b) Use the above formula for s^2 to compute sample variances for both sites.

(c) Which site is nearer to the traffic lights? Is traffic flow more or less even at this site?

1.8 *Physiotherapy training* — *Example* 1.1

(a) Analyze the data in Table 1.2.1 to provide a comparison of the pain experienced by *trained* and *other* women. As part of the analysis,

 (i) compare levels of both *pain rating index* and *pain reaction* between the two groups;

 (ii) consider the possibility that any differences in pain experienced by the two groups is related to group differences in other factors for which information is provided.

(b) Prepare a short report of your findings.

1.9 *The robustness of statistics*

The following set of numbers, extracted from Table 1.3.1, are pre-race glucose levels on a sample of 11 runners.

72 79 95 84 87 85 77 35 79 86 69 .

When placed in order from smallest to largest, it is apparent that one reading is far removed from the remainder.

35 69 72 77 79 79 84 85 86 87 95 .

Could the reading *35* be an error? If so, what effect would it have on statistics computed from the data?

(a) Compute the *arithmetic mean* and the *median* from both the data as recorded and the data modified by the deletion of the reading *35*. Which statistic has been more affected by the removal of the extreme reading from the data set?

(b) Consider the data sets represented by

y 69 72 77 79 79 84 85 86 87 95 .

where y takes values 35, 45, 55, 65, and 75.

Determine both the *mean* and the *median* from the data sets formed by including each value of y.

Plot *mean* versus y and *median* versus y.

These plots establish that the median is less affected by the odd value than is the mean. In statistical jargon, the median is more *robust* against errors in the data than is the mean.

1.10 *Continuation of* 1.9 — *for the more theoretically inclined*

The possibility of an error in the data and its impact on the determination of the average value of a set of numbers has led to interest in the class of statistics defined below.

Suppose $y_1, y_2, \ldots, y_n$ is a collection of responses for the variable y, and the values when ordered from smallest to largest are denoted by $y_{(1)}, y_{(2)}, \ldots, y_{(n)}$. There is a class of statistics which are defined by the formula

$$m = \sum_{i=1}^{n} c_i y_{(i)},$$

where $c_1, c_2, \ldots, c_n$ is a sequence of non-negative numbers which

(i) sum to 1;0; and

(ii) have the property that $c_i = c_{n+1-i}$ for all $i \leq \frac{1}{2}n$.

In practical terms, all members of this class provide weighted averages of the set of values $y_1, y_2, \ldots, y_n$, with the c_i's being the weights.

(a) Establish that the arithmetic mean and the median are both members of this class.

Trimmed means are more general statistics defined by the formula

$$\begin{aligned} \overline{y}_t(a,b) &= \textstyle\sum_{i=a}^{b} c_i y_{(i)} && \text{for } 1 \le a \le b \le n \\ \text{where} \quad c_i &= 1/(b-a+1) && \text{for } a \le i \le b \\ &= 0 && \text{otherwise.} \end{aligned}$$

Winsorized means are variations on trimmed means defined by the formula

$$\begin{aligned} \overline{y}_w(a,b) &= \textstyle\sum_{i=a}^{b} c_i y_{(i)} && \text{for } 1 \le a \le b \le n \\ \text{where} \quad c_i &= 1/n && \text{for } a < i < b \\ &= (a)/n && \text{for } i = a \\ &= (n-b+1)/n && \text{for } i = b \\ &= 0 && \text{otherwise.} \end{aligned}$$

(a) Establish that the arithmetic mean and the median are extreme cases of trimmed means.

(b) Repeat Problem 1.9 using (i) a trimmed mean from which the lowest and highest values are excluded (i.e., $a = 2$ and $b = n-1$); and (ii) a winsorized mean in a is set equal to 2 and b is set equal to $n - 1$.

2

Data: The Factual Information

2.1 Data Collection

2.1.1 "Representative" Samples

In Analytic Statistics, the data collected are used to provide the factual information about the experimental process which generated them. Hence, there is good reason to seek a sample of persons from a population or a sample of outcomes from an experimental process which is *representative* of the population or the outcomes from the process.

What constitutes a *representative* sample? The notion which comes to mind is a sample which is in essence a miniature version of the population, with characteristics matching those found in the population or process being sampled. Thus, if a population is 60% female and 40% male, the sample should have approximately six females for every four males; if a process generates items of which one percent are defective, the sample should contain one defective for every ninety-nine non-defectives.

There are obvious problems associated with the identification and collection of a representative sample. Some of the more important non-statistical problems are:

- the need to have an extensive knowledge of the characteristics of the population being sampled;
- a confidence that all the characteristics which should be matched between sample and population have been identified;
- a sample size which is sufficiently large to accommodate all the restrictions imposed by matching;
- the difficulty of defining a method of sample selection to obtain a representative sample.

From the statistical viewpoint, it is not generally possible to give a precise meaning to this notion of a representative sample and hence not possible to construct statistical methods which can utilize information from a sample which attempts to mimic the characteristics of the population being sampled. An important exception is in the area of opinion polls and sample surveys where the one population is repeatedly sampled to seek information of a similar nature.

The development of statistical methodology is based largely on a different notion of *representative*. The concept associates with the *method of sampling* rather than the individual samples themselves. When selecting a single individual or unit from a population, *representative sampling* is characterized by the condition that each population member has the same chance of being selected.

The implication of this definition is found in the property that the probability of an individual sample member coming from a specific category in a population equals the proportion of population members in that category. For example, suppose a university student population has the following characteristics when partitioned by faculty:

31% of students are in the Arts faculty
19% of students are in the Science faculty
12% of students are in the Applied Science faculty
38% of students are in other faculties.

Representative sampling requires that there be a probability of 0.31 the selected student will come from the Arts faculty, a probability of 0.19 the student will come from the Science faculty, etc.

2.1.2 Basic Methods of Sample Selection

As distinct from the basis for sample selection, it is necessary to consider the actual method by which samples are selected. Furthermore, representative sampling in itself does not define a method of sampling. Primary sampling methods can be placed in the four categories which are listed below.

1. Systematic sampling
2. Random sampling
3. Haphazard sampling
4. Other methods

Systematic sampling is the method of sampling which utilizes a list or arrangement of population members or a sequence of outcomes from an experimental process plus a repetitive rule for sampling which is based on that list or sequence. For example, choose every fifth name on a list, choose the house numbered 7 in each street. Experimental processes operating in a repetitive manner are ideally suited to systematic sampling.

Random sampling is a method for obtaining a sample of fixed size from a population in which any collection of population members of the given size has the same probability of forming the sample as does any other collection of the same size. By this definition, it follows that random sampling is based on representative sampling, i.e., each population member has the same chance of being included in the selected sample. However, this is insufficient to define random sampling. Additionally, it must be assumed that the probability of any one member being included in the sample is not conditional on the presence or absence of any other member in the sample. In other words, the sample members are *independently* chosen.

A mechanism for selecting a sample which meets the conditions of random sampling requires an objective, machine-based system and is described in Appendix 1 of Chapter 13.

As expressed in the preceding paragraph, the notion of random sampling is not defined for experimental processes. There is no *population* to be sampled. Samples with the statistical characteristics of random samples are, however, obtainable if

(a) the chance of any specific response remains unchanged over the period to which the investigation applies; and

(b) an objective method of sample selection applies.

The first of these conditions is, in part, analogous to the requirement of independence in selection for random sampling. Additionally, it imposes the restriction that the experimental set-up is unchanging over the period of the study. Clearly, this is a necessary requirement since a changing trend, e.g., seasonal variations, would imply that the characteristics of the data would depend on when or from where the sample was selected. It is convenient to call processes which satisfy these conditions *repetitive processes.*

It is conventional to refer to the sampling from repetitive processes by a systematic method as random sampling since it leads to the same statistical properties as achieved by random sampling from a population. That practice will be adopted throughout the book.

Experience suggests that an essential condition for random sampling to be achieved is that the choice of sample members is by a machine-based system, i.e., is independent of human choice. Any attempt to simulate random sampling in which the sample members are subjectively chosen is termed *haphazard sampling.* The term applies equally to attempts to choose a representative sample by non-systematic means. There is evidence that haphazard sampling is likely to incorporate the experimenter's beliefs and prejudices. This is particularly unsatisfactory if the conclusions are labelled as *objective* because of their statistical basis. Where there is a practical alternative to haphazard sampling, there is no justification for using the technique.

There are many specialized methods of sampling in common use. For example, in experimental designs (see Chapter 13) the sample members may be non-randomly selected to ensure a homogeneous collection. This makes sense in a situation in which the units are to be allocated between different treatments for the purpose of comparing the effects of the treatments since non-treatment variability would only make the search for treatment differences more difficult.

In medical studies of rare diseases, it is often necessary to include in the sample all cases of occurrence of the disease to ensure a sufficiently large sample for the statistical analysis to be of adequate precision. Such sampling practice, in which the selection is based on the known response of the subject, is known as *retrospective sampling* as distinct from the usual *prospective sampling.*

These and other specialized sampling methods generally require special statistical methods which are not considered in this book. You should, however, be critically aware of the method of sampling employed in your research or research which you are examining. If a non-standard method of sample selection has been employed, you should seek expert advice on the correct method of statistical analysis. Equally, if there is doubt as to the method by which the sample was selected, the conclusions based on analysis of data in that sample should be viewed with suspicion.

2.1.3 Random or Systematic Sampling?

Sampling procedures which have a subjective component (i.e., require a human choice to decide on the inclusion or exclusion of a unit) should be avoided where possible. However both random sampling and systematic sampling remain possibilities.

Systematic sampling is a common choice when sampling outcomes from a process. As noted above, for repetitive processes it has the same statistical characteristics as random sampling.

Where successive outcomes are not independent, systematic sampling is still commonly employed but with the requirement that the statistical model must contain details of the interdependence between successive outcomes. Specialized areas of Statistics known as *Time Series Analysis* and *Analysis of spatial patterns* have been developed to cater for studies in which responses are related.

The intuitive advantage of systematic sampling is that it can be made to ensure an even coverage over the list or region. If population members with similar characteristics are clustered together, then systematic sampling can be expected to provide samples which are more representative of the population than would be provided by random sampling. A practical advantage of systematic sampling lies in the relative ease of defining the method of selecting a sample.

From a statistical viewpoint, these advantages are generally outweighed by the disadvantages, the most serious of which are the following.

1. There may be systematic trends in the responses which could seriously distort the findings if they coincide with the systematic rule in the selection of the sample. For example, selecting house number **1** in each street sampled in a housing survey will increase the probability of including houses on corner blocks; selecting every fifth outcome in

a sequence may lead to serious error if there is a cyclic pattern in the outcomes which repeats every fifth or tenth outcome.

2. To employ a sampling procedure which prevents neighboring units or persons from appearing in the same sample operates against the idea that systematic sampling leads to representative sampling.

   Thus, to take every one thousandth name in a telephone book reduces the probability of related people appearing in the same sample and prevents more than one occurrence of an uncommon name in the sample. Having the capacity for related population members to appear in the same sample is a reflection of the structure of the population.

These deficiencies are important to statisticians since they introduce considerations which generally cannot be accommodated in statistical models. Hence, the powerful methods of inferential Statistics may not be applicable.

Random sampling does possess disadvantages both in possible difficulty of implementation and in its failure to ensure samples are *representative*. Yet it is unique in possessing the all-important property of providing a basis for valid application of statistical methods without any requirement for knowledge of the characteristics which define a *representative* sample.

Primarily for this reason, random sampling is the basic sampling assumption in the vast majority of statistical models employed in scientific investigations. Where systematic sampling is used in practice, it is generally under circumstances where the investigator is prepared to make the assumption that the properties of systematic sampling are identical with those of random sampling. The sampling of repetitive processes is perhaps the most common example.

2.1.4 Complex Sampling Methods

Random sampling is not without deficiencies. It may lead to individual samples which are different in nature from the population being sampled. For example, if a population contains equal numbers of males and females, there is only one chance in four that a sample of size ten will comprise five males and five females. There is a chance of about one in 500 that the sample will contain ten members of the same sex. If there is the possibility of a sex difference in the attribute under investigation, there is clearly much sense in choosing a sample which is equally representative of the sexes.

Random sampling does not have to be abandoned to meet this objective, merely modified. One approach, known as *Stratified Random Sampling*, involves partitioning the population into groups and randomly selecting from within the groups with the restriction that specified proportions must come from each group. In the example introduced above, the sample could be selected to ensure five males and five females were included, but with random sampling being employed within each sex.

Where there are many sub-populations formed in the partitioning of the population, for example suburbs in a large city, a *Multi-stage sampling* procedure may be introduced. At stage 1, a random selection of sub-populations is made (e.g., a random selection of suburbs). From those sub-populations selected at stage 1, stage 2 comprises a random selection of sub-sub-populations (e.g., streets in a suburb). Units (e.g., houses) are randomly selected at the third stage of selection. Obviously, the process could be extended to more stages.

There are many variations which employ random sampling as the basis, but which strengthen the basic method. Coverage of these methods is to be found in the field of *Survey sampling.* (See, for example, Cochran [1977].)

2.1.5 Practical Difficulties in Sampling

It is one thing to speak of the ideals of sample selection and another to be faced with the real-life task of sample selection. What happens to the medical researcher who must rely on volunteers rather than a random selection of population members; the geologist who is given rock specimens taken from the moon; the experimenter who cannot readily identify the population under study, let alone seek a means of collecting a random sample?

Rarely does it make sense to stop the experiment or throw away data because random sampling was not employed in the collection of the data. Equally, it is foolish to pretend that the data were collected by random sampling and proceed to an analysis and interpretation on a possibly faulty assumption.

There is not a single approach which is best for all occasions. The one common sense piece of advice which applies is to clearly state in the report of the study the method of sample selection and to make clear the assumptions which underlie the subsequent analysis.

In some cases it may be wise to do no more than employ Descriptive Statistics to summarize information contained in the data collected and to use the study as a pointer to further investigations. In other cases, there may be strong grounds for arguing that the method of sampling can reasonably be claimed to possess the properties of random sampling which underlie the preferred statistical methods. In such cases, the analysis may proceed as though random sampling had been employed. Experience and scientific honesty are the guideline.

2.2 Types of Data

Data come in many forms and are collected for many purposes. The form of statistical methodology which may be applied and the sensible application of Statistics depend significantly on the nature of the data which are available. In Table 1.2.1, it is obvious that variables such as *Age* and *Time in labor* are measured on a scale while others such as *Marital status* are categorical. Other variables are less obvious. For example, *Pain reaction* is recorded as a whole number value between 1 and 5. Can the numbers be considered to lie on a scale or are they simply indexing a set of ordered categories? In the latter case, the letters A,B,...,E could equally well have been employed to identify the categories.

The reason for selecting a variable can influence its place in the statistical analysis. Again, Example 1.1 can be used to provide an illustration. *Marital status* is a variable on which information was collected, not because it was of interest to the investigator to establish the relative numbers of married and unmarried mothers in the community, but to give supplementary information which might aid the investigation in understanding the reasons why women experience different levels of pain during childbirth.

When faced with the task of selecting statistical methods or interpreting statistical analyses, an understanding of the different types of data and their functions is essential. The information required is summarized in this section of the book.

2.2.1 "Response" and "Supplementary" Variables

In the simplest possible terms, a *Response* variable is one which is being studied because the investigator is primarily interested in characteristics of the pattern of variation or level of response of the variable. Thus, in Example 1.1, both *Pain reaction* and *Pain rating index* would be classified as Response variables because the primary function of the study was to examine pain experienced by women at birth.

The definition extends to cover variables which are being studied in pairs or sets to learn something of an unmeasurable variable about which they provide information. For instance, a biologist may wish to compare the *shape* of leaves in different species of a plant. Shape is not, in itself, measurable, but it is related to length and width measurements on the leaves. Thus, the biologist may utilize length and width variables to provide information on shape and such variables would be referred to as Response variables.

The alternative to a Response variable is a *Supplementary* variable. As the name suggests, the information supplied by such variables supplements that provided on response variables and is collected to better understand the variation observed in the Response variables. In opinion polls, information is sought on characteristics such as Age, Sex, and Occupation to improve the interpretation of the findings on the question which is of primary interest, e.g., support for a political party.

Supplementary variables which are categorical are used in many situations to define groups which are to be compared. The variable *Sex* which partitions a population into males and females generally has this role. In cases where the supplementary variable has this role, it is usually referred to as a *Factor*. The distinction is introduced for the convenience of users—it has no theoretical or methodological significance.

There is the possibility that in one stage of analysis a variable may be a response variable while at another stage or in a later study it may be a supplementary variable. The classification is determined solely by the questions posed by the investigator.

2.2.2 "Random" and "Non-Random" Variables

The manner in which a sample is selected is of great importance in establishing the sensible use to which the information collected from the sample may be put. Clearly, if I were to seek out tall people in a community to comprise my sample and use the mean height of these people as an estimate of the mean height in the community, I would be likely to overestimate the mean height of the community of people. Statisticians have special names for variables according to the way the sample on which they were measured has been collected.

The intuitive notion of a *random variable* is dictated by the manner of sample selection and depends on the requirement that each sample member is selected without regard to the value taken by that variable. If the sample has been randomly selected from a population or comprises a set of independent outcomes from a repetitive process (as defined in Section 2.1) then any variables for which data are obtained from the sample members are random variables.

Where sample members are selected because they possess certain values of a variable then that variable is a *non-random variable*. Thus, to select a sample of people to ensure there are equal numbers of males and females or a good spread of ages would automatically identify *Sex* and *Age* as non-random variables.

Sometimes a variable is a random variable conditional on values taken by other variables as the following situation illustrates:

Suppose a sample of humans is selected to ensure equal numbers of males and females, but with random selection from within the sexes. If, for each sample member, the level of cholesterol in the blood is measured, *Cholesterol* is a random variable within each sex category. In this case, Cholesterol has all the properties of a random variable if statistical analysis is performed within each sex separately.

2.2.3 Categorical and Scaled Variables

A basic division of statistical methods is determined by the type of variable under investigation. The primary distinction is decided by the method of assigning values of the variable. If the value is assigned by counting (e.g., number of insects on a leaf) or by measurement (e.g., diameter of a rivet) then the variable is a *scaled* variable. The alternative is the assignment of values by classification. For example, humans may be classified by race or eye color; a rock may be classified by geological origin; floods may be classified as *minor, moderate* or *severe.* Such variables are *categorical* variables and can be distinguished from scaled variables because no concept of distance or scale can be defined. Even in the flood classification where *moderate* is intermediate between *minor* and *severe,* there is no assumption that it is midway between the other two categories.

Scaled variables can have related numerical properties when applied to a population or process. It is sensible to speak of the mean diameter of rivets. It makes no sense to speak of the mean eye color. The statistical methods appropriate to the analysis of data collected on categorical variables are of limited form. The most common form of analysis is the examination of the data after summarization in frequency tables.

There is a division of scaled variables into two categories—*discrete* and *continuous.* Discrete variables are easily identified by the fact that they can only take whole number values. The number of occurrences of an event over time or space (e.g., cars sold in a week, diseased trees in a sample area in a forest) is a common source of discrete variables. Continuous variables, as the name implies, can assume values over a continuous range. Note that the practical limitation on measuring accuracy does not generally remove the tag *continuous* from such variables.

Categorical variables are classified as *ordered* if the categories have a natural ordering—as in the flood example where the categories are *minor, moderate* and *severe.* Where the ordering applies to all categories, the variable is said to be at *ordinal* level. Where there is no obvious ordering of the categories, the variable is said to be at *nominal* level. An example of a variable at nominal level is the simple classification of humans by blood types. The four categories are O, A, B and AB.

Note that categories at nominal level may have a structure apart from order. For example, if immigrants are classified by country of origin, the categories may be grouped according to the continent in which the countries are found, e.g., Asia, Europe, etc.

There is also a need to make reference to numerical coding of data. A source of confusion arises when categories of a categorical variable are assigned numerical values. This practice is common in the social and medical sciences. Suburbs in a city might be assigned numbers to denote their socio-economic level, perhaps using the numbers 1 to 10. This practice has become particularly prominent since the widespread introduction of computers for data storage. Care must be exercised to use and interpret these data in an appropriate form. If the variable is to assume the status of a scaled variable, then the arbitrariness of the scaling employed must always be borne in mind. This is particularly so when the number of categories and their definition is not obvious.

Problems

2.1 *Sampling experiment for an individual*

(a) Find a convenient spot for observing people passing by—a seat in a park or in a cafe—and record, for 100 sets of five consecutive passers-by, the number of females in each set. Make a score sheet like the following:

No. of females in sample		
0							5
1																				18
2																												26
3																								22
4																							21
5										8
Total		100

Under a model which proposes (i) equal numbers of males and females in the pool of persons from whom the passers-by come; and (ii) independence of classification of sample members, the expected numbers of samples with $0, 1, \ldots, 5$ females is the following:

Number of females	0	1	2	3	4	5
Expected numbers*	3	16	31	31	16	3

(b) When comparing the observed set of frequencies in (a) with the set expected under the conditions described above, it was noticed that the signs of the differences had the following pattern:

Number of females	0	1	2	3	4	5
Frequency						
–observed(o_i)	5	18	26	22	21	8
–expected(e_i)	3	16	31	31	16	3
–diff.($o_i - e_i$)	2	2	−5	−9	5	5
Sign of difference	+	+	−	−	+	+

*These numbers are based on (7.1.1).

It can be seen that higher than expected frequencies were observed at either end and lower than expected frequencies observed in the middle. The expected frequencies are based on a model. Deduce the possible error in the model which would lead to this pattern of signs of differences.

(c) Compare the set of frequencies which you obtained with the expected frequencies provided above. Is there a pattern to the set of differences? If there is, suggest how the model underlying the expected frequencies might be in error.

2.2 *Sampling experiment for a group*

Example 3.2 provides a model for a sampling experiment which illustrates how a pattern can underlie apparent chaos. Individual pieces of data, in the form of lengths of sampled sentences, showed enormous variability. Yet frequency tables based on collections of sentence lengths showed an increasing stability in pattern as they were based on larger sample sizes. This is an essential quality of Analytic Statistics, being able to define a pattern for the structure of data yet not being able to accurately predict responses from individual sample members.

Employ the same format as that described in Example 3.2 and see if your class can identify an author from some measurable characteristic of his or her writing.

Spend some time in designing the sampling experiment. How will the samples be selected?

Make sure there is no ambiguity in the quantity to be measured. (It was necessary in the experiment reported in Example 3.2 to define what constituted a sentence and what constituted a word.)

2.3 *Sampling using the computer*

Most computers have been programmed to generate what are commonly called *random numbers.* These are a sequence of numbers which typically lie between zero and one and are presumed to have the statistical properties that (i) the chance of a number lying within the interval $[a, b]$ where $0 \leq a < b \leq 1$ is equal to the size of the interval, i.e., $b - a$ (for example, the chance that the value will lie in the interval $[0.2, 0.3]$ is 0.1); (ii) the value of any one *random number* is not dependent on the values of any numbers which have gone before it.

The examination of the variation in *random numbers* gives one a feeling for the scope of what is called *sampling variation.*

(a) Compute 200 random numbers and put the first ten in variable C1, the next ten in variable C2, etc.

(b) Construct separate line diagrams for all of the twenty sets of data and note the variation in pattern.

(c) Compute the arithmetic means for the twenty sets of data and present the twenty means in a line diagram. Note that the pattern of variability in the means is quite different to that in the individual data sets.

(d) Combine the data for C1 and C2, C3 and C4, etc. to form ten variables each with twenty values. Construct line diagrams from the data in the ten variables.

(e) Combine the data from C1 to C4, C5 to C8, etc. to give five variables each with forty values. Construct line diagrams from the data in the five variables.

(f) Combine the data into a single variable with 200 observations. Construct a line diagram.

The expectation is that the values will be uniformly distributed between zero and one. Note how the expected distribution becomes more evident as sample size increases.

2.4 *Methods of sampling*

Ideally, the function of shuffling a pack of cards before commencing a game is to ensure that the process of dealing cards from the pack has the characteristics of random selection in respect of any one hand.

(a) Take a pack of cards in which the cards are in the order *Ace*, 2, ..., *King of hearts, Ace*, 2,..., *King of diamonds, Ace*, 2, ..., *King of Clubs, Ace*, 2, ..., *King of Spades.*

(b) Shuffle the cards for one minute. Lay out the top thirteen cards face up. Do you consider the shuffling has been effective? What criteria have you used to reach this decision?

Is it possible to define expected characteristics of a sample obtained by random selection?

Can you suggest an alternative way of achieving random selection of a hand?

3

Statistical Models: The Experimenter's View

3.1 Components of a Model

The idea was introduced in Chapter 1 that Statistics provides a means of using data collected in a study together with other factual information to gain knowledge of the population or process from which the information was collected. There is a formal structure by which Statistics is employed for this purpose and it involves the construction of statistical models.

One perspective on these models is provided by mathematical statisticians which is necessary for the general development of the discipline. Scientists who wish to use Statistics as a tool in investigations require quite a different view of the models. Their perspective must define the models in terms which have a direct experimental meaning.

The description can be presented in three components, although not all components will necessarily be present in any one model. Broadly speaking, the components are determined by

1. Experimental aims
2. The method of sample selection
3. Additional information about the nature of the population or process under investigation.

It would be difficult to envisage a worthwhile investigation or experiment which was planned without the investigator having a reason for undertaking the study. It commonly happens, however, in the course of the study, that statistical analysis is employed to explore information collected without a prior statement of intention by the investigator.

Models employed in this secondary analysis may not have a stated experimental aim and would, therefore, fall within the province of *Exploratory data analysis.* (Tukey [1977] and McNeil [1977] provide introductions to this area of statistical application.) Typically, this involves highlighting structure in the data which might suggest hypotheses for future examination.

Given the primary purpose for the study, it is always possible to convert experimental aims into hypotheses contained within statistical models. Section 3.2 is concerned with the manner in which this is done and the range and scope of such hypotheses.

In Chapter 2, variations in methods of sample selection were discussed. Of particular concern in the construction of statistical models is the presence of any spatial or temporal connection between the sample members which might cause a relationship between responses from different units. In Section 3.3, the manner in which information on the method of sample selection is incorporated into statistical models is discussed.

Finally, there is a need to utilize additional information which is known about the population or process under investigation. Basically speaking, the more relevant information which can be incorporated into the statistical model, the stronger will be the conclusions which can be reached. Much of the book is concerned with the type of information which can be used and the manner in which it is included in the statistical models. Section 3.4 provides some insight into this component of statistical models.

3.2 Hypotheses

Let us consider possible experimental aims in what is perhaps the most popular area of statistical application of our times—the "opinion poll." Suppose interest lies in the electoral support for the "Radical" party which attracted 15% of the vote at the last election. At the time the poll is taken, the percentage support for the party is unknown. Call it p. The type of questions which might be asked about the popularity of the party are:

1. "Has the party maintained the same level of support as it gained at the last election?"

2. Has support for the party grown since the last election?"

3. "What is current level of support for the party?"

The first question identifies a specific value for p, namely 15 and the aim of the statistical analysis could be stated as testing the hypothesis that $p = 15$. There is a qualitative difference between the first and second questions. In the latter there is not a single value of p to be examined—the hypothesis to be tested states that $p > 15$. Thus, acceptance of any value of p in excess of 15 would provide support for an affirmative answer.

The third question offers another variation. Here there is no obvious hypothesis, i.e., no prior conditions are imposed on the value of p. From the statistical viewpoint, this is a problem in *estimation*.

To provide a unified statistical approach to these different experimental questions, the following strategy may be employed. Any statistical model defined will include only one value for p, i.e., will state $p = p_0$ where p_0 is a specified number. Thus, in forming a statistical model for use with question 1, a single model with $p = 15$ will be employed. The statistical aim would be to establish if this model is consistent with the data obtained in the opinion poll.

When considering question 2, there are two facets of interest. The most obvious is the need to establish if there is any model with $p > 15$ which is consistent with the data. However, even if there exists a model with $p > 15$ which is reasonable, such a find does not preclude the possibility that a model exists which has a value of p between 0 and 15 and is consistent with the data. Strong evidence in favor of a growth in support for the political party requires both (i) an acceptable model which includes a value of p in excess of 15, and (ii) no model with p in the range 0 to 15 which is acceptable.

To provide an answer to question 3, models constructed with all possible values for p (i.e., from 0 to 100) would have to be compared with the data and those models identified which are judged to be compatible with the data. As will become apparent later in this book, this invariably means that the statistician cannot provide a single value for p but must give a range of values which are likely to include the actual level of support for the party.

Hypotheses are not necessarily concerned with the determination of a numerical characteristic of a population or process as Examples 3.1 to 3.3 illustrate.

Example 3.1 (Seasonal Variations in Thyrotoxicosis in Tasmania). *A condition of the thyroid gland known as "thyrotoxicosis" was under investigation in Tasmania in the* 1960's. *Monthly data on the number of cases were collected over a three year period and are presented in Table* 3.2.1.

Table 3.2.1. Cases of Thyrotoxicosis in Northern Tasmania from May 1966 to April 1969

Number of Occurrences

	Month											
Year	J	F	M	A	M	J	J	A	S	O	N	D
1966					2	4	2	5	6	8	6	8
1967	5	4	4	2	1	8	8	6	2	2	11	4
1968	4	3	1	6	2	3	4	3	2	5	1	1
1969	1	2	5	4								

Of particular interest was the theory that there existed a seasonal variation in the incidence of the disease. This led to the consideration of a model which included the hypothesis of no seasonal variation. Thus, if M is the average monthly incidence of the disease, the hypothesis merely states that

M takes the same value for all seasons without any statement as to what that level might be.

For the purposes of statistical analysis, the data are summarized in Table 3.2.2. *Also included are the frequencies expected under the assumption of no seasonal variation. These frequencies are simply determined by dividing the total number of occurrences over all seasons* (145) *equally among the four seasons.*

Would you reject the possibility that there is no seasonal variation based on the available information? (*Chapter* 16 *provides a formal method of statistical analysis.*)

Table 3.2.2. Distribution of Thyrotoxicosis by Season (Based on Data in Table 3.2.1)

Season Months[1]	Summer DJF	Autumn MAM	Winter JJA	Spring SON	Total
Observed Frequency	32	27	43	43	145
Expected Frequency[2]	36.25	36.25	36.25	36.25	145

Notes:

[1] Summer in the southern hemisphere is December (D) to February (F), etc.

[2] Expected frequencies constructed under the assumption of no seasonal variation.

Example 3.2 (A Case of Disputed Authorship). *A class of statistics students in the University of Tasmania was given a novel known to be written by one of three authors with the aim of determining which of the three wrote the book. Additionally, the students were given a book of known authorship by each of the authors in question.*

The students chose to compare patterns of sentence length in each book of known authorship with that in the book of unknown authorship in a bid to identify the unknown author.

The data they collected are recorded in Table 3.2.3. *In this case there were three competing hypotheses, each one claiming that the pattern of sentence lengths for the unknown author is identical with a different known author, i.e., Green, Morgan and Greene. The problem was to make a choice between the three.*

The students produced tables of relative frequencies (*Table* 3.2.4) *and plotted the* Relative cumulative frequency *versus* Sentence length *for each known author. The plot for the unknown author was then superimposed and the resulting graphs can be seen in Figure* 3.2.1.

Table 3.2.3. Distributions of Sentence Lengths Based on the Study Reported in Example 3.2

GREEN

	Sentence length							
Group	1–5	6–10	11–15	16–20	21–25	26–30	31–35	>35
1.	12	23	17	9	5	9	1	4
2.	13	18	10	13	7	3	5	11
3.	10	21	16	9	10	3	3	8
4.	13	23	25	5	6	4	0	4
5.	5	25	13	14	6	3	6	8
6.	14	23	14	13	8	4	2	2

MORGAN

	Sentence length							
Group	1–5	6–10	11–15	16–20	21–25	26–30	31–35	>35
1.	20	19	10	7	3	3	3	15
2.	10	17	10	8	8	6	6	15
3.	15	13	12	12	8	3	5	12
4.	10	13	6	7	12	4	6	22
5.	3	16	9	11	9	3	2	7
6.	9	19	12	8	10	5	4	13

GREENE

	Sentence length							
Group	1–5	6–10	11–15	16–20	21–25	26–30	31–35	>35
1.	7	8	17	7	9	7	8	17
2.	3	21	20	13	8	6	4	5
3.	8	22	15	10	6	7	3	9
4.	5	15	15	16	10	6	4	9
5.	6	15	21	9	13	5	4	7
6.	3	10	22	18	14	7	1	5

UNKNOWN

	Sentence length							
Group	1–5	6–10	11–15	16–20	21–25	26–30	31–35	>35
1.	73	150	168	153	116	69	49	76
2.	20	52	42	59	43	25	19	42
3.	17	79	77	76	55	53	21	34
4.	14	49	55	43	50	27	16	24
5.	35	74	78	63	56	25	17	20

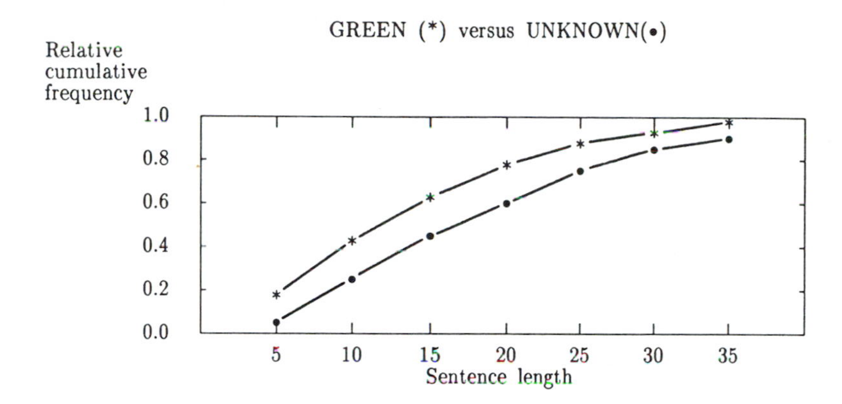

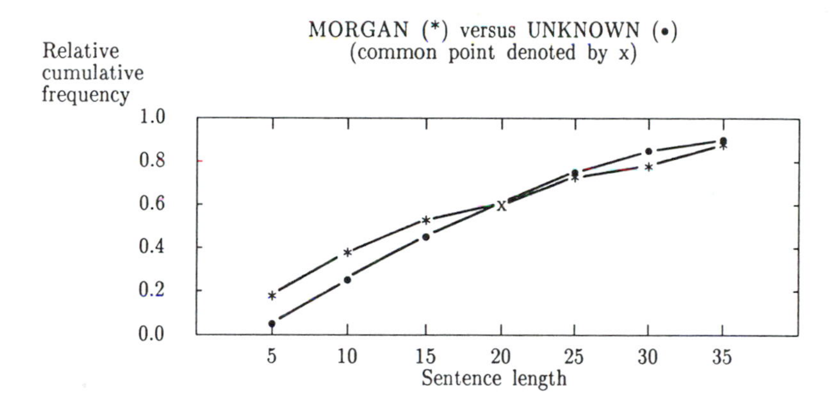

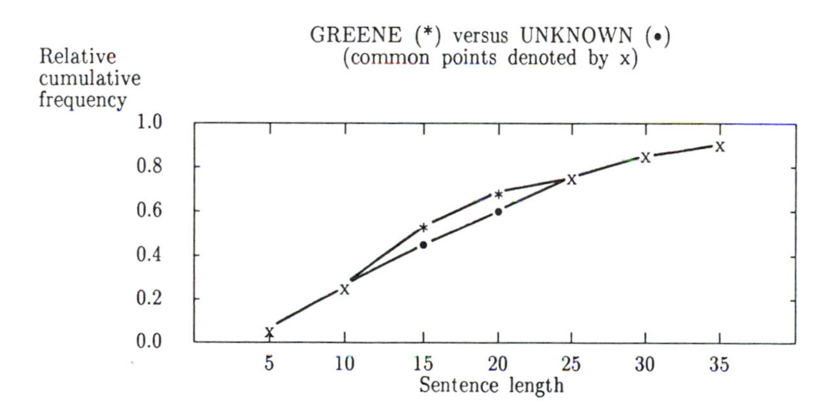

Figure 3.2.1. Relative cumulative frequency versus sentence length (based on data in Table 3.2.4).

Although there was huge variation in observed distribution of sentence length between tutorial groups, the argument was advanced that the large amounts of information collected from each book and the clear cut graphical picture indicates that Greene was the mystery author.

This was one of those rare occasions in which the conclusion from a statistical analysis could be verified. The students were correct!

Table 3.2.4. Relative Percentages of Sentence Lengths Based on Data in Table 3.2.3

Sentence length	1–5	6–10	11–15	16–20	21–25	26–30	31–35	>35	Sample size
Green	14	28	20	13	9	5	4	7	480
Morgan	15	21	13	12	11	5	6	18	460
Greene	7	19	23	15	13	8	5	11	480
Unknown	7	18	19	18	14	9	6	9	2214

Example 3.3 (Examining a Mathematical Model in Blood Flow). *Dr. Kilpatrick of the Department of Medicine at the University of Tasmania has spent some time considering models for blood flow in veins. It has been common to assume that the drop in pressure along a length of vein was linearly related to flow rate. This did not accord with his model and so he undertook an experiment in which a piece of tubing took the place of a vein and fluid of known viscosity was forced along the tube at a range of flow rates. The drop in pressure was recorded for each flow rate. A portion of the data collected is contained in Table* 3.2.5.

The hypothesis examined in this case was one which Dr. Kilpatrick believed to be incorrect, namely that the expected pressure drop is linearly related to the form of the mathemtical structure of the relationship.

Table 3.2.5. Data from the Experiment Reported in Example 3.3

Flow rate	0.11	0.19	0.51	0.75	0.95	1.13	1.45
Pressure drop	1.1	2.1	5.3	6.8	8.4	9.8	12.9

Flow rate	1.65	1.99	2.37	2.69	3.31	3.66	4.25
Pressure drop	14.7	16.7	20.6	23.3	29.5	33.3	39.9

There are many other variations on the form of hypotheses which might be encountered. There are even occasions in which the hypothesis concerns the form of sample selection. Some cases of "rigged experiments" have been uncovered in this way. Example 3.4 *provides an illustration of one in which the author was involved.*

Example 3.4 (In the Footsteps of Mendel). *Gregor Mendel is credited with putting forward a theory of genetic inheritance concerned with single genes. Biology students in a practical session were given the task of providing experimental evidence that Mendel's theory was expressed through corn seeds appearing in an expected ratio of three round seeds to one wrinkled seed. Six groups of students were each asked to classify* 200 *seeds in the experiment. The task was rather boring and since the students knew that they were expected to get about* 150 *round seeds and* 50 *wrinkled seeds they decided to fudge results by making up totals. The figures they obtained are recorded in Table* 3.2.6.

These numbers look reasonable. Certainly they show clear support for the 3:1 *theory.*

The fact is that they are too good. The greatest departure from the theoretical expectation of 150:50 *is shown by group* 4 *with* 153:47. *Under the model which presumes students examined each seed independently, greater sampling variation is expected. Indeed, a result as extreme as* 138:62 *would not be surprising. The chance of six groups obtaining such close results to the expectation as those presented in Table* 3.2.6 *in a properly conducted experiment is so small that the model was queried.*

Specifically, the component which presumed random sampling was queried. Of course, the statistical analysis itself could not prove that the experiment was rigged. It did, however, provide the lever by which a confession was extracted from the students concerned.

Table 3.2.6. Frequencies of Round and Wrinkled Seeds (Data from Example 3.4)

Group	1	2	3	4	5	6
Number of round seeds	152	148	149	153	152	151
No. of wrinkled seeds	48	52	51	47	48	49
Total number of seeds	200	200	200	200	200	200

One feature about the translation of the investigator's question into a statistical hypothesis which often seems contradictory to non-statisticians is the reversal of the question. In Example 3.4, to provide evidence for non-random sampling, the hypothesis was made that random sampling was employed. In Example 3.3, to establish if the relation between pressure drop and flow rate was non-linear, the hypothesis was proposed that it was linear. In Example 3.1, to seek evidence for seasonal variation in the incidence of thyrotoxicosis, the hypothesis included in the model assumed no seasonal variation.

One reason for this apparently contradictory approach by statisticians lies, not in a desire to be contrary, but in the fact that the model defined must have the capacity to define expected outcomes for comparison with observed outcomes.

Consider the situation in the examination of seasonal variation in thyrotoxicosis (Example 3.1). Under the assumption of no seasonal variation, the expected frequencies are identical in all seasons (Table 3.2.2) and are uniquely determined. Under the hypothesis that there are seasonal fluctuations, there is not a unique set of expected frequencies.

As another example, consider the assumption of linearity in Example 3.3. A linear relation between two quantities is well defined whereas a non-linear relation encompasses an infinity of forms.

A second reason for the reversal of the hypothesis lies in the need to exclude hypotheses as possible contenders. For example, to provide strong evidence that a political party has majority support in an electorate, it is necessary to provide evidence against the possibility that it has minority support. The fact that an opinion poll based on a sample of electors yields 52% support for the party may not exclude the possibility that in the electorate at large there is only minority support.

In some circumstances there is an alternative to placing the onus on the investigator to provide a statistical hypothesis. In problems where *estimation* can be employed, experimenters need do no more than note the existence of a parameter of the population or process, and statistical analysis can be employed to provide likely values for the parameter.

3.3 Sampling Assumptions

The vast majority of statistical models require the assumption that the sample which provides the data has been selected by a process of *random selection* (Section 2.1). The importance of this assumption is made apparent in Chapter 5.

Where the sample members are not independently selected, there is a need to assume a structure or mechanism by which observations made on the sample members are connected. Generally, the statistical description is difficult even though the experimental description may be simple. For example, where plants are competing for light, moisture or nutrients, a strong growing plant is likely to be surrounded by weaker plants because of competition. Attempting to model the effects of this competition is not an easy task and frequently leads to the introduction of parameters of unknown value into the model.

Example 3.5 offers some insight into the problems of model construction.

Example 3.5 (Modeling Population Growth in Tasmania). *Figure 3.2.2 displays a plot of the estimated population of Tasmania by year for the period 1900 to 1932. The plot suggests a linear trend in population growth over the time of study. However, there is clearly some structure in the pattern of deviations of the points from the linear trend. Note that when the population is above the line in one year, there is an increased chance it* *will be above the line in the following year. In other words, if the trend line is regarded as defining the average level, one above average year is likely to be followed by another.*

In the absence of this connectedness, it would be sufficient to rely on the line to make a prediction of next year's population. As it stands, a model must be produced which utilizes (at least) this year's population value as well as the average value obtained from the line to predict next year's value.

How might the two pieces of information be combined? If A_{n-1} *is the prediction from the line at time* $n-1$ *and* Y_{n-1} *is the actual population at time* $n-1$*, then perhaps the predicted population at time* n *could be defined as* $A_{n-1}+p(Y_{n-1}-A_{n-1})$ *where* p *is a parameter whose value is to be specified. Perhaps the prediction should be augmented with a component from time* $n-2$*?*

There are many possibilities, but all of them increase the complexity of the model and make statistical analysis qualitatively more difficult as a result of the lack of independence between observations.

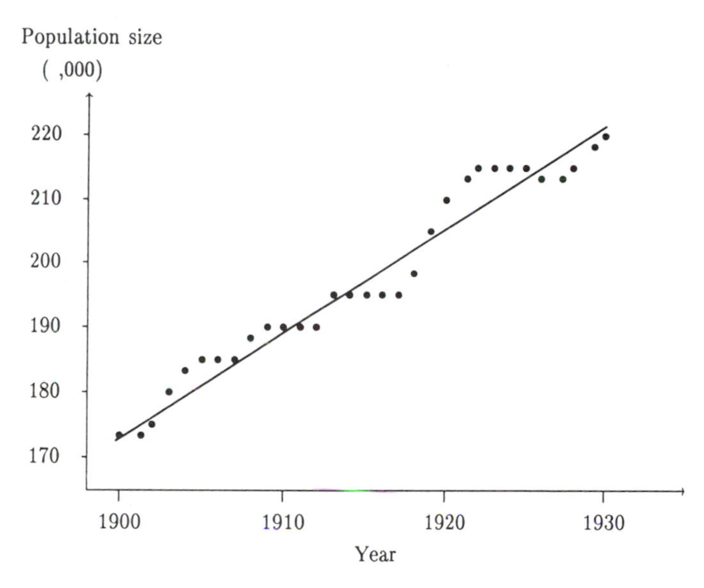

Figure 3.2.2. Population size of Tasmania by year between 1900 and 1932.

The relationship among sample members may not always be obvious. For instance, a researcher in education who selects groups of children from different classrooms for a study must include in the statistical model provision for the fact that children from the same classroom are likely to produce responses which are related because of the common connection of being in the same class. Unlike the case in the population study where each year has a neighbor, in the school case each child is equally close to many neighbors. The structure in the statistical model to reflect this fact is quite different from that employed in the population case.

Translating information about connectedness of sample members into statistical assumptions is a process which requires both expertise and experience.

3.4 Statistical Assumptions

The final component of the model expresses additional information which may take many and varied forms. This will become apparent as the book develops. At this point, only a brief reference will be made since more background information must be provided before the significance of the statistical assumptions can be appreciated.

3.4.1 Distributional Assumptions

The data in Table 3.2.1 recording monthly levels of thyrotoxicosis possess a feature which is typical of sets of observations that are subject to statistical examination, namely *unexplained variation.*

While there may be some identifiable reasons for the variation in incidence of the disease, e.g., seasonal fluctuations, there remains a component of the variation which is not accounted for by available knowledge. For the application of many statistical procedures it is necessary to include in the statistical model a characterization of this unexplained variation.

Interest in this aspect of the model dates back at least to Gauss who coined the name *random error* for the unexplained variation. There was a time when the pattern of variation in responses from scaled variables was thought to be almost universally described by a particular mathematical form which consequently was given the title *Normal distribution.* The basis for this distribution is given in Section 6.5.1.

Nowadays, there are many distributions which are employed and some of these are introduced in Chapter 6. The particular importance of distributions will become apparent with the discussion of probability in Chapter 5.

3.4.2 Design Structure

Where the aim is to compare the effects of different treatments on a collection of subjects or experimental units, the researcher may employ special experimental designs. In such cases, there is a need to introduce into the statistical model a statement of the experimental design. An introduction to this area is found in Chapter 13.

Problems

3.1 *Experimental and statistical hypotheses*

In Example 3.1, the investigator was seeking evidence of seasonal variation in a disease. While he was quite clear about the aim of the study, the best approach to a statistical analysis is not so obvious as is revealed in this problem.

(a) Use the data for Example 3.1 which are provided in Table 3.2.1 to complete the following table of frequencies.

Season	Month	Observed frequency	Expected frequency	
			No seasonal variation	Seasonal variation
Summer[1]	Dec.	13	12.1[2]	10.7[3]
	Jan.			
	Feb.			
Autumn	Mar.			
	Apr.			
	May			
Winter	Jun.			
	Jul.			
	Aug.			
Spring	Sep.			
	Oct.			
	Nov.			

[1]Note that summer occurs in the southern hemisphere in December, January, February, etc.

[2]Assuming the expected frequency is unchanging throughout the year.

[3]Assuming the expected frequency may vary between seasons but not within seasons.

(b) Under what circumstances might it be wiser to base the analysis on the monthly figures rather than the seasonal figures? (Hint: consider the different forms in which seasonal variation might occur.)

(c) What hypothesis would be examined by comparing the observed monthly frequencies with those expected under a model which assumes seasonal variation but no variation within seasons?

3.2 *Simulation*

Computers offer the possibility of simulating experiments. This technique can be used to simulate the genetic experiment described in Example 3.4 and will allow you to gain a feel for the magnitude of sampling variation which is to be expected in such an experiment.

(a) Use your statistical package to generate a sequence of 200 *uniformly distributed pseudo-random numbers.* There should be a simple means of doing this.

Pseudo-random numbers, or more strictly, uniformly distributed pseudo-random numbers, have the property that they are simulating output from an experimental set-up in which each outcome is equally likely to come from any point in a specified range of values—usually zero to one—and the value for any outcome is independent of values of previously generated pseudo-random numbers. Thus, if the range is zero to one, then the odds of the value being below 0.75 are 3:1.

To relate this to Example 3.4, suppose a pseudo-random number below 0.75 is equated with examining a corn seed and finding it is round, whereas a value of 0.75 or higher is equated with finding the seed is wrinkled.

By using the computer to repeatedly generate sets of 200 pseudo-random numbers and determining how many of these numbers are below 0.75 in each set, it is possible to approximate the expected pattern of sampling variation in the biological experiment, given that the assumption of independence of outcomes is observed.

(b) Use your statistical package to count how many of the 200 pseudo-random numbers have values less than 0.75 and print out this figure. (Usually this would be achieved by coding the pseudo-random numbers such that values below 0.75 are assigned the number 1 and values above are assigned the number 2. A frequency table of the coded values then gives the numbers of values below and above 0.75.)

(c) Repeat the whole process six times to provide a table comparable with Table 3.2.6. Does your table show more variability in the numbers of round seeds than are found in Table 3.2.6? Unless, by chance, you have obtained unusual results, there should be evidence of considerably more variability in your results than those observed in Table 3.2.6.

I used the MINITAB statistical package to produce six results and obtained the following figures:

Simulation	1	2	3	4	5	6
No. below 0.75	148	142	151	153	145	166
No. above 0.75	52	58	49	47	55	34

3.3 *The "Normal" distribution*

Construct histograms of (i) the 137 observations of blood cholesterol levels given in Table 1.3.2; and (ii) the 250 readings in Table 1.3.4 of temperatures at which limit thermostats cut out. Notice the similarity in shapes of the two histograms.

The widespread occurrence of this particular pattern of distribution of numbers obtained from measurement on random samples of objects has been noted for over two hundred years. It was titled the *Normal distribution* because of the frequency with which it was encountered. This is the basis of one model which statisticians employ widely. It is discussed in Chapter 6.

4

Comparing Model and Data

4.1 Intuitive Ideas

4.1.1 THE SIGNIFICANCE APPROACH

The use of inferential Statistics to answer experimenter's questions requires the comparison of a statistical model with data from an experiment. The purpose of this chapter is to introduce the important statistical approaches to this task. However, an appreciation of the statistical bases is more easily gained if thought is first given to the way in which the problem might be approached without the benefit of Statistics.

Consider the following data which are taken from Table 1.3.1 and which relate to blood glucose levels in untrained runners before and after a long distance race.

Subject	Blood glucose level		Difference
	Pre-race level	Post-race level	
1	67	100	33
2	46	73	27
3	67	89	22
4	66	95	29
5	70	76	6
6	60	108	48
7	72	77	5
8	76	88	12
9	72	104	32
10	53	108	55
11	66	84	18

The question of interest centers on the change in glucose levels during a race and the hypothesis that glucose levels are expected to rise.

Clearly the data support the hypothesis—there are eleven rises and no falls among the sample members. Yet is this sufficient grounds for accepting the hypothesis? Is it possible that rises and falls is simply a chance happening—like tossing a coin eleven times and obtaining eleven heads?

If eleven rises and no falls provides sufficient grounds for rejecting the hypothesis that rises and falls are equally likely, could this hypothesis be confidently rejected if there had been ten rises and one fall, nine rises and two falls, eight rises and three falls, seven rises and four falls or six rises and five falls?

All of these outcomes have the property that there is a surplus of rises over falls. Yet, as the surplus of rises over falls lessens, there comes an increasing possibility that the difference is due to chance rather than a property of the physiological process.

While individuals may vary in the point at which chance becomes an acceptable explanation, a common basis exists for making a decision. This comprises (i) a ranking of possible outcomes to reflect a decreasing acceptance that chance provides the explanation; and (ii) a point in the ranking beyond which chance is rejected as an acceptable explanation.

Without a statistical basis, the decision is purely subjective. With a statistical basis, there is a mechanism for introducing objectivity into the process of reaching a decision. Can the essence of the subjective approach outlined above be converted into a formal statistical procedure?

There is a statistical approach which adopts this strategy. It is known as the *Significance* approach and is developed and discussed in Section 4.3.

Intuition may suggest that the key to deciding on the acceptability of a model lies in determining how *likely* is the observed outcome under the chosen model. Such a direct application of probability is easily dismissed. Consider, for example, the result that in a sample of 1,000 runners, there are 500 rises and 500 falls in blood glucose level during a race. This is the most likely outcome if rises and falls are expected with equal frequency. Yet, even if the true situation is that rises and falls are expected with equal frequency, there is only about one chance in forty of obtaining this result—not a likely outcome.

Statistics offers three alternatives to this direct reliance on the chance of obtaining a specified outcome under a chosen model. One of the statistical approaches is discussed in Section 4.1.2 and the others in Section 4.3.

4.1.2 The Bayesian Approach

The intuitive argument employed in Section 4.1.1 has rested on the premise that the first step is to define a (single) model and then to consider possible outcomes under that model. In reality, this appears to be contrary to the interest of scientists. They *know* the outcome as a result of having done the experiment. It is the experimenter's hypothesis, or more generally, the *statistical model* describing the experimental set-up which generated the data, which provides the uncertainty. Thus, given the observed outcome, it would seem more sensible to attempt to attach a degree of belief to a model, or, at least, to rank possible models from the most plausible to the least plausible.

It is not possible to attach a degree of belief to the correctness of a model without reference to other possible models. Attempting to attach a probability to a model without specifying the possible alternatives would be like asking a bookmaker to give odds on a horse winning a race without knowing the other horses in the race. It is only if all possible models are identified, that attaching degrees of belief to different models becomes feasible. This is analogous to the horse racing situation in which the bookmaker provides odds on each horse in the race.

Even so, while this approach has application in horse racing, there is disagreement in the statistical profession as to how widely it can be transferred to scientific studies. Obviously, there is a potential problem in turning knowledge and feeling about the correctness of a model into a numerical value. Bookmakers have the advantage that they can rely on precise information in the form of bets laid to set their odds.

A further problem relates to the need to be able to specify all possible models. Again, the racing analogy is relevant. Bookmakers know precisely which horses are in the race. Scientists cannot always identify all possible models. Yet, without complete specification of the possible models, the approach has no basis. In practice, this is perhaps the most serious obstacle to the general application of this approach, which is known as the *Bayesian approach* since it is based on a theorem proposed by Thomas Bayes.

There is little objection to the principle of the Bayesian approach. Where practical objections can be met it is widely recommended. Example 4.1 provides an illustration of the classical use of the Bayesian approach.

Example 4.1 (Screening Test for Breast Cancer). *In the population at large, one woman in eight hundred has undetected breast cancer. Thus, a woman selected from the population by random sampling has one chance in eight hundred of having breast cancer without the fact being known. This constitutes the degree of belief prior to any information being collected from the woman herself. If the woman undergoes a screening test for breast cancer and obtains a negative result, what is the modified probability of her having breast cancer?*

If the test were one hundred percent accurate, the data alone would provide a certain conclusion. Unfortunately, the test is not infallible. What is known is (i) *the probability the test will give a false positive reading is one in ten; and* (ii) *there is a 95% chance of a correct positive reading. Before the screening test, the odds against the woman having breast cancer were* 800:1. *What are they after the positive test is obtained?*

The Bayesian solution to this question relies on a consideration of rules of probability. It is presented in Section 8.5.

4.1.3 Significance Approach or Bayesian Approach?

The Significance approach and the Bayesian approach adopt fundamentally different strategies. In the former, the model is presumed correct and the judgment is based on how well each of the possible data sets supports the model. In the latter, the data are accepted as the fixed component and the possible models are compared to assess the extent to which each model is supported by the data. The choice depends on circumstances.

If there are competing models and there is a reasonable basis for assigning degrees of belief in the models prior to the current study, then the Bayesian approach is likely to be the better. Alternatively, if there is a particular model of interest and the set of alternative models is difficult to formulate, or in the absence of reasonable prior degrees of belief in the possible models, the Significance approach may be needed.

The statistical and scientific community have, through usage, acknowledged the Significance approach as the approach which generally satisfies practical needs. Consequently, this book contains little development of Bayesian methodology. Nevertheless, there are circumstances in which the Bayesian approach is valuable. Readers are strongly urged to read the discussion on the Bayesian approach in Section 8.5 and, where it seems relevant to a study they are undertaking, to seek Bayesian methodology.

4.1.4 The Likelihood Approach

A valid criticism of the Significance approach is that it has no formal mechanism for utilizing information on possible alternative hypotheses to the one under investigation. If an alternative hypothesis is known, there is sense in seeking an ordering of the possible outcomes which reflects lessening support for the stated hypothesis *relative to the stated alternative hypothesis.* This would presumably lead to the greatest chance of detecting that the alternative hypothesis is correct.

Under a wide range of important practical conditions, statistical theory can establish the best ordering of possible outcomes. It does so through a procedure known as the *Likelihood approach,* so named because it depends on how likely an outcome is under different possible hypotheses (or more completely, models). The statistic which defines the ordering is based on a ratio of how likely is an outcome under the stated and the alternative hypotheses. Not surprisingly, therefore, it is called the *Likelihood ratio statistic.* This statistic can be defined so that outcomes are assigned larger values as they become more likely under the alternative hypothesis relative to the stated hypothesis. Further discussion of the Likelihood approach is to be found in Section 4.3.

4.2 The Role of Statistics

Whether or not Statistics is employed in the process, a decision about the acceptability of a model is generally based on only a portion of the information contained in the data. This is easily demonstrated by reference to the blood glucose example used in Section 4.1. The information on change in blood glucose level is contained in the set of differences before and after the race for the eleven runners, i.e.,

Runner	1	2	3	4	5	6	7	8	9	10	11
Change	33	27	22	29	6	48	5	12	32	55	18

The intuitive belief that there tends to be an increase is based only on the fact that there were eleven increases and no decreases observed. The facts that Runner 5 had a lower reading than Runner 2 or that Runner 7 had the lowest reading of all were not considered. These facts seem to be irrelevant to the question. In statistical terms, the decision was based on the value of a *statistic,* namely *the number of increases in the sample of* 11 *readings.*

In the Significance approach, a statistic is always used to define an ordering of the possible data sets from that one (or several) which is judged most consistent with the model to that one (or several) which is judged least consistent with the model. Thus in eleven tosses of a coin where the method of tossing is presumed fair and the coin unbiased, the results which are most consistent with the model comprise those which produce either 5 or 6 heads. As the number of heads tends towards 0 or towards 11 the result becomes less consistent with the model (and implicitly more consistent with either a biased coin or unfair tossing).

Once the ordering has been established there remains only the task of converting each position in the ordering into a probabilistic measure between model and data which can be used by investigators as a guide to the acceptability of the model. The manner in which this measure of agreement is defined and computed is described in Section 4.3.

There is no suggestion in the above discussion as to how the statistic should be chosen. Where there is a choice between two statistics which will lead to different orderings among the possible outcomes, there is obviously the possibility that a different conclusion might be reached dependent on the choice of statistic.

For example, in the study of changes in blood glucose, an alternative to the statistic which records the number of increases in the eleven readings would be the statistic which records the mean increase. An assumption of no average change in glucose level would be best supported by a sample mean difference of zero, and worsening support would be reflected in an increasing magnitude for the sample mean difference.

Which is the better statistic to use, the number of increases in eleven runners or the average increase in the eleven observations? Could there be another statistic which is better than either of these possibilities? These questions will be addressed in general terms in Chapter 8. Note however, that the Likelihood approach has the property that for an important range of statistical models it utilizes a statistic which is guaranteed to use all relevant information. This makes it a very attractive approach.

4.3 Measuring Agreement Between Model and Data

The final steps in the application of Inferential Statistics are (i) the measurement of the closeness of model and data and (ii) the translation of this measure of closeness into something which is meaningful to scientific users. For scientists, this is a critical area in the application of Statistics—the translation of statistical findings into the language of science. As with any translation, there may be aspects which are difficult to grasp or which are liable to misinterpretation. However, with an understanding of the statistical basis, the potential to misuse or misunderstand is greatly diminished.

This section provides an illustrative description of the manner in which the non-Bayesian approaches (Significance and Likelihood) are used to compare models and data and to the interpretation of their findings. The key features which appear in the discussion underlie all applications of these approaches. Rather than attempt this explanation in an abstract way or through a formal approach, it is developed by reference to the example described in Section 4.3.1.

4.3.1 An Example

Wildlife authorities are frequently interested in the size of a population in a given region. Many examples of experiments for estimation of population size are described in the literature—as diverse as hares, lobsters, pheasant, deer and fish.

A widely used method of estimation, known as the *Capture–Recapture method* has the following basis where the task is to estimate the number of fish present in a lake:

> A catch of fish is made from the area under consideration and all fish of the species of interest which are caught are tagged and returned to the water. A second catch is made and the numbers of tagged fish and untagged fish of the species are noted.

By way of illustration let the number of fish in the first catch be 1,200 (all tagged and returned to the lake), and suppose the second catch consists of 1,000 fish of which 40 are tagged. There are three basic questions which the investigator might pose. They are

1. What is the most likely number of fish in the lake?
2. Is 40,000 a reasonable estimate of the number of fish in the lake?
3. What are reasonable upper and lower limits on the size of the fish population in the lake?

4.3.2 Forming a Statistical Model

The statistician faced with seeking answers to these questions sets up a *model* based on the information provided. As noted in Chapter 3, there are potentially three components in the model.

1. The question posed by the investigator.

   (In this case that relates to the number of fish in the lake, i.e., the population size N. Different values of N identify different models.)

2. The sampling assumption.

   (Ideally, the presumption is that any one collection of 1,000 fish in the lake is as likely as any other to form the actual sample collected at the second catch. This, of course, is the presumption of random sampling. In practice, this assumption may be unacceptable—any keen fisherman will tell you that the big fish are harder to catch than the little fish! The problem of establishing a realistic statistical model is one which often confronts the user of Statistics.)

3. Statistical assumptions.

   (There is a place in the statistical model for a third component which includes additional information on the experimental set-up needed in the construction of statistical methodology. In this simple model, the only additional component is a "frequency distribution" which describes the relative proportions of population members in each of the two classes—tagged fish and untagged fish.

Class	1	2
Relative frequency	π_1	π_2

   where π_1 and π_2 are unknown "parameters.")

Models are indexed by different values of the population size N. The minimum value for N is 1,200 since that number of tagged fish was released into the lake. There is no upper limit which can be placed on the population size given the information supplied.

4.3.3 Data

The *sample* which has supplied the *data* comprises 1,000 fish in the second catch and the response which has been measured on each sample member, i.e., each fish, is whether the fish is tagged or untagged. Thus, the *response variable* is *categorical* with two possible categories.

4.3.4 Statistic

The *statistic* which is seen as relevant for the model-data comparison is the *number of tagged fish* in the sample.

4.3.5 Derived Information

The model and statistic which have been selected are a standard choice. Indeed, they will be found in Section 6.2 where a formula is given for calculating the probabilities of obtaining $0, 1, 2, \ldots, 1000$ tagged fish in the sample of 1,000. The formula requires that a value be specified for N. Of course, the correct value of N is unknown—the purpose of the study is to learn something about the number of fish in the lake.

To gain an understanding of the mechanics of the approaches, it is necessary to consider a selection of possible values for N. To establish a likely value, we could make an educated guess based on the fact that (i) the proportion of tagged fish in the second catch is 40/1000; and (ii) the proportion of tagged fish in the lake is $1200/N$. If the proportion of tagged fish in the sample equals the proportion of tagged fish in the lake, the value of N would be 1200/0.04 or 30,000.

Suppose we consider three possibilities, 30,000, 35,000 and 40,000, which, on the basis of our educated guess, represent successively the best estimate and values which are becoming progressively less likely. Figure 4.3.1 displays graphically the probabilities $0, 1, 2, \ldots, 1000$ tagged fish in a sample of 1,000 fish for each of these possible population sizes. (The calculation of these probabilities uses (6.2.1) which is derived using rules of probability introduced in Chapter 5.)

What can be learnt from the graphs in Figure 4.3.1 given that the observed number of tagged fish is 40?

There are two ways of looking at the graphs. One way is to consider each graph separately and base the conclusion on the probabilities of different possible outcomes from no tagged fish up to 1,000 tagged fish. This is the strategy adopted in the Significance approach. The other way is to compare the probabilities of an outcome of 40 tagged fish under different assumed population sizes. This underlies the Likelihood approach.

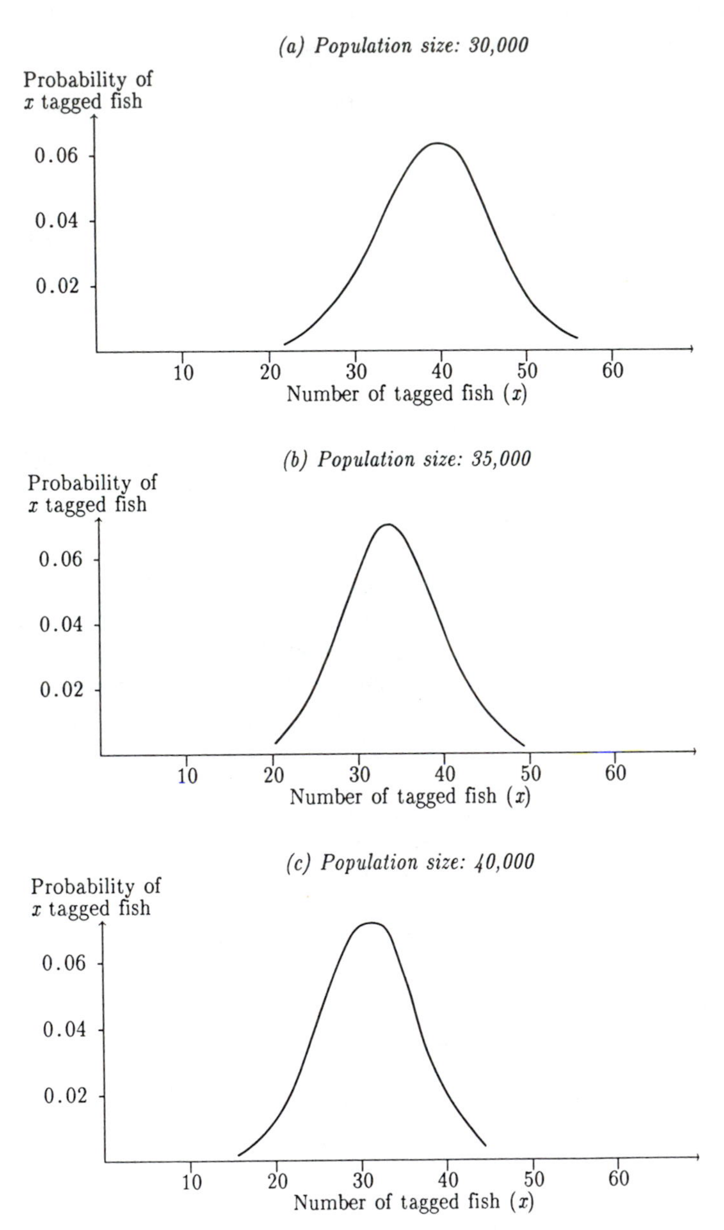

Figure 4.3.1. Distributions of probabilities under three models.

4.3.6 The Significance Approach

The basic requirement is that, for a model with a specified value of the population size N, there should be an ordering of possible outcomes from that outcome (or those outcomes) which is most consistent with the model to that outcome (or those outcomes) which is least consistent with the model.

An intuitively reasonable basis for the ordering would be to use the probabilities associated with the different outcomes. Thus, if the assumed population size is 30,000 (Figure 4.3.1(a)) the outcomes most consistent with the model are those containing 40 tagged fish in the sample of 1,000. The least likely outcome is a sample comprising 1,000 tagged fish in the sample of 1,000.

Changing the assumed population size (N) changes that ordering. When the assumed population size is 40,000, a sample containing 40 tagged fish is not the outcome which is most consistent with the model—that position is now held by outcomes containing 30 tagged fish.

Under the Significance approach, each position on the ordering is assigned a number called a *p-value* or *significance* which is equal to the probability of being at the observed point in the ordering or beyond (i.e., more extreme). This value derives directly from the probabilities of the different outcomes. The p-value has a maximum value of one which is reserved for the outcomes most consistent with the model and decreases towards zero as the outcomes occupy a more extreme position in the ordering.

The appealing feature of the approach is the constancy of interpretation of the p-value (p). The following convention is widely applied throughout scientific disciplines:

- if p exceeds 0.05 the data are said to be consistent with the proposed model and hence the experimental hypothesis contained within it is regarded as reasonable;
- if p is less than 0.05 there is some evidence against the assumed model;
- if p is less than 0.01 there is strong evidence against the assumed model.

Evidence against the model is usually taken as evidence against the experimentally-based hypothesis it contains.

There is a natural and obvious extension of the Significance approach which permits judgment to be passed on a collection of possible models. This is employed when the models are indexed by a parameter. By setting the p-value at a pre-defined level, say 0.05, it is possible to identify all models which have a p-value above the pre-defined level. In most practical applications this leads to a collection of parameter values which lie between two limits. Such limits are called *confidence limits* and the range of values they enclose is called the *confidence interval* for the parameter. The term is prefixed by the level of confidence which can be attached to it. If the pre-defined level of the p-value on which the interval is based is p_0, the interval is called the $100(1-p_0)\%$ *confidence interval.* In particular, if p_0 is set at 0.05, the corresponding interval is called the 95% *confidence interval.*

The above ideas can be given concrete expression in respect to the fish example. A plot of p versus population size N is provided in Figure 4.3.2.

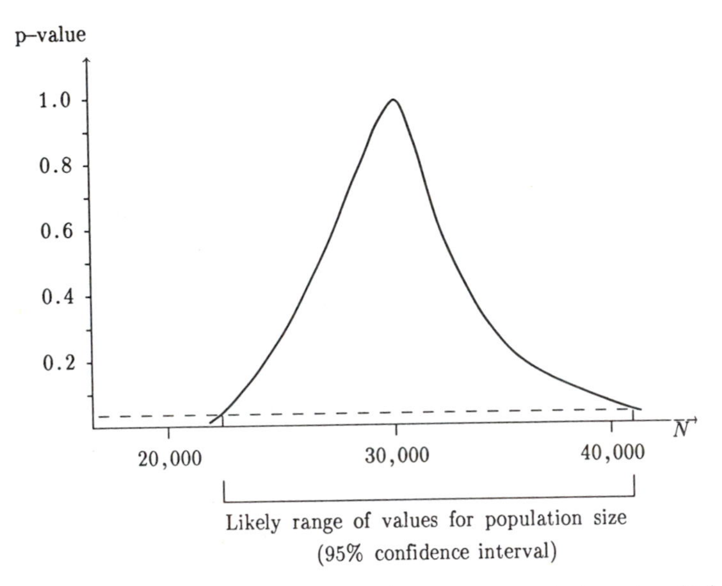

Figure 4.3.2. Plot of p-value (p) versus population size (N).

Based on the information in this figure, the Significance approach coupled with the guideline for interpreting the p-value quoted above, leads to the following answers to the three questions posed at the beginning of the section.

Question What is the most likely number of fish on the lake?

Answer The value of N corresponding to a p-value of unity, namely 30,000.

Question Is 40,000 a reasonable estimate of the number of fish in the lake?

Answer Since the value of 0.06 for p is above the level of 0.05, the outcome (i.e., the data) is consistent with the model and hence 40,000 could be regarded as a reasonable estimate of the number of fish in the lake.

Question What are likely upper and lower bounds on the number of fish in the lake?

Answer Typically, this question is answered by providing the 95% confidence interval. From Figure 4.3.2 this is the range 22,000 to 41,000.

The practical interpretation of p-values is not quite as simplistic as the above set of answers suggests as is revealed in the discussion in Section 8.2.5.

4.3.7 The Likelihood Approach

A deficiency in the Significance approach is the failure to provide a statistic on which the ordering of data sets can be based. The consequence is that the choice is arbitrary and there is no guarantee that it is optimal.

The Likelihood approach provides an objective basis for selecting the statistic which defines the ordering. It does so by establishing how likely is the observed outcome under each possible model (for the present, equate *likelihood* with *probability*) and identifying the model under which the observed outcome is most likely. The likelihood under this model is termed the *maximum likelihood* and serves as a yardstick against which to measure the likelihood of an outcome under other models.

For the fish example, Figure 4.3.3 illustrates graphically the relation between the likelihood of the observed outcome and the population size, N, which indexes the possible models. It is established that the model with $N = 30,000$ is the most likely model. It is termed the *Maximum Likelihood estimate* of N and the corresponding likelihood is termed the *Maximum likelihood.*

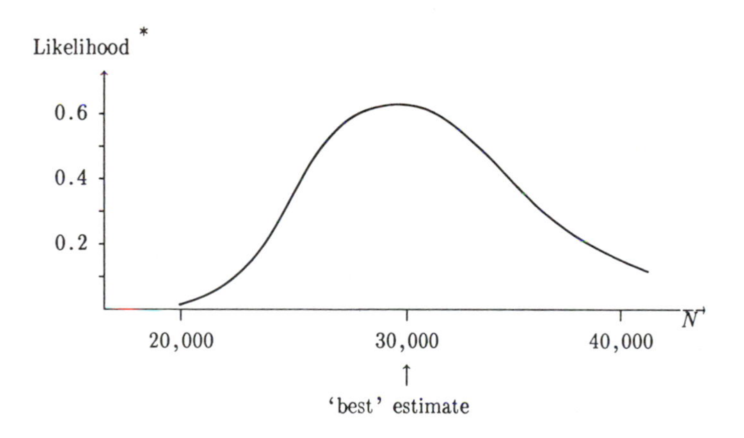

* *computed as probability of 40 tagged fish in sample of 1,000*

Figure 4.3.3. Plot of likelihood versus population size.

To judge whether population sizes other than 30,000 might be regarded as reasonable, the *Likelihood ratio* defined by (4.3.1) is employed.

$$\mathrm{LR}(N) = \frac{\text{Likelihood assuming population size } N}{\text{Maximum likelihood}} \tag{4.3.1}$$

where the Maximum Likelihood is the value of the Likelihood when N equals 30,000.

The Likelihood ratio must take values in the range zero to one, with larger values indicating better agreement between model and data. When N is set equal to the maximum likelihood estimate, namely 30,000, the Likelihood ratio reaches its largest value, and as N increases above or decreases below 30,000 the Likelihood ratio declines. For N equals 30,000, 35,000 and 40,000, the Likelihood ratios are as tabulated.

Population	30,000	35,000	40,000
Likelihood ratio	1.0	0.6	0.2

The Likelihood ratio of 0.2 for 40,000 indicates that an outcome of 40 tagged fish in a sample of 1,000 is five times as likely under the model which assumes N equals 30,000 as it is under the model which assumes N equals 40,000. However, the fact that the observed outcome is less likely given the latter value for N does not eliminate 40,000 as a possible population size. Allowance must be made for the vagaries of chance.

This can be done by considering the Likelihood ratio, (4.3.1), with the value of N fixed at 40,000. The Likelihood ratio is used to order possible data sets. Increasing values of the Likelihood ratio suggest greater support for the population size of 40,000 relative to the population size of 30,000. In fact, this ordering defines a statistic, the *likelihood ratio statistic.* This statistic provides a means for converting the observed outcome to a p-value in the manner described in the Significance approach.

In the Significance approach, the *number of tagged fish in the sample* was used to order the possible outcomes. There was no justification of this choice, it simply seemed the natural basis for the ordering. In fact, the Likelihood ratio approach leads to the same choice, but with sound reasons for the choice.

From the scientists' viewpoint, the Likelihood approach provides answers which have the same interpretation as those provided by the significance approach with the added bonus of defining a statistic which is guaranteed to use all relevant information in the data.

5

Probability: A Fundamental Tool of Statistics

5.1 Probability and Statistics

There is sometimes confusion over the distinction between Probability and Statistics. While both may use probabilistic models, one useful point of separation is in the purpose of the application. *Probability* is concerned with the characteristics of outcomes which are generated by a chosen model whereas *Statistics* is concerned with the characteristics of models which might have generated an observed outcome. In other words, they may be looking at the same experimental situation but from opposite directions.

While Probability has an important role in scientific investigation in its own right, perhaps a more important role is an indirect one, as a tool in Statistics. Since this book is written on the topic of Statistics, it is in this role that Probability is considered.

In the application of Statistics, there are three places in which recourse is made to Probability. The first is in the definition of the statistical model, the second is in the definition of sampling methods and the third is in the construction of a measure of closeness between the data and a proposed model.

The formal development of Probability and its application in statistical theory requires a mathematical level beyond that which can reasonably be assumed for readers of this book. Yet it is important for all users of Statistics to have an understanding of the manner in which the ideas and definitions of Probability are applied in the construction of statistical methodology. To this end, Section 5.2 is devoted to a largely non-mathematical introduction to the probabilistic constructs which appear in the application of Inferential Statistics.

For the mathematically inclined reader, Sections 5.3 and 5.4 provide definitions and rules of probability, introduce the means by which probability is defined in the continuum, and illustrate the formal connection between the definitions of probability and the statistical application of Probability. It must be appreciated however, that the function of this chapter is merely to provide an insight into the role of Probability in Statistics. I hope that readers with the appropriate level of mathematical expertise will be motivated to proceed from this informal introduction to a rigorous development provided by a lecture course or textbook on mathematical Statistics.

5.1.1 The Meaning of Probability

The word *probability* means different things to different people and may change its meaning depending on context. At least three views are useful in a scientist's view of Statistics.

1. The *Equal-likelihood* view. Given a population which may be partitioned into sub populations, the definition of probability can be in terms of proportions and sampling. Under a sampling scheme in which each population member is *equally likely* to be selected, the *probability* of selecting a member from a given sub-population is equal to the proportion of population members in that sub-population. The idea of equal likelihood in selection is embodied in the mechanism of random selection. It is also a concept which has a wide non-statistical perception.

2. Physical property of a system. Consider the statements that a previously unseen coin has a *probability* of one half of coming down heads when tossed, or that the *probability* of a child being albino is one quarter if both normal parents carry a gene for albinism. In these cases, the notion of probability is based on the physical properties of the experimental set-up.

   In assigning the probability to the result of tossing the coin, we implicitly assume a physical model in which the coin is symmetric and the method of tossing the coin is *fair.* In the biological example, we rely on a scientific model of genetic inheritance.

3. Long term frequency. To say that there is a *probability* of 0.01 that an item coming off a production line will be defective, is a consequence of past experience rather than a knowledge of the physical structure of the production line. The person making the statement presumably has knowledge that, of the previous items coming of the production line, one hundredth of them were defective.

None of these approaches provides a meaning for probability in the case where the response is continuous. Consider the case of round-off error and suppose the measurement of a quantity yields the value 23 *to the nearest whole number.* We may interpret this as meaning that the actual value is somewhere between 22.5 and 23.5. What is the probability it lies between 22.5 and 22.6? What is the probability it is exactly 22.6?

With no knowledge to the contrary, we may be prepared to accept that all values between 22.5 and 23.5 are equally likely to be the correct value. If there were a fixed number of values, say 10, then we could say that each value had a probability of 1/10 of being the correct value. But there are an uncountable number of values. Hence, this approach would lead to the probability that each possible value has a probability of zero.

There is some glimmer of hope if we consider the probability that the value lies between 22.5 and 22.6. Since this is one tenth of the total distance between 22.5 and 23.5 and since all values are equally likely, then a value of one tenth would seem reasonable as the probability the real value lies between 22.5 and 22.6.

In summary, it seems possible to base probabilities on collections of possible outcomes rather than individual outcomes. This notion is formalized in Section 5.3.

5.2 Sampling Distributions

A feature of scientific studies in which the processes of Inferential Statistics may be employed is the existence of *chance* or *sampling variation*. This is a term introduced to account for the fact that repetitive sampling of a population or process leads to samples with different compositions. A classic example is found in gambling. The characteristics of a pack of cards and the mechanics of dealing the cards may be well known, but the composition of individual hands will vary in a manner which makes the prediction of the contents of an individual hand impossible.

In scientific experiments, sampling variation tends to camouflage the characteristics of the population or process which are of interest to the scientist who is conducting the experiment. An essential task for Inferential Statistics is to establish whether any discrepancy between model and data could be explained as sampling variation, and, more generally, to quantify the uncertainty which sampling variation introduces. There is a well-defined sequence of steps involved in this process. The function of this section is to trace those steps.

The first step is to characterize the pattern of unexplained variation in the responses of population members or outcomes from an experimental process. This is done within the statistical model through the definition of a *frequency distribution* or a *probability distribution* (Section 5.2.1).

The second step is to provide a formal connection between this description of variation in the population and the pattern of variation expected in the sample. Because of the effects of chance variation, the response of a sample member which has been selected from the population or process cannot generally be predicted. However, it is possible to quantify the relative likelihoods of different possible outcomes. This involves the definition of a *sampling distribution* (Section 5.2.2) and a *sampling distribution of a sample* (Section 5.2.3).

The final step is the construction of the *sampling distribution of a statistic* (Section 5.2.4) which underlies the link between the formal statistical findings in the analysis and the scientific interpretation of these findings.

5.2.1 Frequency Distributions and Probability Distributions

For a population, the characterization of the pattern of variation in a variable is defined by a frequency distribution. This distribution has its simplest form when the variable has a fixed number (k) of categories, in which case the distribution defines the proportion of population members in each category and is commonly represented in a tabular form, i.e.,

Class	1	2	$\cdots$	k
Relative frequency	p_1	p_2	$\cdots$	p_k

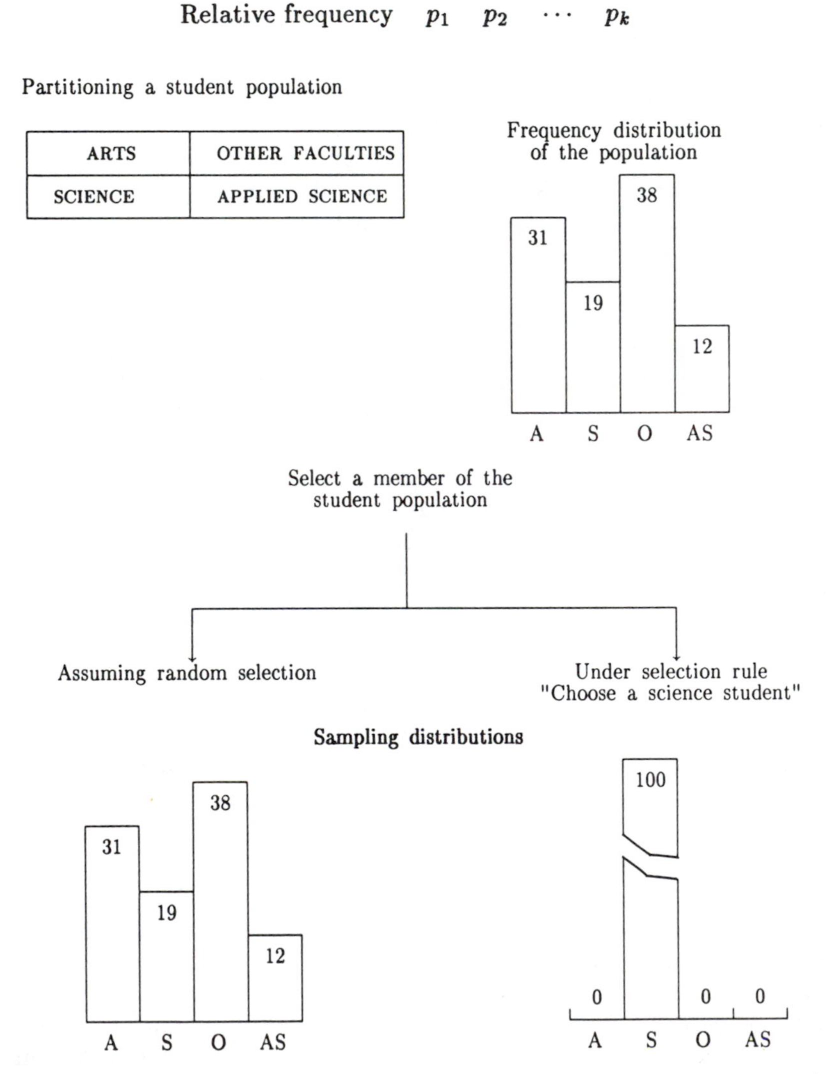

Figure 5.2.1. An example of a frequency distribution and the construction of sampling distributions.

A simple example is provided in Figure 5.2.1 where a categorical variable with four categories is defined. The variable identifies the faculty in which the students in a university are enrolled. The frequency distribution identifies the fact that there is a proportion 0.31 of students in the Arts faculty, 0.19 in the Science faculty, etc. This information may be expressed either in the histogram of Figure 5.2.1 or the following frequency table.

Class	1 Arts	2 Science	3 Other	4 Appl. Science
Relative frequency	0.31	0.19	0.38	0.12

When the attribute under investigation exhibits continuous variation (e.g., birth weights of babies), the number of possible classes is noncountable and the above definition of a frequency distribution is not applicable.

There is, however, a simple way of defining the distribution which both embodies the concepts introduced above and allows a unified statistical development. Under this approach, proportions are assigned to collections of values rather than single values, where the collections are chosen to correspond to intervals on the scale of measurement. The frequency distribution is then defined by a continuous mathematical function in such a way that the area under the curve which represents the function graphically, equates with the proportion of population members in the interval covered. This is illustrated in Figure 5.2.2.

Rather than relying on areas under curves to define probabilities for continuous variables, statisticians utilize the mathematical equivalent, integration. A frequency distribution is defined by a mathematical function f which is assumed to be defined over the whole real line. Should it be impossible for the variable to take values over a region of the real line—birth weights cannot be negative for example—the function is simply assigned the value zero over such a region.

The proportion of population members with values at or below a specified value y_0 is given by (5.2.1)

$$\Pr(y \leq y_0) = \int_{-\infty}^{y_0} f(y)\mathrm{d}y. \tag{5.2.1}$$

Some explanation is required with respect to continuous distributions. A value is typically recorded as a single number even though it is only accurate to within the limits of accuracy of the measuring instrument. Thus, a baby's weight is quoted as 3152 grams rather than 3151.5 to 3152.5 grams.

A frequency distribution for a continuous variable assigns the value zero to the proportion of babies with a birth weight of exactly 3152 grams (on the grounds that no baby will have a birth weight which is 3152.000 ... grams to an infinite degree of accuracy). This has the implication that

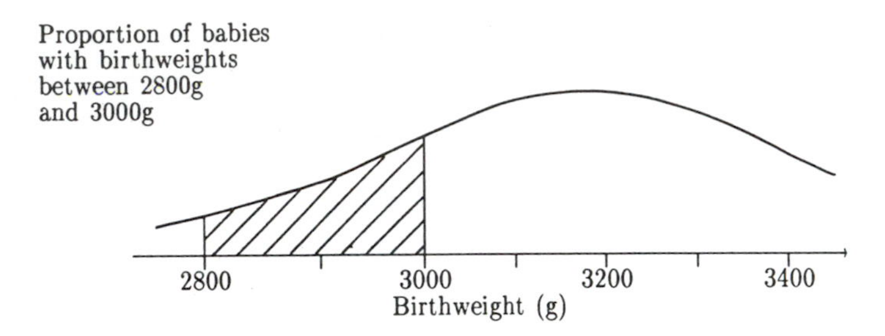

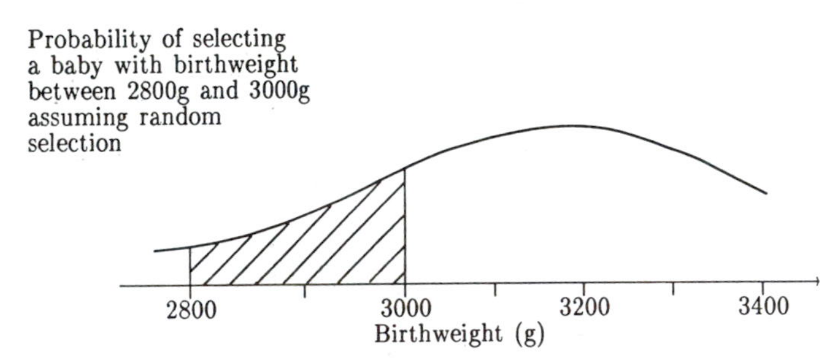

Figure 5.2.2. Examples of frequency and sampling distributions for a continuous response variable.

$$\Pr(y < y_0) = \Pr(y \leq y_0).$$

The proportion of babies in the interval [3151.5, 3152.5] would be defined by (5.2.1) as

$$\Pr(y < 3152.5) - \Pr(y \leq 3151.5) = \int_{3151.5}^{3152.5} f(y)\mathrm{d}y.$$

The above definitions of frequency distributions can be applied equally to the definitions of *probability distributions* of processes. For a categorical variable which may take values in k categories or a discrete scaled variable which may take one of k distinct values, the probability distribution defines the probability p_i that the variable will take the value i $(i = 1, 2, \ldots, k)$. For a continuous variable, (5.2.1) defines the probability that the value will lie at or below the value y_0. Note, however, that values of the function f are not probabilities for a continuous variable even though such a function is called a *probability* distribution.

A third type of variable is recognized in the case of processes. This variable is a scaled variable which takes whole number values but for which the number of possible values is not defined. For example, the number of cars passing an observer in the interval of one minute may be $0, 1, 2, \ldots$ with no means of fixing an upper limit. In this case, the probability distribution is represented by $f(y)$ for $y = 0, 1, 2, \ldots$ where f is typically defined by a mathematical formula.

5.2.2 The Concept of a Sampling Distribution

A requirement for the application of Inferential Statistics is that a population or process be sampled to provide data for the statistical analysis. The characteristics of the data obtained are in part determined by the characteristics of the population or process and in part determined by the method of sampling.

The combined effects of these two contributors is defined by a *sampling distribution.*

For a categorical or discrete scaled variable, this distribution defines the probability that a single sample member will have a specified value for the variable. If there are k possible values, the sampling distribution can be defined by a table of probabilities, i.e.,

Class	1	2	$\cdots$	k
Probability	π_1	π_2	$\cdots$	π_k

If the variable (y) is continuous, there is a sampling distribution, π, which can be employed to define the probability that the sample member will take a value at or below y_0 according to (5.2.2).

$$\Pr(y \leq y_0) = \int_{-\infty}^{y_0} \pi(y) \mathrm{d}y. \tag{5.2.2}$$

To illustrate the construction of a sampling distribution, consider the example portrayed in Figure 5.2.1. If a person is selected from the student population, what is the probability that person will be from the Arts faculty?

The answer depends on the frequency distribution for the student population and on the manner in which the selection is made.

If the person selecting the student adopts the strategy of deliberately selecting a Science student then the probability of selecting a student of the Arts faculty is zero. Alternatively, if the process of random sampling is employed and is based on the complete list of students, the probability of selecting an Arts student is precisely the same as the proportion of Arts students in the university, namely 0.31.

It is this special property of random sampling, namely the equating of the sampling distribution and the frequency distribution of the population being sampled, which makes random sampling so valuable.

If the aim is to learn something of the characteristics of the frequency distribution of the population in the situation portrayed in Figure 5.2.1, clearly random sampling has advantages over the rule *Select a Science student.*

5.2.3 Sampling Distribution of a Sample

For practical application it is necessary to extend the idea to a *sampling distribution of a sample.* This requires the specification of a sampling distribution, π_S, which assigns a numerical value to each possible experimental outcome defined as a sequence of responses $(y_1, y_2, \ldots, y_n)$ from the n members in a sample.

Again there is the distinction between the discrete and the continuous cases. Where the variable is categorical or a discrete, scaled variable, the sampling distribution may be represented in a tabular form and the values taken by the distribution may be interpreted as probabilities. If the variable is continuous, the sampling distribution defines probabilities only through the technique of integration.

Table 5.2.1 illustrates a sampling distribution for a sample of size three for a categorical variable which can take only two possible values labelled D (defective) and N (non-defective). It arises in the context of a production line in which there is a long term frequency of defectives of 1%. The remaining 99% of items are non-defective.

Table 5.2.1. Example of a Sampling Distribution (sampling distribution for a sample of three items from a production line where the outcomes are either defective (D) or non-defective (N) and outcomes are independent)

Sample	DDD	DDN	DND	NDD
Probability	0.000001	.000099	0.000099	0.000099

Sample	DNN	NDN	NND	NNN
Probability	0.009801	0.009801	0.009801	0.970299

Given that an objective method of sample selection is employed, the probability of a single item being defective is then 0.01 and of being non-defective is 0.99. To establish the probability that a selection of three items will all be defective, requires the application of the rules of probability. These are introduced in Section 5.3 and illustrated using the above example.

The definition of the sampling distribution of a sample is considered in Section 5.4.

5.2.4 Sampling Distribution of a Statistic

The final stage in constructing sampling distributions is to determine the sampling distribution of a statistic. Recall that a statistic is a rule for assigning an experimental outcome in the form of a set of responses, $(y_1, y_2, \ldots, y_n)$, to a single number.

For the example relating to items coming off a production line, a possible statistic would be ***the number of defectives in the sample.*** The sample outcome DNN would be assigned the value 1 by this statistic. In fact, the connection between possible sample outcomes and possible values of the statistic is easily listed:

Outcome	NNN	NND	NDN	DNN	NDD	DND	DDN	DDD
Value of statistic	0	1	1	1	2	2	2	3

Where the response variable is categorical, the connection between the set of possible outcomes and the values of a statistic can always be explicitly stated in the above manner.

For example, it is apparent from the above table that one defective in the sample of three will result if the outcome is NND or NDN or DNN (and from no other outcome). Hence, it is plausible that the probability of one defective in a sample of three is the probability of one of NND, NDN or DNN occurring. Using the rules of probability described in Section 5.3, it is possible to construct the sampling distribution for the statistic from the sampling distribution for the sample when the response variable is categorical. Table 5.2.2 provides an illustration and (5.2.3) provides the formal connection where the statistic has possible values $s_1, s_2, \ldots, s_k$, and the sampling distribution of the statistic is denoted by π_{st}.

Table 5.2.2. Relating the Sampling Distribution of a Statistic* to the Sampling Distribution of a Sample

Sampling Distribution of Sample		Sampling Distribution of Statistic	
Sample	Probability	Value of statistic (No. of defectives)	Probability
NNN	0.970299	0	0.970299
NND	0.009801		
NDN	0.009801	1	0.029403
DNN	0.009801		
NDD	0.000099		
DND	0.000099	2	0.000297
DDN	0.000099		
DDD	0.000001	3	0.000001

*Number of defectives in a sample of size 3.

$$\pi_{st}(s_i) = \Sigma\ \pi_S(y_1, y_2, \ldots, y_n) \quad \text{for} \quad i = 1, \ldots, k. \tag{5.2.3}$$

where the summation is over all possible outcomes which are assigned the value s_i by the statistic S.

The sampling distribution of the statistic provides the basis for constructing p-values which, as stated in Section 4.3, provide the basis for interpretation of the results of methods of Inferential Statistics. This connection is developed in Section 5.4.5.

5.3 Probability: Definitions and Rules

5.3.1 The Mathematical View of Probability

To develop a mathematical view of probability which can then serve as the basis for construction of statistical methodology, it is necessary to have a unified structure for probability. The intuitive ideas of probability presented above give little clue to the formalized mathematical structure.

To encompass situations in which the response is continuous, it is necessary to distinguish the set of possible outcomes, which in probability terms is called the *sample space,* and the set of entities to which probabilities can be attached. These entities are called *events.*

The distinction between the set of possible outcomes and the set of possible events is not difficult to appreciate. If a coin, which may land *heads* (H) or *tails* (T) is tossed three times, the set of possible outcomes (sample space) is

HHH HHT HTH THH HTT THT TTH TTT.

Possible events include individual outcomes, e.g., THH, as well as combinations of outcomes. Thus, *Two heads* is an event comprising the outcomes HHT, HTH, and THH; *Less than three heads* is an event which includes all outcomes except HHH; and *First two coins heads* is an event which includes the two outcomes HHH and HHT.

Where the response is continuous and the sample space comprises all possible values in the interval [22.5 to 23.5], an event can be formed as

(i) all outcomes in an interval contained within these limits;

(ii) the *union* of two intervals, which comprises all outcomes appearing in **either** of the intervals; and

(iii) the *intersection* of two intervals, which comprises all outcomes appearing in **both** intervals.

For example, the intervals [22.50, 22.60] and [22.55, 22.65] would comprise events as would their union [22.50, 22.65] and their intersection [22.55, 22.60].

5.3.2 Rules of Probability

The application of probability to construct sampling distributions of statistics requires the use of basic rules to determine the probabilities of events considered collectively. These are presented below using the notation E_1, E_2, $\ldots, E_n$ for n possible events.

1. The probability that both events E_1 and E_2 occur can be determined from either of the following formulae

$$\Pr(E_1 \text{ and } E_2 \text{ both occur}) = \Pr(E_1 \text{ occurs}) \times \Pr(E_2 \text{ occurs if } E_1 \text{ occurred})$$

or equivalently,

$$\Pr(E_1 \text{ and } E_2 \text{ both occur}) = \Pr(E_2 \text{ occurs}) \times \Pr(E_1 \text{ occurs if } E_2 \text{ occurred})$$

which are abbreviated to (5.3.1).

$$\Pr(E_1 \text{ and } E_2) = \Pr(E_1)\Pr(E_2/E_1) = \Pr(E_2)\Pr(E_1/E_2). \qquad (5.3.1)$$

Conditional probability

The probability that E_2 occurs when E_1 is known to have occurred is denoted by $\Pr(E_2/E_1)$ and is referred to as the *conditional probability* that E_2 occurs when E_1 is known to have occurred, and $\Pr(E_1/E_2)$ is referred to as the *conditional probability* that E_1 occurs when E_2 is known to have occurred.

2. The probability that at least one of the events E_1 and E_2 occurs can be determined from (5.3.2).

$$\Pr(E_1 \text{ or } E_2) = \Pr(E_1) + \Pr(E_2) - \Pr(E_1 \text{ and } E_2). \qquad (5.3.2)$$

Independent events

If the probability that E_1 occurs is not affected by the occurrence or non-occurrence of E_2 then E_1 and E_2 are said to be *independent events* with the consequence that (5.3.1) and (5.3.2) may be rewritten as (5.3.3) and (5.3.4), respectively.

$$\Pr(E_1 \text{ and } E_2) = \Pr(E_1)\Pr(E_2) \qquad (5.3.3)$$

$$\Pr(E_1 \text{ or } E_2) = \Pr(E_1) + \Pr(E_2) - \Pr(E_1)\Pr(E_2). \qquad (5.3.4)$$

Mutually exclusive events

If the occurence of E_1 means that E_2 cannot occur, or the occurrence of E_2 means that E_1 cannot occur, the events E_1 and E_2 are said to be *mutually exclusive* events and (5.3.5) and (5.3.6) follow.

$$\Pr(E_1 \text{ and } E_2) = 0 \tag{5.3.5}$$

$$\Pr(E_1 \text{ or } E_2) = \Pr(E_1) + \Pr(E_2). \tag{5.3.6}$$

Extension to more than two events. Since E_1 **and** E_2 and E_1 **or** E_2 are events if E_1 and E_2 are events, the above results may be extended to the case of more than two events. Thus,

$$\Pr(E_1 \text{ and } E_2 \text{ and } \ldots \text{ and } E_k) = \Pr(E_1) \times \Pr(E_2/E_1)$$

$$\times \Pr(E_3/E_1 \text{ and } E_2) \times \ldots \times \Pr(E_k/E_1 \text{ and } E_2 \text{ and } \ldots \text{ and } E_{k-1}) \tag{5.3.7}$$

$$\Pr(E_1 \text{ or } E_2 \text{ or } \ldots E_k) = \sum_{i=1}^{k} \Pr(E_i) - \sum\sum_{i<j} \Pr(E_i \text{ and } E_j)$$

$$+ \ldots + (-1)^k \Pr(E_1 \text{ and } E_2 \text{ and } \ldots \text{ and } E_k). \tag{5.3.8}$$

If the events are pair-wise independent, (5.3.7) simplifies to (5.3.9) and if, for all events, the occurrence of any one of the events precludes the occurrence of any other event, (5.3.8) simplifies to (5.3.10).

$$\Pr(E_1 \text{ and } E_2 \text{ and } \ldots \text{ and } E_k) = \Pr(E_1) \times \Pr(E_2) \times \ldots \times \Pr(E_k) \tag{5.3.9}$$

$$\Pr(E_1 \text{ or } E_2 \text{ or } \ldots \text{ or } E_k) = \Pr(E_1) + \Pr(E_2) + \ldots + \Pr(E_k). \tag{5.3.10}$$

5.4 Random Variables

5.4.1 Random Variables, Probability Distributions and Distribution Functions

From the mathematical viewpoint, the interest in statistics and their sampling distributions lies in the mathematical properties they possess. There is interest in the development of a mathematical framework which allows the connection of these statistical constructs and the probabilistic ideas of sample spaces, events and probabilities of events.

In the jargon of Probability theory the terms *statistic* and *sampling distribution of a statistic* do not appear. A statistic is a special example of a quantity which probabilists and mathematical statisticians refer to as a *random variable* and the sampling distribution of the statistic is the *probability distribution* of the random variable.

Random variable

The basic structure for Probability requires the definition of a sample space (i.e., set of possible outcomes), a collection of events and, attached to each event, a probability that the event will occur.

There is an extensive mathematical structure developed by introducing functions which map the elements in a sample space to points on a line such that intervals on the line correspond to events. If X is a variable which relates to values on the line, then a general theory is based on a structure in which the set of events are formed by the intervals $X \leq x$ for all numbers x on the real line, together with the unions and intersections of such intervals. The mathematical function which defines the transformation of the elements in the sample space to the line is known as a *random variable*. (It is somewhat confusing that a function is given the title of random *variable*. The usual notation for a function is $f(X)$ where X is called a *variable*. In the traditional formulation of random variables, X is the value taken by the function rather than an element in the domain of the function. Hence the symbol X **is** the function while at the same time representing a variable which defines events. If this dual meaning for X is understood, it should be clear which role it is performing in the following discussion of random variables.)

Distribution function

Associated with a random variable is a unique mathematical function Π, which has the property that, for any real number x, the value of Π at x is equal to the probability of the event that the random variable X will take a value which is less than or equal to x, i.e., that (5.4.1) holds.

$$\Pi(x) = \Pr(X \leq x) \qquad \text{for all real } x. \tag{5.4.1}$$

Probability distribution

When the random variable X takes possible values $x_1, x_2, x_3, \ldots$, i.e., there is a countable number of possible values, the event $X \leq x$ occurs if an outcome is transformed into a value x_i which is less than or equal to x. In this case, there exists a unique function, π, which has the property that (5.4.2) holds for all x.

$$\Pr(X \leq x) = \sum_{x_i \leq x} \pi(x_i) \qquad \text{for all real } x. \tag{5.4.2}$$

The function π which satisfies (5.4.2) is termed the *probability distribution* of the random variable X.

When the random variable maps the elements of the sample space into a non-countable set of values, such as occurs when the elements are mapped into an interval on the real line, there exists a unique function π defined on the set of real numbers which satisfies (5.4.3) for any real number x.

$$\Pr(X \leq x) = \int_{-\infty}^{x} \pi(X) dX. \tag{5.4.3}$$

In those cases where (5.4.2) applies, X is said to be a *discrete* random variable while in cases where (5.4.3) applies, X is said to be a *continuous* random variable.

In statistical applications, random variables appear as statistics with probability distribution being represented by the sampling distributions of the statistics. Table 5.2.2 provides an example of a discrete random variable with its accompanying probability distribution. By convention, only the values of the discrete random variable for which $\pi(X)$ is non-zero are recorded.

5.4.2 Properties of Probability Distributions

The probability distribution π of a random variable X has the following properties.

1. (a) If X is a discrete random variable, with non-zero probabilities at $x_1, x_2, x_3, \ldots$
   (i) $0 \leq \pi(x_i) \leq 1$ for all x_i;
   (ii) $\pi(x) = 0$ for all $x \neq x_i$;
   (iii) $\sum_{\text{all } x_i} \pi(x_i) = 1$;
   (iv) $\Pr(X = x_i) = \pi(x_i)$.

   (b) If X is a continuous random variable,
   (i) $\pi(X) \geq 0$ for all X and may exceed unity;
   (ii) $\int_{-\infty}^{+\infty} \pi(X)\, \mathrm{d}X = 1$;
   (iii) $\Pr(X < x) = \Pr(X \leq x) = \int_{-\infty}^{x} \pi(X)\, \mathrm{d}X$.

2. Parameters

   (a) *Mean.* If the values of the probability distribution are thought of as providing the relative chances of the different values of the random variable occurring in an experiment, then they can logically be used as weights to define an average or mean value for the probability distribution. The *mean* or *expected value* of the probability distribution π or its random variable X is defined by (5.4.4).

$$\begin{aligned} M &= \textstyle\sum_{\text{all } x_i} x_i \pi(x_i) && \text{if } X \text{ is discrete} \\ &= \textstyle\int_{-\infty}^{+\infty} X\pi(X)\, \mathrm{d}X && \text{if } X \text{ is continuous.} \end{aligned} \tag{5.4.4}$$

Illustration: *discrete random variable.* Suppose that X is a random variable with the following probability distribution which arises if X is the number of heads in two tosses of a fair coin.

X	0	1	2
$\pi(X)$	1/4	1/2	1/4.

By application of (5.4.4),

$$M = 1/4 \times 0 + 1/2 \times 1 + 1/4 \times 2 = 1.$$

From symmetry conditions this is clearly the correct value, i.e., on average, the number of heads in two tosses of a fair coin is expected to be one.

Illustration: *continuous random variable.* Consider the mean value of the round-off error when a real number, which is equally likely to be anywhere in the range 22.5 to 23.5, is rounded off to 23. Since the error must be in the range -0.5 to $+0.5$, the probability distribution $\pi(X)$ must be set to zero for all values of X outside this range. Within the interval, all values are presumed to be equally likely. Thus, $\pi(X)$ must have the form

$$\begin{aligned} \pi(X) &= c \qquad \text{for } -0.5 \le X \le 0.5 \\ &= 0 \qquad \text{otherwise.} \end{aligned}$$

To satisfy the requirement that $\int_{-\infty}^{+\infty} \pi(X)\,\mathrm{d}X = 1$, c must equal one. Hence, the probability distribution of X is

$$\begin{aligned} \pi(X) &= 1 \qquad \text{for } -0.5 \le X \le 0.5 \\ &= 0 \qquad \text{otherwise.} \end{aligned}$$

By (5.4.4), the mean of the probability distribution is

$$M = \int_{-0.5}^{+0.5} X 1\,\mathrm{d}X = 0.$$

(b) *Moments.* The mean of the probability distribution or random variable, as defined in (5.4.4), is a special case of the collection of parameters known as the *moments of the distribution.* These are defined by (5.4.5).

$$\begin{aligned} E[X^r] &= \textstyle\sum_{\text{all } x_i} x_i^r \pi(x_i) \qquad &\text{if } X \text{ is discrete} \\ &= \textstyle\int_{-\infty}^{+\infty} X^r \pi(X)\,\mathrm{d}X \qquad &\text{if } X \text{ is continuous} \end{aligned} \tag{5.4.5}$$

for $r = 1, 2, 3, \ldots$.

A comparison of (5.4.4) and (5.4.5) reveals the mean to be the moment for which $r = 1$. It is referred to as the *first moment* of the probability distribution.

There is also a set of moments known as the *central moments* of the probability distribution or random variable. These are defined by (5.4.6).

$$\begin{aligned} M_r &= \textstyle\sum_{\text{all } x_i} [x_i - E(X)]^r \pi(x_i) && \text{if } X \text{ is discrete} \\ &= \textstyle\int_{-\infty}^{+\infty} [X - E(X)]^r \pi(X)\, \mathrm{d}X && \text{if } X \text{is continuous} \end{aligned} \tag{5.4.6}$$

for $r = 2, 3, 4, \ldots$.

The second central moment, i.e., M_2, is the *variance* of the probability distribution or, equivalently, of the associated random variable and is, therefore, defined by (5.4.7).

$$\begin{aligned} V(X) &= \textstyle\sum_{\text{all } x_i} [x_i - E(X)]^2 \pi(x_i) && \text{if } X \text{ is discrete} \\ &= \textstyle\int_{-\infty}^{+\infty} [X - E(X)]^2 \pi(X)\, \mathrm{d}X && \text{if } X \text{ is continuous.} \end{aligned} \tag{5.4.7}$$

3. *Expected value of a function of a random variable.* In the above definitions of moments, $\pi(x_i)$ and $\pi(X)$ provide weights which represent the relative frequencies of the values or the relative chances of each value occurring in an experiment. For a wide class of functions, the expected value of a function $g(X)$ of the random variable X, where X has probability distribution π, is defined by (5.4.8).

$$\begin{aligned} \mathrm{E}[g(X)] &= \textstyle\sum_{\text{all } x_i} g(x_i)\pi(x_i) && \text{if } X \text{ is discrete} \\ &= \textstyle\int_{-\infty}^{+\infty} g(X)\pi(X)\, \mathrm{d}X && \text{if } X \text{ is continuous.} \end{aligned} \tag{5.4.8}$$

5.4.3 Properties of Distribution Functions

A distribution function Π for a random variable X has the following properties for every real number x.

1. Probabilistic connections
   (a) $\Pi(x) = \Pr(X \leq x)$
   (b) $1 - \Pi(x) = \Pr(X > x)$
   (c) $\Pi(x) = \Pr(X < x)$ if X is a continuous random variable since $\Pr(X = x) = 0$ for all x when X is continuous.

2. Definition of percentiles.

   For X a continuous variable, the value x which satisfies (5.4.9) is the *Pth percentile* of the random variable or, equivalently, of the probability distribution π.

$$\Pi(x) = P/100. \tag{5.4.9}$$

If X is a discrete variable there is not generally a unique value x which satisfies (5.4.9).

5.4.4 Joint Probability Distributions

To provide the theoretical tools for constructing a sampling distribution of a statistic, it is necessary to define probability distributions and distribution functions for random variables considered collectively.

If $x_1, x_2, \ldots, x_n$ is any set of values which might be taken by random variables $X_1, X_2, \ldots, X_n$, respectively, then (5.4.10) provides a definition for the joint distribution function Π_S of $X_1, X_2, \ldots, X_n$.

$$\Pi_S(x_1, x_2, \ldots, x_n) = \Pr(X_1 \leq x_1 \text{ and } X_2 \leq x_2 \text{ and } \ldots \text{ and } X_n \leq x_n)$$

$$\text{for all real-valued } n \text{ tuples } (x_1, x_2, \ldots, x_n). \tag{5.4.10}$$

Furthermore, where $X_1, X_2, \ldots, X_n$ are all discrete random variables, the joint probability distribution π_S is given by (5.4.11).

$$\pi_S(x_1, x_2, \ldots, x_n) = \Pr(X_1 = x_1 \text{ and } X_2 = x_2 \text{ and } \ldots \text{ and } X_n = x_n). \tag{5.4.11}$$

If $X_1, X_2, \ldots, X_n$ are all continuous random variables, there exists a unique function π_S which satisfies (5.4.12) for all real-valued n-tuples $(x_1, x_2, \ldots, x_n)$. This is the joint probability distribution for $(X_1, X_2, \ldots, X_n)$.

$$\Pi_S(x_1, x_2, \ldots, x_n) = \int_{-\infty}^{x_n} \cdots \int_{-\infty}^{x_2} \int_{-\infty}^{x_1} \pi_S(X_1, X_2, \ldots, X_n) dX_1 dX_2 \ldots dX_n$$

$$\text{for all real-valued } n \text{ tuples } (x_1, x_2, \ldots, x_n). \tag{5.4.12}$$

Conditional probability distributions: Given two random variables X and Y, (5.4.13) defines the *conditional probability distribution for X given $Y = y$*.

$$\begin{array}{lll} \pi_{X \cdot y} = \frac{\pi_S(X,y)}{\pi_Y(y)} & & \text{if } \pi_Y(y) > 0 \\ \quad \text{is undefined} & & \text{if } \pi_Y(y) = 0, \end{array} \tag{5.4.13}$$

where $\pi_S(X, y)$ is the joint probability distribution of X, Y; $\pi_Y(y)$ is the probability distribution for Y evaluated at $Y = y$.

If X and Y are discrete random variables, the conditional probability distribution is an expression of the conditional probability of events of the type $X = x$ given the event $Y = y$ has occurred. If X and Y are continuous random variables, the equivalence with the probabilistic interpretation does not hold since the probability of the event $Y = y$ is zero. There prove to be useful practical applications of conditional probability distributions and they have an important role in the theoretical development of Statistics.

Independence of random variables: Of particular importance is the situation where the events of (5.4.10) and (5.4.11) are independent. In that case, (5.4.10) can be simplified to (5.4.14).

$$\Pi_S(x_1, x_2, \ldots, x_n) = \Pr(X_1 \leq x_1)\Pr(X_2 \leq x_2)\ldots\Pr(X_n \leq x_n)$$
$$\text{for all real-valued } n \text{ tuples } (x_1, x_2, \ldots, x_n) \qquad (5.4.14)$$

which establishes that the joint distribution can be presented in terms of the individual distribution functions according to (5.4.15).

$$\Pi_S(X_1, X_2, \ldots, X_n) = \Pi_1(X_1)\Pi_2(X_2)\ldots\Pi_n(X_n) \qquad (5.4.15)$$

where Π_i is the distribution function for X_i $(i = 1, 2, \ldots, n)$. Furthermore, the probability distribution $\pi_S(X_1, X_2, \ldots, X_n)$ may be expressed as the product of the probability distributions of the individual random variables, as stated in (5.4.16).

$$\pi_S(X_1, X_2, \ldots, X_n) = \pi_1(X_1)\pi_2(X_2)\ldots\pi_n(X_n) \qquad (5.4.16)$$

where π_i is the probability distribution for X_i $(i = 1, 2, \ldots, n)$.

A set of random variables which satisfies (5.4.15) or (5.4.16) are said to be *independent*.

Independence plays an important role in statistical theory. It should come as no surprise that the construction of a joint probability distribution based on (5.4.16) is generally far easier than if the construction must be based on (5.4.10) since the function for independent random variables relies only on the properties of the variables considered individually. Where there is a dependence in the outcomes, the construction of the joint distribution function involves the use of conditional probability.

Functions of random variables: In the context of statistical application, a statistic is a function of an ordered set of random variables, i.e., $S = g(X_1, X_2, \ldots, X_n)$ where X_i represents the response from sample member i $(i = 1, 2, \ldots, n)$. For example, if S is the arithmetic mean of the sample, then $S = (X_1 + X_2 + \ldots + X_n)/n$. An important role for the theory of Probability is to provide the formal mechanism for constructing the probability distribution of S from a knowledge of the joint probability distribution of $(X_1, X_2, \ldots, X_n)$. At an elementary level this can be done by simple probability argument, as is illustrated in Section 5.4.5. For the interested reader, the general development can be found in introductory books on mathematical Statistics—for example, Freund and Walpole [1980].

Linear functions of random variables: Suppose $X_1, X_2, \ldots, X_n$ are random variables with the mean and variance of X_i being denoted by $\mathrm{E}(X_i)$ and $\mathrm{V}(X_i)$, respectively, $(i = 1, 2, \ldots, n)$ and $c_1, c_2, \ldots, c_n$ are constants.

Then any statistic which satisfies (5.4.17) is a linear function of the random variables $X_1, X_2, \ldots, X_n$.

$$S = c_1 X_1 + c_2 X_2 + \ldots + c_n X_n. \tag{5.4.17}$$

Examples of statistics which satisfy (5.4.17) are the sum of n random variables $X_1 + X_2 + \ldots + X_n$, the arithmetic mean $(X_1 + X_2 + \ldots + X_n)/n$ and the difference of two random variables $X_1 - X_2$.

For any statistic S which satisfies (5.4.17), the expected value can be determined from (5.4.18). Furthermore, if the random variables $X_1, X_2, \ldots, X_n$ are independent, then (5.4.19) permits the determination of the variance of S without the requirement to determine the probability distribution of S.

$$\mathrm{E}(S) = c_1 \mathrm{E}(X_1) + c_2 \mathrm{E}(X_2) + \ldots + c_n \mathrm{E}(X_n) \tag{5.4.18}$$

$$\mathrm{V}(S) = c_1^2 \mathrm{V}(X_1) + c_2^2 \mathrm{V}(X_2) + \ldots + c_n^2 \mathrm{V}(X_n). \tag{5.4.19}$$

Products of independent random variables: If $X_1, X_2, \ldots, X_n$ are independent random variables, the expected value of the product $X_1 X_2 \ldots X_n$ is given by (5.4.20).

$$\mathrm{E}(X_1 X_2 \ldots X_n) = \mathrm{E}(X_1)\mathrm{E}(X_2) \ldots \mathrm{E}(X_n). \tag{5.4.20}$$

Covariance

The joint distribution for two random variables is termed a *Bivariate probability distribution.* Corresponding to the *Variance* of a univariate distribution as defined by (5.4.7), is a *Covariance* of a bivariate distribution which is defined by (5.4.21)

$$M_{XY} = \sum_X \sum_Y [X - \mathrm{E}(X)][Y - \mathrm{E}(Y)]\, \pi_S(X,Y) \text{ for } X, Y \text{ discrete}$$
$$= \int_{-\infty}^{+\infty} \int_{-\infty}^{+\infty} [X - \mathrm{E}(X)][Y - \mathrm{E}(Y)]\, \pi_S(X,Y) dX dY \tag{5.4.21}$$

for X, Y continuous.

5.4.5 Constructing Sampling Distributions — An Illustration

An important application of the rules of probability in the area of Statistics is to be found in the construction of sampling distributions of statistics. As introduced in Section 5.2, this is formally a three stage process which begins with frequency distributions for populations or probability distributions for processes, and proceeds through the construction of sampling distributions for samples to sampling distributions for statistics.

The general mathematical development is beyond the scope of this book and is generally developed by utilizing the theory of random variables. However, with a relatively low level of mathematical expertise it is possible to trace the application of the rules of probability to construct the sampling distribution of a statistic in the case where the statistic is a discrete random variable.

The development which is shown below leads to the sampling distribution of the statistic which is presented in Table 5.2.2. The experiment involves the selection of a sample of 3 items from a production line in which the individual outcomes are classified as either defective (D) or non-defective (N). The model contains the assumptions that (i) the probability of an item being defective is 0.01 and the probability of it being non-defective is 0.99; and (ii) the probability of any one item being defective is assumed to be unaffected by the responses of previous items.

Probability distribution. The probability distribution has the simple expression:

Response	D	N
Probability	0.01	0.99

Sampling distribution. Let $E_i = D$ be the event that the ith sample member is defective and $E_i = N$ be the event that the ith sample member is non-defective ($i = 1,2,3$). Under the assumption of random sampling (Section 2.1.2), the sampling distribution is identical with the probability distribution. Thus,

$$\Pr(E_i = D) = 0.01$$
$$\Pr(E_i = N) = 0.99.$$

Equivalently, a random variable, X_i, can be defined which denotes the number of defectives in the ith outcome ($i = 1,2,3$) with a probability distribution π_i where

$$\begin{aligned} \pi_i(X_i) &= 0.99 && \text{for } X_i = 0 \\ &= 0.01 && \text{for } X_i = 1 \\ &= 0.00 && \text{otherwise.} \end{aligned}$$

Sampling distribution of the sample. The possible outcomes from the experiment comprise the eight possible sequences of defectives (D) and non-defectives (N), i.e.,

$$(DDD)\ (DDN)\ (DND)\ (NDD)\ (DNN)\ (NDN)\ (NND)\ (NNN).$$

Let π_S denote the sampling distribution of the sample. Then

$$\pi_S(DDD) = \Pr(E_1 = D \text{ and } E_2 = D \text{ and } E_3 = D).$$

Based on the condition in the model that the component events $E_1 = D$, $E_2 = D$ and $E_3 = D$ are independent, then, by (5.3.9)

$$\begin{aligned}\pi_S(DDD) &= \Pr(E_1 = D) \times \Pr(E_2 = D) \times \Pr(E_3 = D)\\ &= \pi_1(1) \times \pi_2(1) \times \pi_3(1)\\ &= 0.01 \times 0.01 \times 0.01\\ &= 0.000001\end{aligned}$$

which is the value quoted in Table 5.2.1.

Values for π_S for the remaining seven possible outcomes can be computed by following the same steps.

Sampling distribution of statistic. The statistic employed is the *number of defectives in the sample* (s) which can take values 0, 1, 2 and 3 with non-zero probabilities. If π_{st} is the sampling distribution of s, then

$$\begin{aligned}\pi_{\mathrm{st}}(0) &= \pi_S(NNN) = 0.970299\\ \pi_{\mathrm{st}}(1) &= \pi_S(NND) + \pi_S(NDN) + \pi_S(DNN) = 3 \times 0.009801\\ &= 0.029403\\ \pi_{\mathrm{st}}(2) &= \pi_S(NDD) + \pi_S(DND) + \pi_S(DDN) = 3 \times 0.000099\\ &= 0.000297\\ \pi_{\mathrm{st}}(3) &= \pi_S(DDD) = 0.000001.\end{aligned}$$

Construction of p-values. The model underlying the construction of the probabilities states that the probability of a defective is 0.01. Assuming this is the correct value, and under the other conditions imposed, the possible values for the statistic, *number of defectives in the sample of size three,* may be placed in the order 0 1 2 3 to reflect decreasing likelihood of occurrence (based on the values taken by the sampling distribution π_{st}).

By definition, the p-value is the probability of a result which is at least as extreme as that observed. Thus,

- if there are 3 defectives, $p = \pi_{\mathrm{st}}(3) = 0.000001$;
- if there are 2 defectives, $p = \pi_{\mathrm{st}}(3) + \pi_{\mathrm{st}}(2) = 0.000298$;
- if there is 1 defective, $p = \pi_{\mathrm{st}}(3) + \pi_{\mathrm{st}}(2) + \pi_{\mathrm{st}}(1) = 0.029701$;
- if there are 0 defectives, $p = \pi_{\mathrm{st}}(3) + \pi_{\mathrm{st}}(2) + \pi_{\mathrm{st}}(1) + \pi_{\mathrm{st}}(0)$
  $p = 1.000000$.

Interpretation of p-values. Following the conventional guidelines for the interpretation of p-values (Section 4.3.6), the values of the sampling distribution may be assigned the following interpretations:

- if 2 or 3 defectives were observed in a sample, since the value of p is less than 0.01, there would be strong evidence against the hypothesis that the probability of a defective is 0.01;

- if there is one defective in the sample, there is some evidence against the hypothesis;
- if there are no defectives, the result can be said to be consistent with the presumed rate of defectives coming off the production line.

In this example, it was possible and feasible to enumerate all possible outcomes, determine the individual probabilities and sum those probabilities. In general, direct enumeration in this manner is lengthy and alternative methods have been developed.

In some cases, the task is simplified because sets of outcomes all have the same probability of occurrence. The determination of probabilities then reduces to the task of counting the number of possible outcomes. Formulae for this purpose are provided in the appendix to this chapter.

Over and beyond direct enumeration, there is an extensive statistical theory which may be employed to provide the necessary mathematical tools for the construction of sampling distributions of statistics. Those readers who are interested, should seek a book or a lecture course on mathematical Statistics.

Appendix. Combinatorial Formulae

1. NOTATION

The symbol $n!$ pronounced "N factorial" is an abbreviation used by mathematicians for the product

$$n \times (n-1) \times (n-2) \times \ldots \times 3 \times 2 \times 1.$$

Thus, $2! = 2 \times 1 = 2$, $3! = 3 \times 2 \times 1 = 6$, etc. Additionally, by definition,

$$0! = 1.$$

The symbol $\binom{n}{x}$ or equivalently ${}^nC_x$, pronounced "N C X," is an abbreviation for

$$\frac{n!}{x!(n-x)!}.$$

For example,

$$\binom{5}{2} = \frac{5!}{2!(5-2)!} = \frac{5!}{2!3!} = \frac{5 \times 4 \times 3 \times 2 \times 1}{2 \times 1 \times 3 \times 2 \times 1} = 10.$$

2. SELECTION

Given the task of selecting n sample members from a population of size N, how many distinct samples are there?

Assuming that once a sample member is taken from a population it is not replaced during the course of the sampling operation, there are $\binom{N}{n}$ possibilities.

Suppose a population can be partitioned into two sub-populations such that there are N members in the population, N_1 of whom are in sub-population 1 and N_2 of whom are in sub-population 2. How many ways are there of selecting a sample of size n in which there are n_1 members from sub-population 1 and n_2 members from sub-population 2?

Assuming that once a sample member is taken from the population it is not replaced during the course of the sampling operation, there are $\binom{N_1}{n_1}\binom{N_2}{n_2}$ ways.

This result may be extended to the case where there are k sub-populations with numbers $N_1, N_2, \ldots, N_k$ in sub-populations $1, 2, \ldots, k$, respectively. The number of possible selections is then

$$\binom{N_1}{n_1}\binom{N_2}{n_2}\cdots\binom{N_k}{n_k}.$$

3. ARRANGEMENTS

Suppose a sample contains n sample members of which n_1 are of one type and n_2 of another type where $n_1 + n_2 = n$. How many distinct arrangements are there?

There are $\binom{n}{n_1} = \binom{n}{n_2}$ arrangements.

This result may be extended to the case where there are k types in the sample with $n_1, n_2, \ldots, n_k$ members of types $1, 2, \ldots, k$, respectively, where $n_1 + n_2 + \ldots + n_k = n$.

The number of distinct arrangements is then $\frac{n!}{n_1!n_2!\ldots n_k!}$.

Problems

5.1 *Sample space, events, probabilities of events*

The numbers $1, 2, \ldots, 100$ are allocated among 100 persons so that each person is assigned one number. The numbers are assigned so that persons who were born in the same country are assigned consecutive numbers according to the following scheme:

Country of birth	Numbers assigned to persons
Vietnam	1 to 5
China	6 to 11
Poland	12 to 14
Italy	15 to 24
Greece	25 to 39
Australia	40 to 100

(a) If a random selection is made of one member of the group, the *sample space* comprises 100 elements all of which are equally likely to be selected. What is the probability of each of the following *events:*

  (i) selecting a Greek?

  (ii) selecting a Vietnamese or a Chinese?

  (iii) not selecting an Australian?

(b) State the frequency distribution for the group based on the variable *Country of birth.* Note that when random selection is employed, the frequency distribution for the group is identical with the sampling distribution which defines the probability of selecting a member from a particular country of birth.

Consider the random selection of a sample of size 2 from the group under the assumption that the first person who is selected from the group is not returned before the second person is selected.

(c) If the first person selected was person number 10,

  (i) what is the sample space for the second selection?

  (ii) what are the probabilities of the events defined in (a) (i)–(iii) in respect of the second person selected given that the first person selected was Chinese?

(d) Use the relevant equations from Section 5.3 to establish the probabilities of the events:

  (i) the first person selected is Chinese and the second person selected is Greek;

  (ii) the first person selected is Chinese and the second person selected is either Vietnamese or Chinese;

  (iii) the first person selected is Chinese and the second person selected is not Australian.

(e) Use relevant equations from Section 5.3 to establish the probabilities of the events:

  (i) both of the persons selected are Chinese;

  (ii) at least one of the persons selected is Chinese.

5.2 *Independent events and combinatorial formulae*

From long term meteorological information, it is established that the probability of rain falling on any one day in the city of Hobart is approximately 0.25. Furthermore, it has been established that Hobart has a most unusual characteristic in respect of rainfall prediction. The knowledge that rain is falling today is of no value in attempting to predict if rain will fall tomorrow. In probabilisitic terms, the events that *rain will fall today* and *rain will fall tomorrow* are *independent* events.

(a) Use the fact that events corresponding to rain falling on different days are independent plus relevant equations from Section 5.3 to determine the probabilities of the following events:

(i) there is no rain for three successive days;

(ii) there is no rain for three successive weekends;

(iii) on a holiday weekend comprising Saturday, Sunday and Monday, there will be rain on at least two days;

(iv) in a football competition running for 10 days, there will be no rain throughout the competition.

(b) On each of 10 days, the day may be classified as either *wet* (W) or *dry* (D). One possible sequence is $DDWDDDDWWD$.

(i) How many sequences of 10 days have three wet days and seven dry days? (Use the formula in the appendix to this chapter if you wish.)

(ii) What is the probability of the sequence $DDWDDDDWWD$?

(iii) Use the results of (i) and (ii) plus relevant formulae from Section 5.3 to determine the probability of three wet days and seven dry days (in any order) in a sequence of ten days.

(iv) Generalize the result of (iii) to produce a formula for computing the probability of n_1 wet days and n_2 dry days in a sequence of $n = n_1 + n_2$ days.

5.3 *Probability, sample size and the detection of treatment differences*

In a food tasting experiment, five tasters are each given two pieces of fish to taste. One piece of fish has been frozen and the other chemically treated to retard deterioration.

(a) The aim of each taster is to detect the chemically treated fish. Assuming the tasters reach their decisions independently, what is the probability that all tasters will correctly identify the chemically treated fish if

(i) there is no distinguishable difference in taste between fish which have been subjected to the different treatments?

(ii) the expected success rate in detecting the chemically treated fish is 80%.

(b) Suppose there is no discernible difference in taste for the frozen and chemically treated fish, i.e., tasters are only guessing which is the chemically treated fish. Suppose n tasters are used in the experiment and p_n is the probability that all tasters correctly identify the chemically treated fish.

(i) Find an expression for p_n.

(ii) What is the minimum number of tasters who must be employed to ensure that the chance of all tasters correctly identifying the chemically treated fish (by guessing) is less than one in one hundred?

5.4 *Hypergeometric distribution*

In Section 4.3 a *capture-recapture* experiment was described in which a sample of n fish was randomly selected from a lake containing N fish. In the lake there were N_1 *tagged* fish and N_2 *untagged* fish. Use combinatorial formulae from the appendices to Chapter 5 and rules from Section 5.3 to establish that the probability, P, of the sample containing n_1 *tagged* fish and n_2 *untagged* fish, where $n_1 + n_2 = n$, is determined from the expression

$$P = \frac{\binom{N_1}{n_1}\binom{N_2}{n_2}}{\binom{N}{n}}.$$

The following problems are for the more mathematically inclined readers. Problems 5.5 and 5.6 relate to basic properties of probability distributions. Problems 5.7 and 5.8 illustrate the role of the expected value operator in establishing properties of sampling distributions of statistics.

5.5 *Properties of probability distributions — discrete random variables*

(a) Binomial distribution. The random variable X is said to have a *Binomial distribution* if the probability distribution of X is given by

$$\begin{aligned} \pi(X) &= \binom{n}{x} p^x (1-p)^{n-x} \qquad && \text{for } x = 0, 1, 2, \ldots, n \\ &= 0 && \text{otherwise} \end{aligned}$$

for n a positive integer and p a real number in the interval $[0, 1]$.

Prove that (i) $\sum_{x=0}^{n} \pi(x) = 1$; (ii) $\mathrm{E}(X) = np$.

(b) Poisson distribution. The random variable X is said to have a *Poisson distribution* if the probability distribution of X is given by

$$\begin{aligned} \pi(x) &= [\exp(-M)M^x]/x! && \text{for } x = 0, 1, 2, \ldots \\ &= 0 && \text{otherwise,} \end{aligned}$$

for M a positive real number.

Prove that (i) $\sum_{x=0}^{\infty} \pi(x) = 1$; (ii) $\mathrm{E}(X) = M$.

5.6 *Properties of probability distributions — continuous random variables*

(a) Uniform distribution. The random variable X is said to have a *Uniform distribution* if the probability distribution of X is given by

$$\begin{aligned} \pi(x) &= 1/[b-a] && \text{for } a \leq x \leq b \\ &= 0 && \text{otherwise,} \end{aligned}$$

for a and b real numbers with $a < b$.

Prove that (i) $\int_{-\infty}^{+\infty} \pi(X)\mathrm{d}X = 1$; (ii) $\mathrm{E}(X) = \frac{1}{2}(b+a)$.

(b) Exponential distribution. The random variable X is said to have an *Exponential distribution* if the probability distribution of X is given by

$$\begin{aligned} \pi(x) &= \tfrac{1}{M}\exp(-x/M) && \text{for } x \geq 0 \\ &= 0 && \text{otherwise,} \end{aligned}$$

for M a positive real number.

Prove that (i) $\int_{-\infty}^{+\infty} \pi(X)\mathrm{d}X = 1$; (ii) $\mathrm{E}(X) = M$.

5.7 *Expected values of functions of random variables*

(a) Let X be a continuous random variable with expected value M and variance V.

If $g(X) = c$ and $h(X) = cX$, prove that (i) $\mathrm{E}[g(X)] = c$; (ii) $\mathrm{E}[h(X)] = cM$; (iii) $\mathrm{V}[h(X)] = c^2V$.

Hence, prove that (i) $\mathrm{E}[a + bX] = a + bM$; and (ii) $\mathrm{V}[a + bX] = b^2V$.

5.8 *Expected value and variance of the arithmetic mean*

Let $Y_1, Y_2, \ldots, Y_n$ be independent random variables each with mean M and variance σ^2. Determine the expected value and variance of (i) $Y = Y_1 + Y_2 + \ldots + Y_n$; and (ii) $\overline{y} = (Y_1 + Y_2 + \ldots + Y_n)/n$.

6

Some Widely Used Statistical Models

6.1 The Binomial Model

6.1.1 Model Identification

A repetitive experimental process is envisaged in which there are only two possible types of outcome. These types may take many and varied forms—*success* and *failure*, *yes* and *no*, *diseased* and *healthy*, *large* and *small*, etc. The requirements for valid application of the model are (i) the outcome from one application of the process is not influenced by the results of other applications of the process; (ii) the probability of a type 1 outcome (or a type 2 outcome) is constant over the time period or region of the study; and (iii) an objective method of sample selection has been employed.

6.1.2 Probability Distribution

There is a simple probability distribution represented by

Type	1	2
Probability	π_1	π_2

where $\pi_1 + \pi_2 = 1$.

6.1.3 Applications

The Binomial model is probably the most widely used model in Statistics. Some illustrations are given below.

1. In all areas of scientific and technological research there are situations in which interest is in the estimation of the success rate of a treatment.

2. There may be interest in examining an hypothesis about a proposed success rate. For example, a geneticist may wish to examine the hypothesis that a gene segregation is in the ratio of 3:1. This is equivalent to testing the hypothesis that $\pi_1 = 3/4$.

3. The Binomial model is a component in more complex models where the aim is to compare success rates for two or more processes or treatments.

4. The Binomial model is also used as an approximation to the Two-state population model described in Section 6.2.

6.1.4 Data Reduction

Commonly, the data are not presented or recorded as a sequence of outcomes, but are summarized as either the number of successes (x) and number of failures ($n - x$) in the sample of size n, or as the number of trials (n) to obtained exactly x successes.

6.1.5 Statistics

The statistics which are typically employed relate directly to the data reduction described above. They are

either the number of successes (x) in a fixed number of outcomes (n) for which the sampling distribution, known as the *Binomial distribution,* is given by (6.1.1)

$$\pi_{st}(x) = \binom{n}{x}\pi^x(1-\pi)^{n-x} \quad \text{for} \quad x = 0, 1, 2, \ldots, n \tag{6.1.1}$$

where π is the probability of success at one trial;

or the number of trials (n) to obtain exactly x successes for which the sampling distribution, known as the *Negative Binomial distribution,* is given by (6.1.2).

$$\pi_{st}(n) = \binom{n-1}{x-1}\pi^x(1-\pi)^{n-x} \quad \text{for } n = x, x+1, x+2, \ldots. \tag{6.1.2}$$

6.2 The Two-State Population Model

6.2.1 Model Identification

A population of N members can be partitioned into two sub populations (which for convenience can be identified as Groups 1 and 2) and from this population a sample can be selected by a process of random selection. An illustration of the model is given in Chapter 4 in respect of the Capture-recapture example where a fish population is subdivided into two sub populations comprising *tagged* and *untagged* fish.

6.2.2 Frequency Distribution

The frequency distribution has the simple form

Group	1	2
Relative frequency	π_1	π_2

where $\pi_1 + \pi_2 = 1$.

6.2.3 Applications

1. Questions may relate to the estimation of π_1 (or equivalently π_2). In a herd of animals or a forest plantation, the proportion of diseased individuals might be of interest. Note that information collected in opinion polls and surveys does not usually fit this model since random selection is generally not employed.

2. In mass production processes where the items are packed in boxes of known size, the Two-state population model provides the basis for developing quality control methods. Each box of objects constitutes a population and the items sampled from the box are grouped according to whether they are defective or non-defective.

3. In the estimation of population size, the model may be used in the manner illustrated in the Capture-recapture example presented in Chapter 4.

4. The model may be employed as part of a more complex model when comparisons are to be made between different populations to establish if the ratio π_1/π_2 varies between populations, and, if so, by how much.

6.2.4 Data Reduction

The data are typically recorded as simply the number of sample members in each of Groups 1 and 2, which are denoted by x_1 and x_2, respectively, where $x_1 + x_2 = n$ is the sample size.

6.2.5 Statistic

The statistic most commonly employed is the number of sample members from Group 1 (or equivalently, from Group 2). Under the Two-state population model the sampling distribution of the statistic is known as the *Hypergeometric distribution* and is defined by (6.2.1).

$$\pi_{st}(x_1) = \frac{\binom{N_1}{x_1}\binom{N_2}{n-x_1}}{\binom{N}{n}} \quad \text{for} \quad x_1 = 0, 1, 2, \ldots, \min(n, N_1) \qquad (6.2.1)$$

where N_1 and N_2 are numbers of population members in Groups 1 and 2, respectively, with $N_1 + N_2 = N$; $\min(n, N_1)$ is the smaller of n and N_1.

6.3 A Model for Occurrences of Events

6.3.1 Model Identification

A process exists which generates events in either time or space (e.g. customers entering a shop for service, cars passing an observer, misprints on a page of writing, natural disasters over a region).

The mechanism of the process is assumed unchanging over the time or region of interest. Thus, the average rate of occurrence of the process per unit of time or space is assumed constant. This might be acceptable for requests for car parts from a supplier but would unlikely to be correct for telephone calls to a switchboard at different hours of a day.

The generation of any one event is assumed to neither increase nor decrease the probability of occurrence of a future event. Thus, the process would not apply to the occurrence of infectious diseases but may have application if a disease is non-infectious.

The sampling process is assumed objective, commonly systematic, over time or space.

6.3.2 Probability Distributions

The assumptions in the model define two probability distributions.

1. *Poisson distribution.* If the conditions of the model defined in Section 6.3.1 are met, then (6.3.1), which is known as the *Poisson distribution,* defines (a) the probability of finding n occurrences of an event in a time period; or (b) the expected proportion of spatial units in which n occurrences of the event occurred.

$$\pi(n) = \frac{\exp(-M) \cdot M^n}{n!} \quad \text{for} \quad n = 0, 1, 2, \ldots, \qquad (6.3.1)$$

where M is the average number of occurrences expected per unit of time or space.

2. *Exponential distribution.* The probability of waiting a time period t or having to travel a distance t before the next occurrence of the event is

defined by (6.3.2) which is known as the *Exponential distribution.*

$$\begin{aligned} \pi(t) &= \frac{1}{M}\exp\left(-\frac{t}{M}\right) && \text{for } t \geq 0 \\ &= 0 && \text{otherwise,} \end{aligned} \tag{6.3.2}$$

where M is the mean time between occurrences of events.

6.3.3 Applications

1. The most common use for this model is to examine the hypothesis that the events occur independently. Analysis based on this model has provided support for the theory that radioactive emissions occur independently, allowed assessment as to whether equipment faults are independent or are related happenings, and permitted a judgment as to whether a disease in forests is infectious or non-infectious.

   The assumptions in the model lead to a pattern of occurrence of events which is termed a *random* pattern. There is no descriptive definition of a random pattern possible. Rather it is defined by what it is not! A *random* pattern is intermediate between two *non-random* patterns. At the one extreme is the *systematic* pattern in which occurrences of events are spaced evenly over time or space. At the other extreme is the *clustering* pattern in which many occurrences of the event occur together and there are intervals devoid of occurrences of the event in between. Commonly the model is employed in order to seek evidence of either a systematic or a clustering pattern.

2. Where the model is assumed correct, it may be employed in the estimation of the average number of occurrences of the event per unit of time or space.

3. When two or more processes are to be compared, the model may be included as part of a larger model.

4. As well as providing a base for examining the number of occurrences in a fixed interval of time or space, the model is also employed in studying the distances between events in time or space. It may be used to define the pattern of lifetimes of light bulbs or electronic components, or the time between arrival of radioactive particles at a Geiger counter.

5. The model provides a useful approximation to the Binomial model in cases where the sample size (n) is large and the probability of success (π) is small. This is typically the case in the study of rare events.

6.3.4 Data

The data may be recorded either as the number of occurrences of events in equal-sized periods of time or space $(n_1, n_2, \ldots, n_k)$, or as the successive intervals between occurrences of the event $(t_1, t_2, \ldots, t_k)$.

6.3.5 Statistics

1. The statistics employed for the examination of the data set $(n_1, n_2, \ldots, n_k)$ vary according to the circumstances and will be presented in later chapters.

2. *Gamma distribution.* If T is the waiting time for the occurrence of the nth event, then the sampling distribution of T is known as the Gamma distribution and is defined by (6.3.3).

$$\begin{aligned} \pi_{st}(t) &= \frac{1}{(n-1)!} M^n t^{n-1} \exp(-Mt) \quad \text{for } t \geq 0 \\ &= 0 \qquad\qquad\qquad\qquad\qquad\qquad \text{otherwise.} \end{aligned} \tag{6.3.3}$$

6.4 The Multinomial Model

6.4.1 Model Identification

An obvious extension of the Binomial model is obtained by allowing more than two categories in the response. The Multinomial model allows for any specified number of distinct outcomes which may be identified by the numbers $1, 2, \ldots, k$.

The assumptions in the model are the same as those for the Binomial model.

6.4.2 Probability Distribution

The distribution is simply

Type	1	2	$\cdots$	k
Probability	π_1	π_2		π_k

where $\pi_1 + \pi_2 + \cdots + \pi_k = 1$.

6.4.3 Applications

The Multinomial model is widely employed to examine hypotheses about the structure of probability distributions. Typical examples are provided below.

1. Example 3.1 provides the following table of frequencies for occurrence of a disease known as Thyrotoxicosis.

Type Season	1 Summer	2 Autumn	3 Winter	4 Spring
Frequency	32	27	43	43
Expected probability	π_1	π_2	π_3	π_4

The hypothesis under investigation states that the probability of occurrence of the disease is the same for all seasons, i.e. $\pi_1 = \pi_2 = \pi_3 = \pi_4$.

2. A more complicated model is required in applications designed to study the relation between responses. The following frequency table is taken from Example 17.1.

Age	Length of Stay	
	Short	Long
Young	364	721
Old	1579	1723

This is not obviously of the form to which the Multinomial model might apply. It becomes more familiar if presented in the following manner.

Type Length of Stay Age	1 Short Young	2 Short Old	3 Long Young	4 Long Old
Frequency	364	1579	721	1723
Probability	π_1	π_2	π_3	π_4

The question of interest is whether the ratio of long to short stays differs between young and old tourists. The hypothesis that the pattern is the same is defined by

$$\frac{\pi_1}{\pi_3} = \frac{\pi_2}{\pi_4}.$$

Note that this hypothesis makes no assumption as to the actual values of the individual probabilities.

6.4.4 Data Reduction

The data are typically summarized in a frequency table as is illustrated above. The table has the general form displayed in Table 6.4.1.

Table 6.4.1. Data Reduction Under the Multinomial Model — Frequency Table

Table of Frequencies

Group	1	2	$\cdots$	k
Frequency	f_1	f_2		f_k

6.4.5 The Multinomial Distribution

The probability of obtaining exactly f_1 sample members in Group 1, f_2 sample members in Group 2, etc. in a sample of n members is defined by (6.4.1). It is termed the *Multinomial distribution.*

$$\pi(f_1, f_2, \ldots, f_k) = \frac{n!}{f_1! f_2! \cdots f_k!} [\pi_1^{f_1} \pi_2^{f_2} \cdots \pi_k^{f_k}] \qquad (6.4.1)$$

where π_i is the probability of a sample member being from Group i ($i = 1, 2, \ldots, k$);

$$\pi_1 + \pi_2 + \cdots + \pi_k = 1$$

$$f_1 + f_2 + \cdots + f_k = n.$$

6.4.6 Statistics

There is no obvious or natural statistic to employ in the comparison of model and data in this case. Discussion of the choice of statistics which are in common use will be found in Chapter 14.

6.5 The Normal Distribution Model

6.5.1 The Theory of Errors

There are many circumstances in which there is not an obvious set of conditions on which to base the frequency or probability distribution. This is particularly so when the response is continuous.

However, as early as the eighteenth century it was noted that data sets from many diverse sources, when represented in histograms, showed a striking similarity in appearance. This pattern is illustrated by the two histograms in Figure 6.5.1.

Attempts were made in the early part of the last century to construct models which might characterize this commonly observed pattern. Gauss proposed that each observation, y, be portrayed as the sum of two components, an average or *expected value,* M, and a *random error,* e. Thus, $y = M + e$. Mathematicians sought to characterize the pattern of variation in e.

In 1837, a German mathematician named Hagen did produce a satisfactory characterization. The assumptions on which the mathematical form he derived were based, stated that

(i) the random error is the sum of an infinitely large number of what he termed *elementary errors,* each of which contributes in an arbitrarily small way to the sum;

(ii) the elementary errors make independent contributions; and

(iii) each elementary error is equally likely to be positive or negative.

Hagen proved that there is only one type of mathematical function which satisfies these conditions and that is the function defined by (6.5.1).

$$f(t) = \frac{1}{\sqrt{2\pi\sigma^2}} \exp\left[-\frac{1}{2\sigma^2}t^2\right] \quad \text{for} \quad -\infty < t < \infty. \tag{6.5.1}$$

(a) Cut–out temperatures of limit thermostats
(Histogram based on data in Table 1.3.7)

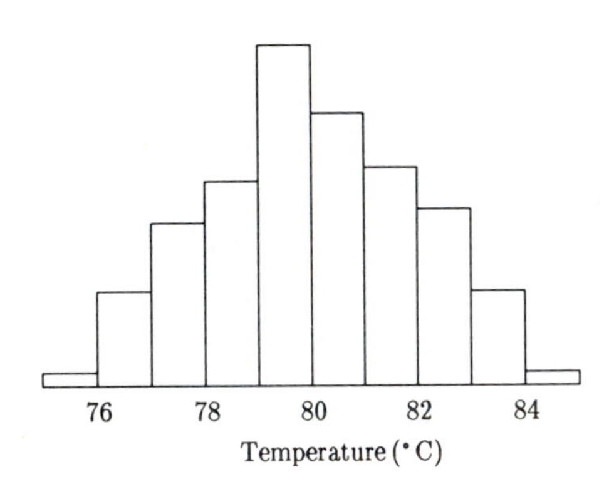

(b) Blood cholesterol levels

(Histogram based on data in Table 1.3.2)

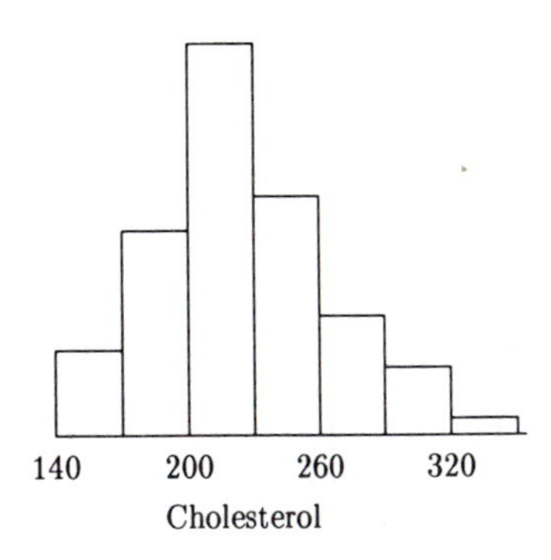

Figure 6.5.1. Patterns of variation consistent with the normal distribution.

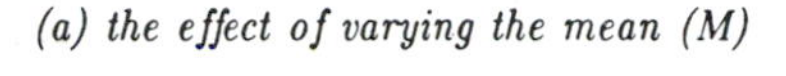

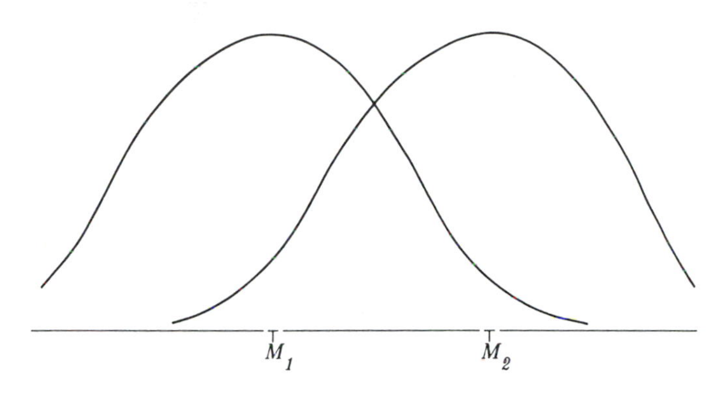

(b) the effect of varying the variance (σ^2)

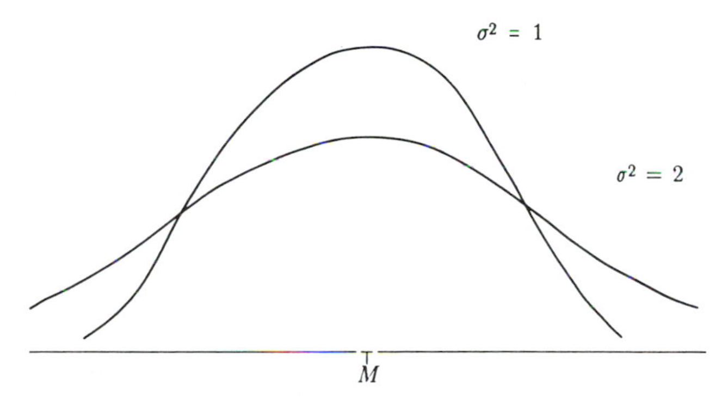

Figure 6.5.2. The shape of the normal distribution and the effect of varying the parameters M and σ^2.

The graph of this function has a characteristic bell-shape which can be seen in Figure 6.5.2. The quantity σ^2 is a constant for any curve but every value of σ^2 produces a different curve.

Because this curve provides an approximation to the pattern observed in so many and diverse histograms based on real data sets, the function became known as the *Normal distribution.* An alternative name which appears is the *Gaussian distribution* in honor of the interest Gauss showed in the distribution.

6.5.2 Model Identification

The model has application when (i) the quantity under investigation can take many values and (ii) it can reasonably be assumed that an observed outcome is the end result of contributions from many factors, with no one factor having a large effect.

Random sampling is assumed.

6.5.3 Frequency or Probability Distribution

The general form of the Normal distribution is expressed by (6.5.2).

$$f(x) = \frac{1}{\sqrt{2\pi\sigma^2}} \exp\left[-\frac{1}{2\sigma^2}(x-M)^2\right] \qquad -\infty < x < \infty, \qquad (6.5.2)$$

where M and σ^2 are properties of the particular experimental process.

6.5.4 The Parameters of the Normal Distribution

While it is common to speak of *the* Normal distribution, it is in fact a family of distributions with the members distinguished by the pair of values for M and σ^2.

The parameter M is the *mean* or *expected value* of the distribution and the parameter σ^2 is the *variance*. The effects of changing these parameters are illustrated in Figure 6.5.2.

Notation: $N(M, \sigma^2)$ is the usual abbreviation for the Normal distribution with mean M and variance σ^2.

$N(0, 1)$ is often called the *standard Normal distribution.*

6.5.5 Properties of the Normal Distribution

1. *Relationship between distributions.* If y is a random variable with a $N(M, \sigma^2)$ distribution and z is a random variable with a $N(0, 1)$ distribution, then, for any real number y_0,

$$\Pr(y < y_0) = \Pr\left[z < \frac{y_0 - M}{\sigma}\right].$$

2. *Useful probability statements.* If the scale of the variable is measured in units of σ, the following is true where a Normal distribution is presumed to describe the pattern of variation for an attribute in a population:

- approximately 68% of the values will be within 1 unit of the mean;
- approximately 95% of the values will be within 2 units of the mean;
- approximately 99.7% of the values will be within 3 units of the mean.

Expressed in probabilistic terms these statements are respectively

$$\Pr(-\sigma < y - M < \sigma) \cong 0.68;$$
$$\Pr(-2\sigma < y - M < 2\sigma) \cong 0.95;$$
$$\Pr(-3\sigma < y - M < 3\sigma) \cong 0.997,$$

where y has a $N(M, \sigma^2)$ distribution.

6.5.6 Applications

The Normal distribution has two distinct applications. On the one hand it is a widely used approximation to frequency and probability distributions for many attributes (for the reason given in Section 6.5.1). On the other hand it provides a good approximation to the sampling distributions of many statistics (see Section 7.3.5). In the latter role it is great importance in the development of statistical theory and in providing simple computational formulae.

The two applications are illustrated below.

1. **Approximating a frequency or probability distribution.** As with many attributes measured on mass-produced items, the *cut-off temperature* of limit thermostats, which provided the data in Table 1.3.7, is likely to satisfy the conditions imposed by the Theory of Errors, namely that the observed values are under the control of many factors, each of which makes a small contribution. The data collected in the study which are summarized in Table 6.5.1 superficially fit the pattern expected under a Normal distribution in being symmetric with the maximum concentration of values in the center of the range of the observations. Based on the best fitting Normal distribution, proportions expected in each of the ten categories defined in Table 6.5.1 are calculated following steps described in Table 6.5.1. Visual comparison of the two sets of frequencies suggests they show good agreement. The comparison could be carried further by the application of a formal statistical test as described in Section 16.2.12.

2. **Approximating the sampling distribution of a statistic — the Normal approximation to the Binomial.** The Binomial distribution is defined by (6.1.1) as

$$\pi_{st}(x) = \binom{n}{x} \pi^x (1-\pi)^{n-x} \qquad \text{for} \quad x = 0, 1, 2, \ldots, n$$

where x is the number of successes in n independent trials with the probability of success at each trial being π. For large values of n, the computation of sums of Binomial probabilities using this formula is time consuming. Instead, an approximation is available which is based on a Normal distribution. If the aim is to compute the probability that the Binomial statistic, x, will take a value in the range x_1 to x_2, i.e. $\Pr(x_1 \leq x \leq x_2)$, and y is a statistic with a Normal distribution having the same mean and variance as the Binomial distribution, i.e. $M = n\pi$ and $\sigma^2 = n\pi(1-\pi)$, the Normal approximation to the Binomial is defined by (6.5.3).

$$\Pr\left(x_1 - \frac{1}{2} \leq y \leq x_2 + \frac{1}{2}\right) = \Pr(x_1 \leq x \leq x_2). \tag{6.5.3}$$

Table 6.5.1. Examining the Fit of a Normal Distribution to a Data Set (Data from Table 1.3.7)

Class	Upper Class Limit obs.	Upper Class Limit stand.	Proportion Below Limit	Proportion Expected in Class	No. in Class exp.	No. in Class obs.
i	y_i [1]	z_i [2]	p_i [3]	$p(i)$ [4]	$np(i)$ [5]	f_i [1]
1	75.95	−1.83	0.0336	0.0336	8.4	4
2	76.95	−1.37	0.0853	0.0517	12.9	16
3	77.95	−0.90	0.1841	0.0988	24.7	28
4	78.95	−0.44	0.3300	0.1459	36.5	32
5	79.95	+0.03	0.4880	0.1580	39.5	52
6	80.95	+0.49	0.6879	0.1999	50.0	41
7	81.95	+0.96	0.8315	0.1436	35.9	33
8	82.95	+1.42	0.9222	0.0907	22.7	25
9	83.95	+1.89	0.9706	0.0484	12.1	15
10			1.0000	0.0294	7.4	4

Explanations:

[1]From Table 1.3.7. Note that the upper class limit for class 10 is infinity.

[2]$z_i = (y_i - M)/\sigma$ where values for M and σ^2 are the sample mean and variance respectively which have values 79.90 and 4.6086.

[3]$p_i = \Pr(Z < z_i)$ from $N(0,1)$ distribution. Note that p_0 and p_{10} are necessarily zero and one respectively to satisfy requirements of a probability distribution.

[4]$p(i) = p_i - p_{i-1}$.

[5]n is the sample size which is 250.

Figure 6.5.3 illustrates both why the Normal distribution provides a good approximation to Binomial probabilities and why the adjustment $\frac{1}{2}$ appears in the limits for the Normal distribution in (6.5.3). The area of each rectangle is seen to be well approximated by the area under the Normal distribution within the same limits. For example, the probability of an outcome of zero under the Binomial model is approximately equal to the area under the Normal curve between −0.5 and 0.5; the probability of an outcome in the range zero to nine is approximately equal to the area under the Normal distribution curve between −0.5 and 9.5.

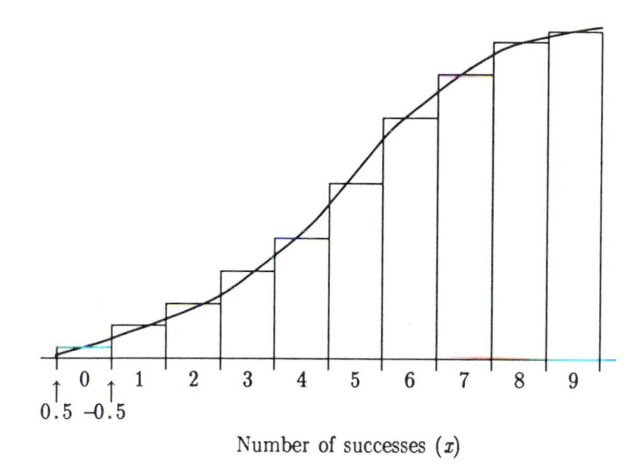

Figure 6.5.3. Normal approximation to the Binomial distribution.

The simplicity of the calculation based on the Normal distribution can easily be illustrated. In an experiment by the man who first proposed the genetic theory of inheritance, Gregor Mendel, a sample of 7324 seeds were counted. According to his theory, three quarters of these should be round. Thus, $n = 7324$ and $\pi = \frac{3}{4}$. The observed number of round seeds was 5474. What is the probability of 5474 or fewer round seeds under the Mendelian theory?

To apply the Binomial model directly would require the evaluation of $\Pr(x = 0) + \Pr(x = 1) + \cdots + \Pr(x = 5474)$, i.e. 5475 terms, using (6.1.1).

To employ the Normal approximation, (6.5.3) is used with x_1 and x_2 set equal to 0 and 5474, respectively. From the information provided, $M = 7324 \times \frac{3}{4} = 5493$ and $\sigma^2 = 7324 \times \frac{3}{4} \times \left(1 - \frac{3}{4}\right) = 1373.25$. The area required is between -0.5 and 5474.5. A value of 0.31 is obtained with little computational effort—see Table 7.1.1 for computational details.

The Normal approximation is used as an approximation to many and varied sampling distributions of statistics as will become apparent in later chapters of the book.

6.5.7 Statistics

There are many statistics which have practical application when based on a sample presumed to have arisen under a Normal distribution model. Three of these statistics are described below. In their definitions, $y_1, y_2, \ldots, y_n$ are presumed to constitute independent random variables with $N(M, \sigma^2)$

distributions. (From an experimental viewpoint, the variables $y_1, y_2, \cdots, y_n$ are presumed to supply the responses from n sample members which have been randomly selected from a population.)

1. *The arithmetic mean.* The statistic $\bar{y} = (y_1 + y_2 + \cdots + y_n)/n$, the arithmetic mean of a sample, also has a sampling distribution which is Normal with mean M and variance σ^2/n. Hence, the statistic $z = [\bar{y} - M]/[\sigma/\sqrt{n}]$ has a $N(0,1)$ sampling distribution.

2. *The t-statistic.* The statistic $t = [\bar{y} - M]/[s/\sqrt{n}]$, where s is the sample standard deviation, has a sampling distribution which is a member of a family of distributions known collectively as the *t-distribution* which is defined in Section 7.4.

3. *The Chi-squared statistic.* The statistic $X^2 = \sum_{i=1}^{n} \frac{(y_i - M)^2}{\sigma^2}$, i.e. the sum of squares of the *standardized* observations, has a sampling distribution which is a member of a family of distributions that are collectively known as the *Chi-squared distribution* (see Section 7.5).

6.6 The Logistic Model

6.6.1 Motivation

Examples 6.1 and 6.2 describe completely different situations. Yet, from a statistical viewpoint, a common model can be expected to apply. Read the descriptions of the examples and look for the common theme.

Example 6.1 (Susceptibility of Household Flies to Insecticide). *Table* 6.6.1 *contains data taken from an experiment in which samples of household flies were subjected to varying levels of an insecticide prepared from naturally occurring ingredients. The proportion of flies killed is seen to rise as the concentration of active ingredient in the insecticide is increased—scarcely a surprising result. What is of interest is the relation between the expected proportion of insects killed (P) and the level of active ingredient (x).*

Table 6.6.1. Data Collected in the Study Reported in Example 6.1 (Data supplied by Dr. D. Ratkowsky, C.S.I.R.O.)

Active ingredient Concentration (x)[1]	0.0	1.0	1.5	2.0	2.5	3.0	3.5	4.0
Total number	20	20	20	20	20	20	20	20
Number dead	0	3	5	4	9	11	16	20
Proportion dead	0.00	0.15	0.25	0.20	0.45	0.55	0.80	1.00

[1] Values for x are recorded on a logarithmic scale base 10.

The fact that some insects are killed by a given concentration of insecticide while others are not, suggests that the insects have differing Levels of susceptibility (u) *to the insecticide. The proposition is that the insects in a population show a continuous range from those which are very resistant to those which are very susceptible to the insecticide. It is only those insects whose Level of susceptibility is above a* threshold (u_T) *which are killed by the insecticide. Furthermore, Level of susceptibility of an insect changes with changing concentration of the insecticide—as the concentration of the insecticide* (x) *increases, so the proportion* (P) *of insects with levels of susceptibility above the threshold increases.*

There is an obvious advantage in being able to describe the relationship between P and x in order that (i) *the expected proportion of insects killed by an insecticide with a stated level of active ingredient can be determined; or* (ii) *that the minimum level of active ingredient required to kill a fixed proportion of insects can be determined. The latter problem is of particular interest and has prompted the introduction of the term* LD50 *as the level of active ingredient which leads to exactly* 50% *of insects being expected to die.*

Example 6.2 (Loyalty to a Political Party). *At the time of an election, voters vary in their level of support for a political party. However, the only information which is observable is the proportion of voters who actually vote for the party.*

It might reasonably be postulated that there is an underlying continuous variable, which might be titled Degree of support *for the political party* (u), *and that in the population of voters, there is a distribution of values of this underlying variable from low to high levels. If a voter has a value for the* Degree of support *variable which is above a fixed threshold* (u_T), *the voter will vote for the party. Otherwise they will not vote for the party. Under a fixed set of conditions, a proportion P of voters in the electorate will vote for the party.*

Consider the effect of a change in taxation policy by the party on the proportion of voters who would support it. All other factors remaining constant, as the amount of taxation concessions (x) *increases, so the proportion of voters who would vote for the party* (P) *would increase. What is the expected form of the relation between P and x?*

6.6.2 A Consideration of Statistical Models

Examples 6.1 and 6.2 are typical of many and diverse situations in which data have been collected which suggest a form of relation between an expected proportion (P) and a supplementary variable (x) with the graphical form displayed in Figure 6.6.1.

An explanation for this common form of relationship commences from the following basis.

> Suppose there exists a continuous response variable u which has the property that the only information available on the value of u for a population member is whether the value is above or below a threshold value u_T. In the population at large, there will be a proportion of values, P, which will be above the threshold.
>
> Consider the effect of changing the conditions under which the population is held. In particular, suppose the change in conditions can be defined in terms of a variable x, and that when x takes the value x_0, the proportion of responses for u which are above the threshold is $P(x_0)$.

What additional conditions might be added which lead to the form of relation between $P(x)$ and x displayed in Figure 6.6.1?

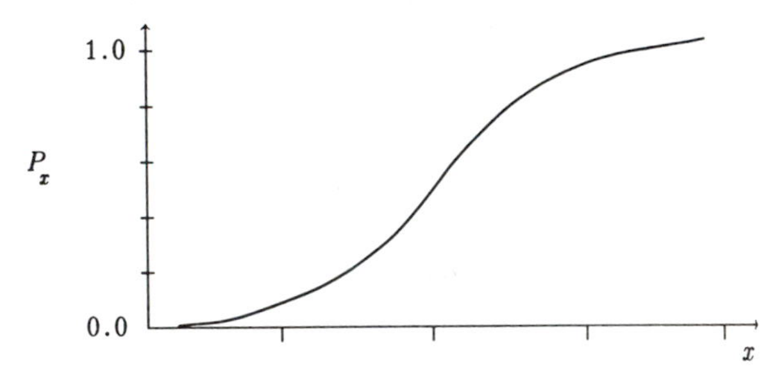

Figure 6.6.1. The S-shaped logistic curve.

To answer this question, it is firstly necessary to establish a connection between $P(x)$ and the probability distribution of the response variable u. If π_x is the probability distribution of u given x, then (6.6.1) provides a definition of $P(x)$ in terms of π_x.

$$P(x) = \int_{u_T}^{\infty} \pi_x(u)\, du. \tag{6.6.1}$$

Figure 6.6.2 illustrates the manner in which $P(x)$ is constructed by displaying the relationship when u is set equal to the threshold value u_T.

The form of the probability distribution π_x cannot be established from direct empirical evidence since values of u are not measurable. The only clue to a possible form for π_x is the nature of the variable. For reasons advanced in Section 6.5, the Normal distribution might be expected to provide a reasonable approximation in many situations, and the early development of statistical models was almost exclusively based on this assumption. The relation between $P(x)$ and y under the assumption of Normality which is based on (6.6.1) is known as the *Probit transformation.*

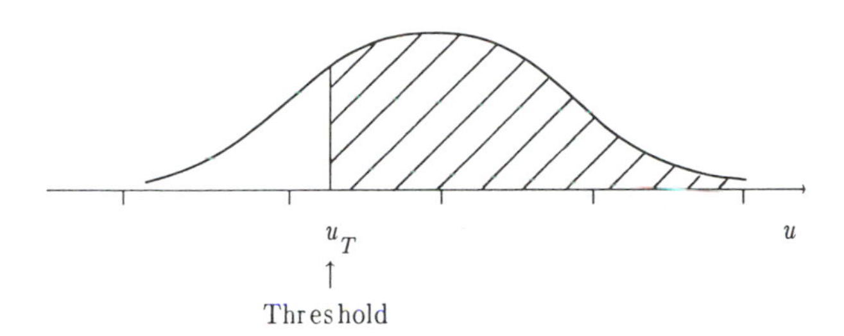

Figure 6.6.2. Relating the expected proportion (P) and latent variable u in a logistic model.

In addition to the nature of the frequency distribution, there is need to define the manner in which the distribution π_x changes with changing values in the supplementary variable x. The simplest and most widely employed statistical models are based on the assumption that (i) the effect of changing the value of x is to cause the distribution to be shifted along the axis without change in shape; and (ii) the translation is defined by the relation between the mean or expected value of the distribution, M_x, and x which is expressed in (6.6.2).

$$M_x = A + Bx. \tag{6.6.2}$$

Under the assumption that the probability distribution of u given x, namely π_x, is $N(M_x, \sigma^2)$, and M_x is related to the variable x according to (6.6.2), the expected proportion above a fixed threshold u_T, $P(x)$, is related to x in the manner displayed graphically in Figure 6.6.1. Particularly in biological assay work, the model appears to fit well much of the data which are collected and, under the name of the *Probit model,* it has been widely used.

With time, there has come the realization that by varying conditions in the Probit model there is scope for much broader application. However, for technical and computational reasons, there are difficulties in realizing this broader application. There is an alternative model, the *Logistic model,* which has similar properties to the Probit model from the user's point of view, but which has the capacity to generalize to meet the broader range of applications plus the added bonus of superior properties from the statistical viewpoint.

6.6.3 The Logistic Model

For a continuous random variable u, the probability distribution defined by (6.6.3) is termed the *Logistic distribution,* and if this distribution is inserted into (6.6.1), the relation between P and u_T, as defined in (6.6.4), is termed the *Logit transformation.*

$$\pi(u) = \frac{\exp\{(u-\alpha)/k\}}{k[1+\exp\{(u-\alpha)/k\}]^2} \quad \text{for} \quad -\infty < u < \infty \tag{6.6.3}$$

where α and k are parameters which are fixed for a given distribution.

$$P = \frac{1}{1+\exp[(u_T-\alpha)/k]}. \tag{6.6.4}$$

The mean or expected value of the *Logistic distribution is* α, and, by employing the assumption $\alpha = A + Bx$ suggested in (6.6.2), the expected proportion of population members with values above u_T is related to $P(x)$ according to (6.6.5).

$$P(x) = \frac{1}{1+\exp\{[u_T-(A+Bx)]/k\}} \quad \text{for} \quad -\infty < x < \infty. \tag{6.6.5}$$

An equivalent expression for (6.6.5) is provided in (6.6.6).

$$\log\{(1-P(x))/P(x)\} = C + Dx \quad \text{for} \quad -\infty < x < \infty \tag{6.6.6}$$

where $C = (u_T - A)/k$, $D = -B/k$.

6.6.4 Applications

Central to the original development of methodology based on the Probit distribution was the desire to establish the value of X which would yield a specified value of P. In biological assay, this application is so well established that the level of X at which $P(x)$ equals 0.5 has the special title of the LD50.

Since the parameters which appear in (6.6.6) derive from the underlying variable u, there is scope for examining hypotheses relating to changes in these parameters under different treatments or different conditions.

The Logistic model can be generalized by permitting a division of the underlying scale into more than two regions. For example, in a food tasting experiment, it is common to use sevel divisions which range from *very bad* to *very good.* Each taster classifies the food into one of the seven categories. By assuming there is an underlying continuous variable, *Degree of acceptability,* there are six thresholds which define the seven categories. For the ith threshold, $u_T(i)$, $(i = 1, 2, \ldots, 6)$ there is a parameter P_i which is defined by the Logit transformation, (6.6.4).

Problems

Some of the problems in this set require the use of a statistical package which can generate pseudo-random numbers from Poisson and Normal distributions.

6.1 *Computing Binomial probabilities*

(a) Use (6.1.1) to compute the probability of 0, 1, 2, 3, 4, and 5 successes in 5 trials assuming π equals 0.5. Hence, reproduce the set of expected frequencies obtained in Problem 2.1.

(b) Use the set of observed frequencies given in Problem 2.1 to compute the average number of females in a sample of 5 persons ($\overline{y}_0$). Hence, determine the average proportion of females in a sample of 5 persons ($p_0 = \overline{y}_0/5$). Use this figure as a value of π to compute Binomial probabilities using (6.1.1) with n equal to 5.

6.2 *Computing Poisson probabilities*

The data in Table 1.2.2 (Example 1.2) displays a set of observed frequencies from 576 sites and a set of expected frequencies.

(a) Compute the arithmetic mean, $\overline{y}_0$, for the data using the additional information that the eight readings in the ≥ 4 category are 4,4,4,4,4,4,4 and 7.

(b) Use $\overline{y}_0$ as the value for M in (6.3.1) and compute probabilities for $n = 0, 1, 2, 3$ and for $n \geq 4$. Multiply each probability by the sample size, 576, to reproduce the expected frequencies in Table 1.2.2.

(c) Use a statistical package to generate 576 pseudo-random values from a Poisson distribution with mean equal to $\overline{y}_0$. Group the data in a frequency table and visually judge if this generated set shows greater or less similarity to the expected (Poisson) frequencies than does the observed table of frequencies.

6.3 *Computing Normal probabilities*

(a) Using the formulae and explanation in Table 6.5.1 plus the data in Tables 1.3.2 and 1.3.3, complete the following table and obtain a set of expected frequencies under the assumption of a Normal distribution model.

(b) Display graphically (e.g. in histograms) the set of observed and expected frequencies in a manner which allows a visual assessment of the Normality assumption to be made.

Class	Upper class limit obs.	Upper class limit stand.	Proportion below limit	Proportion expected in class	No. in class exp.	No. in class obs.
i	y_i	z_i	p_i	$p(i)$	$np(i)$	f_i
1	159.5					5
2	179.5					13
3	199.5					17
⋮						

6.4 *Sampling variation*

(a) Use a statistical package to generate 1,000 observations from a Normal distribution with mean zero and variance one.

(b) Construct a histogram based on the set of numbers generated.

(c) Determine the proportion of numbers in the ranges

(i) -1.0 to $+1.0$;

(ii) -2.0 to $+2.0$;

(iii) -3.0 to $+3.0$.

Do these proportions appear consistent with the expected proportions stated in Section 6.5.5?

(d) Determine the mean and standard deviation of the 1,000 observations. Note how close to the expected values—zero for the mean and 1 for the standard deviation—they are.

(e) Partition the 1,000 observations into 10 sets of 100 observations each and construct a histogram for each set of 100 observations. Note that while the characteristic pattern of the Normal distribution is retained, there is more evidence of sampling variability when the smaller sized samples are employed.

(f) Produce tables of the means and the standard deviations for each set of 100 observations.

(g) Partition the first 100 observations into 10 sets of 10 observations each. Produce tables of the means and the standard deviations for each set of 10 observations.

Note the greater variability in values of both statistics when computed from the smaller sample sizes.

The following questions require mathematical derivations of formulae for sampling distributions of statistics using the rules of probability.

6.5 *Binomial and Negative Binomial distributions*

Derive (6.1.1) and (6.1.2) from direct probability arguments using the rules of probability given in Section 5.3 based on the assumptions in Section 6.1.1.

6.6 *The Binomial approximation to the Hypergeometric distribution*

Establish that as N_1 and N_2 tend to infinity while n remains constant, the probabilities defined in (6.2.1) having a limiting form which are defined by (6.1.1). (The significance of the result is that statistical methods developed for the Binomial model may be employed when the Two-state Population model applies provided the population size is large compared to the sample size.)

6.7 *The Multinomial distribution*

Based on the assumptions in the Multinomial model (Section 6.4), use the rules of probability stated in Section 5.3 to derive the Multinomial distribution (6.4.1).

7

Some Important Statistics and Their Sampling Distributions

7.1 The Binomial Distribution

The Binomial distribution arises in situations in which the Binomial model (Section 6.1) applies.

7.1.1 STATISTIC

The typical description is expressed as the number of successes, x, in n repetitive trials, i.e. trials in which outcomes are independent and the probability of success, π, remains constant from trial to trial.

7.1.2 SAMPLING DISTRIBUTION

$$\pi_{st}(x) = \binom{n}{x}\pi^x(1-\pi)^{n-x} \qquad \text{for} \quad x = 0, 1, 2, \ldots, n, \tag{7.1.1}$$

where π is the probability of success in a single trial and as such must take values between zero and one, n is the sample size and as such must be a positive whole number.

7.1.3 PROPERTIES

1. MEAN = $n\pi$ VARIANCE = $n\pi(1-\pi)$.

2. If x_1 and x_2 are two Binomial statistics based on sample sizes n_1 and n_2, respectively, and with a common probability of success of π, then, provided x_1 and x_2 derive values from independent outcomes, $x = x_1 + x_2$ is also a Binomial statistic with sample size $n = n_1 + n_2$ and probability of success π.

3. If $\pi = 0.5$ the distribution is symmetric, otherwise it is skewed. This is illustrated in the histograms presented in Figure 7.1.1.

4. The distribution for $\pi = \pi_0$ is the mirror image of the distribution for $\pi = 1 - \pi_0$ for a common value of n.

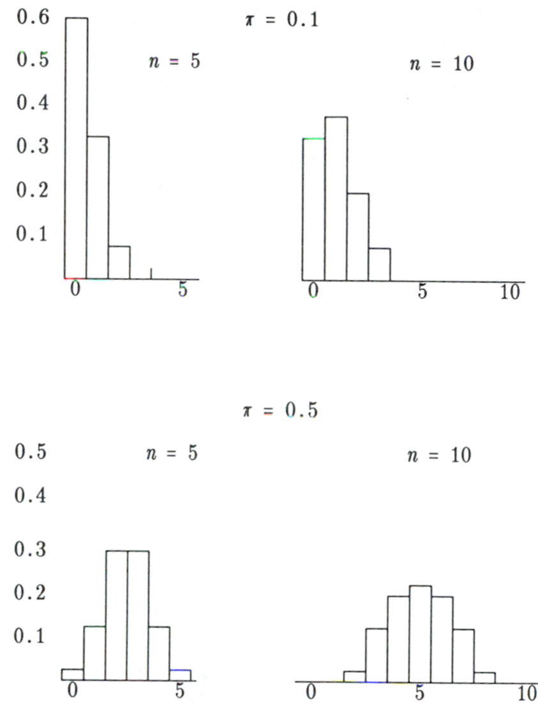

Figure 7.1.1. Graphical representations of Binomial distributions.

7.1.4 Calculation of Probabilities

Values may be required for $\Pr(x = c)$, $\Pr(x \leq c)$, $\Pr(c_1 \leq x \leq c_2)$, etc. Statistical computing packages and statistical tables are sources of $\Pr(x = c)$, i.e. values of (7.1.1), and cumulative probabilities, $\Pr(x \leq c) = \sum_{x=0}^{c} \pi_{st}(x)$. Variations on these basic forms require the application of rules of probability. The common variations are presented and illustrated below on the assumption that values can be obtained for $p_1 = \Pr(x \leq c_1)$ and $p_2 = \Pr(x \leq c_2)$.

Probability	Example for $\pi = 0.3 \quad n = 10$
$\Pr(x > c_1) = 1 - p_1$	$\Pr(x > 4) = 1 - 0.8497 = 0.1503$
$\Pr(x \geq c_1 + 1) = 1 - p_1$	$\Pr(x \geq 5) = 1 - 0.8497 = 0.1503$
$\Pr(c_1 < x \leq c_2) = p_2 - p_1$	$\Pr(2 < x \leq 5) = 0.9527 - 0.3828 = 0.5699$

For large values of n, an approximation based on the Normal distribution may be required (see 3 below for relevant formulae and an illustration).

1. *Computer-based calculations.* Value of $\Pr(x = c)$ and $\Pr(x \leq c)$ are widely available as values of the *probability distribution* and (*cumulative*) *distribution functions*, respectively.

2. *Tables.* Statistical tables are available for a range of values of n and π.

Table 7.1.1. Normal Approximations to Binomial Probabilities (see Section 7.3.4 for methods of computing Normal probabilities)

Binomial Probability	Normal Approximation	
$\Pr(x \leq x_0)$	$\Pr\left[z \leq \frac{x_0 + \frac{1}{2} - n\pi_0}{\sqrt{n\pi_0(1-\pi_0)}}\right]$	
$\Pr(x \geq x_0)$	$\Pr\left[z \geq \frac{x_0 - \frac{1}{2} - n\pi_0}{\sqrt{n\pi_0(1-\pi_0)}}\right]$	
$\Pr(x = x_0)$	$\Pr\left[z \leq \frac{x_0 + \frac{1}{2} - n\pi_0}{\sqrt{n\pi_0(1-\pi_0)}}\right]$ $-\Pr\left[z \leq \frac{x_0 - \frac{1}{2} - n\pi_0}{\sqrt{n\pi_0(1-\pi_0)}}\right]$	
$\Pr(x_1 \leq x \leq x_2)$	$\Pr\left[z \leq \frac{x_2 + \frac{1}{2} n\pi_0}{\sqrt{n\pi_0(1-\pi_0)}}\right]$ $-\Pr\left[z \leq \frac{x_1 - \frac{1}{2} - n\pi_0}{\sqrt{n\pi_0(1-\pi_0)}}\right]$	
$\Pr(\lvert x - n\pi_0\rvert \geq \lvert x_0 - n\pi_0\rvert)$	$2\Pr\left[\leq \frac{x_0 + \frac{1}{2} - n\pi_0}{\sqrt{n\pi_0(1-\pi_0)}}\right]$	for $x_0 < n\pi_0$
	$2\Pr\left[z \geq \frac{x_0 - \frac{1}{2} - n\pi_0}{\sqrt{n\pi_0(1-\pi_0)}}\right]$	for $x_0 > n\pi_0$
	1	$x_0 = n\pi_0$

3. *Approximations using the Normal distribution.* This approximation is based on the Central Limit Theorem (Section 7.3.5). Table 7.1.1 lists the range of probabilities for the Binomial distribution which are required in practice and their Normal distribution approximations. (The reason for adjustment of limits by one half which appears in the formulae for the Normal distribution is explained in Section 6.5.6.)

The approximation applies in large samples, although the circumstances under which it has practical value depend on the value of π. There are various rules-of-thumb suggested for identifying the conditions under which the approximation is adequate. Simple rules which rely only on the product $n\pi(1-\pi)$ provide poor guidance for extreme values of π. I have developed the formula presented in (7.1.2) to provide a guide to the minimum sample size, $n_{\min}$, under which the Normal approximation to the Binomial provides an adequate approximation.

$$n_{\min} = \frac{2}{3\{\pi(1-\pi)\}^2}. \tag{7.1.2}$$

A selection of values of $n_{\min}$ for different values of π, computed using (7.1.2) are presented below.

π	0.5	0.3 or 0.7	0.1 or 0.9	0.01 or 0.99
$n_{\min}$	11	16	83	6803

When $\pi = \frac{1}{2}$ the distribution is symmetric and the Normal distribution provides a good approximation for small sample sizes. As the value of π tends towards zero or one, the distribution becomes more asymmetric for a fixed sample size. However, asymmetry decreases as sample size increases.

Illustration. Consider the Binomial distribution with n and π equal to 20 and 0.5, respectively. (This model would have application, for example, when a fair coin was tossed twenty times and π was the probability of the coin coming down heads.)

The following questions could be answered using the formulae in Table 7.1.1.

1. *What is the probability of at least five heads?* This requires the determination of $\Pr(x \geq 5)$. Using the Normal approximation,

$$\Pr(x \geq 5) \cong \Pr\left[z \geq \frac{5 - \frac{1}{2} - 20(\frac{1}{2})}{\sqrt{20(\frac{1}{2})(1-\frac{1}{2})}}\right] = \Pr[z \geq -2.46] = 0.9931.$$

2. *What is the probability of exactly five heads?* This requires the determination of $\Pr(x = 5)$. Using the Normal approximation,

$$\begin{aligned}\Pr(x = 5) &\cong \Pr\left[z \le \frac{5 + \frac{1}{2} - 20(\frac{1}{2})}{\sqrt{20(\frac{1}{2})(1 - \frac{1}{2})}}\right] - \Pr\left[z \le \frac{5 - \frac{1}{2} - 20(\frac{1}{2})}{\sqrt{20(\frac{1}{2})(1 - \frac{1}{2})}}\right] \\ &= \Pr[z \le -2.01] - \Pr[z \le -2.46] \\ &= 0.0222 - 0.0069 \\ &= 0.0153.\end{aligned}$$

3. *What is the probability of between 5 and 10 heads?* This requires the determination of $\Pr(5 \le x \le 10)$. Using the Normal approximation,

$$\begin{aligned}\Pr(5 \le x \le 10) &\cong \Pr\left[z \le \frac{10 + \frac{1}{2} - 20(\frac{1}{2})}{\sqrt{20(\frac{1}{2})(1 - \frac{1}{2})}}\right] \\ &\quad - \Pr\left[z \le \frac{5 - \frac{1}{2} - 20(\frac{1}{2})}{\sqrt{20(\frac{1}{2})(1 - \frac{1}{2})}}\right] \\ &= \Pr[z \le 0.22] - \Pr[z \le -2.46] \\ &= 0.5871 - 0.0069 \\ &= 0.5802.\end{aligned}$$

4. *What is the probability that the number of heads differs from the mean value (10) by at least five?* This requires the determination of $\Pr(|x - 10| \ge |5 - 10|)$. Using the Normal approximation,

$$\begin{aligned}\Pr(|x - 10| \ge |5 - 10|) &\cong 2\Pr\left[z \le \frac{5 + \frac{1}{2} - 20(\frac{1}{2})}{\sqrt{20(\frac{1}{2})(1 - \frac{1}{2})}}\right] \\ &= 2\Pr[z \le -2.01] \\ &= 2\,(0.0222) \\ &= 0.0444.\end{aligned}$$

4. *Approximation using the Poisson distribution.* As n tends towards infinity, π towards zero and $n\pi$ towards a constant M, then the probability that the Binomial statistic takes the value x_0 tends towards the probability that the Poisson statistic with mean $n\pi$ takes the value x_0 which is defined by (7.2.1).

7.2 The Poisson Distribution

7.2.1 Statistic

The number of occurrences, n, of an event in a fixed period of time or in a fixed region of space where the process generating the events is expected to produce an average of M events in the given time period or region of space. The occurrence of an event is presumed to neither increase nor decrease the chance of a further event in the following period, i.e. the events are said to be *independent* occurrences. The mechanism which generates the events is presumed to be unchanging over the period or area of sampling and over the period or area to which conclusions will apply.

7.2.2 Sampling Distribution

$$\pi_{st}(n) = \frac{\exp(-M) \cdot M^n}{n!} \qquad \text{for} \quad n = 0, 1, 2, \ldots \tag{7.2.1}$$

where M is a parameter which is the average number of occurrences expected per unit of time or space.

The effect of varying the value of M is illustrated in Figure 7.2.1.

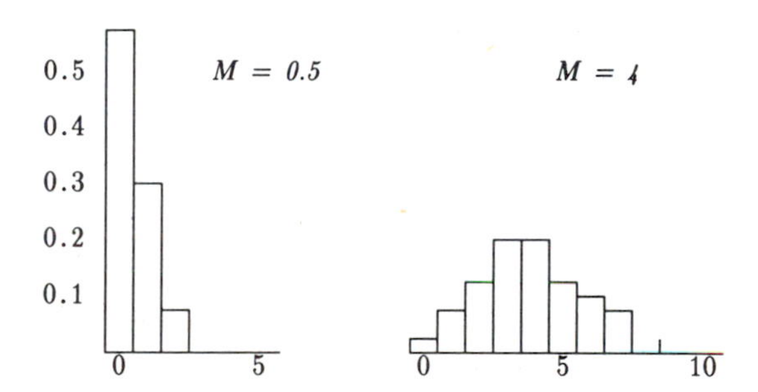

Figure 7.2.1. Sampling distributions for Poisson statistics.

7.2.3 Properties

1. MEAN $= M$ VARIANCE $= M$.

2. If n_1 and n_2 are statistics based on independent outcomes and having sampling distributions which are Poisson distributions with means M_1 and M_2, respectively, the sampling distribution of $n = n_1 + n_2$ is Poisson with mean $M = M_1 + M_2$.

7.2.4 Calculation of Probabilities

Values may be required for $\Pr(x = c)$, $\Pr(x \le c)$, $\Pr(c_1 \le x \le c_2)$, etc. Statistical computing packages and statistical tables are sources of

$\Pr(x = c)$, i.e. values of (7.2.1), and cumulative probabilities, $\Pr(x \leq c) = \sum_{x=0}^{c} \pi_{st}(x)$. Variations on these basic forms require the application of rules of probability, e.g. $\Pr(x > c) = 1 - \Pr(x \leq c)$. The common variations are presented and illustrated below on the assumption that values can be obtained for $p_1 = \Pr(x_1 \leq c_1)$ and $p_2 = \Pr(x_2 \leq c_2)$.

Probability	Example for $M = 3$
$\Pr(x > c_1) = 1 - p_1$	$\Pr(x > 4) = 1 - 0.6472 = 0.3528$
$\Pr(x \geq c_1 + 1) = 1 - p_1$	$\Pr(x \geq 5) = 1 - 0.6472 = 0.3528$
$\Pr(c_1 < x \leq c_2) = p_2 - p_1$	$\Pr(2 < x \leq 5) = 0.9161 - 0.4232 = 0.4929$

For large values of n, an approximation based on the Normal distribution may be required.

1. *Computer-based calculations.* Values of $\Pr(x = c)$ and $\Pr(x \leq c)$ are widely available as values of the *probability distribution* and (*cumulative*) *distribution functions,* respectively.

2. *Tables.* Statistical tables are available which provide probabilities and cumulative probabilities for a range of specified values of the mean M.

3. *Approximation using the Normal distribution.* For M in excess of 9, the Normal distribution may be used to provide adequate approximations to both individual and cumulative Poisson probabilities. (Table 7.1.1 can be adapted to define the relationship between Poisson and Normal probabilities. The only changes required are (i) to identify the probabilities in the first column as Poisson probabilities for a distribution with a value M_0 for the parameter M; and (ii) to replace $n\pi_0$ in the numerator of an expression for the Normal approximation by M_0 and to replace $n\pi_0(1 - \pi_0)$ in the denominator by M_0.

7.3 The Normal Distribution

7.3.1 Statistic

The variable is presumed to show continuous variation over the range of all possible positive and negative values. In practice it may also be used to approximate the distribution for

(i) discrete variables which take many values—an illustration using the Binomial statistic is provided in Section 6.5.6; and

(ii) variables which can only take values within a limited range—for example, *weight* or *length* which can only take positive values. In this case, a Normal distribution could only have relevance if the probability of a value lying outside the permissible range is suitable close to zero—a situation which arises frequently in practice.

There are many statistics which have sampling distributions well approximated by a Normal distribution when based on large sample sizes.

7.3.2 Sampling Distribution

The sampling distribution of a Normally distributed statistic, y, is defined by (7.3.1).

$$\pi_{st}(y) = \frac{1}{\sqrt{2\pi\sigma^2}} \exp\left[-\frac{1}{2\sigma^2}(y - M)^2\right] \qquad -\infty < y < \infty \qquad (7.3.1)$$

where M and σ^2 are parameters of the distribution, with M being permitted to take any real value and σ^2 being any positive real number. The effects of changing values of M and σ^2 are illustrated in Figure 6.5.2.

Notation. The following notation is used throughout the book.

$N(M, \sigma^2)$ denotes a Normal distribution with mean M and variance σ^2.

$N(0,1)$ is referred to as the *standard Normal distribution.*

z_α is the value of a statistic z which has a $N(0,1)$ sampling distribution and satisfies $\Pr(z > \mathrm{z}_\alpha) = \alpha$. For example, $\mathrm{z}_{0.05}$ is the value which satisfies $\Pr(z > \mathrm{z}_{0.05}) = 0.05$.

7.3.3 Properties

1. MEAN = M VARIANCE = σ^2.

2. If y_1 and y_2 are independent statistics having sampling distributions $N(M_1, \sigma_1^2)$ and $N(M_2, \sigma_2^2)$, respectively, and c_1 and c_2 are real numbers, then

   (a) $y_1 + y_2$ has a $N(M_1 + M_2,\ \sigma_1^2 + \sigma_2^2)$ distribution;

   (b) $c_1y_1 + c_2y_2$ has a $N(c_1M_1 + c_2M_2,\ c_1^2\sigma_1^2 + c_2^2\sigma_2^2)$ distribution; in particular, $y_1 - y_2$ has a $N(M_1 - M_2,\ \sigma_1^2 + \sigma_2^2)$ distribution.

3. (a) $\Pr(y > M - a) = \Pr(y < M + a)$.

   If $a > 0$, then

   (b) $\Pr(M - a < y < M + a) = 1 - 2\Pr(y \leq M - a)$;

   (c) $\Pr(M \leq y \leq M + a) = \frac{1}{2} - \Pr(y > M + a)$;

   (d) $\Pr(|y - M| \geq a) = \Pr(y \geq M + a \text{ or } y \leq M - a) = 2\Pr(y \leq M - a)$.

4. *Properties of the standard normal distribution.* If z is a statistic which has a standard Normal distribution, i.e. a $N(0,1)$ distribution, then

   (a) $\Pr(z < -z_0) = \Pr(z > z_0)$.

   If z_0 is a number greater than zero, then

   (b) $\Pr(-z_0 \le z \le z_0) = 1 - 2\Pr(z > z_0)$; and
   (c) $\Pr(0 \le z \le z_0) = \frac{1}{2} - \Pr(z > z_0)$.
   (d) $\Pr(|z| \ge z_0) = \Pr(z \ge z_0 \text{ or } z \le -z_0) = 2\Pr(z > z_0)$.

5. *Linking probabilities under different Normal distributions.* If y and z are statistics which have Normal distributions $N(M, \sigma^2)$ and $N(0,1)$, respectively, then

$$\Pr(y \le y_0) = \Pr(z \le z_0) \quad \text{where} \quad z_0 = \frac{y_0 - M}{\sigma}.$$

7.3.4 Calculation of Probabilities

Notation. See the standard notation presented in Section 7.3.2 which is used throughout the book.

The use of statistical packages and tables for the determination of probabilities is described and illustrated below.

Two basic applications arise when the distribution of y is $N(M, \sigma^2)$. They are

(i) given a value y_0 for y, determine $p_0 = \Pr(y \le y_0)$; and

(ii) given a value $p_0 = \Pr(y \le y_0)$, determine the value of y_0.

Variations on the first of these tasks are considered in Table 7.3.1 and illustrations based on formulae in the table are given below.

An important application of the inverse operation is to find a value $\mathrm{z}_{\frac{1}{2}\alpha}$ which satisfies $\Pr(|z| \ge \mathrm{z}_{\frac{1}{2}\alpha}) = \alpha$. Since statistical packages typically provide $p_0 = \Pr(z \le \mathrm{z}_0)$, this equation must be transformed using (7.3.2) to determine the value of $\mathrm{z}_{\frac{1}{2}\alpha}$. For example, if $\alpha = 0.05$, it is necessary to set $p = 0.975$ to determine the value for $\mathrm{z}_{0.025}$.

$$\Pr(z \le \mathrm{z}_{\frac{1}{2}\alpha}) = 1 - \frac{1}{2}\alpha. \tag{7.3.2}$$

1. *Computer-based calculations.* All major statistical packages provide the option of obtaining either y_0 from p_0 or the inverse operation from the relation $p_0 = \Pr(y \le y_0)$ where y has a $N(M, \sigma^2)$ distribution.

2. *Tables of Normal probabilities.* There is widespread availability of Normal probability tables which will provide probabilities for $N(0,1)$ distributions (for example, Tables A.1.1–A.1.5). The tables are provided only for the

standard Normal distribution. Hence, to make use of the tables, it is necessary to employ Property 5 in Section 7.3.3 and convert a value y_0 from a $N(M, \sigma^2)$ distribution to $z_0 = (y_0 - M)/\sigma$ for entry in the tables. Note that the tables vary widely as to the probability which is provided. They variously provide $\Pr(z \leq z_0)$, $\Pr(z > z_0)$, $\Pr(0 \leq z \leq z_0)$, and $\Pr(|z| \geq |z_0|)$. Table 7.3.1 allows for conversion from the tabulated form to the required form.

Table 7.3.1. Probabilities of Events Based on the Normal Distribution

$N(0,1)$ distribution	$N(M,\sigma^2)$ distribution	Probability	
$\Pr(z \leq z_0)$	$\Pr(y \leq y_0)$	p	
$\Pr(z \geq z_0)$ $\Pr(z > z_0)$	$\Pr(y \geq y_0)$ $\Pr(y > y_0)$	$1-p$	
$\Pr(z \geq -z_0)$	$\Pr(y \geq 2M - y_0)$	p	
$\Pr(0 \leq z \leq z_0)$	$\Pr(M \leq y \leq y_0)$	$p - \frac{1}{2}$	for $z_0 > 0$ or $y_0 > M$
$\Pr(\lvert z\rvert \geq \lvert z_0\rvert)$	$\Pr(\lvert y - M\rvert \geq \lvert y_0 - M\rvert)$	$2p$	if $z_0 \leq 0$ or $y_0 \leq M$
		$2(1-p)$	otherwise
$\Pr(z_1 \leq z \leq z_2)$	$\Pr(y_1 \leq y \leq y_2)$	$p_2 - p_1$ *	

*$p_1 = \Pr(z \leq z_1) = \Pr(y \leq y_1)$ where $z_1 = \frac{y_1 - M}{\sigma}$
$p_2 = \Pr(z \leq z_2) = \Pr(y \leq y_2)$ where $z_2 = \frac{y_2 - M}{\sigma}$.

Illustration. Data are provided in Table 1.3.7 on the cut-off temperatures for thermostats. With a nominal mean cut-off temperature (M) of 80 and a value for σ of 2.1,

(a) *What proportion of thermostats is expected to have cut-off temperatures below* 83°C ($y_0 = 83$)? From Table 7.3.1 property 1, $z_0 = (83 - 80)/2.1 = 1.43$.

$$\Pr(y < 83) = \Pr\left(z < \frac{83 - 80}{2.1}\right) = \Pr(z < 1.43) = 0.9236.$$

Thus, approximately 92% of thermostats would have cut-off temperatures below 83°C.

(b) *What proportion of thermostats would have cut-off points which are at least five degrees from the mean value* (80°C)? This question uses property 5 of Table 7.3.1 with $y_0 - M$ equal to 5.

$$\Pr(|y - 80| \geq 5) = \Pr\left(|z| \geq \frac{5}{2.1}\right) = \Pr(|z| \geq 2.38)$$

$$= 2(1 - 0.9913)$$
$$= 0.0174.$$

Thus, about 1.7% of thermostats would have cut-off temperatures which are at least five degrees from the mean.

7.3.5 The Central Limit Theorem

There is an important theorem which is a generalization of the result deriving from the Theory of Errors (Section 6.5.1). It is known as *The Central Limit Theorem* and is expressed as follows.

> Suppose S_n is a statistic which can be expressed as the sum of n independent variables, $X_1, X_2, \ldots, X_n$, where X_i has mean M_i and variance σ_i^2 $(i = 1, 2, \ldots, n)$. If $M(n)$ and $\sigma^2(n)$ are the mean and variance of S_n, then as n tends towards infinity, the probability distribution of $[S_n - M(n)]/\sigma(n)$ tends towards a limiting form which is a standard Normal distribution.

This result has importance both in statistical theory and in application. It ensures that, for any statistic which satisfies the conditions of the Central Limit theorem, there is scope for using probabilities based on the standard Normal distribution to approximate probabilities based on the probability distribution of S_n. Equation (7.3.3) provides the connection.

$$\Pr(S_n \leq s_0) \cong \Pr\left[z \leq \frac{s_0 - M(n)}{\sigma(n)}\right] \qquad \text{for} \quad -\infty < s_0 < \infty \qquad (7.3.3)$$

where z has a $N(0,1)$ distribution. Following directly from (7.3.3) is (7.3.4).

$$\Pr\{M(n) - z_{\frac{1}{2}\alpha}\sigma(n) \leq S_n \leq M(n) + z_{\frac{1}{2}\alpha}\sigma(n)\} \cong 1 - \alpha. \qquad (7.3.4)$$

In practice, the application of the Central Limit Theorem is commonly required in situations in which the standard deviation $\sigma(n)$ is unknown. In large samples it is generally acceptable to replace $\sigma(n)$ by its sample estimate for application of (7.3.3) or (7.3.4).

For any statistic to which the Central Limit Theorem applies, there will be a minimum value of n for which (7.3.3) is acceptable as a basis for a Normal approximation to probabilities defined in terms of the sampling distribution of S_n. For example, Table 7.1.1 illustrates the use of the Central Limit Theorem in the computation of Binomial probabilities based on the minimum sample size defined in (7.1.2).

Two applications of the Central Limit Theorem which are of particular importance are described below.

1. Number of successes in n independent trials.

Suppose that x_i is a random variable which takes the value 1 if the outcome at the ith of n independent trials yields a success and the value 0 if it yields a failure. If the probability of success is π and the probability of failure is $1 - \pi$ then, from properties of the Binomial distribution, x has a mean of π and a variance of $\pi(1 - \pi)$.

If $x = x_1 + x_2 + \cdots + x_n$, then x is a statistic with a Binomial distribution (Section 7.1) with a mean $n\pi$ and a variance $n\pi(1 - \pi)$. By the Central Limit Theorem, $[x - n\pi]/\sqrt{[n\pi(1 - \pi)]}$ has a distribution which tends towards a standard Normal distribution as n tends towards infinity. The use of this result to approximate sums of Binomial probabilities is known as the Normal approximation to the Binomial distribution. It is employed in Section 7.1.4.

2. The arithmetic mean $\overline{y}$.

By definition, the arithmetic mean for a random sample with responses provided by variables $y_1, y_2, \ldots, y_n$ is

$$\begin{aligned}\overline{y}_n &= (y_1 + y_2 + \cdots + y_n)/n \\ &= \frac{y_1}{n} + \frac{y_2}{n} + \cdots + \frac{y_n}{n}.\end{aligned}$$

If y_i is a random variable with mean M and variance σ^2, then by properties of random variables introduced in Section 5.4, y_i/n is also a random variable with mean M/n and variance σ^2/n^2, and hence $\overline{y}_n$ is a random variable with mean M and variance σ^2/n. Therefore, by the Central Limit theorem, $[\overline{y}_n - M]/[\sigma/\sqrt{n}]$ is a statistic which has a sampling distribution which is well approximated by a standard Normal distribution when n is large. In fact, for the vast majority of experimental situations, the Normal distribution provides a good approximation to the sampling distribution of the arithmetic mean if n exceeds 30.

7.3.6 Applications

The two major applications of the Normal distribution are described in Section 6.5.6.

7.4 The t-Distribution

7.4.1 Statistic

The t-statistic exhibits continuous variation and is defined over the set of real numbers. It is typically denoted by the letter t.

It has an important application when a statistic (x) with a sampling distribution which is a Normal distribution with mean M_x and variance σ_x^2

is converted to its *studentized form* which standardizes the statistic in the manner defined in (7.4.1).

$$t = [x - M_x]/s_x \tag{7.4.1}$$

where s_x^2 is the statistic which provides the unbiased Maximum Likelihood estimate of σ_x^2.

The statistic t defined by (7.4.1) has a sampling distribution which is a member of a family of distributions known collectively as the *t-distribution.*

7.4.2 Sampling Distribution

The sampling distribution of t is defined by (7.4.2).

$$\pi_{\text{st}}(t) = K(\nu)\left[1 + \frac{t^2}{\nu}\right]^{-\frac{1}{2}(\nu+1)} \qquad \text{for} \quad -\infty < t < \infty \tag{7.4.2}$$

where ν is a parameter which is known as the *degrees of freedom;* $K(\nu)$ is a constant which is a function of ν.

The *t-distribution* is a family of distributions distinguished by the values of the *degrees of freedom,* ν. The value taken by ν must be a positive whole number. Specific rules for determining the value of ν accompany the various applications of the t-statistic. However, the rules have the following (informal) basis. A sample comprising n independent responses can be thought of as providing n pieces of information. If there are p distinct parameters to be estimated in establishing the value of the statistic x in (7.4.1) then there are assumed to be p restrictions placed on the information. The amount of information which remains is $n - p$ and is titled the *degrees of freedom.*

Notation. The following notation is employed throughout the book:

$t(\nu)$ is a t-statistic with a sampling distribution which is a t-distribution with ν degrees of freedom. Where there is unlikely to be confusion, the abbreviated form t is used.

$\mathrm{t}_\alpha(\nu)$ is the value for a t statistic with a sampling distribution which is a t-distribution with ν degrees of freedom and satisfies $\Pr\{|t(\nu)| \geq \mathrm{t}_\alpha(\nu)\} = \alpha$. For example,

$\mathrm{t}_{0.05}(\nu)$ is the value satisfying $\Pr\{|t(\nu)| \geq \mathrm{t}_{0.05}(\nu)\} = 0.05$.

7.4.3 Properties

1. MEAN $= 0$ VARIANCE $= \nu/(\nu - 2)$ for $\nu > 2$.
2. The distribution is symmetric. Thus, $\Pr(t < t_0) = \Pr(t > -t_0)$.
3. If t_0 is positive, then
   (a) $\Pr(-t_0 < t < t_0) = 1 - 2\Pr(t > t_0)$

(b) $\Pr(0 < t < t_0) = \frac{1}{2} - \Pr(t > t_0)$.

The following property applies whether t_0 is positive or negative and provides a means of computing the probability that t will have a magnitude in excess of t_0 irrespective of sign.

4. $\Pr(|t| > |t_0|) = \Pr(t > |t_0|) + \Pr(t < -|t_0|) = 2\Pr(t > |t_0|)$.

7.4.4 Calculation of Probabilities

Notation. Standard notation presented in Section 7.4.2 is used throughout the book.

Both statistical computing packages and tables are widely available for determining either probabilities from numerical values of t-statistics or the converse. The application of both methods is described and illustrated below.

1. *Computer-based calculations.* Statistical computing packages typically have an option to provide a value of $p = \Pr\{t(\nu) \leq t_0\}$ for a specified value t_0 from a $t(\nu)$ distribution. By reference to Table 7.4.1, probabilities of related events may be computed

Table 7.4.1. Probabilities of Events Based on the t-Distribution

$\Pr(t \leq t_0)$	p	
$\Pr(t > t_0)$	$1-p$	
$\Pr(t \geq -t_0)$	p	
$\Pr(0 \leq t \leq t_0)$	$p - \frac{1}{2}$	if $t_0 > 0$
$\Pr(t_0 \leq t \leq 0)$	$\frac{1}{2} - p$	if $t_0 < 0$
$\Pr(\lvert t\rvert \geq \lvert t_0\rvert)$	$2\Pr(t \leq -\lvert t_0\rvert)$	
$\Pr(t_1 \leq t \leq t_2)$	$\Pr(t \leq t_2) - \Pr(t \leq t_1)$	

There may also be need for the inverse operation, i.e. to obtain the value of t_0 which is associated with a specified probability. This operation is required when the aim is to find a value $\mathrm{t}_\alpha(\nu)$ which satisfies $\Pr\{|t(\nu)| \geq \mathrm{t}_\alpha(\nu)\} = \alpha$. This is transformed into a form which can be solved using a statistical package by use of (7.4.3).

$$\Pr\{t \leq \mathrm{t}_\alpha(\nu)\} = 1 - \frac{1}{2}\alpha. \tag{7.4.3}$$

2. *Tables of probabilities.* There is widespread availability of tables for the t-distribution (for example, A.2.1). Typically, the tables provide values of $\mathrm{t}_\alpha(\nu)$ for selected values of α, i.e. provide the value of $t(\nu)$ which satisfies $\Pr\{|t(\nu)| > \mathrm{t}_\alpha(\nu)\} = \alpha$. Take care to establish the probability which is provided, however, since some tables provide the probability $\Pr\{t(\nu) > \mathrm{t}_\alpha(\nu)\}$

which is equal to $\frac{1}{2}\alpha$. Where α is provided, the term *two-tailed* is applied and where $\frac{1}{2}\alpha$ is provided, the term *one-tailed* is applied. Commonly, the graph of a t-distribution is shown with a shaded region corresponding to probability provided. Check if the shading is at one end or both ends of the graph to establish if a one-tailed or two-tailed probability is provided. Equation (7.4.3) and Table 7.4.1 may be used to convert a value of α to the probability of a related event.

Illustration. For a problem in which the degrees of freedom are determined to be ν, there are two tasks which are likely to arise. They are

(i) the computation of a probability given a value t_0; and

(ii) the computation of a value of t_α for a specified value of α.

The sample calculations below are for (i) $t_0 = 3.17$, $\nu = 10$ and (ii) for $\alpha = 0.05$, $\nu = 10$.

1. *Computer-based calculations.* The probability usually provided is $\Pr\{t(10) \leq t_0\}$, which, for $t_0 = 3.17$ is $\Pr(t(10) \leq 3.17) = 0.995$. Using Table 7.4.1, it is possible to convert this probability to

$$\Pr\{t(10) > 3.17\} = 1 - \Pr\{t(10) \leq 3.17\} = 0.005; \text{ or}$$

$$\Pr\{|t(10)| \geq 3.17\} = 2\Pr\{t(10) \leq -3.17\} = 0.010.$$

The determination of the value $t_{0.05}(10)$ requires the use of (7.4.3) to compute a value for p of $1 - \frac{1}{2}(0.05) = 0.975$. The value of 2.23 satisfies the equation $\Pr\{|t(10)| \geq 2.23)\} = 0.05$.

2. *Table-based calculations.*

(a) If the tables provide values of $\mathrm{t}_\alpha(\nu)$ based on $\Pr\{|t(\nu)| \geq \mathrm{t}_\alpha(\nu)\} = \alpha$, then $\mathrm{t}_{0.05}(10)$ may be read from the body of the tables as the value for 10 degrees of freedom and a probability of 0.05.

(b) If the tables provide values of $\mathrm{t}_\alpha(\nu)$ based on $\Pr\{t(\nu) \geq \mathrm{t}_\alpha(\nu)\}$, then $t_{0.05}(10)$ must be read from the body of the tables as the value for 10 degrees of freedom and a probability of $\frac{1}{2}(0.05) = 0.025$. In either case, the value obtained is 2.23.

7.4.5 Applications

1. *Examining the mean of a population or process.* The statistic $t = (\overline{y} - M)/s_x$ has a t-distribution with $n-1$ degrees of freedom if s_x is replaced by $[s/\sqrt{n}]$, where $\overline{y}$ and s^2 are respectively the arithmetic mean and variance from a sample of size n selected by random sampling from a population for which the frequency distribution is a Normal distribution with mean M.

This t-statistic may be used to examine an hypothesized value for M or to provide a range of values within which M is likely to lie, i.e. a confidence interval for M (see Chapter 11).

2. *Comparing two means.* By constructing a studentized form of the difference in two sampling means using (7.4.1), questions concerning the magnitude of the difference between two population means can be examined (see Chapter 12).

There are numerous and diverse applications of the t-statistic based on the studentized form of statistics as defined by (7.4.1).

7.5 The Chi-Squared Distribution

7.5.1 Statistic

The Chi-squared statistic is so named because it is often represented as χ^2, where χ is a letter in the Greek alphabet. It is a statistic which exhibits continuous variation and is defined for all positive real numbers.

7.5.2 Sampling Distribution

The sampling distribution defined by (7.5.1) is a member of a family of distributions known collectively as the *Chi-squared distribution.*

$$\begin{aligned} \pi_{st}(x) &= \mathrm{K}(\nu)x^{\frac{1}{2}(\nu-2)}\exp\left(-\tfrac{1}{2}x\right) && \text{for } x > 0 \\ &= 0 && \text{otherwise,} \end{aligned} \quad (7.5.1)$$

where ν is a parameter which is known as the *degrees of freedom* and may take any whole number positive value; $\mathrm{K}(\nu)$ is a constant which depends on ν for its value.

Notation. When reference is made to Chi-squared statistics and distributions, the following notation is employed.

$\chi^2(\nu)$ is a statistic with a sampling distribution which is Chi-squared with ν degrees of freedom. This is abbreviated to χ^2 when reference to a specific value of ν is unnecessary.

$\mathrm{X}^2_\alpha(\nu)$ is a number satisfying $\Pr\{\chi^2(\nu) > \mathrm{X}^2_\alpha(\nu)\} = \alpha$.

7.5.3 Properties

1. MEAN $= \nu$ VARIANCE $= 2\nu$.

2. If χ_1^2 and χ_2^2 are two Chi-squared statistics with degrees of freedom ν_1 and ν_2, respectively, which are independent in a probabilistic sense, then $\chi^2 = \chi_1^2 + \chi_2^2$ has a Chi-squared distribution with $\nu = \nu_1 + \nu_2$ degrees of freedom.

7.5.4 Calculation of Probabilities

Notation. The notation presented in Section 7.5.2 applies throughout the book.

Both statistical computing packages and tables for Chi-squared distributions are widely available. Their application is described and illustrated below.

1. *Computer-based calculations.* To determine $p = \Pr\{\chi^2(\nu) > X_0^2\}$, it is usually necessary to obtain a value for $1-p = \Pr\{\chi^2(\nu) \leq X_0^2\}$, i.e. a value of the distribution function at X_0^2.

For example, if $X_0^2 = 7.96$ and $\nu = 16$, $\Pr\{\chi^2(\nu) \leq 7.96\} = 0.05$ and, therefore, $p = 1 - 0.05 = 0.95$.

To obtain a value $X_\alpha^2(\nu)$ which satisfies $\Pr\{\chi^2(\nu) > X_\alpha^2(\nu)\} = \alpha$, it is usually necessary to supply a value for $1-\alpha$ since the computation is based on the distribution function, i.e. $\Pr\{\chi^2(\nu) \leq X_\alpha^2(\nu)\}$.

2. *Tables* are available (for example, A.3.1) which give values of $X_\alpha^2(\nu)$ for a range of values of ν and either (i) selected values of α, i.e. values satisfying $\Pr\{\chi^2(\nu) > X_\alpha^2(\nu)\} = \alpha$; or (ii) selected values of $p = 1 - \alpha$, i.e. values satisfying $\Pr\{\chi^2(\nu) \leq X_\alpha^2(\nu)\} = 1 - \alpha$.

7.5.5 Applications

The Chi-squared distribution has an important role as an approximation to many distributions. It provides a large sample approximation to the sampling distribution of the *Likelihood ratio statistic* (Sections 4.3.7 and 8.4) which is widely employed in the construction of statistical tests.

7.6 The F-Distribution

7.6.1 Statistic

The F-statistic, named in honor of a famous statistician, R.A. Fisher, exhibits continuous variation and is defined for all positive numbers. It arises in statistical application as the ratio of two independent Chi-squared statistics.

7.6.2 Sampling Distribution

The sampling distribution defined by (7.6.1) is a member of a family of distributions which are collectively known as the *F-distribution.*

$$\begin{aligned} \pi_{st}(x) &= K(\nu_1, \nu_2) x^{\frac{1}{2}(\nu_1 - 1)} (\nu_2 + \nu_1 x)^{-\frac{1}{2}(\nu_1 + \nu_2)} && \text{for } x > 0 \\ &= 0 && \text{otherwise} \end{aligned} \quad (7.6.1)$$

where ν_1 and ν_2 are parameters which are known as the *degrees of freedom* and which may take any positive whole number values; $K(\nu_1, \nu_2)$ is a constant for fixed ν_1 and ν_2.

Notation. Where reference is made to the F-statistic and the F-distribution, the following notation is employed.

$F(\nu_1, \nu_2)$ is a statistic with an F-distribution having (ν_1, ν_2) degrees of freedom. Where there is unlikely to be ambiguity, this is abbreviated to F.

$\mathrm{F}_\alpha(\nu_1, \nu_2)$ is a number satisfying $\Pr\{F(\nu_1, \nu_2) > \mathrm{F}_\alpha(\nu_1, \nu_2)\} = \alpha$.

7.6.3 Properties

1. MEAN $= \nu_2/(\nu_2 - 2)$ for $\nu_2 > 2$.
2. Where $\chi_1^2(\nu_1)$ and $\chi_2^2(\nu_2)$ are independent Chi-squared statistics, the statistic defined by (7.6.2) has an $F(\nu_1, \nu_2)$ distribution where ν_1 and ν_2 are known as the *degrees of freedom of the numerator* and *degrees of freedom of the denominator*, respectively.

$$F = \left[\frac{\chi_1^2(\nu_1)}{\nu_1}\right] / \left[\frac{\chi_2^2(\nu_2)}{\nu_2}\right]. \qquad (7.6.2)$$

7.6.4 Calculation of Probabilities

Notation. The notation introduced in Section 7.6.2 is used throughout the book. Note that ν_1 and ν_2 are commonly referred to as the *degrees of freedom of the numerator and denominator* respectively in F-tables and in references accompanying statistical computing packages.

Both statistical computing packages and F-tables are widely available to supply probabilities and F-values. Their use is described and illustrated below.

1. *Computer-based calculations.* The probability most commonly required is $p = \Pr\{F(\nu_1, \nu_2) > F_0\}$. Typically, the quantity provided by the computer is a value for the distribution function, i.e. $1 - p = \Pr\{F(\nu_1, \nu_2) \leq F_0\}$. Hence, the value obtained must be subtracted from one to obtain p.

For example, if $F_0 = 1.72$, $\nu_1 = 3$ and $\nu_2 = 7$, $\Pr\{F(3,7) \leq 1.72\}$ is equal to 0.75 and, therefore, $p = 1 - 0.75 = 0.25$.

To determine a value $\mathrm{F}_\alpha(\nu_1, \nu_2)$ satisfying $\Pr\{F(\nu_1, \nu_2) > \mathrm{F}_\alpha(\nu_1, \nu_2)\} = \alpha$, the distribution function of the F-statistic is commonly employed. This necessitates inserting the value of $p = 1 - \alpha$ into the computer instructions.

2. *Tables* are available (for example A.4.1) which give values of $\mathrm{F}_\alpha(\nu_1, \nu_2)$ for a range of values of ν_1 (degrees of freedom of the numerator) and ν_2 (degrees of freedom of the denominator) and selected values of α. Less commonly, tables are available which give values of $\mathrm{F}_\alpha(\nu_1, \nu_2)$ for a range of values of ν_1 and ν_2 and selected values of $p = 1 - \alpha$.

7.6.5 Applications

The F-statistic has its most important role in the statistical method known as *Analysis of variance* (Section 14.3.4).

7.7 Statistics Based on Signs and Ranks

7.7.1 Statistics

Given a set of numbers which form the data from an experiment, statistics based on the data may be grouped according to whether they utilize only (i) the signs of the numbers, (ii) the relative magnitudes of the numbers; or (iii) the signs plus the magnitudes of the numbers.

The three components of the data are illustrated in Table 7.7.1 where the information of interest in the table comprises (i) the signs of the differences, (ii) the ranks of the differences based on magnitudes and (iii) the signed differences.

Table 7.7.1. Times Taken at First and Second Attempts of a Maze

Mouse	1	2	3	4	5	6	7	8	9
Attempt									
First	93	162	115	94	162	132	137	140	111
Second	72	95	46	118	121	105	130	78	101
Difference	21	67	69	−24	41	27	7	62	10
Sign	+	+	+	−	+	+	+	+	+
Rank*	3	8	9	4	6	5	1	7	2

*based on magnitude.

The question that prompted the experiment which provided the data in Table 7.7.1 was the following.

> "Do mice forget the correct passage through the maze after a time period of seven days?"

The set of *differences* contains all the information which can be obtained from the experiment about the changes in time taken for the mice to complete the maze when comparing the two attempts. Why then should statisticians be interested in statistics based on *signs* and *ranks* which use only a portion of the information? The primary reasons are

(a) valid application of the methods based on signs and ranks requires fewer statistical assumptions and, hence, lessens the chance of an incorrect conclusion because of an error in the model; and

(b) a lower sensitivity to errors in the data.

In pre-computer days a third reason would have been added—simplicity of calculation.

7.7.2 Distribution-Free Models

Sign and rank statistics are examples of statistics for which statistical methods can be developed which are based on *distribution-free models.* These are statistical models in which no statement of the specific mathematical form is required in the specification of the frequency or probability distribution. Such models are extremely important in the practical application of Statistics since they offer the possibility of analysis in situations in which there is insufficient experimental knowledge to define the mathematical form of the frequency or probability distribution for the response variable.

The function of this and the subsequent section is to illustrate the manner in which sampling distributions are constructed for statistics based on distribution-free models. The statistics considered in this section are based on signs and ranks. In Section 7.8, a statistic is considered which is based on the signed numerical measurements.

Tests based on signs and ranks vary according to the nature of the experimental problem and the manner in which the experiment is designed. However, the conceptual bases can be demonstrated by reference to the experiment which yielded the data in Table 7.7.1. The manner in which sign-based statistics are constructed is illustrated in Section 7.7.3 and the manner in which rank-based statistics are constructed is illustrated in Section 7.7.4.

7.7.3 The Construction of a Sign-Based Statistic

If a sequence of n outcomes from the process yields numerical values $y_1, y_2, \ldots, y_n$ for a continuous response variable y, then these values may be converted to a sequence of n pluses (+) and minuses (−) according to whether each value is above or below the median M. (To simplify the discussion, assume that no population member has a response which is exactly equal to M.)

From a statistical viewpoint this is equivalent to defining a new variable x which is categorical and has two states identified as + and − both with an accompanying probability of one half.

The relation between the measured response y and the categorical response x is illustrated in Table 7.7.1 where y is the *difference* and x is the *sign* of the difference.

Practical application occurs when the value of the median M is unknown and an hypothesized value M_0 is introduced into the model. In the example underlying Table 7.7.1, M_0 was assigned a value of zero to reflect the assumption that, on average, mice took the same time to complete the maze at both attempts.

If n_+ and n_- are statistics denoting, respectively, the numbers of pluses and minuses in the sample, a test of the hypothesis $M = M_0$ can be based on the statistic which is the difference in the magnitude of the number of pluses and minuses in the sample, namely $d = |n_+ - n_-|$.

If the hypothesized value for the median, M_0, is the correct value, then outcomes become more extreme as d increases in size. If the observed value of d is d_0, the p-value is simply the probability of a difference of at least d_0, i.e. $\Pr(d \geq d_0)$.

Based on the experiment which provided the data in Table 7.7.1, the sample size is 9 and the probability of an outcome at or beyond the observed point in the ordering is easily determined from first principles using the rules of probability. There are a total of 2^9 or 512 possible outcomes all of which are equally likely under the assumption that (i) a plus and a minus have the same probability; and (ii) individual responses are independent.

The observed outcome, comprising the sequence $+ + + - + + + + +$, has 8 increases and one decrease. Thus $n_+ = 8$ and $n_- = 1$ and, hence, d_0 equals 7.

Those outcomes which are at least as extreme as the observed results are (i) other outcomes in which there are 8 increases and one decrease; (ii) outcomes in which there are 8 decreases and one increase; (iii) the outcome in which there are nine increases; and (iv) the outcome in which there are nine decreases. There are a total of 20 outcomes out of the possible 512 which are at least as extreme as that observed. Hence, the p-value is $20/512 = 0.04$.

Adopting the commonly used guidelines, since the observed p-value is below 0.05, there is some evidence against the hypothesis of no change in the median time taken by mice in the maze between first and second attempts.

In fact, there is no need to resort to direct probabilistic arguments to determine the sampling distribution of the statistic since the Binomial model is applicable in the manner explained below.

Suppose M is the true value for the median in a population from which a random sample of n members yielded the measurements $y_1, y_2, \ldots, y_n$, and $y_{(1)}, y_{(2)}, \ldots, y_{(n)}$ is the set of n measurements after ordering from smallest to largest. (For example, if the set of nine differences in Table 7.7.1 provide the values for $y_1, y_2, \ldots, y_n$, then the values for $y_{(1)}, y_{(2)}, \ldots, y_{(n)}$ are

$$-24 \quad 7 \quad 10 \quad 21 \quad 27 \quad 41 \quad 62 \quad 67 \quad 69,$$

respectively.)

The statement that there are i numbers below M and $n - i$ numbers above M can be expressed by the relation $y_{(i)} < M < y_{(i+1)}$. Furthermore, if x is the categorical variable which takes as its response the number of observations below M, then (7.7.1) follows from the Binomial distribution (7.1.1) with $\pi = \frac{1}{2}$.

$$\Pr(y_{(i)} < M < y_{(i+1)}) = \Pr(x = i) = \binom{n}{i}\left(\frac{1}{2}\right)^n. \qquad (7.7.1)$$

The relation expressed in (7.7.1) can be employed to determine a p-value between a model which includes an hypothesized value of M_0 for the median and data in which there is an observed number j of observations below M_0 and $n - j$ above M_0.

If M_0 is the correct value for the median M, then the construction of the p-value requires the summation of probabilities of outcomes for values of $i = 0, 1, 2, \ldots, n$ which satisfy $\Pr(x = i) \le \Pr(x = j)$. If j is less than $\frac{1}{2}n$, these values are $i = 0, 1, 2, \ldots, j,\ n - j,\ n - j + 1, \ldots, n$. Thus,

$$p = \sum_{i=0}^{j}\binom{n}{i}\left(\frac{1}{2}\right)^n + \sum_{i=n-j}^{n}\binom{n}{i}\left(\frac{1}{2}\right)^n = 2\sum_{i=0}^{j}\binom{n}{i}\left(\frac{1}{2}\right)^n.$$

If j is equal to or greater than $\frac{1}{2}n$, these values are $i = 0, 1, 2, \ldots, n - j$, $j, j + 1, \ldots, n$.

$$p = \sum_{i=0}^{n-j}\binom{n}{i}\left(\frac{1}{2}\right)^n + \sum_{i=j}^{n}\binom{n}{i}\left(\frac{1}{2}\right)^n = 2\sum_{i=0}^{n-j}\binom{n}{i}\left(\frac{1}{2}\right)^n.$$

Equation (7.7.1) also has a role in the construction of a confidence interval for M as is explained in Section 8.3.6.

7.7.4 The Construction of a Rank-Based Statistic

In terms of utilization of information in data, rank-based statistics lie one step above sign-based statistics. In addition to taking account of the signs of the numbers, they also depend on the relative magnitudes of the numbers.

The significance of this additional information can be seen by reference to the following two sets of numbers. The first is the set of differences from Table 7.7.1 in which the one negative difference is fourth in the set of rankings. The second varies from the first in that the one negative difference has been modified to make it the smallest magnitude of all differences.

Mouse	1	2	3	4	5	6	7	8	9
Difference									
observed	21	67	69	−24	41	27	7	62	10
modified	21	67	69	−1	41	27	7	62	10

In each of the above data sets, there are eight positive and one negative numbers. Thus, the statistic based on signs would have the same value for both data sets. However, when the magnitudes of the numbers are ranked from smallest to largest, the negative value has a rank of 4 in the observed data set and a rank of 1 in the modified data set.

This difference appears in the rank-based statistic and would reflect a greater support for an hypothesized median difference of zero in the observed data set than in the modified data set.

To construct a test of the hypothesis that the median of a population is zero, the measurements obtained from a random sample drawn from the population are employed, namely $y_1, y_2, \ldots, y_n$. To simplify development, it will be assumed that no two measurements have the same magnitude and no observation has a value of zero.

The test is based on *ranks* which are formed when the *magnitudes* of the observations are ranked from smallest to largest and assigned the numbers $1, 2, \ldots, n$. Rank-based tests can be constructed by utilizing the sum of the ranks of the positive numbers (R_+) or the sum of the ranks of the negative numbers (R_-). Which is these is used is immaterial since the sum $R_+ + R_-$ is constant (and equals $1 + 2 + \cdots + n = \frac{1}{2}n(n+1)$).

Given the model which includes the hypothesis that the true median is zero, the outcomes said to be most consistent with the model are those for which $R_+ = R_- = \frac{1}{2}\left[\frac{1}{2}n(n+1)\right] = n(n+1)/4$. Outcomes which have sums for R_+ and R_- which are increasingly distant from $n(n+1)/4$ are said to show lessening support for the model and, hence, for the hypothesis that the median difference is zero.

The sampling distributions of R_+ and R_- are not defined unless the model contains further distributional information. The usual assumption which is added states that the response variable has a symmetric frequency distribution. This condition has the effect of ensuring that, when the model is correct and the median difference is zero, (i) each rank is as likely to be associated with a positive as with a negative response; and (ii) the probability that one rank is positive (or negative) is independent of the sign associated with any other rank.

Consequently, if the ranks are listed in sequence $(1, 2, \ldots, n)$ the corresponding sequence of signs is equally likely to be any one of the 2^n possible sequences ranging from n pluses to n minuses. Hence, the p-value is simply the proportion of sequences which are associated with rank sums (R_+ or R_-) which are more extreme than the observed sum.

The determination of p is given concrete expression below using the data from the mouse experiment. It can be seen from the data in Table 7.7.1 that only one of the nine mice had a negative response and that response was fourth in the ranks.

To construct the sum of ranks, consider the sequence $(1, 2, \ldots, 9)$. The statistic R_- uses only those ranks for which the difference is negative. To identify the negative ranks in the sequence, it is convenient to replace ranks associated with a positive response by zero. In the observed data set this results in the sequence $(0, 0, 0, 4, 0, 0, 0, 0, 0)$. The observed value of R_- is then the sum of these ranks, namely 4. Equivalently, to construct R_+ from the positive ranks it is convenient to replace the negative ranks in the sequence by zero. Thus, in the observed data set the sequence $(1,2,3,0,5,6,7,8,9)$ provides the ranks from which R_+ is constructed and leads to a value of 41 for R_+.

Under this scheme, sequences which are at least as extreme as the observed sequence are

(i) those which lead to a value of 4 or less for R_- which are

$(0,0,0,0,0,0,0,0,0)$ $(1,0,0,0,0,0,0,0,0)$ $(0,2,0,0,0,0,0,0,0)$
$(1,2,0,0,0,0,0,0,0)$ $(0,0,3,0,0,0,0,0,0)$ $(1,0,3,0,0,0,0,0,0)$
$(0,0,0,4,0,0,0,0,0)$;

(ii) those which lead to a value of 41 or more for R_+ which are

$(1,2,3,4,5,6,7,8,9)$ $(0,2,3,4,5,6,7,8,9)$ $(1,0,3,4,5,6,7,8,9)$
$(0,0,3,4,5,6,7,8,9)$ $(1,2,0,4,5,6,7,8,9)$ $(0,2,0,4,5,6,7,8,9)$
$(1,2,3,0,5,6,7,8,9)$.

The total number of possible outcomes which are at least as extreme as the observed sequence is 14 out of a total number of possible sequences of 2^9 or 512. Hence, the p-value is $14/512 = 0.03$. This compares with the value of 0.04 from the application of the sign-based statistic.

In practice, it is not necessary to resort to direct computations of this type. Statistical packages provide p-values as part of the computations, or alternatively, tables or formulae providing large sample approximations are widely available.

7.8 Statistics Based on Permutations

7.8.1 Permutation Tests

Neither the Sign statistics nor the Rank statistics introduced in Section 7.7 utilize all the information contained in a set of values for a scaled variable. There is a class of statistical procedures based on distribution-free models which rely on the counting strategy employed with Sign and Rank statistics but which use the numerical values of the responses.

The procedures have application in the comparison of groups or processes. The basic premise is that the frequency distributions for all groups are identical and consequently each response is equally likely to have come from any of the defined groups. By way of illustration, consider the simplest situation in which there are two groups, labelled A and B and data are presumed to take the form of independent, random samples from the two groups, with readings $y_1, y_2, \ldots, y_m$ from members of group A and y_{m+1}, $y_{m+2}, \ldots, y_{m+n}$ from members of group B.

To construct a permutation test which examines the hypothesis that groups A and B have identical frequency distributions for a response variable y, choose a statistic which can order data sets such that larger values of the statistic imply less support for the hypothesis.

In practice, the statistic should be based on the type of departure which is of interest to the experimenter—if, for example, there was interest in establishing if group A responses tend to be higher than group B responses, then the difference in sample means, $d = \overline{y}_A - \overline{y}_B$, would be a sensible choice, with increasing values of d implying lessening support for the hypothesis of identical frequency distributions.

Having defined a statistic, a value for the statistic is computed for each possible permutation (ordering) of the $m + n$ observations under the assumption that the first m observations are from group A and the last n from group B.

If the value of the statistic for the permutation which corresponds to the actual data set is s_0, then the proportion of permutations with values of the statistic equal to or greater than s_0 can be interpreted using the guidelines adopted for a p-value in a test of significance.

The crucial properties of the set of permutations are

(i) all possible permutations are equally likely under the stated hypothesis that the groups have identical frequency distributions; and

(ii) permutations show less support for the hypothesis as the value of the statistic increases.

7.8.2 Approximate Permutation Tests

A major practical limitation to the wide scale adoption of permutation tests is the vast number of permutations which are possible except in the cases of very small sample sizes. Even with the advent of high speed computers, it is commonly impractical to examine all possible permutations. Unlike the sign-based tests and the rank-based tests, it is rarely possible to construct sampling distributions by analytic techniques. The consequence has been a lack of exploitation of a very flexible and powerful statistical procedure.

There is, however, a simple solution to this limitation. It involves the random sampling of all possible permutations to select a sample of permutations to provide an estimate of p. The precision of the estimate can be calculated and is determined by the number of samplings employed. Where the possible number of permutations may be in the millions, a selection of 200 permutations is adequate for most practical purposes.

Given the power of desk top computers, the implementation of approximate permutation tests is a rapid operation in most practical situations. Furthermore, most general-purpose statistical packages can be simply instructed to perform the sampling and associated computations.

The steps in application are explained and illustrated below using data taken from Example 12.2 which records the level of ascorbic acid in the bodies of two groups of scientists involved in a study. The data comprise 12 readings from members of the supplement group (one observation in the original data set, 35.9, is excluded because it is thought to be an error) and 11 readings from members of the control group.

Supplement group

18.3 9.3 12.6 15.7 14.2 13.1 14.3 16.2 18.1 19.4 15.5 11.7

Control group

24.9 16.0 26.3 25.5 19.3 16.8 15.7 24.6 19.9 9.4 17.4

The experimenter was seeking to establish if the average level of ascorbic acid in persons taking the supplement treatment differs from the average level in persons who are not taking the supplement and who are represented in the sample as the *Control* group.

Calculation based on the data establishes that the means for the Supplement and Control groups are 14.87 and 19.62, respectively. Is it reasonable to suppose that the magnitude of the observed difference in means, $19.62 - 14.87 = 4.75$, can be attributed to sampling variation rather than a treatment difference?

To answer this question, an approximate permutation test may be employed by following the steps below.

1. Choose a suitable statistic, s.

The method is valid for any choice of statistic, but is more likely to detect an incorrect hypothesis if the chosen statistic reflects the type of departure which is consistent with the possible alternative hypotheses. Thus, to examine the difference in two population means it makes sense to use the difference in sample means.

2. Apply the statistic, s, to the data to produce the observed value s_0.

For the sample data, the statistic chosen was the magnitude of the difference in means between the first 12 numbers which constitute the Supplement group and the remaining 11 numbers which constitute the Control group. The value obtained is 4.75 which is the value assigned to s_0.

3. Use the pseudo-random number generator provided by a convenient Statistical package to generate a set of pseudo-random numbers equal in number to the number of sample members. Use the computer to place the sample members in the same order as the set of random numbers.

Each ordering is a possible arrangement of the sample members and, hence, the data. Using the set of numbers produced by the pseudo-random number generator allows for a random selection from among the possible arrangements.

In the sample data there are 23 observations. Hence, 23 pseudo-random numbers are to be generated. Using the MINITAB *statistical package, the set* 0.144, 0.295, *etc. in Table* 7.8.1 *was generated and associated with sample members* 1, 2, ... *The random numbers were then ordered and the sample members with their responses carried with the random numbers in the ordering process to generate a randomly selected arrangement of the sample members.*

Table 7.8.1. Approximate Randomisation Test Example of Step 3

Original Form			After Ordering		
Sample member	Random number	Data y_i	Sample member	Random number	Data y_i
1	0.144	18.3	23	0.015	17.4
2	0.295	9.3	16	0.035	25.5
3	0.498	12.6	18	0.063	16.8
4	0.815	15.7	1	0.144	18.3
5	0.313	14.2	20	0.251	24.6
6	0.998	13.1	2	0.295	9.3
7	0.580	14.3	19	0.299	15.7
8	0.973	16.2	5	0.313	14.2
9	0.879	18.1	10	0.339	19.4
10	0.339	19.4	13	0.372	24.9
11	0.637	15.5	12	0.411	11.7
12	0.411	11.7	14	0.476	16.0
13	0.372	24.9	3	0.498	12.6
14	0.476	16.0	7	0.580	14.3
15	0.949	26.3	11	0.637	15.5
16	0.035	25.5	21	0.656	19.9
17	0.825	19.3	4	0.815	15.7
18	0.063	16.8	17	0.825	19.3
19	0.299	15.7	9	0.879	18.1
20	0.251	24.6	22	0.936	9.4
21	0.656	19.9	15	0.949	26.3
22	0.936	9.4	8	0.973	16.2
23	0.015	17.4	6	0.998	13.1

4. Compute a value of the statistic for the new arrangement of the data and record if its value equals or exceeds s_0.

For the ordered data in Table 7.8.1, *the magnitude of the difference in means between the first twelve numbers and the last eleven is* 0.55. *This value is less than the value of* 4.75 *for the observed permutation.*

5. Repeat step 3 and 4 at least 200 times and determine the proportion p_{est} of times the observed value is equaled or exceeded.

The quantity p_{est} is an estimate of $p = \Pr(s > s_0)$, the proportion of permutations which equal or exceed s_0. (In theory the procedure should be defined to avoid repetition of any permutation. However, in practical applications, the number of possible permutations is usually so large as to cause the possibility of repetition to be of little practical consequence.)

6. Interpretation: the quantity p_{est} can be interpreted according to the guidelines applicable for a p-value in a test of Significance.

Problems

Problems 7.1 to 7.11 are practical exercises in computing probabilities based on sampling distributions of statistics. It is presumed that a statistical computing package is available for at least the basic computations.

7.1 *Computation of Binomial probabilities based on* (7.1.1)

(a) Obtain the values of $\Pr(x = c)$ for $c = 0, 1, 2, \ldots, 10$ for the Binomial distributions with $n = 10$ and (i) $\pi = 0.1$ (ii) $\pi = 0.25$ (iii) $\pi = 0.5$ (iv) $\pi = 0.75$ (v) $\pi = 0.9$.

(b) For each of the sets of probabilities computed in (a) construct graphical representations using the form of presentation employed in Figure 7.1.1. Note that (i) the distributions become more symmetric as π approaches 0.5 and (ii) the distributions for $\pi = 0.75$ and $\pi = 0.9$ are mirror images of the distribution for $\pi = 0.25$ and $\pi = 0.1$, respectively.

(c) For the Binomial distribution with $n = 10$ and $\pi = 0.5$

(i) compute $\Pr(x < 2)$;

(ii) calculate $\Pr(|x - 5| \geq 3)$.

7.2 *Normal approximation to the Binomial*

For the Binomial distribution with $n = 100$ and $\pi = 0.3$

(a) use (7.1.2) to establish that the Normal approximation to the Binomial distribution is reasonable.

(b) Use the Normal approximation to the Binomial and formulae given in Table 7.1.1 to compute approximate values for

(i) $\Pr(x \leq 38)$ (ii) $\Pr(x \geq 38)$ (iii) $\Pr(x = 38)$

(iv) $\Pr(27 \leq x \leq 38)$ (v) $\Pr(|x - 30| \geq 8)$.

7.3 *Computation of Poisson probabilities based on* (7.3.1)

(a) For $M = 1$ and $M = 10$ compute $\Pr(x = c)$ for all values of c for which $\Pr(x = c) > 0.01$. Graph the sets of probabilities in the form employed in Figure 7.2.1. Note that the distribution is more symmetric when $M = 10$.

(b) For $M = 10$ compute cumulative probabilities for $x = 0, 1, \ldots, 15$.

(c) For $M = 10$ compute $\Pr(x > 3)$.

7.4 *Normal approximation to the Poisson*

Use the Normal approximation to the Poisson distribution with $M = 10$ (Section 7.2.4) to compute approximate values of $\Pr(x = 8)$ and $\Pr(x > 3)$. Compare the approximate values with the exact values obtained in Problem 7.3 (a) and (c).

7.5 *Computations of probabilities for the Normal distribution*

(a) For the Normal distribution with mean 10 and variance 4 determine

(i) $\Pr(y \leq 8)$ (ii) $\Pr(y \geq 8)$ (iii) $\Pr(8 \leq y \leq 10)$
(iv) $\Pr(y \geq 12)$ (v) $\Pr(8 \leq y \leq 12)$ (vi) $\Pr(|y - 10| \geq 2)$.

(b) For the Normal distribution with mean 10 and variance 1 determine

(i) $\Pr(y \leq 9)$ (ii) $\Pr(y \geq 9)$ (iii) $\Pr(9 \leq y \leq 10)$
(iv) $\Pr(y \geq 11)$ (v) $\Pr(9 \leq y \leq 11)$ (vi) $\Pr(|y - 10| \geq 1)$.

(c) For the Normal distribution with mean 25 and variance 9 determine

(i) $\Pr(y \leq 22)$ (ii) $\Pr(y \geq 22)$ (iii) $\Pr(22 \leq y \leq 25)$
(iv) $\Pr(y \geq 28)$ (v) $\Pr(22 \leq y \leq 28)$ (vi) $\Pr(|y - 25| \geq 3)$.

7.6 *Normal distribution — application*

The weights of bars of chocolate are produced according to the specification that, over a long production run, they should have a Normal distribution with mean 200 grams and standard deviation 0.8 grams.

(a) If bars which are less than 198 grams are rejected as being underweight, what is the expected rejection rate?

(b) If the company decides to raise the mean weight so that the rejection rate is only one in ten thousand, what must be the new mean if the standard deviation remains unchanged?

7.7 *Normal distribution — application*

The number of cars (n) crossing a bridge per day averages 1,280 with a standard deviation of 35.6. The distribution of n is known to be well approximated by a Normal distribution.

Assuming a $N(1280, [35.6]^2)$ distribution,

(a) determine the 1st, 5th, 10th, 25th, 50th, 75th, 90th, 95th and 99th percentiles of the distribution.

(b) Complete the following sentences by inserting upper and lower limits which are symmetric about the mean.

On 50% of days the number of cars crossing the bridge will lie between ____ and ____.

On 90% of days the number of cars crossing the bridge will lie between ____ and ____.

On 2% of days the number of cars crossing the bridge will lie outside the limits ____ and ____.

7.8 *The t-distribution*

(a) Determine (i) $\Pr\{t(15) \leq 2.1\}$. Hence, compute

$$\text{(i) } \Pr\{t(15) > 2.1\} \qquad \text{and} \qquad \text{(ii) } \Pr(|t(15)| \geq 2.1).$$

(b) Find $t_{0.05}(15)$.

7.9 *Chi-squared distribution*

(a) Find $\Pr\{\chi^2(10) \geq 12.6\}$.

(b) Find $X^2_{0.025}(8)$.

7.10 *Application of the Chi-squared distribution*

If $u_1, u_2, \ldots, u_n$ are independent random variables with sampling distributions which are Chi-squared distributions with one degree of freedom, Property 2 (Section 7.5.3) establishes that the probability distribution of $u_s = u_1 + u_2 + \cdots + u_n$ is $\chi^2(n)$. This result has application in the following problem.

For any statistic S which yields a value S_0 in an experiment, the quantity $p_0 = \Pr(S \geq S_0)$, is, by definition, itself the value of a statistic p. From a probabilistic viewpoint, the value of p is equally likely to lie in the interval $[0, 1]$. Technically, the statistic p is said to be "uniformly distributed" on the interval $[0, 1]$. Furthermore, statistical theory establishes that the statistic $u = -2\log(p)$ has a Chi-squared distribution with two degrees of freedom if p is uniformly distributed on the interval $[0, 1]$.

An application of this result can be illustrated with reference to the findings in Example 3.4 where the model was brought into question because the experiment was thought to be *rigged.* The steps in a statistical examination of this claim are developed below.

(a) For each of the six groups considered separately, assume the observed *number of round seeds* is a value of a Binomial statistic, x, with $n = 200$ and $\pi = 0.75$. Use the Normal approximation to the Binomial to complete the following table:

Group	1	2	3	4	5	6
$p_0 = \Pr(\lvert x - n\pi \rvert \geq \lvert x_0 - n\pi \rvert)$	p_{10}	p_{20}	p_{30}	p_{40}	p_{50}	p_{60}

Assuming the groups worked independently to obtain the results and the Binomial model with $n = 200$ and $\pi = 0.75$ applies in all cases, the six p-values obtained are independent values for a statistic which is uniformly distributed on the interval $[0, 1]$.

(b) Compute values $u_{i0} = -2\log(p_{i0})$ for $i = 1, 2, \ldots, 6$ and, hence, $u_{s0} = u_{10} + u_{20} + \cdots + u_{60}$.

(c) If the students deliberately selected results which were close to the expected results, the p-values will cluster towards the upper end of the range and u_{s0} will take a value near zero. Hence, the measure of agreement is computed as $p = \Pr\{\chi^2(12) \leq u_{s0}\}$ with a low value of p offering evidence the fit of the data to the model is too good. Compute a value of p. Is there strong evidence the results were rigged?

7.11 *The F-distribution*

(a) Determine $\mathrm{F}_{0.01}(3, 20)$.

(b) Evaluate $\Pr\{F(2,5) \geq 6.7\}$.

(c) There exists the relationship $\Pr\{|t(\nu)| \geq \mathrm{t}_\alpha(\nu)\} = \Pr\{F(1,\nu) \geq \mathrm{t}_\alpha^2(\nu)\}$, i.e. that $\mathrm{F}_\alpha(1,\nu) = \mathrm{t}_\alpha^2(\nu)$ for all combinations of ν and α. Verify this relationship for $\nu = 10$ and $\alpha = 0.05$.

7.12 *Deriving means and variances for discrete distributions*

Using (5.4.4) and (5.4.7),

(a) establish that the mean and variance of (7.1.1) are $n\pi$ and $n\pi(1-\pi)$, respectively.

(b) establish that the mean and variance are equal for the Poisson distribution.

7.13 *Mean and variance of the Uniform distribution*

If x is a random variable with the probability distribution

$$\pi(x) = 1 \quad \text{for } 0 \leq x \leq 1$$
$$= 0 \quad \text{otherwise,}$$

determine the mean and variance of x.

7.14 *Relationships between Normal distributions*

If z has a $N(0,1)$ distribution verify that $y = M + z\sigma$ has a $N(M, \sigma^2)$ distribution.

8

Statistical Analysis: The Statisticians' View

8.1 The Range in Models and Approaches

8.1.1 Perspectives on Statistics

The function of this book is to provide users of Statistics with a view of the discipline which reflects their interests—an *application* perspective. There are, however, at least three other ways in which statisticians look at their discipline. There is the *philosophical* perspective which is concerned with the logical structure of Statistics and the purpose it serves; the *mathematical* perspective which views Statistics from its mathematical foundation and the mathematical building blocks used in statistical theory; and the *methodological* perspective which partitions Statistics according to types of statistical methods.

To a greater or lesser degree, professional statisticians have the capacity to view Statistics from all four perspectives. However, the vastness of the discipline and the different situations in which statisticians are employed, forces most to concentrate on a single perspective.

Ideally, scientists seeking statistical advice or support for a specific project should have access to a statistician who can look at Statistics through the eyes of a user. A well trained and experienced statistical consultant is able to provide assistance far beyond the mechanical prescription and application of statistical methods. Help may be provided in the area of experimental design; the translation of experimental questions into a statistical framework; the selection of methods of analysis given practical limitations; and the interpretation of results in a manner which is meaningful to scientists.

While scientists might hope for this level of support, there remains a chronic shortage of skilled statistical consultants. In fact, with the rapid growth in the use of Statistics in scientific investigation, the situation is worsening. Consequently, scientists are likely to be seeking assistance from statisticians who may have difficulty in viewing Statistics through the eyes of the user. Good communication between the statistician and the scientist in such circumstances requires the scientist to have an understanding of the statistician's perspective. Beyond this reason, scientists can better employ Statistics as a tool in their research if they have the overview offered by the different perspectives. This chapter provides that overview and the relevance to users of Statistics.

8.1.2 The Role of Statistics in Science

There has been, and remains, considerable controversy over the central role for Statistics in scientific investigation. One view is that Statistics should provide a vehicle for quantifying uncertainty in the beliefs and knowledge of the investigator. This is the Bayesian viewpoint and is expanded upon in Section 8.5. This fits in with the progress of Science towards an ultimate goal of perfect understanding by a sequence of steps, each of which is aimed at reducing uncertainty.

By way of contrast, there is the decision theoretic viewpoint in which an investigation is seen as a mechanism for making a decision between competing scientific hypotheses. The scientist is presumed to be in a position to quantify not only degrees of belief in the competing hypotheses, but also the relative importance of making incorrect decisions about each hypothesis. The ability to supply this information is largely limited to situations in which there is repetition, for example, industrial areas of quality control, biological testing programs, and medical diagnosis.

A third view of Statistics presents a technique for providing an objective assessment of the acceptability of a model using data from an experiment. There is no mechanism for including the prior belief of the scientists. Under this approach, scientists are left to combine the statistical findings with their own beliefs using non-statistical criteria.

To many statisticians, the three approaches are complementary rather than competitive. Where scientists' beliefs can be quantified, it makes sense to use this additional information. If decision making is the purpose of the investigation and the relevant information is available, decision theory provides a sensible basis for statistical analysis. Where the necessary information is not available, the lesser level of statistical involvement offered by the third option is recommended.

If usage is to be taken as a guide, neither the Bayesian nor the decision theory approaches have a widespread application. In part this is due to the lack of information necessary for their implementation. Additionally, they force users into acceptance of a particular format for the problem—either the modification of prior belief in the Bayesian case or decision making in the Decision Theory case. Scientists show an inclination to the more flexible approach which the third option provides. There is an opportunity to tailor the statistical approach to the perceived needs of the scientists. This flexibility is illustrated by the following three annotated examples.

1. *Simple hypothesis in the absence of an alternative.* Mendelian theory of genetic inheritance proposes that round to wrinkled seeds of the garden pea are expected to occur in the ratio 3:1. Are the data

consistent with this theory?

(*In this example, there is a question pertaining to a specific biological model which translates into the hypothesis that the expected ratio of round to wrinkled seeds is* 3:1. *The function of statistical analysis is to assess the acceptability of that model without any alternative biological model being provided.*)

2. *Composite hypothesis in the presence of an alternative.* A supplier of home insulation claims that the average saving in heating costs to homes which install insulation to his specifications is at least 40%. Given data from homes in which is insulation has been installed, does his claim seem reasonable?

   (*This problem varies from that described in Example* 1 *in two ways. Firstly, the supplier provides no single figure for the increase, the hypothesis being that the percentage gain is equal to or greater than* 40. *One task of statistical analysis is to establish if any one or more of the values in the range* 40 *to* 100 *is consistent with the data. Secondly, there is reason to examine the alternative possibility, namely that the percentage gain is less than* 40. *In this case, the aim is to establish if all values below* 40 *are unacceptable in light of the available data.*)

3. *Estimation.* An opinion poll is to be used to estimate the percentage support for a political party in an electorate.

   (*In this case, there is no claim put forward by the investigator. All possible values in the range* 0 *to* 100 *are possibilities. The task of statistical analysis is to provide an estimate of the most likely value or the likely range of values which will contain the correct value.*)

The first two examples come within the province of *Hypothesis Testing* which is introduced and examined in Section 8.2. The third example is classified as a problem in *Estimation* and is explored in Section 8.3.

8.2 Hypothesis Testing

8.2.1 Null and Alternative Hypotheses

As the name suggests, hypothesis testing is an approach to statistical analysis which examines a statistical hypothesis contained in a statistical model. While there may be more than one hypothesis which could be defined, the assumption is that only one hypothesis, called the *null hypothesis,* is to be examined. Other possible hypotheses, collectively called the *alternative hypothesis,* may be identified to indicate the manner or direction in which the null hypothesis might be in error.

Note that in hypothesis testing, no judgment is passed on the acceptability of the alternative hypothesis except by implication—if the null hypothesis is judged unacceptable, then the alternative hypothesis may be taken to be acceptable by default. The converse is certainly not true—acceptability of the null hypothesis does not imply the alternative hypothesis is unacceptable or even that the null hypothesis is more likely to be true than the alternative hypothesis.

8.2.2 Simple and Composite Hypotheses

A second classification of hypotheses is according to whether they are *simple* or *composite*. In this context, the word *hypothesis* rather confusingly relates the complete statistical model, not merely to the component which is of interest to the experimenter.

A *simple hypothesis* refers to a model in which there are unique numerical values assigned to all parameters which appear in the model. Where this condition is not met, the term *composite hypothesis* is employed. The distinction is illustrated by the following examples.

Example 1 – simple hypothesis. In the case of tossing a coin which is presumed to be unbiased, the Binomial model might be employed and would include the assumption that the probability of heads is one half. Thus, the sole parameter in the model, π, is assigned a unique value, namely $\pi = \frac{1}{2}$.

Example 2 – composite hypothesis. In a quality control study, the company requirement is that the percentage of defective products coming off the production line should be no more than two percent. Again, a Binomial model might be presumed with π being the proportion of defectives expected from the production line. The specification is that $\pi \leq 0.02$. There is not a single value specified for π. Hence, the term *composite hypothesis* is applied.

Example 3 – composite hypothesis. Table 1.3.4 contains data from a study of thermostats. The specifications for the thermostats stated that they should be activated when the temperature reaches a nominal 80°C. The experimental hypothesis which appears in the statistical model is $M = 80$, where M is the average temperature at which the thermostats are activated. Thus, the parameter M which appears in the model is assigned a unique value. However, if the model includes the assumption that the pattern of variation expected in the activation temperatures follows a Normal distribution, there are two parameters, M and σ^2, for which values must be provided to uniquely define the model. If the experimenter has no expected value or hypothesized value for σ^2, the model contains a parameter which is not uniquely defined. Theoretical statisticians would identify this as a *composite* hypothesis since there is a parameter in the model which is not completely specified even though that parameter is not of interest to the experimenter.

8.2.3 Tests of Significance

The basic idea of Significance testing was introduced in Chapter 4. It involves the comparison of the model with the data to provide a measure of closeness which is termed the *significance* or *p-value.*

The comparison depends on the definition of a statistic, s, which has a sampling distribution completely defined by the model containing the null hypothesis. Each possible data set identifies with a value of the statistic in such a way that increasing values of the statistic imply decreasing agreement between model and data, i.e. decreasing support for the model.

p-value: definition. Given an observed value s_0 for the statistic, the *significance* or *p-value* is defined to be the probability the statistic s takes a value equal to or greater than s_0, i.e. $p_0 = \Pr(s \geq s_0)$.

p-value: illustration. The construction of a p-value is easily grasped through an example. Consider the Binomial model suggested for the coin tossing example above. Assume a fair coin, i.e. $\pi = \frac{1}{2}$, and let the sample size be ten. Intuition suggests that the most likely outcome is that in which there are five heads and five tails. Worsening agreement comes with increasing discrepancy between heads and tails. Thus, a sensible statistic to use would be that one which is defined as the difference between the number of heads and tails in the sample ignoring the sign of the difference. Then

$$\begin{aligned} s &= 0 \text{ if there are five heads;} \\ &= 2 \text{ if there are four heads or six heads;} \\ &= 4 \text{ if there are three heads or seven heads;} \\ &= 6 \text{ if there are two heads or eight heads;} \\ &= 8 \text{ if there is one head, or nine heads;} \\ &= 10 \text{ if there are no heads or ten heads.} \end{aligned}$$

Based on the Binomial distribution as defined in (7.1.1) and the rules of probability provided in Chapter 5, the following table of p-values can be produced.

s_0	0	2	4	6	8	10
$p_0 = \Pr(s \geq s_0)$	1.00	0.75	0.34	0.11	0.02	0.00

p-value: interpretation. The widely used guidelines for interpreting the p-value offers the following interpretation:

- if p_0 exceeds 0.05, the data are presumed to be consistent with the proposed model and, hence, the experimental hypothesis contained in the model is presumed to be reasonable;
- if p_0 lies between 0.05 and 0.01, there is some evidence against the model and it is presumed that the possible flaw in the model is in the experimental hypothesis it contains;
- if p_0 is less than 0.01, there is strong evidencc against the model and, hence, against the experimental hypothesis it contains.

Many professional journals accept reports of statistical analyses in which the magnitude of the p-value is related to these yardsticks with the implication that the guidelines stated above apply. The wisdom of such a universal classification is queried in Section 8.2.5.

The discussion to date has not addressed either of the following questions.

1. Is there a suitable statistic which can be used for the comparison of the model and the data?
2. If there is more than one statistic which might be used, which is the best statistic to employ?

Answers to these questions require information on the nature of possible alternative models if the model under investigation is incorrect. There is no mechanism within the framework of Significance testing for the inclusion of such information. A more general formulation of the problem is called for and that is the topic of Section 8.2.4.

8.2.4 Hypothesis Testing

Suppose that one of two models, M_0 and M_1, is presumed to provide the correct description of the experimental set-up which is under investigation. Model M_0 contains an hypothesis which the experimenter proposes, namely the ***null hypotheses,*** and M_1 differs from M_0 only to the extent that it replaces the null hypothesis with an ***alternative hypothesis.*** A statistic s is proposed on which to base a decision about the acceptability of M_0 and, hence, the null hypothesis. In the manner of a Test of significance, increasing values of s are presumed to reflect lessening agreement between M_0 and the data.

A formal rule is introduced which states that

- M_0 is *accepted* if the statistic s has a value which is less than s_α;
- M_0 is *rejected* if the statistic s has a value which is equal to or greater than s_α,

where s_α is chosen such that when M_0 is correct, $\Pr(s \geq s_\alpha) = \alpha$.

By following this rule, two types of error are possible.

Either M_0 may be rejected when it is correct

or M_0 may be accepted when it is incorrect.

The probabilities of these errors may be determined from a knowledge of the sampling distribution of s under M_0 and M_1, respectively. Specifically,

$$\Pr(\text{reject } M_0 \text{ when } M_0 \text{ is true}) = \Pr(s \geq s_\alpha \text{ when } M_0 \text{ is true}) = \alpha$$
$$\Pr(\text{accept } M_0 \text{ when } M_1 \text{ is true}) = \Pr(s < s_\alpha \text{ when } M_1 \text{ is true}) = \beta.$$

From the experimenter's point of view, the aim is to make the probabilities of error, i.e. α and β, as small as possible. It should come as no surprise, however, that any attempt to decrease α by changing the cut-off point between the acceptance and rejection regions will result in an increase in β and vice versa.

The approach which has been adopted for statistical development of hypothesis testing is to fix the value of α and use the resulting value of β as a measure of the worth of the statistic s as a basis for comparing the model and the data. This approach offers scope for comparing different statistics. If s and t are two statistics which satisfy the condition

$$\Pr(s \geq s_\alpha \text{ when } M_0 \text{ is true}) = \Pr(t \geq t_\alpha \text{ when } M_0 \text{ is true}) = \alpha$$

then s is the better choice as a test statistic to decide if M_0 is likely to be incorrect if it has the smaller value of β, i.e. if

$$\Pr(s < s_\alpha \text{ when } M_1 \text{ is true}) < \Pr(t < t_\alpha \text{ when } M_1 \text{ is true}).$$

This comparative approach then has potential to define the best statistic from among a collection of possible statistics as that statistic which provides a division into *acceptance* and *rejection* regions which leads to the smallest probability of accepting the null hypothesis when it is false subject to the condition that all statistics have the same probability of rejecting the null hypothesis when it is true.

The approach can be extended to the case where the alternative hypothesis is composite by considering a composite hypothesis as being a collection of simple hypotheses. The approach can then be applied for each simple hypothesis.

The following technical terms are in common use to describe properties of tests which are based on the concept introduced in this section.

Type 1 error rate (α) — the probability of incorrectly rejecting the null hypothesis under the chosen partitioning of outcomes into acceptance and rejection regions.

Type 2 error rate (β) — the probability of incorrectly accepting the null hypothesis under the chosen partitioning of outcomes into acceptance and rejection regions.

Power of a test $(1 - \beta)$ — the probability of correctly rejecting the null hypothesis.

Power curve — a graph of the Powers of a test defined as a function of a parameter in situations in which the alternative hypothesis is defined by a continuously varying parameter, e.g. $M > 30$.

Uniformly more powerful test — if two competing tests which have the same Type 1 error rate are being compared and one has a higher power than the other for all simple hypotheses defined under the alternative hypothesis, then that one is said to be uniformly more powerful than the other.

8.2.5 Interpretation of p-Values

Central to the process of hypothesis testing is the determination and interpretation of a *p-value.* In Section 8.2.3 there are guidelines for the interpretation of p as they are conventionally employed.

As noted in Chapter 4, these guidelines are simplistic and should be used within the following constraints.

1. *Interpreting a large value of p.* If p exceeds 0.05 it has been emphasized that the statistical analysis is not supporting *acceptance* of the model, merely establishing that the observed outcome is *reasonable* given the model provided. The strength of acceptance which this provides for the model and the experimental hypothesis it contains depends on the sensitivity of the test to the detection of an error in the model. In technical terms, it depends on the power of the test.

As a general rule, increasing sample size equates with greater power. All other things being equal, a p-value in excess of 0.05 when the sample size is large implies more confidence in the model than when the same sized p-value derives from data based on a smaller sample.

Rather than testing a specified value of the parameter, the scientist may seek the collection of values of the parameter which are reasonable given the available data. This approach involves the definition of a *confidence interval* for the parameter and is discussed in Section 8.3.

In circumstances where there are a limited number of alternative models, the possibility exists for (i) employing Bayesian methods; or (ii) providing a p-value for each model to establish if any one or more models can be rejected.

2. *Interpreting a small value of p.* As a general rule, a value of p less than 0.01 is regarded as providing strong grounds for rejecting the model and the experimental hypothesis it contains. Implicitly, this is taken as evidence for accepting the alternative hypothesis.

There are two sound reasons for not automatically converting this *statistical* rejection into an *experimental* rejection of the model. Firstly, there is nothing in the construction of a test of significance which gives a universal guarantee that a low p-value ensures the observed outcome is more likely under the alternative hypothesis than under the null hypothesis. Secondly, the sensitivity of most tests of significance to an error in the null hypothesis increases with increasing sample size. This can produce a situation in which an error in the null hypothesis, which a scientist would regard as inconsequential, has a high probability of being detected.

The question of what to accept when the null hypothesis is rejected lies outside the scope of tests of significance. Presumably a scientist will look for a plausible or likely alternative in this case.

Where the hypothesis assigns a value or a set of values to a parameter, the use of confidence intervals (Section 8.3) provides a precaution against scientific misinterpretation of the analysis. Given the likely range of values for a parameter, scientists can draw their own judgment as to the acceptability of the original hypothesis.

8.3 Estimation

Investigators do not always undertake a study to examine a predefined hypothesis or have a question which can be represented as a statistical hypothesis. Consider the following problems:

1. A pollster may survey public opinion to judge support for a government action. There may be no reason for making a guess at the level of support as would be required for the application of hypothesis testing. Rather, the pollster may wish to provide a single number as an estimate of percentage public support.

2. A chemical analyst may wish to provide an estimate of the accuracy of a given assay procedure.

3. An agronomist may wish to provide an estimate of the average grain yield for a variety which is under study.

In each case there is a collection of possible models to be considered, with the members of the collection distinguished by a parameter. It is the value of the parameter which is of interest to the experimenter.

8.3.1 Approaches to Estimation

There are two basic approaches. One possibility is to seek a single number which is, in some sense, the *best* estimate of the quantity under investigation. The statistical process by which this is obtained is known as *Point estimation.* The other approach is to seek the collection of values which are judged to be consistent with the data set. Typically, this collection of values comprises all numbers between two limits which graphically could be represented as an interval on a line and consequently the approach is termed *Interval estimation.*

From a user's point of view, point estimation may appear to be the more desirable approach since the end result is a single number rather than a collection—it may appear that a statement "*The estimated support for the government is* 55%" is more valuable than the statement "*The support for the government is likely to lie between* 45% *and* 65%." In fact, precisely the reverse is true. To quote a single value gives no indication of the reliability of the figure and can be quite misleading.

In some areas of application, an attempt is made to overcome this deficiency in point estimation by appending a measure of reliability termed the *standard error of the estimate.* In relative terms, the larger the standard error of the estimate, the less reliable is the estimate. However, the numerical value of the standard error does not always have an interpretation. Even where it does, the practical use to which this interpretation can be put is to permit the construction of an interval estimate, as is demonstrated in Section 8.3.3.

The particular importance of point estimation is in theoretical Statistics where it has an important role in the characterization and construction of estimators as is explained in Section 8.3.2.

8.3.2 Point Estimation

The process of point estimation involves the definition of a statistic which provides a formula for obtaining an estimate of a parameter.

Estimator and **estimate**: A statistic which is employed to provide an estimate of a parameter is termed an *estimator* and the value obtained by applying the statistic to a set of data is termed an *estimate* of the parameter of interest.

Theoretical statisticians study the properties of estimators with the ultimate goal of providing a basis for defining the *best* estimator for a parameter. Attempts to provide a formal definition of the *best estimator* have probabilistic bases. Intuitively, the best estimator is the statistic with the highest probability of providing an estimate which is close to the true value of the parameter being estimated.

To give formal definition to these ideas, consider that there exists a statistical model which includes a parameter M for which the correct value is M_0.

There is no obvious way of formalizing the notion that there should be a *high* probability of an estimate *close* to M_0. Statisticians have chosen particular paths to follow in quantifying the notion in order to allow a general theoretical development. However, it should be borne in mind that no approach can lay claim to being the natural or the best approach.

Mean square error: The strategy whic has been perhaps the most fruitful involves the use of the average squared distance between the estimator (S) and the true value, M_0. This, called the *Mean square error,* is designated by $E(S - M_0)^2$ and is defined by (8.3.1).

$$E(S - M_0)^2 = \sum_{\text{all } S_i} (S_i - M_0)^2 \pi_{st}(S_i) \qquad \text{if } S \text{ is discrete}$$
$$= \int_{-\infty}^{\infty} (S - M_0)^2 \pi_{st}(S) dS \text{ if } S \text{ is continuous.} \tag{8.3.1}$$

where π_{st} is the sampling distribution of S. By this criterion, the *best* estimator of M_0 is that statistic which has the smallest mean square error.

Even though the mean square error is well defined, it is not generally possible to develop the statistical theory by which the statistic with the smallest mean square error can be determined, or to provide a general algorithm by which a value of the best estimator can be computed.

Unbiased estimators: The more fruitful approach has been to restrict possible estimators to those whose sampling distributions have a mean equal to M_0, i.e. for which $E(S) = M_0$. Such statistics are termed *unbiased* estimators of the parameter M. For example, the arithmetic mean of a sample obtained from a population by random sampling is an unbiased estimator of the population mean, i.e. $\overline{y} = \Sigma y_i / n$ is an unbiased estimator of M_0.

Minimum variance estimators: The effect of restricting the possible estimators of a quantity to those which provide unbiased estimation is, in effect, replacing the *mean square error* criterion with the criterion of *minimum variance* since mean square error reduces to variance if $E(S)$ equates with M_0.

Biased estimators: If $E(S)$ does not equal M_0, the statistic is said to be a *biased* estimator of M_0. Intuition is not always a good guide in the identification of which statistics provide unbiased estimators of a parameter as the following example reveals.

Since the variance is the average squared deviation of an observation from the mean (M_0) of a probability distribution of a variable, y, i.e. $\sigma^2 = E[(y - M_0)^2]$, it might be expected that an unbiased estimate of σ^2 is provided by the average squared deviation in the sample, $\Sigma(y_i - \overline{y})^2/n$. In fact, this estimator is biased and it is the sample variance $s^2 = \Sigma(y_i - \overline{y})^2/(n-1)$, which proves to be an unbiased estimator. To further confuse matters, while s^2 is an unbiased estimator of σ^2, its square root, the sample standard deviation, is a biased estimator of σ!

Relative efficiency: There may be practical reasons for not using the best statistic according to the mean square error or minimum variance criteria.

In such circumstances, one may wish to know the relative loss of precision from using a less-than-optimal estimator. The way this is most commonly done is to state how many more sample members would be required under the less optimal method to achieve the same level of precision in estimation. The information is expressed as the *relative efficiency* of statistic S_2 compared to statistic S_1. If S_1 is the more efficient estimator, then a relative efficiency of 0.8 would imply that 10 sample members would be required using S_2 to every 8 needed if S_1 provided the estimation for the same level of precision to be achieved.

8.3.3 Interval Estimation

Suppose a collection of models are distinguished only by the values of a parameter M. Based on a test of significance which employs a statistic S and is applied to each model separately, a subset of models can be defined which have p-values above a specified level p_0. For most practical applications, this collection can be identified with a range of values of M forming an interval between a lower limit $M_1(S)$ and an upper limit $M_2(S)$. An interval so formed is known as a *confidence interval* for the parameter M and is prefixed by a number determined from the value of p_0. If p_0 is set at 0.05, the interval is known as a 95% *confidence interval.* More generally, the interval is known as a $100(1 - p_0)\%$ *confidence interval.*

If the two intervals are based on the same statistic, a 99% confidence interval has a higher probability of including the true value of M than does a 95% confidence interval. The price to pay for the higher confidence will be a wider interval. Traditionally, the 95% confidence interval is the one most commonly quoted, although the choice is arbitrary.

Since the construction of a confidence interval depends on a test of significance, the interval obtained will depend on the choice of statistic used in the construction of the test of significance. Thus, for a given collection of models, differing only in the value of M, and a given data set, each statistic employed will provide a confidence interval and, in general, these confidence intervals will differ.

Which is the best statistic to employ to construct the confidence interval?

Because of the method of construction, all statistics will provide intervals which have the property that the interval has a probability of $1-p_0$ of covering the true value of the parameter being estimated. (This follows directly from the connection with tests of significance.) In theory, the *best* statistic is the one which can provide the confidence interval with the smallest average width.

The limits of a confidence interval, $M_1(S)$ and $M_2(S)$, are themselves values of statistics and, thus, are subject to variation. Hence, the placement of the confidence interval is subject to chance. To state that a 95% confidence interval has a probability of 0.95 of including the correct value of M derives from the variability in the statistics. The correct value of

M is regarded as something fixed although unknown. This contrasts with the Bayesian approach which operates on the premise that a probability distribution can be attached to M.

8.3.4 Estimation and the Normal Distribution

There proves to be a vast number of statistics which have sampling distributions that are well approximated by Normal distributions when the sample size, n, is large. This has important implications for the construction of confidence intervals as is explained below.

Suppose x_n is such a statistic based on a sample of size n which is an unbiased estimator of the parameter M. If σ_n^2 is the variance of x_n and the conditions of the Central Limit Theorem apply, the following probability statement is true for large n.

$$\Pr\left(-1.96 < \frac{x_n - M}{\sigma_n} < 1.96\right) \cong 0.95,$$

or, more generally,

$$\Pr\left(-z_{\frac{1}{2}\alpha} < \frac{x_n - M}{\sigma_n} < z_{\frac{1}{2}\alpha}\right) \cong 1 - \alpha.$$

These statements may be rearranged to yield the following:

$$\Pr(x_n - 1.96\sigma_n < M < x_n + 1.96\sigma_n) \cong 0.95$$

and

$$\Pr(x_n - z_{\frac{1}{2}\alpha}\sigma_n < M < x_n + z_{\frac{1}{2}\alpha}\sigma_n) \cong 1 - \alpha.$$

Hence, by definition, the intervals defined by (8.3.2) are, respectively, approximate 95% and $100(1-\alpha)\%$ confidence intervals for M.

95% confidence interval	$x_n - 1.96\sigma_n$ to $x_n + 1.96\sigma_n$	(8.3.2)
$100(1-\alpha)\%$ confidence interval	$x_n - z_{\frac{1}{2}\alpha}\sigma_n$ to $x_n + z_{\frac{1}{2}\alpha}\sigma_n$.	

There are two further results with important applications which apply when the value of σ_n^2 is unknown but there is a value of s_n^2, the unbiased estimator of σ_n^2 derived from the Maximum Likelihood estimator of σ_n^2.

1. For most practical applications (8.3.3) provides good large sample approximations to the 95% and $100(1-\alpha)\%$ confidence intervals for the statistic M.

95% confidence interval	$x_n - 1.96s_n$ to $x_n + 1.96s_n$	(8.3.3)
$100(1-\alpha)\%$ confidence interval	$x_n - z_{\frac{1}{2}\alpha}s_n$ to $x_n + z_{\frac{1}{2}\alpha}s_n$.	

2. In situations where the sampling distribution of x_n is a Normal distribution, exact confidence limits for M are defined by (8.3.4).

$$\begin{array}{ll} \text{95\% confidence interval} & x_n - t_5(\nu) \text{ to } x_n + t_5(\nu)s_n \\ 100(1-\alpha)\% \text{ confidence interval} & x_n - t_\alpha(\nu)s_n \text{ to } x_n + t_\alpha(\nu)s_n \end{array} \quad (8.3.4)$$

where ν is a parameter which depends on the statistic x_n.

8.3.5 Linking Estimation and Hypothesis Testing

By the definition provided in Section 8.3.3, a $100(1 - p_0)\%$ confidence interval for a parameter M contains all values of M which yield a p-value above p_0 in a test of significance. Hence, if the guideline is adopted that a value of M is reasonable when the corresponding p-value exceeds 0.05, a decision about the acceptability can be based on a 95% confidence interval for M. Similarly, if a p-value less than 0.01 is considered to provide stong evidence against the hypothesized value of M, a decision can be based on a 99% confidence interval.

Thus, a confidence interval can serve a role not only in estimation but also in hypothesis testing.

8.3.6 Levels of Confidence for Statistics Taking Discrete Values

A $100(1 - \alpha)\%$ confidence interval for a parameter M which is based on a statistic S is defined by a pair of statistics, $M(S_1)$ and $M(S_2)$, which satisfy (8.3.5).

$$\Pr\{M(S_1) \leq M \leq M(S_2)\} = 1 - \alpha. \quad (8.3.5)$$

In equations (8.3.2) to (8.3.4), these statistics were assumed to exhibit continuous variation. The consequence is that, for a specified level α, it is possible to find statistics to meet the condition imposed by (8.3.5).

There are, however, some confidence intervals of practical importance for which the pair of statistics $M(S_1)$, $M(S_2)$ cannot take infinitely many values with the consequence that (8.3.5) can only be satisfied for a finite set of values of α.

An important area of application of confidence intervals where there are limited choices for $M(S_1)$, and $M(S_2)$ occurs when there is a set of n scaled responses for the variable y and the confidence interval is based on the ordered values. There is a set of *order statistics* which can be defined—$y_{(1)}$ is the smallest response in the set of n, $y_{(2)}$ is the second smallest, etc. A pair of order statistics $y_{(i)}, y_{(j)}$ where $j > i$ provides a practical means of defining a confidence interval for a quantile (Q) of the frequency or probability distribution of the response variable y.

Using the relation developed in Section 7.7.3 for the median of the distribution, $\Pr(y_{(i)} < M < y_{(i+1)}) = \binom{n}{i}(\frac{1}{2})^n$, it follows that

$$\Pr(y_{(i)} < M < y_{(j)}) = \sum_{k=i}^{j-1}\binom{n}{k}\left(\frac{1}{2}\right)^n$$

for any pair i and j which satisfy $0 \leq i < j \leq n$.

This leads to (8.3.6) as the basis for a confidence interval for the median which is free of any distributional assumptions about y.

$$\Pr(y_{(i)} < M < y_{(n+1-i)}) = \sum_{k=i}^{n-i}\binom{n}{k}\left(\frac{1}{2}\right)^n = 1 - 2\sum_{k=0}^{i-1}\binom{n}{k}\left(\frac{1}{2}\right)^n . \quad (8.3.6)$$

Since the Binomial statistic can only take values $0, 1, 2, \ldots, n$, it is not generally possible to find a value of i which provides the exact level of confidence sought.

For example, if $n = 10$, the following confidence intervals can be constructed.

i	lower limit	upper limit	confidence level
1	$y_{(1)}$	$y_{(10)}$	0.9980
2	$y_{(2)}$	$y_{(9)}$	0.9785
3	$y_{(3)}$	$y_{(8)}$	0.8926

A confidence level of 95% cannot be achieved using any of the pairs of statistics defined in (8.3.6). In practice, two avenues are open to partially solve this dilemma.

The first is to present the smallest interval which is *at least* equal to the nominal level of confidence. This is the interval $(y_{(2)}, y_{(9)})$ in the above example.

The alternative is to define an approximate confidence interval using the mathematical process of linear interpolation which is based on the following development.

If the required level of confidence is $1 - \alpha$ and intervals available are $1 - \alpha_1$ and $1 - \alpha_2$ where $1 - \alpha_1 < 1 - \alpha \leq 1 - \alpha_2$ then the proportional distance of the nominal level $(1 - \alpha)$ between the two obtainable levels is given by (8.3.7).

$$d = [(1 - \alpha) - (1 - \alpha_1)]/[(1 - \alpha_2) - (1 - \alpha_1)]. \quad (8.3.7)$$

Thus, to define an approximate lower limit for a $100(1-\alpha)\%$ confidence inteval, $S_1(\alpha)$, it seems sensible to place the lower limit proportionately the same distance between the values defined by the statistics $S_1(\alpha_1)$ and

$S_1(\alpha_2)$ which are lower limits at confidence levels $1 - \alpha_1$ and $1 - \alpha_2$, respectively. This is achieved by using (8.3.8).

$$S_1(\alpha) = (1 - d)S_1(\alpha_1) + dS_1(\alpha_2), \tag{8.3.8}$$

where d is defined by (8.3.7).

This same approach can be used to define an approximate upper limit $S_2(\alpha)$ in terms of $S_2(\alpha_1)$ and $S_2(\alpha_2)$.

Application of the method is illustrated in Section 11.2.2.

8.4 Likelihood

Central to a large amount of statistical theory and methodology is the concept of *Likelihood.* This concept can be applied in situations in which there is a family of probability distributions defined which have the same mathematical form but differ in values of one or more parameters. Consider the distribution

$$\begin{aligned} \pi(y/\alpha) &= 1 - \alpha && \text{for } y = 0 \\ &= \alpha && \text{for } y = 1 \\ &= 0 && \text{otherwise,} \end{aligned}$$

where α is a fixed value in the interval $[0, 1]$.

From a different perspective, this equation may be seen as describing the variation in α when the variable y is observed to take the value y_0. In this capacity it is a function L defined by (8.4.1).

$$\begin{aligned} L(\alpha/y_0) &= (\alpha)^{y_0}(1 - \alpha)^{1-y_0} && \text{for } 0 \le \alpha \le 1 \\ &= 0 && \text{otherwise.} \end{aligned} \tag{8.4.1}$$

Where a mathematical equation used to define a probability distribution is instead regarded as a function of one or more parameters of the probability distribution for a fixed value of the variable, it is termed a *Likelihood function.* The application of a likelihood function was introduced and discussed in Section 4.3.

An important practical role of Likelihood arises when L is based on the joint distribution of n variables, $y_1, y_2, \ldots, y_n$ whose values $y_{10}, y_{20}, \ldots, y_{n0}$ constitute the data in an experiment. If there is a single parameter, M, of the joint probability distribution of $y_1, y_2, \ldots, y_n$ and if the set of values $y_{10}, y_{20}, \ldots, y_{n0}$ is denoted by the symbol D, then from both an intuitive and a theoretical viewpoint it is reasonable to regard values of the likelihood function $L(M/D)$ as providing relative degrees of support for different values of the parameter M when the data set D is observed.

By way of illustration, if $y_1, y_2, \ldots, y_n$ are independent and identically distributed random variables all with the probability distribution defined by (8.4.1) then, by (5.4.16), the joint probability distribution of $y_1, y_2, \ldots, y_n$

is the product of the individual probability distributions. Hence, the Likelihood function $L(\alpha/y_{10}, y_{20}, \ldots, y_{n0})$ is defined by (8.4.2).

$$L(\alpha/y_{10}, y_{20}, \ldots, y_{n0}) = (\alpha)^{Y_0}(1-\alpha)^{n-Y_0} \quad \text{for } 0 \le \alpha \le 1 \qquad \text{(8.4.2)}$$
$$= 0 \quad \text{otherwise}$$

where $Y_0 = y_{10}+y_{20}+\cdots+y_{n0}$. In the interval $[0,1]$, the Likelihood function defined by (8.4.2) has a minimum value of zero at $\alpha = 0$ and $\alpha = 1$ and rises to a maximum at $\alpha = Y_0/n$. The value Y_0/n is known as the *Maximum likelihood estimate* of the true value of the parameter α. Correspondingly, the statistic $\overline{y} = (y_1+y_2+\cdots+y_n)/n$ is known as the *Maximum Likelihood estimator* of the parameter α.

The two uses to which Likelihood is put are (i) to establish the value of the parameter which has maximum Likelihood; and (ii) to compare Likelihoods for different values of the parameter. The latter is achieved by forming Likelihood ratios. For a broad range of statistical applications, Likelihood ratios provide the basis for hypothesis testing and play key roles in the Bayesian approach and Decision theory.

If $L(M, D)$ is a likelihood function for a parameter M given data D, and M_* is the value of M which maximizes the likelihood function when the observed data set is D_0, then (8.4.3) defines the *Maximum likelihood ratio statistic* under the hypothesis $M = M_1$.

$$LR(D) = \frac{L(M_1, D)}{L(M_*, D)}. \qquad \text{(8.4.3)}$$

In (8.4.3), M_1 and M_* are fixed values and variation is presumed to result from changes in the data D. $LR(D)$ is, by definition, a statistic because each possible data set is assigned a numerical value by (8.4.3). The significance of the statistic is that increasing values correspond to relatively greater support for M_1 relative to M_*.

Statistical theory establishes that, for many statistical models of practical importance, the Maximum likelihood ratio statistic provides the basis for the best test of the hypothesis $M = M_1$ against the alternative $M \neq M_1$.

The Likelihood-based procedures are extremely important in the development of statistical methodology, and those readers who possess the necessary mathematical background would benefit from an exploration of the properties and application of Likelihood.

8.5 The Bayesian Approach

A complete introduction to the Bayesian approach could in itself fill a book. This section presents the Bayesian viewpoint free of any critical comparison with the non-Bayesian approach which dominates this book. Those who find the approach has relevance to their scientific work are urged to seek further information from texts which are written entirely from the Bayesian viewpoint (for example, Box and Tiao [1973]).

8.5.1 The Essence of the Bayesian Approach

An essential feature of the Bayesian approach is that it provides a unified approach to Analytic Statistics. There is no need to distinguish Hypothesis Testing and Estimation, Point and Interval Estimation; Null hypotheses and Alternative hypotheses, Tests of significance and Likelihood ratio tests. Bayesian analysis works from the same basis for all types of scientific studies.

Central to Bayesian philosophy is the requirement that elements in a statistical model which are not known exactly, can be assigned a probabilistic structure. In particular, parameters appearing in the model can be assigned probability distributions. Furthermore, such probability distributions based on pre-experimental knowledge must be completely specified. Various techniques exist for coping with the situation in which the investigator is unable to supply probabilities for possible hypotheses or is unable to supply probability distributions for parameters in the model.

By the nature of its operation, Bayesian analysis has the capacity to automatically make use of all relevant information in the data and all relevant information deriving from other sources. The assumptions being made are visible to the user and it is, in theory, possible to establish the effect of varying any component in the model. This property is made easy by the standard form of result from Bayesian analysis.

8.5.2 Bayes' Theorem

The statistical base is provided by a relation known as *Bayes' theorem* which is defined by (8.5.1). It is based on probabilistic notions which are developed below.

Suppose that an experiment or investigation is conducted in which the data come from one of the possible outcomes, $D_1, D_2, \ldots, D_r$ and the experimental set-up which generated the data is described by one of the possible models, $M_1, M_2, \ldots, M_p$. Which model generated the data is presumed unknown to the investigator, although probabilities can be assigned to the models based on information available prior to the current experiment, which indicates the relative levels of belief the investigator assigns to the models.

From the probabilistic viewpoint, both D_i $(i = 1, 2, \ldots, r)$ and M_j $(j = 1, 2, \ldots, p)$ can be interpreted as events and, hence, the following sets of probabilities may be defined:

$\Pr(D_i)$	–	the probability that the experiment produces the event D_i;
$\Pr(M_j)$	–	the probability that the correct model for the experimental set-up is M_j *prior* to any knowledge of the outcome from the experiment;
$\Pr(D_i/M_j)$	–	the probability that outcome D_i will occur if model M_j is the correct model; and
$\Pr(M_j/D_i)$	–	the probability that model M_j is the correct model given the information that the experiment produced the outcome D_i.

Given that the observed data set is D_i, the task of Bayesian analysis is to determine the set of probabilities, $\Pr(M_j/D_i)$ for $j = 1, 2, \ldots, p$. Thus, the aim is to modify the prior probabilities, $\Pr(M_j)$, $j = 1, 2, \ldots, p$, in the light of information obtained from the experiment. Equation (8.5.1) provides the mechanism for doing that. It is known as Bayes' theorem and derives from (5.3.1).

$$\Pr(M_j/D_i) = \frac{\Pr(D_i/M_j) \times \Pr(M_j)}{\Pr(D_i)}. \tag{8.5.1}$$

Commonly, information is not available directly for the evaluation of $\Pr(D_i)$ and (8.5.2) is employed.

$$\Pr(D_i) = \sum_{j=1}^{p} [\Pr(D_i/M_j) \times \Pr(M_j)]. \tag{8.5.2}$$

8.5.3 An Illustration of the Use of Bayes' Theorem

Example 4.1 provides the information required for an application of Bayes' theorem. The data are provided by the results of a screening test for breast cancer on a woman. The possible outcomes are a *negative* finding, D_n, or a *positive* finding, D_p. The two possible models are distinguished by the actual state of the woman—either free from cancer, M_-, or having the disease, M_+.

The information supplied is listed below.

$\Pr(M_+) = 0.0013$	*The probability of a woman having breast cancer prior to the screening.*
$\Pr(M_-) = 0.9987$	*The probability of a woman being free of breast cancer prior to the screening.*
$\Pr(D_p/M_-) = 0.10$	*The probability of a false positive.*
$\Pr(D_n/M_-) = 0.90$	*The probability of a correct negative.*
$\Pr(D_p/M_+) = 0.95$	*The probability of a correct positive.*
$\Pr(D_n/M_+) = 0.05$	*The probability of a false negative.*

Using (8.5.2) the probabilities of positive and negative findings may be calculated.

$$\begin{aligned}\Pr(D_n) &= \Pr(D_n/M_-)\Pr(M_-) + \Pr(D_n/M_+)\Pr(M_+)\\ &= 0.90 \times 0.9987 + 0.05 \times 0.0013\\ &= 0.8989\end{aligned}$$

$$\begin{aligned}\Pr(D_p) &= \Pr(D_p/M_-)\Pr(M_-) + \Pr(D_p/M_+)\Pr(M_+)\\ &= 0.10 \times 0.9987 + 0.95 \times 0.0013\\ &= 0.1011.\end{aligned}$$

Application of Bayes' theorem then yields the following probability of a woman having breast cancer *after the screening test gives a negative reading.*

$$\Pr(M_+/D_n) = \frac{\Pr(D_n/M_+) \times \Pr(M_+)}{\Pr(D_n)} = \frac{0.05 \times 0.0013}{0.8989} = 0.000072.$$

Before the test, the odds against breast cancer were about 800 to 1 (0.9987 : 0.0013). As a result of a negative test, the odds have lengthened to about 14,000 to 1 (0.999928 : 0.000072). Thus, the odds have lengthened by a factor of about 18 (14,000 : 800).

8.5.4 Generalizing Bayes' Theorem

Use of the representation given in (8.5.1) for Bayes' theorem is limited by the fact that both the number of possible outcomes and the number of possible models are presumed to be finite. If the response variable is continuous or if the models are indexed by a parameter which shows continuous variation, (8.5.1) cannot be applied.

These possibilities can however be accommodated if a statistical rather than a probabilistic view is taken of the formula in (8.5.1). Instead of considering the elements in the formula to be probabilities of events, this view represents them as values of probability distributions. This leads to the formulation of Bayes' theorem given in (8.5.3) where D_0 denotes the data and θ is the parameter whose values distinguish the set of possible models.

$$\pi(\theta/D_0) = \frac{\pi_s(D_0/\theta) \times \pi(\theta)}{\pi_s(D_0)} \qquad (8.5.3)$$

where

$\pi(\theta)$ is the probability distribution assumed for θ prior to information collected from the current investigation—the *prior distribution* for θ;

$\pi(\theta/D_0)$ is the probability distribution for θ after the information from the current experiment has been included—the *posterior distribution* for θ;

$\pi_s(D_0)$ is the value from the sampling distribution of the sample when the data from the current experiment are employed; and

$\pi_s(D_0/\theta)$ is the value from the conditional sampling distribution of the sample given θ when the data from the current experiment are employed.

If the response variable is discrete or categorical with a finite number of possible outcomes and if there are only a finite number of possible models, then (8.5.3) can be utilized in exactly the same manner as (8.5.1) with values of probability distributions being equal to probabilities of events. Probabilities are then available to define the relative degrees of belief in the different models.

Where θ is a continuous variable, both the prior and posterior probability distributions are continuous and probabilities can only be expressed for selected sets of values of θ, in particular for intervals. For example, the mean of a population may be presumed to have a probability distribution which is Normal with mean (θ) equal to 25.0 and to have a known variance of 4.0 prior to an experiment being conducted. From properties of the Normal distribution, it could then be determined that the probability of the correct value of θ lying in the interval 21 to 29 is approximately 0.95.

Using (8.5.3) it is possible to define the posterior probability distribution for θ if the prior probability distribution for θ and the sampling distribution of the sample are known. For example, if (i) the prior probability distribution for θ has a Normal distribution with assumed mean θ_0 and known variance σ_0^2, and (ii) the population from which the sample is randomly selected has a Normal distribution with the same variance, then the posterior distribution for θ, given the sample yields an arithmetic mean of $\overline{y}_0$ from a sample of size n, is Normal with mean $(n\overline{y}_0 + \theta_0)/(n+1)$ and variance $\sigma_0^2/(n+1)$. By way of illustration, if the mean and variance of the prior distribution are $\theta_0 = 25.0$ and $\sigma_0^2 = 4.0$, and the sample mean is $\overline{y}_0 = 18.6$ from a sample of $n = 20$, then the posterior distribution of θ is Normal with a mean of 18.9 and a variance of 0.19. It can be seen that the sample information has modified the prior distribution by decreasing the mean of the distribution from 20 to 18.9 and greatly increasing the precision with a decrease in variance from 4.0 to 0.19. Where there was a probability of about 0.95 that the value of θ lies between 21.0 and 29.0 prior to the data from the experiment, after the incorporation of the data, there is a probability of about 0.95 that the value of θ lies in the interval 18.0 to 19.8.

8.5.5 Practical Considerations

Two points about the Bayesian approach require special consideration. The first is the determination of prior probability distributions. Where these can be directly based on factual knowledge, as in the example on screening for breast cancer, there are few, if any, grounds for controversy. Equation (8.5.1) provides a vehicle for probabilistic calculations and interpretation which are universally accepted. However, this situation has only limited applicability.

Where such an approach is not tenable, two options have been widely suggested. Both are controversial and their relative and absolute acceptance depends very much on the emphasis one places on the range of criteria by which they are judged.

The first approach seeks to quantify prior belief objectively. In practice, the emphasis is placed on situations in which the investigator claims initial ignorance. Superficially, this might seem an easy situation to handle—simply give all possible outcomes equal probability. Unfortunately, there prove to be quite serious difficulties where the response is continuous. Generally, there will not be precise upper and lower bounds on the range of possible values and, in theory, they may be plus and minus infinity.

A more recent approach makes no pretense at objective definition of priors. Instead it relies on how you, the investigator, perceives relative prior probabilities by setting up comparisons in which the answers can be expressed as probabilities. An essential requirement of this approach is that all levels and manner of belief can be quantified in this way and that the probabilities so formed are strictly comparable.

A second, and perhaps more serious restriction on the practical application of Bayesian methods, is the requirement that an exhaustive set of possible models must be defined. As explained in Chapter 4, it is not sensible to assign a probability to a model without knowledge of all possible alternatives. There are situations in which scientists would appear to pose legitimate questions as to the acceptability of a single model without having regard to alternatives. At least in the classical Bayesian approach, such problems have no solution. However, it should be noted that many Bayesians would not regard the examination of a single model in the absence of completely identified alternatives as a valid statistical problem.

8.6 Choosing the Best Method

The selection of statistical methods cannot be made prescriptive. Whether on statistical or non-statistical grounds, there remain value-judgments to be made which rest with the individual scientist and possibly an associated statistical consultant. The *best* statistical method for use in a scientific study simply does not exist in any universal sense.

8.6.1 The Statistical Approach

Given that a scientist has a well defined problem and a clear and accurate description of the experimental conditions of the study, there arises a decision as to the statistical approach to adopt. The choice between a Bayesian and a non-Bayesian approach will generally affect the nature of the problem, the method of analysis and the nature of the interpretation of the findings. In principle, the Bayesian approach offers greater integration of Statistics into the scientific process, a unified form of statistical analysis, and interpretation which fits naturally into the idea of scientific investigation steadily advancing towards a higher level of understanding. In return for these apparently desirable objectives, a greater amount of information is required from the investigator, and there is a need to accept that all forms of uncertainty relevant to the problem can be quantified and are comparable. In practice, non-Bayesian approaches are found to have far greater application.

8.6.2 Theoretical Considerations

Another area where statisticians have a central role is in establishing the *best* method from a theoretical viewpoint. Having decided on the appropriate framework within which to construct a mathematical theory of Statistics, conditions have been proposed and widely accepted for identifying the optimal method. Unfortunately, three problems exist in practice

- there may be no method which meets the optimality conditions;
- there may be no means of finding the optimal solution even though statistical theory says that one exists; or
- the conditions may depend on information which is not available or which is not known with certainty.

Notwithstanding the difficulties which may arise, statistical theory plays a central role in the construction of optimal methods. All other things being equal, the test which is more powerful is the one to be recommended; the point estimator with the smaller mean square error is the one to be recommended; the interval estimator which, on average, gives the shorter confidence interval is the one to be recommended.

8.6.3 Experimental Considerations

As a general rule, the greater the amount of relevant information which can be used by a statistical method, the more confidence which can be placed in conclusions reached. In this way, scientists conducting experiments have a key role in providing information for use in selecting the best statistical method. This is an area where research experience and a knowledge of the type of information which can be used in the determination of statistical methods both play an important role. To some extent, the scientist's experience can be replaced by having a consultant statistician as an advisor. One of the qualities of such a consultant is the capacity to seek relevant information from the scientist—to know the right questions to ask.

8.6.4 Data and Computational Considerations

Computers are playing an ever increasing role in the selection of the best statistical method for a specific investigation. In a direct sense, the method which is judged to be the best on other grounds may be automatically eliminated if there are not facilities for its computer implementation. This is particularly the case when the necessary calculations are complex or when there is a large amount of data. The indirect effect of computers in selecting the best method has been in the practical enhancement of model selection and data checking procedures, points taken up in detail in Chapter 10.

The need to ensure the availability of suitable computing programs or packages, accessibility to such programs at both an operational and interpretive level, and the resources to undertake the computations are important considerations when establishing the best statistical method to employ. This is a task which should be undertaken at the design stage of the experiment.

Problems

8.1 *p-value and hypothesis testing*

The data in Table 1.3.4 comprise 250 readings of cut-off temperatures of limit thermostats. The specified cut-off temperature is 80°C. It is clear from the data that not all thermostats have this cut-off temperature. From the statistical viewpoint, the specified cut-off temperature can be viewed as the hypothesized mean cut-off temperature and the statistical problem is to test the hypothesis that the actual mean, M, has the value 80. If $\bar{y}$ is the statistic which is the mean of the sample of readings, then $d = |\bar{y} - 80|$ provides a basis for judging whether the hypothesis $M = 80$ is reasonable. Increasing values of d reflect a greater difference between the hypothesized mean in the statistical model and the sample mean obtained from the data. Hence, if $d_0 = |\bar{y}_0 - M|$ is the value of d obtained when the observed sample mean is $\bar{y}_0$,

$$p_0 = \Pr(d \geq d_0) = \Pr(|\bar{y} - 80| \geq d_0)$$

is a measure of agreement between the data and the model which includes the hypothesis $M = 80$.

The Central Limit Theorem (Section 7.3.5) may be invoked to establish that the sampling distribution of $\overline{y}$ based on the sample of 250 observations is well approximated by a $N(80, \sigma^2/250)$ distribution if the hypothesis $M = 80$ is correct. The parameter σ^2 is unknown but it is reasonable to employ the value of the sample variance, s_0^2 as the value of σ^2.

(a) Using the data in Table 1.3.4, compute values of $\overline{y}_0$, s_0^2 and $d_0 = |\overline{y}_0 - 80|$.

(b) Using Table 7.3.1 for guidance, obtain the value of

$$p_0 = \Pr(|\overline{y} - 80| \geq d_0)$$

where $\bar{y}$ has a $N(80, \sigma_0^2/250)$ distribution.

(c) Based on the rules for interpretation of p-values given in Section 8.2.3, do the data support the claim that the expected cut-off temperature for the limit thermostats is 80°C.

8.2 *Acceptance and rejection regions*

(a) Suppose x is a statistic which has a $N(M, 6)$ sampling distribution. There is an hypothesis H_0: $M = 10$ and an alternative hypothesis H_1: $M \neq 10$. Construct an *acceptance* region for H_0 based on the statistic x which is symmetric about 10 and for which there is a Type 1 error rate of 0.05.

(b) What is the probability of obtaining a value of x in the acceptance region if the true value of M is 15?

8.3 *Type 1 and Type 2 errors*

Of the persons being admitted to a hospital the belief is that, in the long term, 50% are not in need of hospitalization. A person undertaking studies in health care costs plans an experiment in which 30 successive persons admitted to a hospital are to be examined to ascertain if they need hospitalization. A long-term hospitalization rate of 50% will be regarded as reasonable if somewhere between 12 and 18 persons in the sample of 30 are judged as being in need of hospitalization. Otherwise, the rate will be seen as unreasonable.

Use the Normal approximation to the Binomial to determine the following:

(a) the *Type 1 error rate* for the above scheme;

(b) the *Type 2 error rate* for the above scheme if the true long term hospitalization rate is 75%.

8.4 *Power curve*

A variable x has a $N(M,1)$ distribution. The null hypothesis is $H_0: M = 0$ and the alternative hypothesis is $H_1: M \neq 0$. Suppose the acceptance region for the null hypothesis is $[-1.96, 1.96]$.

(a) What is the Type 1 error?

(b) If the true value of M is M_* find the Type 2 error for $M_* = 0.5$, 1.0, 1.5, ..., 3.0 and hence, the power of the test at these values of M.

(c) Use the values obtained in (b) to construct a power curve for the test.

8.5 *Unbiased estimators*

Let $y_1, y_2, \ldots, y_n$ be n independent random variables which all have an expected value of M and the same variance σ^2. Use (5.4.17)–(5.4.19) and the definitions contained in (5.4.4)–(5.4.7) to establish the following results.

(a) If $\bar{y} = (y_1 + y_2 + \cdots + y_n)/n$, then $\bar{y}$ is an unbiased estimator of M. (This establishes that the sample mean is an unbiased estimator of the population mean.)

(b) If $s_*^2 = \sum_{i=1}^n (y_i - M)^2/n$, then s_*^2 is an unbiased estimator of σ^2.

(c) The variance of $\bar{y}$ is σ^2/n.

(d) If $s^2 = \sum_{i=1}^n (y_i - \bar{y})^2/(n-1)$, then s^2 is an unbiased estimator of σ^2.

Hint: write $(y_i - \bar{y})^2$ as $[(y_i - M) - (\bar{y} - M)]^2$ and expand the square.

8.6 *Constructing a confidence interval*

If $y_1, y_2, \ldots, y_n$ are independent random variables all with a $N(M, \sigma^2)$ distribution, and s^2 is the sample variance as defined by (1.3.6), then $(n-1)s^2/\sigma^2$ is a statistic which has a $\chi^2(n-1)$ distribution. Hence,

$$\Pr\left\{ \mathrm{X}^2_{\frac{1}{2}(1-\alpha)}(n-1) \leq \frac{(n-1)s^2}{\sigma^2} \leq \mathrm{X}^2_{\frac{1}{2}\alpha}(n-1) \right\} = 1 - \alpha.$$

By rearranging the inequalities, construct a $100(1-\alpha)\%$ confidence interval for σ^2. (This is an additional illustration to that provided in Section 8.3.4 of the technique of constructing a confidence interval for a parameter which is based on the sampling distribution of an estimator of the parameter.)

8.7 *Maximum Likelihood estimation*

Let $y_1, y_2, \ldots, y_n$ be n independent random variables which are all $N(M, 1)$.

(a) Based on (7.3.1) and (5.4.16), deduce that the joint probability distribution of $y_1, y_2, \ldots, y_n$ is given by

$$\pi_s(y_1, y_2, \ldots, y_n) = [2\pi]^{-\frac{1}{2}n} \exp\left\{-\frac{1}{2}\sum_{i=1}^{n}(y_i - M)^2\right\}.$$

(b) Let y_{i0} be an observed value of y_i $(i = 1, 2, \ldots, n)$. Replace y_i by y_{i0} $(i = 1, 2, \ldots, n)$ in the expression obtained in (a) and consider the expression to define a function $L(M)$ of M (i.e. consider the function to be a *Likelihood function* rather than a joint probability distribution).

Using the fact that the value of M which maximizes the Likelihood function will be the same value as that which maximizes the logarithm of the Likelihood function, establish that the Likelihood function takes its maximum value when $M = (y_{10} + y_{20} + \cdots + y_{n0})/n$. (This implication is that the sample mean, $\bar{y}$, is the Maximum Likelihood estimator of the population mean M when the frequency or probability distribution is Normally distributed and the sample is selected by random sampling.)

9

Examining Proportions and Success Rates

9.1 Experimental Aims

Consider a situation in which there is a single response variable which has only two possible values. This may arise in the study of a population which is divided into two groups, in the study of a treatment for which there are only two possible results or in the study of a process in which there are only two possible outcomes.

For the population, a frequency distribution may be defined with the following structure:

Group	1	2
Proportion	π_1	π_1.

The experimental questions can be related to the values of π_1 and π_2 which represent the proportions of population members in groups 1 and 2, respectively. Because π_1 and π_2 have the connection that $\pi_1 + \pi_2 = 1$, questions posed about π_1, π_2 or the ratio π_1/π_2 can utilize the same statistical methodology. The type of questions which may be examined by statistical analysis are described in Section 9.1.1.

For a treatment or a process, a probability distribution may be defined with the following structure:

Outcome	1	2
Probability	π_1	π_2.

The questions which may be considered relate to the probabilities of the outcomes and are described and illustrated in Section 9.1.2.

9.1.1 Proportions

The type of problem to which the methods of this chapter relate is well illustrated by a population in which people holding an opinion on an issue are divided into two groups—those supporting and those opposing a proposition. An illustration is provided by Example 9.1.

Example 9.1 (Determining Support for a Hotel Development). *An application for a multi-story hotel development on the waterfront in a city was vigorously opposed by a section of the community. A committee was formed to attempt to determine if a majority of the city's population was opposed to the development. With the help of an army of willing volunteers, a sample comprising* 10% *of the total population was approached. Of the* 7,500 *persons surveyed,* 7,412 *were prepared to give a firm response*—3928 *were against the proposal and* 3484 *in favor. Clearly, in the sample, there is majority support. Can the committee be confident this would be reproduced in the population at large?*

The various statistical techniques employed examine variations on the one question, namely "*What are the relative proportions of the population members in the two classes?*" In statistical terms, the questions relate to the two parameters of the frequency distribution, π_1 and π_2.

The only factual information about the values of π_1 and π_2 is presumed to come from a sample which is randomly selected from the population in which each population member is classified as either *favoring* or *opposing* the proposition. (The *undecideds* are excluded from the statistical analysis, although they should not be ignored in the investigator's conclusions for reasons explained in Section 13.4.)

The nature of the problem is illustrated in Figure 9.1.1. The information which is known comprises (i) the total number of persons in the population (N); (ii) the sample size (n); (iii) the number of sample members who support the government (x_1) and (iv) the number who do not support the government (x_2).

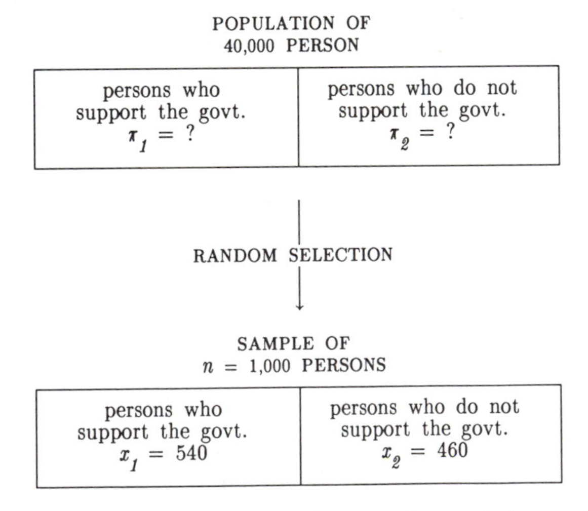

Figure 9.1.1. A statistical view of an opinion poll.

Questions which might arise in the problem illustrated in Figure 9.1.1 are

1. *Can the government be confident that it has majority support in the population?*

2. *What is the likely level of support for the government in the population?*

To answer the first question in the affirmative, it is necessary to find support for at least one value of π_1 in excess of 0.50 and to reject the possibility that π_1 could be less than 0.50. The recommended statistical approach to answering this question would be through the construction of a confidence interval for π_1 (with level of confidence 95% unless otherwise indicated) to establish if any values less than 0.50 are contained in that interval.

The simplest solution to the second question would be to provide a point estimate for π_1, i.e. a single number which is the most likely value to be correct. As explained in Section 8.3.1, this is not a solution which is recommended by statisticians. It is preferable to provide a likely range of values of π_1 in the form of a confidence interval. By so doing, the reliability of the estimation procedure is immediately apparent to the recipient of the information.

As noted in the introduction to this section, any question posed in terms of π_1 may equally well have been posed in terms of π_2 or the *odds* that a population member is in favor rather than opposed to the proposal, namely $\lambda = \pi_1/\pi_2$. The odds would be of interest if the question posed was the following:

Is it likely that those in favor outnumber those against by at least two to one?

In fact, this question is equivalent to the question

Is it likely that the proportion in favor is at least two thirds?

Rejecting the hypothesis $\pi_1 < 2/3$ is equivalent to rejecting the hypothesis $\lambda < 2$.

9.1.2 Probabilities and Success Rates

The second type of scientific problem which is considered in this chapter arises when an experimental process rather than a population is under investigation. The requirement is that there must be only two possible types of outcomes from the process. These may have any pair of a vast array of labels—*success* and *failure, defective* and *non-defective, yes* and *no,* and so on. Some illustrations are provided by the following examples.

Example 9.2 (Examining the Mendelian Theory of Genetic Inheritance). *Under the Mendelian theory of genetic inheritance, three quarters* *of the pods of pea plants are expected to be round and one quarter wrinkled. In a sample of* 7324 *plants,* 5474 *were found to be round. Are the data consistent with the expected ratio of* 3:1?

Example 9.3 (Examining the Rate of Cot Deaths). *The inexplicable death of apparently healthy babies is a worldwide phenomenon. It is officially referred to as the "Sudden Infant Death Syndrome" or* SIDS. *Tasmania has a rate which is significantly above the average of* 4.4 *per* 1,000 *births for other parts of Australia. Extensive research is being undertaken in Tasmania and much data have been collected. The following data come from two local government districts, for the years* 1975 *to* 1980 *inclusive.*

District	No. of live births (n)	No. of SIDS (x_1)
Hobart	3939	24
Green Ponds	66	0

(*Note that* SIDS *babies are included in the live birth numbers since the deaths did not occur at birth.*)

From these figures it can be seen that Green Ponds has an observed rate of zero per 1,000 *live births which is obviously below the national figure of* 4.4. *The observed rate in Hobart is* 6.1 *deaths per* 1,000 *live births, which is above the national figure. However, these figures take no account of sampling or chance variation. It is possible to observe no occurrences of an event in a series of trials even when there is a non-zero probability of the event occurring. What is the likely long term rate in Green Ponds? Is there strong evidence to suggest that the long term rate is lower than the national value of* 4.4?

What is the likely long term rate in Hobart? Can it be reasonably concluded that the region has a rate higher than the national average?

Example 9.4 (Custody of Children in Divorce Cases). *A committee receiving evidence on inconsistency of judges in the matter of which parent is granted custody of children in divorce cases was presented with the facts that, on average, the mother was granted custody in* 70% *of cases. However, one judge who was projected as having an anti-woman bias was reported to have granted custody to the father in* 18 *of his last* 30 *cases. Committee members raised a number of questions regarding the significance of this evidence. These questions include the following.*

1. *Could such a deviation from the average be explained as a chance happening?*

2. *How reliable are figures based on* 30 *cases?*

The model employed in the study of success rates or probabilities differs from that required for proportions in populations in that there is not a finite pool from which sample members are being selected. Hence, there is no reference to a population size.

The type of questions which arise are, however, of the same nature as those described above with reference to proportions. If the outcomes are labelled *success* and *failure* and π_0 is the probability of success, then the range of statistical questions are:

1. Are the data consistent with the assumption that the probability of success is π_0?

2. What is the best estimate of the probability of success?

3. Within what limits is the probability of success likely to lie?

4. Are the data consistent with the assumption that the probability of success is at least as large as π_0?

Additional forms of question could be added if the possible values for π are known to be limited to two possibilities, π_1 and π_2.

5. Which hypothesis, $\pi = \pi_1$ or $\pi = \pi_2$, is better supported by the data?

6. How much greater is the support for π_1 than π_2?

9.2 Statistical Analysis

9.2.1 Data

The variable under study must have only two possible outcomes. The data required for the analysis are the number of sample members in each of the two possible categories, denoted by x_1 and x_2. Note that since $x_1 + x_2$ is equal to the sample size (n), supplying values for x_1 and n or supplying values for x_2 and n is equivalent to supplying values for x_1 and x_2. Commonly, the data are summarized in a frequency table of the following form.

Outcome	1	2	Total
Frequency	x_1	x_2	n

When sampling from a population, the population size N is also required unless the population size is known or expected to be much larger than the sample size. (A commonly used rule-of-thumb is that a value for N is required if the population size is less than twenty times the sample size, n.)

9.2.2 The Statistical Model

The model presumed is either the Two-state population model (Section 6.2) in the study of a population, or the Binomial model (Section 6.1) in the study of an experimental process.

A basic requirement is for outcomes to fall into only one of two possible categories.

An essential requirement is that sample members should be independently selected. Lack of indepedence might be expected in two types of situations. The first is where the data comprise repeated measurements on the one process. In this case, the outcome at one time might be expected to bear a relation to the outcome at the next. For example, in an economic investigation, data formed by noting if an indicator was below or above average on 50 consecutive days are unlikely to meet the independence requirement.

The second situation in which dependence between responses may occur is where there is a spatial relationship between sample members and the response from one sample member affects that from a neighbor. For example, given that data are to be collected on the incidence of an infectious disease in a community, the presence of the disease in a person is likely to increase the chances of a person from the same home or work place being afflicted; plants competing for nutrients, light or water are likely to produce related responses.

The assistance of a consultant statistician is recommended where correlations in either time or space are suspected. Detecting lack of independence is important. To use the methods in this chapter when the independence condition is violated could lead to serious error in the statistical conclusions.

9.2.3 Computational Considerations

Form of data: The data, as collected, comprise a list of responses from the n sample members. The list may be in the form of coded numbers, perhaps 0 and 1; or letters, perhaps y for yes and n for no; or words, *old* and *young*. Sometimes the binary variable must be constructed. For example, a list of weights may require conversion to *above average* and *below average;* a categorical variable with multiple states may require two or more states to be combined to reduce the number of categories to two. On other occasions, only a subset of the sample may be required. For example, if the options are *agree, disagree* and *undecided,* a reduced sample may be formed from which the *undecideds* are excluded.

Statistical packages are generally well adapted to both constructing variables and to counting the number of sample members in each class.

9.2.4 Methods Applicable under the Binomial Model

1. *Point estimation.* Given that there are x_1 successes in n trials or outcomes, then $p_1 = x_1/n$ which is the proportion of successes in the sample, provides the best estimator of the expected probability of successes, π_1.

2. *Interval estimation.* Inferential methods require calculations based on the sampling distribution of p_1 or $x_1 = np_1$ which has a Binomial distribution. However, in practice, methods which rely on the construction of a confidence interval for π require either an approximation to the Binomial distribution or an alternative mathematical form.

The formulae to be used for the construction of confidence intervals depend on the values taken by n and x_1. Computationally simplest is the use of the Normal approximation to the Binomial. As explained in Section 7.1.4, this approximation provides a good approximation only under restricted conditions. The recommendation introduced in that section requires that the value of n should at least equal to $n_{\min}$ as defined in (9.2.1) for the Normal approximation to apply.

$$n_{\min} = \frac{2}{3\{p_1(1 - p_1)\}^2} \tag{9.2.1}$$

where p_1 is the proportion of successes observed in the sample. If the sample size does exceed $n_{\min}$, then the 95% confidence interval for π_1 and, more generally, a $100(1 - \alpha)\%$ confidence interval can be computed from Table 9.2.1.

Table 9.2.1. Confidence Limits for π When n Exceeds $n_{\min}$ as Defined in (9.2.1)

Level of Confidence	Lower Limit	Upper Limit
95% Confidence limits	$p_1 - 1.96s_p$	$p_1 + 1.96s_p$
$100(1 - \alpha)\%$ confidence interval	$p_1 - z_{\frac{1}{2}\alpha}s_p$	$p_1 + z_{\frac{1}{2}\alpha}s_p$

where p_1 is the observed proportion of successes;
$s_p = \sqrt{\{p_1(1 - p_1)/n\}}$.

Illustration. The method has application in Example 9.4. The requirement in this case is to provide a confidence interval for the expected probability that the father is granted custody in a divorce case.

The data in this experiment comprise the sample size ($n = 30$) and the number of successes in the form of the number of cases in which the father was granted custody ($x_1 = 18$).

The first step is to check if the sample size reaches the minimum value prescribed by (9.2.1) for the Normal approximation to the Binomial to have application. In fact, $n_{\min}$ is calculated to be 11.6 which is less than the sample size of 30. Hence, the Normal approximation is reasonable.

Use of the formulae provided in Table 9.2.1 yields values for p_1 and s_p of 0.60 and 0.0894, respectively and hence, 95% confidence limits for the expected probability that the father will be granted custody of 0.42 and 0.78. In other words, the father is expected to be granted custody in between 42% and 78% of cases. This is equivalent to limits of 22% and 58% for the percentage of cases in which the mother is expected to be granted custody.

The reliability of the findings can be judged by the width of the confidence interval. It is possible to estimate the sample size required to achieve a $100(1-\alpha)\%$ confidence interval of a prechosen size. If the width is to be no greater than d, the minimum sample size required is

$$n = 4(z_{\frac{1}{2}\alpha})^2 p_1(1-p_1)/d^2.$$

Because p_1 is necessarily unknown prior to the study, either an estimate may be used or a value of 0.5 may be employed since this provides an upper limit on the value of n once the value of d is fixed. For example, if a 95% confidence interval is assumed, d is set equal to 0.1 and p_1 takes the value 0.5, the minimum sample size required is approximately 400.

Where the sample size, n, is less than the minimum specified by (9.2.1) it is necessary to use the limits provided in Table 9.2.2.

Table 9.2.2. Confidence Limits for π When n Does Not Exceed $n_{\min}$

Level of Confidence	Lower Limit	Upper Limit
95% C.I.	$\dfrac{\nu_2}{\nu_2 + \nu_1 F_{0.025}(\nu_1, \nu_2)}$	$\dfrac{\nu_1' F_{0.025}(\nu_1', \nu_2')}{\nu_2' + \nu_1' F_{0.025}(\nu_1', \nu_2')}$
$100(1-\alpha)\%$ C.I.	$\dfrac{\nu_2}{\nu_2 + \nu_1 F_{\frac{1}{2}\alpha}(\nu_1, \nu_2)}$	$\dfrac{\nu_1' F_{\frac{1}{2}\alpha}(\nu_1', \nu_2')}{\nu_2' + \nu_1' F_{\frac{1}{2}\alpha}(\nu_1', \nu_2')}$

where $\nu_1 = 2(n - x_1 + 1)$ and $\nu_2 = 2x_1$
$\nu_1' = 2(x_1 + 1)$ and $\nu_2' = 2(n - x_1)$.

Note: it is possible for either ν_2 or ν_2' to be zero in which case the value of F is set equal to one.

Illustration. Consider the investigation in Example 9.3. The initial aim is to compute 95% confidence intervals for the rates per thousand of deaths from SIDS for both Green Ponds and Hobart local government areas. The connection between a rate per thousand (r) and the probability of a baby dying from SIDS (π) is $r = 1000\pi$. The formulae available allow for confidence limits for π. Multiplying these by 1,000 gives confidence limits for r.

For the Green Ponds data, n is 66 and x_1 is 0. Attempting to calculate a value for $n_{\min}$ from (9.2.1) indicates a value of infinity. Hence, the Normal approximation to the Binomial is not applicable.

Turning to Table 9.2.2, the following values are obtained for parameters.

$$\nu_1 = 2(66 - 0 + 1) = 134 \qquad \nu_2 = 2 \times 0 = 0$$
$$\nu_1' = 2(0 + 1) = 2 \qquad \nu_2' = 2(66 - 0) = 132$$

Since $\nu_2 = 0$, it follows from the footnote to Table 9.2.2 that $F_{0.025}(134, 0) = 1$. From a statistical package or tables (see Section 7.6.4), $F_{0.025}(2, 132) = 3.794$. Hence, from the formulae given in Table 9.2.2 for lower and upper 95% confidence limits,

$$\text{lower limit} = \frac{0}{0 + 134 \times 1} = 0$$
$$\text{upper limit} = \frac{2 \times 3.794}{132 + 2 \times 3.794} = 0.054.$$

Converted to rates per thousand, these become 0 and 54.

Applying the same procedure to the Hobart data, the value of $n_{\min}$ is computed to be 18,179 and hence, the Normal approximation is not acceptable. Confidence limits on the probability of a baby dying from SIDS based on the formulae in Table 9.2.2 are 0.0039 and 0.0090. This leads to limits on the rate per thousand of 3.9 and 9.0.

In theory, comparison of the Green Ponds or Hobart results with the national average of 4.4 deaths per thousand live births should take into account the sampling variation inherent in the national figure. However, that figure is based on a sample size in excess of one million and hence the sampling variability would be extremely small. For both local government areas, the national figure falls within the confidence interval and, on that basis, both could be said to be consistent with the national average. However, the wide range of values included in the confidence interval for Green Ponds reduces the practical value of such a statement.

3. *Hypothesis testing.* It is generally wise to construct a confidence interval for π even when the aim is to examine an hypothesized value or range of values for π. The width of the interval conveys something of the level of confidence which can be placed in the findings. Thus, to find that a value of $\pi = 0.5$ is consistent with the data when the 95% confidence interval extends from 0.4 to 0.9 might suggest that the finding has limited value and perhaps more data are required. Equally, to reject the value of $\pi = 0.5$ when

the 95% confidence interval is 0.485 to 0.495 might be misleading in that the value of 0.5 could possibly be *close enough* for practical purposes. Without reference to the confidence interval, the investigator and those reading the report of the study are not in a position to make these judgments.

However, there do occasionally arise situations in which the direct application of hypothesis testing is appropriate.

Three variations on the null hypotheses are presented below and the methodology appropriate for each variation is given. In each case, π_0 is presumed to be a numerical value specified by the investigator, x is the statistic denoting number of successes in n trials when the probability of success is π_0. Furthermore, x_1 is the observed number of successes in the sample and $p_1 = x_1/n$ is the sample proportion of successes.

(a) Null hypothesis: $\pi \geq \pi_0$ Alternative hypothesis: $\pi < \pi_0$

If $p_1 \geq \pi_0$ there exists a value of π in the null hypothesis, namely $\pi = p_1$, which is in perfect agreement with the data and hence, the null hypothesis is acceptable. Otherwise it is necessary to use the Binomial distribution with parameters π_0 and n, to compute the p-value as

$$p = \Pr(x \leq x_1)$$

since it is decreasing values of x_1 which reflect lessening agreement with the null hypothesis.

See Section 7.1.4 for methods for computing p.

(b) Null hypothesis: $\pi \leq \pi_0$ Alternative hypothesis: $\pi > \pi_0$

If $p_1 \leq \pi_0$ there exists a value of π in the null hypothesis, namely $\pi = p_1$, which is in perfect agreement with the data and hence, the null hypothesis is acceptable. Otherwise it is necessary to use the Binomial distribution with parameters π_0 and n, to compute the p-value as

$$p = \Pr(x \geq x_1)$$

since it is increasing values of x_1 which reflect lessening agreement with the null hypothesis.

See Section 7.1.4 for methods of computing p.

(c) Null hypothesis: $\pi = \pi_0$ Alternative hypothesis: $\pi \neq \pi_0$

Under the null hypothesis the most likely outcome is $x_1 = n\pi_0$. Where the Normal approximation to the Binomial can be applied, i.e. when n exceeds n_{min} as defined in (9.2.1), the symmetry of the Normal distribution establishes that p can be computed from (9.2.2).

$$p = \Pr(|x - n\pi_0| \geq |x_1 - n\pi_0|). \tag{9.2.2}$$

Computational details for the application of (9.2.2) are provided in Section 7.1.4.

Where the sample size is too small for the application of the Normal approximation to the Binomial, calculations must be based directly on the Binomial probabilities

$$\pi_{st}(x) = \binom{n}{x}\pi_0^x(1-\pi_0)^{n-x} \quad \text{for} \quad x = 0, 1, 2, \ldots, n.$$

The Significance approach is applied in its basic form with p being determined from (9.2.3).

$$p = \Sigma\pi_{st}(x) \tag{9.2.3}$$

where the summation is over all values of x from zero to n for which $\pi_{st}(x) \leq \pi_{st}(x_1)$.

For the situation where n does not exceed $n_{\min}$, it is necessary to resort to computation of the p-value from first principles. The steps are explained below under the heading **Computations — Small samples**.

Illustration — Normal approximation to the Binomial: The Mendelian experiment reported in Example 9.2 could reasonably be seen as a problem in which the hypothesis testing approach might be used. In this case there is a single hypothesis of interest with no specific alternative stated. If the ratio is not exactly 3:1 then the theory is incorrect. The information provided is that $n = 7324$ and $x_1 = 5474$. From (9.2.1), the value of $n_{\min}$ is 18.7 and hence, the Normal approximation to the Binomial is applicable.

The hypothesis is that the expected proportion of round seeds is 3/4, i.e. $\pi_0 = 3/4$ is the null hypothesis and the alternative is that $\pi_0 \neq 3/4$. The computation required is

$$p = \Pr(|x - n\pi_0| \geq |x_1 - n\pi_0|)$$

where $n = 7324$, $\pi_0 = 0.75$, $x_1 = 5474$.

Reference to Table 7.1.1 indicates that the corresponding probability for the standard Normal distribution is

$$p \cong 2 \times \Pr\left(z \leq \frac{5474.5 - 5493}{\sqrt{1373.25}}\right) = 2 \times \Pr(z \leq -0.50).$$

Evaluation of this probability either by reference to a statistical package or Normal distribution tables yields an approximate value of 0.6 for p. Since the p-value is above 0.05, the data can be said to be consistent with the assumption of a 3:1 ratio for round to wrinkled seeds.

Computations — small samples: To compute a value using (9.2.3), note that the Binomial distribution has the property that the probabilities rise smoothly to a single peak and then fall smoothly from that point onwards. Thus, there will only be one or two intervals in the tails of the distribution for which probabilities will have to be computed and these may be obtained using available sources of cumulative probabilities, for example, statistical packages.

To determine the relevant interval or intervals requires an examination of at least a portion of the set of probabilities for the Binomial distribution with parameters n and π_0. The search is aided by the fact that the maximum probability occurs close to the value $n\pi_0$.

By way of illustration of the process, consider the situation in which a production line is assumed to produce, on average, 1% defective, but that in a batch of 100, 3 are observed to be defective. Could this indicate a problem with the production line? The values given are

$$n = 100, \qquad \pi_0 = 0.01, \qquad x_1 = 3.$$

Checking the acceptability of the Normal approximation to the Binomial using the value $p_1 = 3/100$, (9.2.1) provides the value of 787 for $n_{\min}$.

Since the actual sample size is only 100, the Normal approximation is judged inadequate. The maximum probability for a Binomial distribution with $n = 100$ and $\pi_0 = 0.01$ will occur at about 100×0.01 or $x = 1$. Hence, probabilities can be expected to decline beyond the observed value of $x = 3$. This suggests the probabilities in the range 0 to 3 should be examined. For interest, the value of $x = 4$ is also shown.

x_0	0	1	2	3	4
$\Pr(x = x_0)$	0.366	0.370	0.185	0.061	0.015

The probability of the observed number of defectives, namely 3, is 0.061. The p-value is the sum of all probabilities with values equal to or less than 0.061. Given that probabilities decline for values of x beyond 3 and no probabilities for lesser values of x fall below the yardstick, the p-value should be computed as

$$p = \Pr(x \geq 3) = \Pr(x = 3) + \Pr(x = 4) + \cdots + \Pr(x = 100).$$

Such a computation could be performed by noting that

$$\begin{aligned} p &= \Pr(x \geq 3) \\ &= 1 - \Pr(x < 3) \\ &= 1 - [\Pr(x = 0) + \Pr(x = 1) + \Pr(x = 2)] \\ &= 1 - (0.366 + 0.370 + 0.185) \\ &\cong 0.08. \end{aligned}$$

4. *Decision making.* Where the experimenter has two or more competing hypotheses and there is no reason to select one of the possibilities as the null hypothesis, the aim may be to make a decision as to which is the best choice. In its most general form, this is a problem requiring the application of decision theoretic methods and as such is not covered in this book. Rather than decision making in this situation, the aim may be to assign relative degrees of belief to the different hypotheses, possibly incorporating beliefs held prior to the current experiment. In this case, the Bayesian approach should be considered.

9.2.5 Methods Applicable Under the Two-State Population Model

The situation is one in which there is a population of size N which is partitioned into two groups, conveniently labelled *Group* 1 and *Group* 2 which contain proportions π_1 and π_2 of the population, respectively, where $\pi_1 + \pi_2 = 1$. Interest is in the value of π_1 (or equivalently π_2 or π_1/π_2). A sample of size n is available and the statistic used in inferential methods is the number of Group 1 members in the sample (x_1). As stated in Section 6.2, the sampling distribution of x_1 is known as the *Hypergeometric distribution.*

For most practical applications, the Hypergeometric distribution is well approximated by the Binomial distribution and hence, the methodology of Section 9.2.4 may be employed. For this approximation to apply, the population size N should be at least twenty times the value of the sample size n. Interpretation of the methods as described in Section 9.2.4 requires only the substitution of references to the *probability of success* by *proportion of Group 1 members* in the population, and substitution of references to *number of successes* by *number of sample members from Group 1.*

Where the sample size (n) exceeds twenty times the population size (N), alternative methodology must be employed. Some of that methodology is presented in this section. The methods which are presented cover most cases in which statistical analysis is likely to be of value. Where the conditions are not met, seek advice from a consultant statistician.

1. *Point estimation.* Given that there are x_1 members from Group 1 in the sample of n members, then $p_1 = x_1/n$, which is the proportion of Group 1 members in the sample, provides the best estimator of the expected proportion of Group 1 members in the population, π_1.

2. *Interval estimation.* The only method considered relies on the Normal approximation to the Hypergeometric distribution. The approximation is reasonable if the sample size n exceeds the value of $n_{\min}$ defined in (9.2.4).

$$n_{\min} = \frac{2}{3\{p_1(1-p_1)(N-n)/(N-1)\}^2} \qquad (9.2.4)$$

where p_1 is the proportion of Group 1 members observed in the sample. If the sample size does exceed (9.2.4), then the 95% confidence interval for π_1 and, more generally, a $100(1-\alpha)\%$ confidence interval can be computed from Table 9.2.3.

Table 9.2.3. Confidence Limits for π When n Exceeds $n_{\min}$ as Defined in (9.2.4)

Level of Confidence	Lower Limit	Upper Limit
95% Confidence limits	$p_1 - 1.96s_p$	$p_1 + 1.96s_p$
$100(1-\alpha)\%$ confidence interval	$p_1 - z_{\frac{1}{2}\alpha}s_p$	$p_1 + z_{\frac{1}{2}\alpha}s_p$

where p_1 is the observed proportion of successes; and
$s_p = \sqrt{p_1(1-p_1)\left(\frac{N-n}{N-1}\right)/n}$.

Note that the limits for π_1 as given in Table 9.2.3 may be readily converted into limits for π_2 or the ratio π_1/π_2 according to the following formulae. If L_1 and L_2 are the limits of a confidence interval for π_1, then

(a) the corresponding limits for π_2 are $1 - L_2$ and $1 - L_1$; and

(b) the corresponding limits for π_1/π_2 are $L_1/(1-L_1)$ and $L_2/(1-L_2)$.

Illustration. The method can be demonstrated using the information in Example 9.1. There is a problem in determining the population size. While the estimated size of the community is 75,000, it is apparent from the sample that not every person is willing or able to give a clear preference. About 1% of the sample were undecided. Assuming that the percentage who were undecided in the population equals the percentage who were undecided in the sample, the value of N would be 74120. This value of N is clearly less than twenty times the sample size of 7,412. Hence, the Binomial approximation cannot be applied. From (9.2.4), $n_{\min}$ is computed to be 13 and the Normal approximation may be employed.

Using the formulae in Table 9.2.3, the 95% confidence interval for the proportion of people in the population opposing the hotel development is

$$\frac{3928}{7412} - 1.96s_p \quad \text{to} \quad \frac{3928}{7412} + 1.96s_p$$

where

$$s_p = \sqrt{\frac{3928}{7412} \times \frac{3484}{7412} \times \frac{74120 - 7412}{74120 - 1}/7412} = 0.0055.$$

This yields numerical limits of 0.52 and 0.54.

Such limits are based on the assumption of a random selection of sample members from the population and interviewing or opinion collecting methods which were objective. Given that the poll was conducted by a group whose members were opposed to the development, this aspect should be carefully examined in establishing the worth of the findings.

The confidence interval excludes any values which suggest majority population support for the hotel development, i.e. values of π_1 less than 0.50. Assuming the conditions for valid application of the statistical analysis were met, the opponents of the hotel development could properly claim that the statistical analyses provides evidence opposing the belief that the hotel development has majority support. However, from an objective viewpoint, one might say that a realistic assessment is that the proposal has split the population 'down the middle' rather than supplied the opponents with an overwhelming victory.

The example also provides a cautionary note on the size of sample needed to obtain accurate estimation in such cases. It has taken a sample size of seven and one half thousand to obtain an interval with a width of two percentage points. Generally, it is neither practical nor sensible to seek such precision through huge samples. The other possibility is to vary the model by changing the method of sample selection. This point is discussed briefly in Section 13.4.

3. *Hypothesis testing.* If Hypergeometric probabilities are available, then the same approach as that described in Section 9.2.4 based on Binomial probabilities may be used. If the Normal approximation to the Hypergeometric is satisfied, i.e. (9.2.4) is met, a test of the hypothesis $\pi_1 = \pi_0$ versus the alternative that $\pi_1 \neq \pi_0$ may be based on the p-value computed from (9.2.5) where the probability is determined from the standard Normal distribution using one of the methods listed in Section 7.3.4.

$$\begin{aligned} p &= 2 \times \Pr\left[z \leq \frac{x_1 + \frac{1}{2} - n\pi_0}{\sqrt{\{x_1(n-x_1)(N-n)/n(N-1)\}}}\right] \quad \text{if } x_1 < n\pi_0 \\ &= 2 \times \Pr\left[z \geq \frac{x_1 - \frac{1}{2} + n\pi_0}{\sqrt{\{x_1(n-x_1)(N-n)/n(N-1)\}}}\right] \quad \text{if } x_1 \geq n\pi_0. \end{aligned} \tag{9.2.5}$$

Problems

9.1 *Example 9.4 — Normal approximation to the Binomial*

Reproduce the calculations in Section 9.2.4 based on data in Example 9.4 by (i) using (9.2.1) to establish that the Normal approximation to the Binomial may be applied in the computation of confidence limits for the expected probability that the father will be given custody in the divorce case, and (ii) employing the relevant formulae in Table 9.2.1 to compute 95% confidence limits for the expected probability that the father will be given custody.

9.2 *Example 9.3 — exact confidence limits*

Reproduce the calculations in Section 9.2.4 based on data in Example 9.3 by (i) using (9.2.1) to establish that the Normal approximation to the Binomial does not provide an adequate approximation in the computation of confidence limits for the rate of deaths from SIDS, and (ii) employing the relevant formulae in Table 9.2.2 to compute 95% confidence limits for the expected rate of deaths from SIDS.

9.3 *Survey of national park visitors*

Visitors to selected national parks were classified into three groups for the purpose of determining the level of user satisfaction with the facilities and services offered. The three groups were identified as *Hikers*, *Campers* and *Others* (mainly picnickers). Of the visiting parties who entered one of the parks in the six month survey period, a sample was chosen and one member in each selected party was given a questionnaire and asked to return it to the ranger on completion. Among the questions asked were the following.

1. *Were the facilities offered adequate for your purpose?*
2. *Was the information supplied within the park adequate for your needs?*

The following tables of frequencies provide a summary of the information given in answer to these questions by those parties who supplied answers to both questions.

	Hikers (120)		Campers (58)		Others (411)	
	Facilities		Facilities		Facilities	
Info.	Yes	No	Yes	No	Yes	No
Yes	82	26	28	23	255	115
No	2	10	1	6	33	8

(a) Determine the proportions of sample members in each of the groups who found the information provided in the parks adequate.

(b) Use (9.2.1) to establish that the sample sizes for the *Hikers* and the *Other* groups are sufficiently large to allow the Normal approximation to the Binomial to be used while the minimum size is not met for the *Campers*.

(c) Hence, for both the *Hikers* and the *Other* groups, use the formulae in Table 9.2.1 to calculate 95% confidence intervals for the expected proportions of visitors who find the information provided adequate. (Note that the confidence interval for the smaller group, the *Hikers*, is about 1.6 times the width of the interval for the *Others* group. This reflects the difference in sample size.)

9.4 *Normal approximation versus exact distribution* (*continuation of* 9.3)

For the *Campers*, the minimum sample size for the application of the Normal approximation to the Binomial is computed from (9.2.1) to be 60. The actual sample size is 58. Therefore, the approximation is not recommended. What would be the error if the Normal approximation were employed?

(a) Compute approximate 95% confidence limits using the Normal approximation to the Binomial for the expected proportion of *Campers* who regard the information supplied as adequate using formulae in Table 9.2.1.

(b) Compute exact confidence limits using formulae in Table 9.2.2. Note the asymmetry of the latter limits about the sample proportion.

9.5 *Hypothesis testing and likelihood ratios*

Two genetic models are proposed. The first hypothesizes an expected ratio of success to failure of 1:1 and the second a ratio of 9:7. In 100 independent trials, with constant probability of success at all trials, there are 55 successes observed.

(a) Is this result consistent with both hypotheses? (Use (9.2.1) to establish that the Normal approximation to the Binomial may be employed and (9.2.2) to compute a p-value in each case.)

(b) Based on the Binomial model, if π denotes the probability of success, the *Likelihood* that π takes the value π_0 when there are x successes in n trials is given by

$$L(\pi_0) = \binom{n}{x} \pi_0^x (1 - \pi_0)^{n-x}$$

Given an outcome of 55 successes in 100 trials, compute the likelihoods that π takes the values 1/2 and 9/16. Hence, compute the *likelihood ratio*, $L(1/2)/L(9/16)$ to determine the relative likelihoods of the two hypotheses.

9.6 *Unusual happenings — chance or fate?*

Lotto is a popular form of gambling in Australia. It comprises the drawing of six numbered balls from a barrel with the winner(s) being the person(s) who correctly predicts the six numbers drawn. It was recently announced that winning tickets in consecutive *Lotto* draws came from different households in one street of 10 houses. Is this merely coincidence or does it signify some special powers of the householders in the street to foresee results of Lotto draws?

There are two aspects to this question. Consider firstly the statistical analysis based on the following model:

1. Each household in the street entered one ticket in each *Lotto* draw.
2. The choice of numbered combinations on tickets were independent from house to house.
3. At each *Lotto* draw, six selections of numbers were made from a barrel without replacement of numbers previously drawn. The barrel contained the numbers $1, 2, \ldots, 40$ at the beginning of the draw.

   (a) Based on this model, use the rules of probability given in Chapter 5 to establish that the probability of a ticket containing the winning numbers is $1/\binom{40}{6}$ (note that the ordering in which the balls are drawn is unimportant).

   (b) Hence, use the relevant method described in the *Hypothesis testing* subsection of Section 9.2.4 to determine that 2 successes in 10 attempts provides strong evidence to reject the model and hence, the hypothesis that the two wins in the one street is merely a chance happening.

   (c) While the statistical argument and conclusions are correct **given the model proposed**, there is a flaw in the reasoning. Can you explain what it is?

10

Model and Data Checking

10.1 Sources of Errors in Models and Data

10.1.1 Errors in Data

Even the most careful and thorough experimental practice cannot eliminate the possibility of an error appearing in the data collected from an experiment. It may arise because of a simple mistake in recording or transcribing a result, be due to a fault in a recording device, or perhaps result from an undetected fault in an experimental unit or an unknown affliction suffered by a subject used in the experiment. In many circumstances, it is possible to detect serious errors in the data. In other circumstances, it can be determined in advance that the nature of the experimental design will make detection either impossible or difficult.

The process of data checking with the aim of detecting errors is discussed in Section 10.4. It is an important part of the process of statistical analysis since the data provide the factual information about the experimental set-up and hence, the base for judging the adequacy of a model of that set-up.

10.1.2 Errors in Statistical Models

As described in Chapter 3, a statistical model has three components—the hypothesis deriving from the experimenter's question, sampling assumptions, and statistical assumptions. Ideally, the only possible source of error is the hypothesis deriving from the experimenter's question. If such is the case, any conclusion regarding the acceptability or otherwise of the model can be directly transferred to a conclusion concerning the experimental hypothesis.

In reality, there is usually uncertainty about other components of the model and the possibility of errors in the data. Hence, there is the possibility an incorrect conclusion will be reached from the statistical analysis because of a mistake which is unconnected with the question of interest to the experimenter.

To minimize the risk of a serious mistake arising from incorrect statistical assumptions in a model, it is wise to apply whatever procedures are practical to check the acceptability of those assumptions in the statistical model which do not directly relate to the questions raised by the experimenter. These checking procedures range from simple visual checking by the experimenter to formal statistical methods. Basic approaches for this model checking phase of the analysis are described in Section 10.2.

10.1.3 The Role of Computers

A significant advancement in the application of Statistics in scientific experimentation is the improved facilities for model and data examination. This improvement has been made possible by the wide availability of statistical computing packages.

The role of these packages is to provide the means of (i) presenting and summarizing data in simple visual forms which highlight any obviously incorrect features; and (ii) applying Data Analytic and Inferential methods of Statistics designed to detect specific forms of errors in the model and the data. Reference will be made to specific applications when individual methods are presented in later chapters. Some general approaches are discussed in following sections of this chapter.

10.1.4 Statistical Experience

It is impossible to avoid the fact that the quality of model and data checking which an analyst is capable of providing depends on the experience and expertise of that analyst. This is true both in general statistical terms and in respect of the particular area of scientific study in which the experimental problem falls.

To a limited extent the process can be quantified and automated. **For those who lack experience it is wise to follow the guidelines which accompany specific methods and to employ basic procedures suggested in Section 10.4 for data checking.** Even so, it must be recognized that such guidelines are not a substitute for experience. **If a consultant statistician or an experienced researcher is available to assist you, their wisdom and experience should take precedence over general guidelines.**

10.1.5 The Choice of Methods

Because of the subjective elements in the process of model and data checking and gaps in statistical research, universal agreement does not exist among statisticians as to the best set of procedures to employ in a given problem. There may be

(a) differing emphasis placed on information supplied about the experimental set-up;

(b) variation in the choice of statistical procedures; and/or

(c) disagreement as to the relative weights to be attached to the various findings.

Nevertheless, it is to be expected that the varying approaches would all lead to the detection of serious errors in a model and that, where lesser errors exist, different decisions as to model selection would not produce great differences in experimental conclusions.

The suggested guidelines which relate to the selection and application of model and data checking procedures are just that—*guidelines*. They are the result of an extensive review of statistical literature and of research by the author. They are constrained by the fact that any recommended method should be easily accessible to users and, in particular, they are constrained to what is readily available in most statistical packages.

Even though there are subjective elements in the suggested procedures, they are presented in this book in a precise form and with, hopefully, clear and unambiguous interpretations. Readers should feel free to vary the suggested application and/or interpretation of procedures in the light of their own experience or that of more experienced colleagues or statistical consultants.

10.2 Detecting Errors in Statistical Models

10.2.1 Strategies in Model Construction

To some extent, the approach to checking models for errors is dependent on the approach adopted in model construction. Given the experimental information available, the two extremes in the model construction process are to create (i) the most detailed and complex model which might be consistent with experimental knowledge; or (ii) the most conservative model in which the aim is to include the minimal number of assumptions necessary for meaningful statistical analysis. In the former case it would generally follow that extensive model checking is essential to ensure that gross model violations have not occurred. In the latter case, minimal model checking may suffice.

In practice, while the use of models with minimal assumptions has an appeal, there arise many situations of practical importance in which suitable methodology based on such models does not exist. Consequently, scientists who wish to use methods of Inferential Statistics must become familiar with standard practices in model checking and the paths to follow when the checking procedure suggests possible flaws in the model.

A confounding factor is the increasing role of Data analysis in the model construction process. Often this has the effect of blurring the distinction between model construction and model checking. Particularly where complex experiments are constructed, it is necessary for the construction of statistical models to rely on Data Analytic techniques. However, for those being introduced to the process of statistical analysis, it is desirable for the construction and checking phases to be kept distinct in order to clearly identify the building blocks of each.

10.2.2 Errors in the Sampling Assumptions

Incorrect sampling assumptions may arise because the model of sample selection is not of a form which meets the condition of random sampling from a population or the condition of independence of outcomes from a repetitive process. Such violations are difficult to detect unless there is some indication of the manner in which the assumption has been violated.

As a general rule, it is easier to establish the failure of the assumption through a critical examination of the method of sample selection than through formal statistical analysis. While there are statistical tests available, only those which are designed to detect specific alternatives to randomness or independence are of sufficient power to have practical worth. Such tests tend to have specialized application and consequently are not considered in this book.

There are important areas of application where statistical models include assumptions of dependence between outcomes. This is particularly so where indicators are studied over time. Not uncommonly, the response at one point in time is related to the outcome at the next point in time. This is illustrated by the examination of stock market indicators or sales figures of a company.

In agriculture, ecology and geography, applications commonly arise in which there are expected to be spatial relations between the responses. Neighboring plants or animals may compete for resources, leading to the fitter and stronger developing at the expense of the weaker; infectious diseases tend to arise in clumps rather than in a random pattern; faster economic growth in one area will tend to attract population from neighboring areas.

Attempting to quantify the structure of the relationship between responses generally demands complex modeling and the checks on the adequacy of such models is equally difficult. These are specialized areas not covered in this book.

10.2.3 Checking for Errors in Statistical Assumptions

In most methods which appear in subsequent chapters in this book, checking of statistical assumptions in models will be an integral part of the methods. In this subsection, the basic elements to be checked are introduced. Checking procedures are presented in Section 10.3.

1. *Component equations.* An essential feature of any model used in Infer-

ential Statistics is its inability to predict with certainty the responses of individual outcomes from an experiment. In all but the simplest model, there are two types of elements which contribute to a particular response and to the variation between responses. One group of contributors are identifiable—if two sample members received different treatments or came from different populations this is an identifiable source of difference in response. The other group is unidentifiable or at least unexplained in the model.

Where the response is numeric and measured on a scale, it is common practice to include in the model an equation which defines the manner in which the identified and unidentified sources are assumed to contribute to the observed responses.

The simplest equation is defined by (10.2.1) and applies when considering a population in which none of the variation in responses between sample members is explained.

$$\text{RESPONSE} = \begin{matrix}\text{POPULATION}\\ \text{MEAN}\end{matrix} + \text{DEVIATION}. \tag{10.2.1}$$

The *population mean* is the identified component of the response and the *deviation* is the collective contribution from all those factors which cause an individual population member to have a response which is different from the population average.

Such an equation is conveniently referred to as a *component equation* and has the general form represented by (10.2.2).

$$\text{RESPONSE} = \begin{matrix}\text{IDENTIFIED}\\ \text{COMPONENT}\end{matrix} + \begin{matrix}\text{UNIDENTIFIED}\\ \text{COMPONENT.}\end{matrix} \tag{10.2.2}$$

Different component equations are formed by introducing structure into the *identified component.* By way of illustration, consider an agricultural experiment in which the weight of a corn cob is presumed to depend on (i) the variety of corn, (ii) the temperature in which it is growing and (iii) other *unidentified* components. If the effects of the varietal and temperature factors are additive, the *component equation* is described by (10.2.3).

$$\begin{matrix}\text{WEIGHT} & = & \begin{matrix}\text{VARIETAL}\\ \text{COMPONENT}\end{matrix} & + & \begin{matrix}\text{TEMPERATURE}\\ \text{COMPONENT}\end{matrix} & + & \begin{matrix}\text{UNIDENTIFIED}\\ \text{COMPONENT}\end{matrix}\\ y & = & c_1 & + & c_2 & + & c_3\end{matrix} \tag{10.2.3}$$

Equation (10.2.4) is employed in Example 18.1 and portrays the fact that a student's performance in a university mathematics examination (y) is assumed to depend on two components. It is related in a linear fashion to high school performance in mathematics (x) with a contribution from *other sources* (e) added on. The equation establishes a linear *trend* for the relation between university and high school mathematics performance, but with individual performances deviating from the trend line by an unexplained amount e.

$$y = A + Bx + e. \tag{10.2.4}$$

In the process of model checking, there are two aspects related to component equations which must be examined. One is the structure of component equations themselves and the other is the characterization of the pattern of variation in the unidentified component.

The examination of the structure of the component equation must assess the acceptability of

(i) the *additivity* assumption in (10.2.2) which specifies that the response is the *sum* of the two components; and

(ii) the assumed form of the identified component which is represented in (10.2.3) by the sum of the varietal and temperature components and in (10.2.4) by the mathematical equation defining the trend line.

Methods for checking the structure of the component equation are presented in Section 10.3.

2. *Distributional assumptions.* The characterization of the pattern of variation in the unidentified component is contained in the *distributional assumptions* in the model. Where a particular mathematical form is presumed for a distributional assumption, there is generally a need to assess the adequacy of the assumption. By far the most commonly assumed form is the Normal distribution and it is the one form for which a specific checking procedure is provided in Section 10.3.

Distribution-free procedures, despite their name, commonly require models which are not strictly 'distribution-free'; they are merely free of the specification of a specific mathematical form. There may be a requirement that a distribution is symmetric, or where two or more treatments are to be compared, there may be an assumption that the distributions under different treatments are identical even though the precise form is not stated.

10.2.4 Residuals and Fitted Values

The examination of either the structure of the component equation or distributional assumptions in a model generally requires the division of each response into the component parts—an identified component and an unidentified component. While the information is not available to allow exact values of the components to be computed from the data, it is possible to provide estimates of the components. The estimate of the identified component is termed the *fitted value* and the estimate of the unidentified component is termed the *residual.* The manner in which these estimates are defined ensures that the sum of the estimates is equal to the response, i.e., that (10.2.5) holds.

$$\text{RESPONSE} = \frac{\text{FITTED}}{\text{VALUE}} + \text{RESIDUAL}. \tag{10.2.5}$$

It is beyond the scope of this book to provide the statistical theory necessary for the determination of formulae used in the construction of fitted values and residuals. However, in many simple cases the theory leads to estimators which would be predicted by intuition. An illustration of the construction of fitted values and residuals is provided below which embodies the ideas.

Illustration. There are 22 observations in Table 1.3.1 for the *pre-race Sodium* variable. In this case, the data come from two groups which may differ in mean value. Hence, (10.2.6) defines a plausible component equation.

$$y_{ij} = G_i + e_{ij} \tag{10.2.6}$$

where

y_{ij} is the observation on pre-race Sodium for the jth sample member ($j = 1, 2, \ldots, 11$) from the ith group ($i = 1, 2$) as shown in Table 10.2.1;

G_1 **and** G_2 are population parameters which are the average pre-race Sodium levels expected for persons who are untrained ($i = 1$) and trained ($i = 2$), respectively; and

e_{ij} is the deviation of the observation of runner j in group i from the expected level (G_i) and is due to unexplained sources.

Since there is no identified source of variation between observations from the same group, the fitted values for all sample members *within* a group must be identical. The statistic which provides the fitted values $\{f_{1j0}\}$ for Group 1 is, as intuition would predict, the sample mean for the Group 1 (the *Untrained* group). This is traditionally denoted by $\overline{y}_{1.}$. The set of residuals $\{r_{1j0}\}$ are values of the statistics $y_{1j} - \overline{y}_{1.}$ ($j = 1, 2, \ldots, 11$). The set of fitted values for members of Group 2 (the *Trained* group), $\{f_{2j0}\}$, are also identical and are equal to the value computed for the mean of Group 2, $\overline{y}_{2.}$. The corresponding set of residuals, $\{r_{2j0}\}$ are values of the statistics $y_{2j} - \overline{y}_{2.}$ ($j = 1, 2, \ldots, 11$). Based on these formulae, the numerical values for fitted values and residuals were computed from the *Pre-race Sodium* readings in Table 1.3.1. They are presented in Table 10.2.1.

Table 10.2.1. Fitted Values and Residuals — Pre-Race Sodium (Data from Table 1.3.1)

Untrained Group ($i = 1$) ($\overline{y}_{1.} = 142.7$)				Trained Group ($i = 2$) ($\overline{y}_{2.} = 139.1$)			
Subject (j)	y_{1j}	f_{1j}	r_{1j}	Subject (j)	y_{2j}	f_{2j}	r_{2j}
1	141	142.7	−1.7	1	140	139.1	0.9
2	141	142.7	−1.7	2	147	139.1	7.9
3	141	142.7	−1.7	3	140	139.1	0.9
4	142	142.7	−0.7	4	137	139.1	−2.1
5	142	142.7	−0.7	5	137	139.1	−2.1
6	146	142.7	3.3	6	137	139.1	−2.1
7	146	142.7	3.3	7	138	139.1	−1.1
8	143	142.7	0.3	8	141	139.1	1.9
9	142	142.7	−0.7	9	137	139.1	−2.1
10	144	142.7	1.3	10	136	139.1	−3.1
11	142	142.7	−0.7	11	140	139.1	0.9

The basic methods of model and data checking based on residuals and fitted values are described in Section 10.3.

10.3 Analyzing Residuals

10.3.1 Notation

Subscripts are employed both to identify levels of identified contributors to the response and to distinguish sample members who have the same identified contribution.

If there are n sample members, there can be defined n response variables $y_1, y_2, \ldots, y_n$. The general representation of the component equation has the form portrayed in (10.3.1).

$$y_i = M_i + e_i \qquad \text{for} \quad y = 1, 2, \ldots, n \tag{10.3.1}$$

where M_i is the identified contribution to the ith response; e_i is the unidentified or unexplained contribution to the ith response.

When the sample members are placed in groups and there is reason to believe that variations in response exists between groups, it is convenient to replace the single subscript identifying a sample member by a pair of subscripts, one to identify the group and the other to distinguish sample members within the group. Thus, the equation has the form of (10.3.2).

$$y_{jk} = G_j + e_{jk} \qquad \text{for} \quad j = 1, 2, \ldots, g; \; k = 1, 2, \ldots, n_j \tag{10.3.2}$$

where G_j is the contribution from group j. (For example, if there were $n = 6$ sample members, $g = 2$ groups with $n_1 = 2$ sample members in Group 1 and $n_2 = 4$ sample members in Group 2, the following subscripting would be used

Sample member (i)	1	2	3	4	5	6
Group (j)	1	1	2	2	2	2
Member within group (k)	1	2	1	2	3	4

It can be seen that the pair of subscripts j, k provides an alternative means of identifying a sample member to a simple numbering from 1 to 6.)

In models where there is more than one contributor to the identified component, a more complex form of subscripting is necessary. This may be illustrated in the situation for which (10.2.3) applies. In that case, the response is the weight of a corn cob (y) which is presumed to be dependent on the variety of corn (V), the growth temperature (T) and the collective contribution of unidentified factors (e). The conventional representation of the component equation is given in (10.3.3) where the subscript i identifies the variety of corn and the subscript j identifies the growth temperature and k distinguishes plants which are of the same variety and have grown in the same temperature.

$$y_{ijk} = M + V_i + T_j + e_{ijk} \tag{10.3.3}$$

where

M represents the average weight;

V_i is the increase (or decrease) from the average as a result of using variety i ($i = 1, 2, \ldots, n_v$);

T_j is the increase (or decrease) from the average as a result of growing the plant in temperature j ($j = 1, 2, \ldots, n_t$); and

e_{ijk} is the component of response which is attributed to unidentified sources ($k = 1, 2, \ldots, r$).

Based on (10.3.3), the experimental situation is represented below for 2 varieties, 3 temperatures and 2 replications of each combination of variety and temperature.

Variety	i	Temperature (°C)		
	j	1	2	3
$X1$	1	y_{111}	y_{121}	y_{131}
		y_{112}	y_{122}	y_{132}
$X2$	2	y_{211}	y_{221}	y_{231}
		y_{212}	y_{222}	y_{232}

Fitted values and residuals employ the same subscripts as the corresponding response variables.

10.3.2 Computation of Fitted Values and Residuals

The quantity M_i in (10.3.1) is the mean or expected value of the variable y_i and the fitted value, f_i, for sample member i is the estimated value for M_i. In practical applications, $M_1, M_2, \ldots, M_n$ are expressed in terms of a smaller number of parameters and the fitted value f_i is computed by replacing the parameters in the expressions for M_i with estimates obtained from the data. For example, the component equation $y_i = A + Bx_i + e_i$ includes the identified component $M_i = A + Bx_i$. The estimated value for M_i relies on estimates being obtained for the parameters A and B.

Residuals are computed using (10.3.4).

$$\text{RESIDUAL} = \text{RESPONSE} - \text{FITTED VALUE}. \qquad (10.3.4)$$

10.3.3 Standardized Residuals

For some purposes it is convenient to scale the residuals. Most commonly the scaling is performed to permit the magnitudes of the scaled or *standardized* residuals to be individually examined in the course of model and data checking. Some computer packages provide the option of displaying either unstandardized residuals or standardized residuals (or both). When examining residuals which are provided, be sure you are aware of the form in which they are supplied.

A common application for standardized residuals is in cases where they are expected to behave like values from a standard Normal distribution. In such cases, standardized residuals with magnitudes in excess of 2.0 are often regarded as indicators of possible errors in the model or the data on the grounds that there is only about one chance in twenty of obtaining a value from a standard Normal distribution which has such a large magnitude. While the reasoning is flawed by virtue of the fact that it is the residual of largest magnitude which is being considered, nevertheless, the rule has proven itself to be a useful guide when used in combination with other procedures.

Certainly, if an observation has a large standardized residual, the analyst and the investigator should carefully check that no recording or translation errors have been made in respect of that response and experimental records should be checked for any note to the effect that something odd occurred or was observed in respect to the sample member which provided the response.

10.3.4 Normal Probability Plots

There is an important graphical procedure used in the examination of residuals. While the basis of the procedure is applicable for many assumed distributional forms, the discussion is in terms of the Normal distribution since this is the most important application.

1. *Expected values of Normal order statistics* (EOS). Consider the selection of a random sample from a population for which a response variable, y, is defined which has a frequency or probability distribution which is a Normal distribution. A set of statistics can be defined, the first being the smallest value in the sample $y_{(1)}$, the second being the second smallest in the sample, $y_{(2)}$, etc. If the sample has n members, there are n such statistics. Each statistic has a sampling distribution and hence, a mean or expected value. For the statistic $y_{(i)}$ $(i = 1, 2, \ldots, n)$ denote the mean by E_i.

Statistical theory provides the means of computing the expected value of the ith largest value in a sample of size n when the mean and variance of the Normal distribution are known. The *Expected values for Normal order statistics* for a $N(0, 1)$ distribution are denoted below by $\mathcal{N}_i$ $(i = 1, 2, \ldots, n)$.

2. *Normal probability plot — definition.* The graph of the points $(y_{(i)}, \mathcal{N}_i)$ for $i = 1, 2, \ldots, n$ is called a *Normal probability plot.*

The name also applies if the set of ordered responses is replaced by the set of ordered residuals or standardized residuals.

3. *Normal probability plot — application.* If the set of responses or residuals have the characteristics of a random sample drawn from a Normal distribution with fixed mean and variance, then the Normal probability plot has an expected trend line which is linear. This trend line is defined by the point $(E_i, \mathcal{N}_i)$ for $i = 1, 2, \ldots, n$.

Non-linearity in trend suggests a non-Normal distribution. Figure 10.3.1 illustrates the three trend lines which most commonly arise:

- in plot (a) the Normality assumption would be judged acceptable since the trend line is linear;
- in plot (b) the curved trend line indicates a skewed distribution of the type displayed; and
- in plot (c) the "S" shaped curve suggests a distribution which has longer tails than would be expected for the Normal distribution.

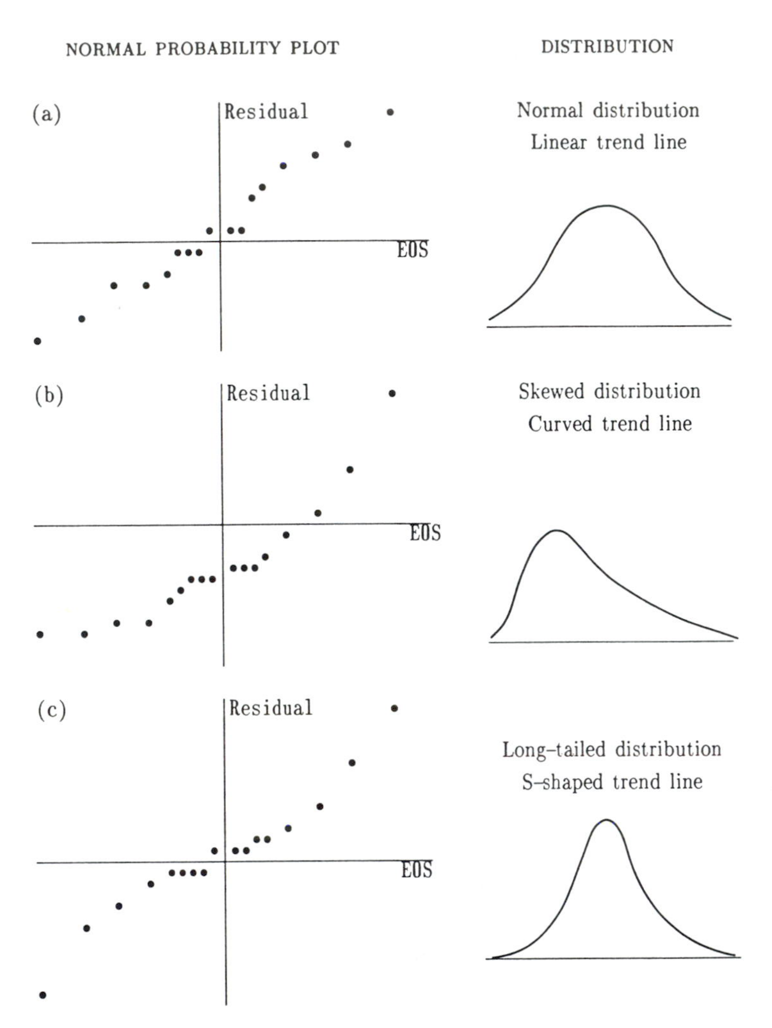

Figure 10.3.1. Interpreting Normal probability plots.

The points will generally not lie on the trend line since the plot is $y_{(i)}$ versus $\mathcal{N}_i$ rather than E_i versus $\mathcal{N}_i$. Hence, there is an element of subjectivity in deciding whether a departure from linearity is an indication of an incorrect trend line or merely sampling variation.

In practice, the observations will not generally all be drawn from the one population or process and it is necessary to employ the set of residuals or standardized residuals rather than the set of observations for constructing the Normal probability plot.

4. *Normal probability plots — computation.* Most general purpose statistical packages offer the option of constructing a Normal probability plot as a

standard part of the statistical analysis. Alternatively, tables of expected values of Normal order statistics are available.

5. *Normal probability plots — interpretation.* Linearity of the trend line is the indication of Normality. Figure 10.3.1 portrays the most common forms of departure from linearity and the meanings to be attached. Note the difference between a non-linear trend as illustrated in Figure 10.3.1 (b) and (c) and an error in the data which is illustrated in Figure 10.4.1.

To use a Normal probability plot in judging the acceptability of the assumption of Normality, it is necessary for the analyst to make a subjective judgment as to whether the departure from linearity is due to a non-linear trend or is due to sampling variability. Particularly with small sample sizes, the distinction may be difficult and it must be accepted that an assessment of the acceptability of the Normality assumption is limited when the sample size is small. A useful component to the Normal probability plot in small samples is the magnitude of standardized residuals. If there is evidence of a curved trend line, check for the presence of a standardized residual with a magnitude in excess of 2.0 as confirmation that the Normality assumption may be suspect.

10.3.5 The Plot of Residuals versus Fitted Values

In those experimental situations where methods of Inferential Statistics can be employed, the values of the unexplained component in a response are determined by chance. If the component equation is correct, there should be no relation between the contribution from identified sources and the component determined by chance. The existence of a relation between the components indicates the inclusion of part of the identified contribution in the unexplained component.

The existence of a relationship between the identified and unexplained components in the component equation is commonly reflected by a pattern in the plot of fitted values versus residuals or fitted values versus standardized residuals, with the latter plot being preferred because it has more general application.

An example of a plot which reflects an incorrect component equation is provided in Figure 10.3.2 (a).

There is also a role for the plot of fitted values versus residuals in examining distributional assumptions. Figure 10.3.2 (b) illustrates a situation in which the variability of standardized residuals increases with increasing sizes of fitted values. This suggests an incorrect variability assumption in the model.

A plot of fitted values versus standardized residuals is recommended as a standard plot of model checking whenever sets of fitted values vary between sample members.

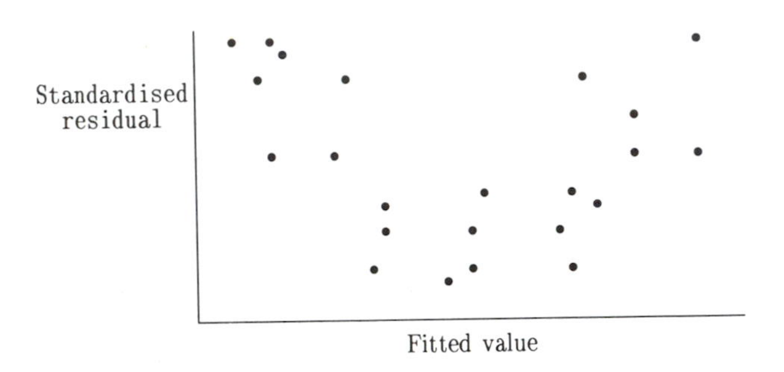
(a)Evidence of an incorrect component equation
Note the presence of a curved trend line.
If the model is correct there should be no relation evident
Standardised residual
Fitted value

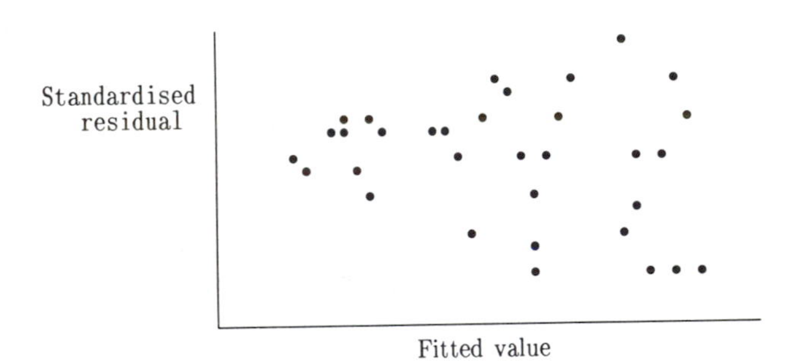
(b) evidence of unequal variance
Note the increasing spread of residuals with increasing fitted values
Standardised residual
Fitted value

Figure 10.3.2. Plots of fitted values versus residuals identifying potential errors in the statistical model.

10.3.6 Checking Assumptions in the Model — A General Philosophy

While assumptions in a statistical model may be stated independently of one another, two factors intervene to introduce a dependence into the detection of errors in assumptions. The first is the fact that frequently an error in one assumption in the model will be accompanied by an error in another assumption of the model, and secondly, a checking procedure which is sensitive to an error in one assumption will generally also be sensitive to an error in another assumption.

The consequence is that model and data checking should be seen as an integrated operation in which the *collective* results from all checking procedures are considered in deciding on the acceptability of a model, and interpretation of specific methods is dependent on the finding of other methods.

Some commonly used model checking procedures are introduced in the following two subsections. They have been selected because of their wide applicability and ease of application. Use additional or alternative methods on the advice of a consultant statistician or where there is published evidence that the method has merit. Be wary of indiscriminately using methods. Each additional arbitrarily chosen method increases the likelihood of falsely declaring a model incorrect—given enough tests, pattern can be detected in any set of numbers which shows variation.

10.3.7 Checking Distributional Assumptions

The distributional assumptions may include a statement

- of the mathematical form of the distribution.
- that two or more distributions are identical.
- that a distribution is symmetric.

Where differences between responses are in part attributable to identified components, checks on distributional assumptions should be made on residuals since it is variation in the unexplained component which is characterized by the distributional assumptions.

Procedures fall into two categories, graphical and statistic-based, and these are discussed separately below.

1. *Graphical procedures.* The simplest procedure is the construction of a line diagram (Section 1.3.5) if the sample size is less than 50 or a histogram (Section 1.3.6) or Box-plot (Section 1.3.10) if the sample size is at least 50. These pictorial presentations offer a visual picture of the pattern of variation in the data. If there are two or more groups, construct a graph for each group separately (but use the same scale of measurement to allow for group comparison).

Look for evidence of asymmetry and for evidence that there is a greater spread of values for one group than for another. Carry a mental picture of these plots with you when interpreting more specific or more sophisticated checking procedures.

Specific checks on assumed distributional forms are provided by probability plots. The Normal probability plot is widely available as an option in statistical packages. Use it in the manner described in Section 10.3.4 whenever the Normality assumption is included in the model or when values of a statistic are obtained which are expected to exhibit variation defined by a Normal distribution.

In specialized areas, other distributions may commonly arise and facilities may be provided for constructing probability plots based on other distributional forms. The method parallels that described in Section 10.3.4 for the Normal probability plot, the only difference in application being the use of expected values of order statistics from a different distributional form.

2. *Statistic-based procedures.* There are formal statistical tests of distributional assumptions. They are not used in the methods described in this book because they are not widely available in statistical packages, require special tables for the computation of p-values, or have limited application. The earlier remark about choosing procedures is relevant. Do not indiscriminately apply procedures. Use a procedure if it is readily available and there is evidence that it will detect the type of departure from the distributional assumption which is likely in the specific experimental situation.

A particular note is appropriate in respect of the Normality assumption. This is the most widely employed distributional assumption and is often employed without strong grounds for its use. If the Shapiro–Wilk U-test of Normality is available, it is strongly recommended as a supplement to the examination of a Normal probability plot.

Tests of equality of distributions which are based on residuals generally concentrate on comparisons of variability. Under the assumption of Normality these tests take the form of testing for equality of variance. If a test exists on the available statistical package, use that test. Generally the validity of these tests depends on correctness of the Normality assumption. Since inequality of variance is commonly associated with failure of the Normality assumption this makes the tests of dubious value as formal tests of the hypothesis of equality of variance. However, they do have a role in the general process of detecting errors in the model.

10.3.8 Errors in the Component Equation

For the purposes of model checking, there are two general features of the component equation which demand examination. They are

1. The additivity of the explained and unexplained components, i.e., the assumption that the component equation can be represented as $y = M + e$.
2. The correctness of the structure of the explained component M.

Approaches for checking each of these assumptions are presented below.

1. *Checking for non-additivity of explained and unexplained components.* Where additivity of explained and unexplained components is incorrectly assumed, there will be a relation between values taken by the two components. This may be reflected by a pattern in the plot of fitted values versus residuals. However, an arbitrary search for non-additivity is unlikely to be

rewarding. It is wise to seek a reason for the lack of additivity in the nature of the experiment.

Commonly, non-additivity is a result of a multiplicative rather than an additive model being appropriate. The means of identifying that situation, specific model checking and the experimental relevance are discussed below under the heading of *Multiplicative component equations.*

Sometimes recommended is the practice of arbitrarily seeking a transformation of the scale of measurement in an effort to achieve additivity on another scale of measurement. Comments on that approach are presented under the heading *Transformations and additivity.*

Transformations and additivity. The reason for such intense interest in additive component equations rests both with statisticians and with scientists. The fact is that additivity in the component equation derives from a requirement to compare parameters by examining *differences* rather than some other form of comparison.

That *differences* associate with *additivity* is easily demonstrated. Consider the additive equation $y = M + e$. If y_1 and y_2 are two responses based on different values of M, M_1 and M_2, then

(i) $y_1 - y_2 = M_1 - M_2 + e_1 - e_2$;

(ii) $y_1/y_2 = (M_1 + e_1)/(M_2 + e_2)$.

The additive equation permits the difference in identified components, $M_1 - M_2$, to be separated from the unexplained components, but the ratio M_1/M_2 cannot be separated from the unexplained components.

Mathematical theory establishes that it is only the difference which can be expressed free of the unexplained component when an additive model is employed. This has significance if M_1 and M_2 represent population means. The statistical task of constructing estimates or tests relating to their difference is relatively easy. Other forms of comparison are either difficult or impossible.

If an experimenter decides that the appropriate form of comparison of the two means is in terms of their ratio, M_1/M_2, then the desirable form for the component equation is the multiplicative form $y = M \times e$. Then the ratio of responses may be expressed in terms of the ratio M_1/M_2 as defined in (10.3.5).

$$\frac{y_1}{y_2} = \frac{M_1}{M_2} \times \frac{e_1}{e_2}. \tag{10.3.5}$$

While statisticians must accept the form of comparison which scientists judge to be the most relevant in an experimental situation, there is a vast array of statistical methodology which is valid only under the assumption of an additive component equation. Hence, statisticians have an interest in attempting to transform the relevant component equation to a scale of measurement in which it is additive and to apply the statistical analysis to the responses on the transformed scale of measurement.

For example, if the required component equation is multiplicative, i.e. $y = M \times e$, then additivity can be attained if the responses are transformed to a logarithmic scale of measurement as is illustrated in (10.3.6).

$$\log(y) = \log(M) + \log(e). \tag{10.3.6}$$

The key to the use of transformations in this situation lies in the interpretability of the findings from an experiment viewpoint. The use of a logarithmic transformation when ratios are of interest is of practical value and is discussed separately below. Other special cases arise which have practical relevance.

There has been extensive statistical development of *power transformations* to convert an observed response y to y^p such that additivity holds on the transformed scale, i.e. (10.3.7) holds.

$$y_i^p = M_1' + e_i' \tag{10.3.7}$$

where M_i' is the value of the parameter of interest (M_i) for the ith sample member on the transformed scale of measurement.

There is no natural relation between M_i' and M_i such as exists for multiplicative models. Note that $y_i = (M_i' + e_i)^{1/p}$ and there is no connection between $M_i' - M_j'$ and a contrast between the mean responses under treatments i and j, namely M_i and M_j. The limited interpretation available lies in the fact that rejection of the hypothesis $M_1' = M_2' = \ldots = M_t'$ is equivalent to rejection of the hypothesis $M_1 = M_2 = \ldots = M_t$.

Multiplicative models. As noted above, a multiplicative model is appropriate when it is experimentally wise to express comparisons in terms of ratios rather than differences. It is commonly applicable when the magnitude of the unexplained component is expected to increase as the response increases. The situation would arise, for example, in the study of weights of humans where the sample comprises a range from babies to adults. The variation of weights in babies would typically be much less than the variation of weights in adults.

Where an additive equation has been incorrectly assumed, the indications that a multiplicative model is more appropriate may be identified by the following conditions being met:

(i) a plot of fitted values versus residuals displaying greater variability in residuals with increasing size of fitted values, in the manner illustrated in Figure 10.3.2 (b);

(ii) a response variable which cannot take negative values and for which the ratio of largest to smallest response in the sample is in excess of ten; and

(iii) a Normal probability plot in which the trend line is curved in the manner illustrated in Figure 10.3.1 (b).

The experimental situation in which this commonly arises is in the comparison of mean response between groups where the mean responses are more sensibly compared by studying ratios than differences. (See Section 12.3 for discussion and illustration of this point.)

2. *Errors in the component equation.* The general effect of an error in the component equation is to cause part of what should be in the explained component to appear in the unexplained component. For example, if the assumed component equation is $y = A + Bx + e$ when it should be $y = A+Bx+Cx^2+e$, the element Cx^2 will incorrectly appear in the unexplained component.

General detection of this type of error is through the presence of a trend in the plot of fitted values versus residuals. An illustration is provided in Figure 10.3.2 (a). The nature of the trend will depend on the form of the component equation and the type of departure.

Formal statistical tests which are sensitive to specific types of departure are available and are indicated where they have application.

Additivity of identified components. From an experimental viewpoint, a special form of departure can arise when there are two or more factors contributing to the explained component of the equation and they are assumed to contribute in an additive fashion. Equation (10.3.3) provides the simplest example. In that example the response is defined as the weight of a corn cob (y) and it depends on both the variety used (V) and the temperature at which it is grown (T).

The implication of assuming that the varietal and temperature components are *additive* is that the expected difference in weights between cobs from plants of different varieties is independent of the temperature at which the plants are grown. This can be demonstrated by showing that the difference in weights between two plants grown at the same temperature is independent of the temperature component, T_j, i.e. $\overline{y}_{1j.} - \overline{y}_{2j.} = V_1 - V_2 + \overline{e}_{1j.} - \overline{e}_{2j.}$. This is a restriction which may not apply in practice. Suppose, for example, that variety X1 ($i = 1$) has been selected because of its good performance at 20°C ($j = 1$) while variety X2 ($i = 2$) has been selected because of its good performance at 30°C ($j = 3$). Then the difference in varietal effects in the component equation will not be temperature independent. This fact is represented in (10.3.8) where V_i is interpreted as an average varietal contribution to be adjusted by an amount $(VT)_{ij}$ at the jth temperature. The effects of the varietal and temperature components are no longer additive. In statistical jargon there is an **interaction** between variety and temperature.

$$y_{ijk} = M + V_i + T_j + (VT)_{ij} + e'_{ijk}. \qquad (10.3.8)$$

The methods used for detecting this form of non-additivity are

(i) a plot of *residuals* versus *fitted values* (see Section 10.3.5);

(ii) tests designed to detect specific forms of non-additivity; and

(iii) a general purpose test for the presence of non-additivity available for a range of *designed experiments* (see Chapter 13).

The application of the methods is illustrated as relevant methodology is introduced.

10.4 Checking Data

10.4.1 Common Sense and Experimental Knowledge

A good experimenter or investigator maintains vigilance throughout the course of the study for evidence of odd features, accidents, malfunctions in equipment and faults in the work practices of assistants which may affect the data.

Statistical consultants may bring valuable experience to the examination of data after collection, but their experience cannot substitute for the intimate knowledge of the scientific content and the day-to-day operation which the investigator can provide.

Any reason for concern in respect to the data set should be noted and, where possible, checked once the data are available. Such information should always be brought to the attention of a statistical advisor.

Quite distinct from being diligent in noting factors of concern in the course of collecting the data, investigators should bring their special knowledge to play in the examination of the data collected. Avoid the temptation to simply feed the data into the computer and allow automated checking procedures to take over the task of data checking.

Ignoring the process of visual checking by the person who collected the data is unwise on at least two grounds. Firstly, the investigator will have generally input much time and effort in the design and execution of the study. To present a flawed report on the study because a small amount of time was not made available for data checking makes no sense. Secondly, no automated process of data checking can incorporate the knowledge possessed by the experimenter. Consequently, serious blunders may be missed. Not only may this result in embarrassment and perhaps ridicule for the investigator in the short term, but in the longer term, support for research and even job opportunities may be lost.

The possible forms of checking are of course many and varied, depending greatly on the particular experiment or investigation being undertaken. New or young research workers can gain enormously from advice from senior colleagues. Comparisons can be made with earlier or parallel studies in the field. Where responses are outside expected ranges or have characteristics which differ greatly from comparable studies, search for the reason before undertaking costly statistical analyses.

10.4.2 Statistical Methods

The statistical tools available for data examination are of four types, which are

1. Checks on data for information which is outside possible bounds. For example, a proportion which is outside the range zero to one; a categorical response which is not within the list of possible answers.

2. The summarization of data either in the form of pictures, tables or values of statistics which are then available to the investigator for subjective judgment. For example, the maximum and minimum values may be supplied for a set of numbers.

3. The examination of data based on a statistical model but using subjective examination of the results. This is the general area of Data Analysis as it is defined in this book. Normal probability plots fall into this category.

4. The examination of data based on a statistical model using techniques of statistical inference.

10.4.3 Data Checking in Practice

The following guidelines are offered for checking data.

1. The person who conducted the experiment should, if practical, check the data and should identify any observations which may be in error either from the data checking or from knowledge of the operation of the experiment. If possible, a second person (assistant, colleague) should also check for obvious errors in the data.

2. If the data comprise more than 50 pieces of information, it is strongly recommended that (i) computer-based checking be implemented to identify responses which are outside defined bounds; and (ii) appropriate data summarization be employed which aids in the identification of errors in the data. As a minimal requirement, for numeric data, the maximum and minimum values should be provided for each variable, and for categorical data the set of categories which appeared for each variable should be listed.

3. If they are available, Data Analytic methods should be employed which exploit knowledge of the assumed statistical model to define expected pattern in the data and to detect to errors in the data. Of great value is the Normal probability plot. As illustrated in Figure 10.4.1, a possible error in the data is identified as a value which deviates from the trend of the remaining points.

A standardized residual with a magnitude in excess of 2.0 can be taken as evidence of a possible error in the data only if the point corresponding to the high residual is considered inconsistent with the trend line. This is the case in Figure 10.4.1. When a standardized residual has a magnitude in excess of 2.0 but appears to lie on the trend line, the problem usually lies with the distributional assumption rather than the data.

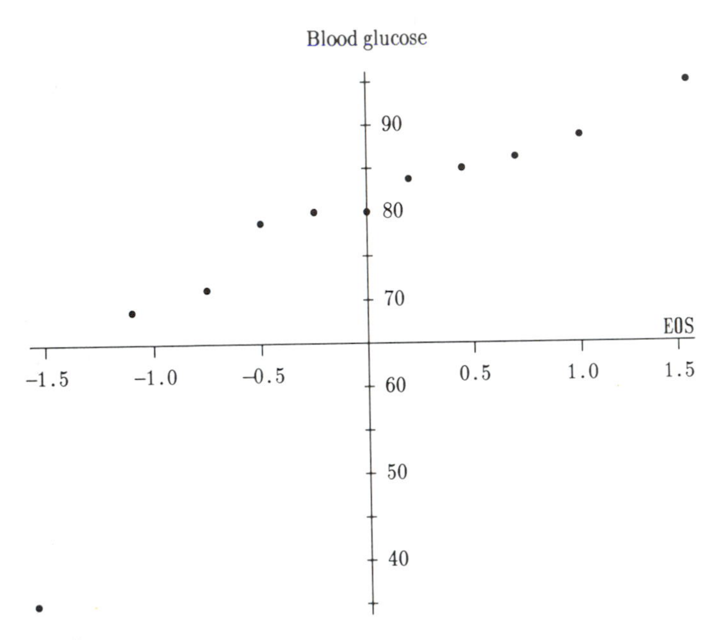

Figure 10.4.1. Normal probability plot displaying an error in the data. (*Plot based on pre-glucose readings for Trained runners in Table 1.3.1.*)

10.4.4 Influence

While the major thrust in the examination of data is the detection of errors in the data, there is another aspect which is of considerable importance in helping scientists decide on the importance they might attach to the findings of a statistical analysis.

While statistical methods make use of all relevant information in a set of data, some pieces of data can play a much more important role than others in the findings of the analysis because of particular placement in respect of other data items and the statistical model employed. A simple example will make the point clear.

Figure 10.4.2 presents a set of points which come from a medical experiment. The experimenter, on the basis of statistical analysis, drew the conclusion that there is a relation between values taken by the two variables—as one increases in value, the other tends to increase also. There are ten points in the graph. Yet, it is clearly the position of just one of those points relative to the remainder which is responsible for the conclusion reached. Remove that single point and a quite different conclusion is reached. That one point is said to be a point of high **influence.**

It is becoming common practice for statistical packages to identify points which have high influence. Obviously, responses which are in error may have high influence, although that will not always be the case.

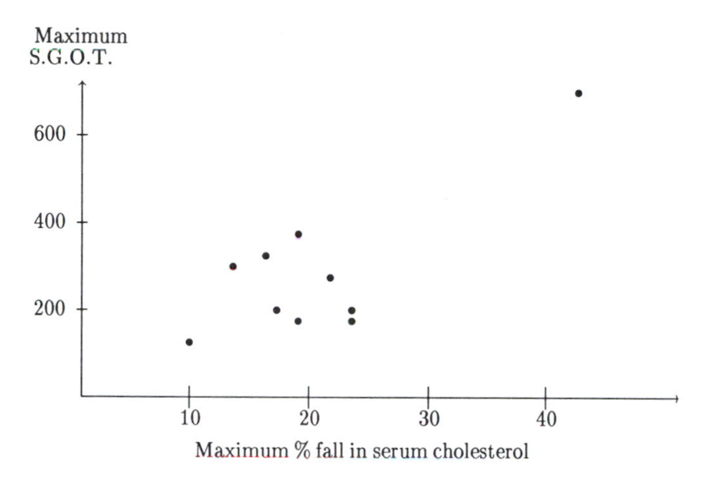

Figure 10.4.2. Scatter plot displaying a point of high influence.

10.5 Data Modification or Method Modification?

The process of model and data checking are neither independent nor guaranteed to identify from which source a suspected error arises. Thus, it may be unclear whether the correct course of action is to alter the data or modify the model when a warning is sounded in the course of model and data checking. The situation is made more complex by the fact that a fault in the model can sometimes be overcome by transforming the data to a new scale of measurement. In this section, an attempt is made to place the different options in perspective. It is impossible to provide rules which have application in individual situations. However, scientists who are employing statistical techniques should be aware of the broad options available.

10.5.1 Data Modifications

Where there is evidence that one or more responses may be incorrect, the scientist may be in the fortunate situation of having indisputable evidence that the value or values are incorrect. If the correct value or values are known then the obvious course of action is to replace the offending data items. If not, then the responses should be treated as missing values, i.e. deleted from the data set. In such cases the fact should appear in the report on the study.

The more difficult situation is one in which there is sound evidence from the data checking process that a response may be incorrect—it is an *odd* value. Three basic courses of action are open in respect of the data. The offending item may be

- excluded from the statistical analysis which proceeds as though the response had never been obtained;
- retained in the data set and the analysis performed without regard to the fact that the checking procedure suggested it may be incorrect;
- modified prior to or during the course of the statistical analysis.

More than one of these options may be invoked in separate analyses. This is a sensible course of action from the viewpoint of establishing the effect of different values for the response. Presumably, if the same conclusion is reached in all analyses, the possible error in a response is unimportant in respect of the experimental question which prompted the analysis. There are practical disadvantages—there is extra computation and more detail is required in the presentation of results, particularly if the analyses give different conclusions.

10.5.2 Method Modification

Some statistical techniques are less sensitive to errors in data than are their competitors. Such statistical procedures are termed *robust* against the presence of errors in the data. Where a robust method is available, is computationally feasible and there is little loss in power or sensitivity through its use, it is strongly recommended.

If robust methodology is not available or not feasible, there is not a simple solution. An apparently odd value may be a sign of an incorrect model. For example, a distribution which is asymmetric or long-tailed will commonly yield samples in which one response is far removed from the remainder. If the *odd* value is removed incorrectly, it must be accepted that the consequence might be to provide a statistical conclusion which is seriously in error or which changes the scope of the statistical inference. For example, if the study is of energy intakes and the largest intake is excluded because it is 'surprisingly large', the conclusions from the analysis may apply to a sub-population from which persons with very large energy intakes have been excluded. This may or may not be acceptable to the investigator.

To ignore the fact that the odd value is inconsistent with the proposed model and proceed with the statistical analysis as though the data checking revealed nothing unusual, will in general invalidate the findings. It is a course of action which is not recommended.

A possible alternative to the use of a robust procedure is to alter the statistical model so that the data item is no longer unusual.

Problems

10.1 *Data checking*

In Table 1.3.1, 22 readings are provided for the pre-race levels of *Glucose.*

(a) Using the model, and following the steps in the illustration given in Section 10.2.4, compute fitted values $\{f_{ij0}\}$ and residuals $\{r_{ij0}\}$ for the pre-race Glucose level and reproduce the information in the tabular form displayed in Table 10.2.1.

(b) Use the residuals to compute sample variances for both groups from the statistics

$$s_i^2 = \sum_{j=1}^{n_i} r_{ij}^2/(n_i - 1) \qquad \text{for } i = 1, 2$$

where n_i is the number of observations in the ith group.

(c) Under the model which presumes that the unexplained component in the model is expected to have the same distribution in both groups, a set of standardized residuals (Section 10.3.3) may be computed from the statistics

$$r_{s_{ij}} = r_{ij}/s \qquad i = 1, 2; \quad j = 1, 2, \ldots, n_i$$

where r_{ij} is the residual for the jth runner in group i;

$$s^2 = [(n_1 - 1)s_1^2 + (n_2 - 1)s_2^2]/(n_1 + n_2 - 2)^*$$

$^*s^2$ is the sample variance from the complete sample and is formed as a weighted average of the sample variances of the two groups.

Compute the set of standardized residuals and note which reading is identified as being a potential error, i.e. which reading has a standardized residual with a magnitude in excess of 2.0.

(d) Construct a Normal probability plot (Section 10.3.4) based on the 22 standardized residuals computed in (c). Note the evidence for an error in the data (refer to Section 10.4.3 and Figure 10.4.1).

(e) Repeat steps (a) to (d) after excluding the odd value from the data set. Is there evidence of any other possible error in the data? (If a statistical package is being employed, the odd value may be excluded by (i) deleting it from the data and modifying the computing program to establish that the number of observations in the *Trained* group is reduced to ten; (ii) replacing the odd value by the *missing value* symbol; or (iii) introducing a dummy variable which takes the value 1 for the subject which has the odd value and the value 2 elsewhere and then restricting analysis to subjects which have a value 2 for the dummy variable. (The last approach has the advantage that no change need be made to the data and is valuable if the original data is required for other analyses.)

10.2 (*Continuation of Problem* 10.1)

(a) Compute values for the variable which is the change in Glucose levels before and after the race (i.e. the difference between pre- and post-levels) for the twenty-two runners.

If there was an error in the pre-race Glucose level of runner 8 in the *Trained* group, there must also be an error in the change in Glucose levels (assuming, of course, that the post-race reading was not also in error by the same amount.)

Following the steps employed in Problem 10.1 for detecting an odd value, examine the data for the change in Glucose levels for evidence of an odd value.

10.3 *Model checking and transformations*

The following data were collected in an experiment to compare the amount of combustible material which was present on areas of land twelve months after clearing programs were employed to reduce the fire risk in the area. The experiment comprised dividing the land into sixteen plots which had approximately the same type of vegetation and density of ground cover. Four clearing programs were used with each of the sixteen plots being allocated one of the four treatments. Twelve months after treatment, the amount of combustible material on each plot was estimated and the figures are recorded below.

TREATMENT

Burning ($i = 1$)	Slashing & clearing ($i = 2$)	Chemical & burning ($i = 3$)	"Bradley" ($i = 4$)
110	102	2193	18
4915	184	34	93
493	314	43	86
221	1626	289	259

Suppose the component equation to be included in the model is given by

$$y_{ij} = M + T_i + e_{ij}$$

where

y_{ij} is the amount of combustible material observed in plot j of treatment i;

M is the expected or mean level of combustible material;

T_i is the expected or mean increase from the average level of combustible material as a result of using method i; and

e_{ij} is the unexplained component of the response from the jth plot which received the ith treatment.

(a) Obtain an estimate of the parameter M using the statistic $\overline{y}_{..}$ and an estimate of the parameter T_i ($i = 1, 2, 3, 4$) using the statistic $\overline{y}_{i.} - \overline{y}_{..}$ (where $\overline{y}_{i.}$ is the sample mean for the observations receiving treatment i ($i = 1, 2, 3, 4$) and $\overline{y}_{..}$ is the sample mean for the whole data set). Hence, compute the set of fitted values $\{f_{ij0}\}$ by replacing M and T_i ($i = 1, 2, 3, 4$) by their sample estimates in the above component equation. Compute the set of residuals $\{r_{ij0}\}$ from the formulae $r_{ij} = y_{ij} - f_{ij}$ for $i = 1, 2, 3, 4$; $j = 1, 2, 3, 4$.

(b) Produce a Normality probability plot (Section 10.3.4) based on the residuals. Note the curved trend line.

(c) Plot the fitted values versus the residuals and notice the increased variability in the residuals as the fitted values increase in size.

(d) Compute the mean and standard deviation for the set of four responses from each group. Plot mean versus standard deviation and notice that increasing mean associates with increasing variability.

The evidence from results in (b), (c) and (d) suggests that a *multiplicative model* (Section 10.3.8) might be appropriate.

(e) Transform the original data to a logarithmic scale and repeat steps (a)–(d) on the presumption that the additive component equation applies on the logarithmic scale. Is additivity reasonable on the logarithmic scale?

10.4 *Robust statistics and the estimate of the population mean*

(a) Generate a sample of size 3 from a $N(0, 1)$ distribution.

(b) Determine the mean and median of the sample and store them for later analysis.

(c) Add 3 to the first number in the sample to simulate an error introduced into the data.

(d) Compute the mean and median for the amended data and store the values for later analysis.

(e) Repeat (a)–(d) a further 99 times to produce four sets of numbers, namely 100 means from original data, 100 medians from original data, 100 means from amended data and 100 medians from amended data.

(f) Construct line diagrams for the four sets of data using the same scale for all sets.

(g) Compute the means and variances for the four sets.

Statistical theory predicts that the sample median is less efficient (as judged by the greater variability in the medians) than the sample mean as an estimate of the population mean when the data are drawn from a population with a Normal distribution and are free of error. However, the sample median is less affected by the presence of an error in the data. Does this agree with your findings?

Could there be statistics which are more efficient estimators than the median in the absence of an error in the data but are less affected by an error than is the sample mean? You may wish to pursue this possibility, perhaps by incorporating an automated process for detection of an error in the data and either deleting or modifying the apparent error before computing an estimate of the population mean. (Consider the possibility of using trimmed means which are defined in Problem 1.11.)

11

Questions About the "Average" Value

11.1 Experimental Considerations

11.1.1 The Quantity Under Investigation

One valuable feature of Statistics is that it demands of the user a precise statement of the aims of the investigation. Part of this statement relates to a clear expression of the quantity being studied. To a statistician, the word *average* has a clear and unambiguous meaning. Popular usage of the word is not so well defined, as the following example reveals.

Example 11.1 (The "Average" Cost of a Medical Consultation). *When the Government decided to introduce a universal health scheme, there was a need to set a fee for a standard consultation of patients with their medical practitioners. There was general agreement between the Government and the medical profession that the fee should reflect the average fee charged prior to the introduction of the scheme.*

The Government's idea of average fee was the most common fee charged (the mode) since this reflected the typical fee. The Medical Association opted for the arithmetic mean. Given a lack of consensus, a committee was established to make a recommendation. Their conclusion was that the median fee should be employed.

The reasons for the differing suggests are not difficult to see if reference is made to Figure 11.1.1.

p.s. the Government definition prevailed!

The example illustrates how different people who are claiming to be talking about the same quantity are in fact referring to different things. A primary consideration in any scientific investigation is to clearly identify the quantity or quantities under investigation. Persons reading a report of the investigation may not agree with the choice made by the investigator, but at least the source of the disagreement is plain and the conclusions reached may be evaluated in the light of the choice which was made.

While the most commonly occurring value, or *mode,* is an unusual choice for the average value, in many investigations there is no obvious reason for choosing between the arithmetic mean and the median. Factors to be considered in making the choice are considered in Section 11.2, as is methodology for answering questions about the mean and the median.

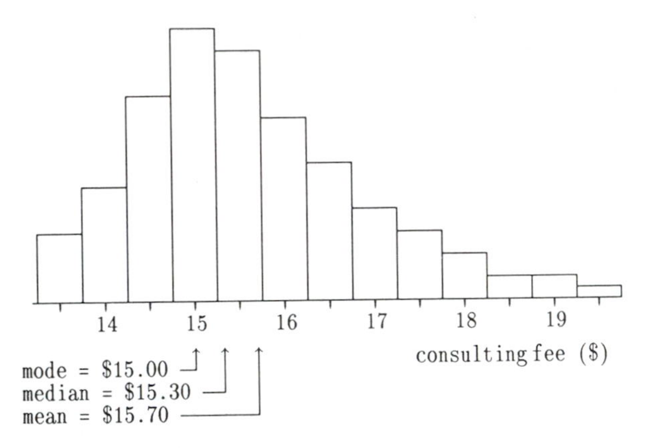

Figure 11.1.1. Distribution of medical charges.

11.1.2 The Experimental Aim

Having decided on what is to be investigated, there are typically three reasons for conducting the investigation when interest lies in a numerical quantity. The first is purely *descriptive*, and involves simply summarizing information in the sample into a single statistic, typically the sample mean or median. The remaining two reasons are *analytic* and require the definition of a statistical model. The presumption is that the sample is taken from a larger collection and the purpose of the investigation relates to properties of the larger collection.

Given that the "average" value in the larger collection is an unknown quantity M, the aim is either to *estimate* the value of M using sample information or to judge the acceptability of an *hypothesized value* for M. The following examples illustrate the range of problems in which analytic methods are required.

Example 11.2 (Laser-Induced Temperature on a Tungsten Surface). *The set of* 8 *readings recorded below are from a laser study in which a clean tungsten surface was heated using a ruby laser with a pulse amplitude of* 1.2. *The readings are surface temperature* (°C) *on the tungsten surface.*

745.1 736.2 726.1 742.9 741.9 753.1 731.8 740.4.

The nominal mean surface temperature produced is 750 *degrees. Are the data consistent with this mean?*

Example 11.3 (Percentage Change in Property Prices Over Time). *The regional manager of a real estate company was examining the rate of return from property investments in his city. He collected data on purchase and subsequent sale prices of* 28 *properties during a ten-year period. Based on the purchase price* (P)*, the sale price* (S) *and the number of months which elapsed between purchase and sale* (n)*, he computed the percentage monthly increase in value* (p) *according to the formula*

$$p = 100(S - P)/(Pn).$$

The values obtained are recorded in Table 11.1.1 *and represent the complete information from the records of one real estate company in the city.*

Table 11.1.1. Percentage Monthly Increase in Property Values (Data from Example 11.3)

0.05	0.99	0.13	1.45	0.74	5.21	1.04	1.51	1.68
0.93	3.01	0.88	1.72	0.73	0.69	1.33	1.01	0.83
1.44	1.81	0.80	0.91	0.95	1.14	0.26	0.97	1.30
0.74								

The aim of the study was to estimate the average monthly percentage increase in property value for the city.

Given the nature of the quantity being measured it is to be expected that the distribution of values of p will be skewed with a small proportion of very large values. This is borne out in the data collected. The consequence is that the mean will be dominated by the small proportion of large values and is perhaps not the best measure of the average increase in this context. When the statistical model and the analysis of the data are considered later in the chapter, it is the median percentage increase which will be investigated.

Example 11.4 (Energy Intakes of 15 Year Old Boys and Girls). *Part of the data collected in a major study of eating habits of Tasmanian school children are contained in Table* 1.3.5. *They comprise the estimated energy intakes of* 15 *year olds, presented separately for boys and girls. One purpose of the survey was to provide the information necessary for the estimation of median intakes for the two groups.*

Example 11.5 (Cholesterol Levels of Workers at an Industrial Site). *The data in Table* 1.3.2 *come from a study of workers who were exposed to high levels of some heavy metals in a zinc works and were shown to have levels of one or more heavy metals in their bodies far above the average for the population.*

The data were collected because of some medical research which suggested that persons who had above average levels of heavy metals in their bodies were likely to have about average levels of cholesterol.

Information available from a pathology laboratory in the community of which these workers were members produced a figure of 220 *mgm% for the mean level of cholesterol. It was also noted that, so far as records allowed, this figure was based on persons without exposure to high levels of heavy metals.*

The investigator wished to know if the mean level of cholesterol was higher in persons who have above normal levels of heavy metals in their bodies.

11.1.3 Sample Selection

For valid application of the commonly used methods of analysis, the sample members are presumed to be selected by either random selection from a population or systematic selection from a repetitive process in which successive outcomes are independent. There is need for concern if the sample members are connected in time or space since it is frequently found that responses from neighboring units are related. Thus, the economist who obtains sales figures for a company from 12 successive months, or the educator who uses a group of children from the same class, is unlikely to meet the conditions required in standard models. In such cases assitance should be sought from a consultant statistician, preferably before the investigation is begun.

11.2 Choosing and Applying Statistical Methods

11.2.1 Choosing a Method

The choice of a statistical method is relatively straightforward. The decision is primarily dependent on two factors—the quantity under investigation and the sample size.

If the *median* is the quantity of interest, the recommended method derives from a sign-based statistic and is described in Section 11.2.2.

Where the *mean* is the quantity of interest, a distribution-free method based on the Normal distribution is described in Section 11.2.3 which may be employed with large sample sizes ($n > 30$). For small sample sizes, a distribution-based method which utilizes the t-statistic should be considered. It is described in Section 11.2.4.

It is for the experimenter to decide if the mean or median is the more relevant quantity to study. The decision will rest on such factors as (i) the traditional choice in the field of study; (ii) the nature of the variable being studied; (iii) the availability of methodology and computing facilities; and (iv) additional or complementary questions which might be asked.

For brevity, no statement is made of sampling assumptions in defining the models, the universal assumption being that sampling is either by random selection from a population or systematic selection from a repetitive process in which successive outcomes can be assumed independent.

11.2.2 Examining the Median — Sign-Based Procedures

As explained in Section 7.7, simple statistical procedures based on the signs of responses and derived values may be employed to examine questions about median response in a population or process.

Model: The procedures are based on the most general model which can be employed. It imposes no conditions on the form of the probability or frequency distribution other than to specify a value M_0 as the median value.

The development of inferential procedures is based on the Binomial model and for this reason the assumption is made that the sample size is small compared with the population size in cases where the investigation involves sampling a population.

Statistic: The statistic employed in conjunction with this model is the Binomial statistic which equates with the number of observations above M_0 in the table below.

	above M_0	below M_0	Total
Number of observations	x	$n' - x$	n'

where n' is equal to the sample size, n, less the number of observations which are equal to M_0.

As explained in Section 7.7.3, if the correct value has been chosen for M_0 then x has a sampling distribution which is Binomial with parameters $\pi = 1/2$ and n'. Hence, the most likely outcome is represented by half of the observations being above and half below the hypothesized value. The least likely outcomes are represented by all observations being above or all observations being below M_0.

Measure of agreement: If x_0 is the observed number of responses above the hypothesized median and x is a Binomial statistic with parameters n' and $\pi = 1/2$, then, following the development outlined in Section 7.7.3, the p-value is defined by (11.2.1).

$$\begin{aligned} p &= 2\Pr(x \leq x_0) && \text{if } x_0 < \tfrac{1}{2}n' \\ &= 2\Pr(x \geq x_0) && \text{if } x_0 > \tfrac{1}{2}n' \\ &= 1 && \text{if } x_0 = \tfrac{1}{2}n' \end{aligned} \qquad (11.2.1)$$

Model and data checking: Check the data for the existence of responses which are impossibly high or low.

If the sample size is in excess of 30, plot either a line diagram (Section 1.3.5) or a histogram (Section 1.3.6) and visually check if the plot is approximately symmetric. If there is no obvious asymmetry in the plot, the more powerful distribution-free procedure described in Section 11.2.3 is recommended. That procedure is developed for examining the mean but will have relevance to the median if the frequency or probability distribution is symmetric since, in that case, the mean equals the median.

Where there is obvious asymmetry, as illustrated in Figure 11.2.1, the methods of this subsection should be employed.

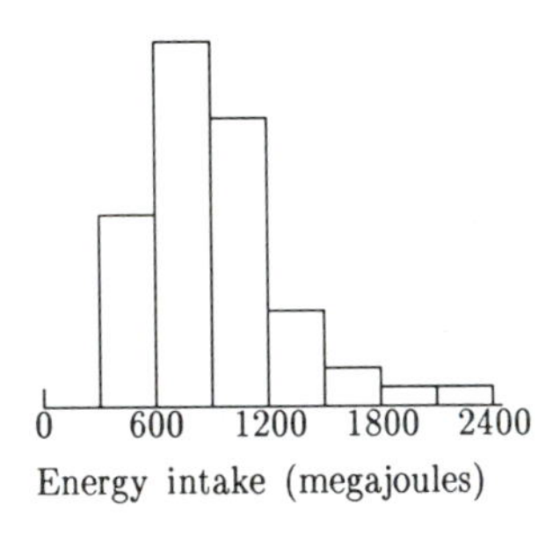

Figure 11.2.1. Histogram illustrating a skewed distribution (based on Girls' data in Table 11.1.2).

Estimation: The sample median provides a point estimate of the true value of M.

As explained in Section 8.3.6, confidence intervals can be based on pairs of order statistics, $y_{(i)}$ and $y_{(n-i+1)}$, by utilizing the Binomial distribution.

Because there are only $\frac{1}{2}n$ pairs of order statistics, it is generally not possible to obtain a confidence interval with the chosen level of confidence. The solution commonly suggested is to employ the confidence interval which has the smallest level of confidence above the required level. Alternatively, the method of linear interpolation described in Section 8.3.6 may be employed. The details are given below.

If the required level of confidence for the interval is $100(1-\alpha)\%$, the general computational requirement is to produce the two sets of confidence limits based on order statistics defined by (11.2.2).

$$\begin{array}{ll} y_{(i)} \text{ and } y_{(n-i+1)} & \text{with level of confidence } 100(1-\alpha_2) \\ y_{(i+1)} \text{ and } y_{(n-i)} & \text{with level of confidence } 100(1-\alpha_1) \end{array} \qquad (11.2.2)$$

where i is chosen so that $1-\alpha_1 < 1-\alpha \leq 1-\alpha_2$.

The approximate $100(1-\alpha)\%$ confidence limits for M, based on (8.3.8), are then provided by (11.2.3).

$$\begin{array}{ll} \text{lower limit} & (1-d)y_{(i+1)} + dy_{(i)} \\ \text{upper limit} & (1-d)y_{(n-i)} + dy_{(n-i+1)} \end{array} \tag{11.2.3}$$

where i, α_1 and α_2 are determined from (11.2.2);

$$d = [(1-\alpha) - (1-\alpha_1)]/[(1-\alpha_2) - (1-\alpha_1)].$$

Numerical calculations are presented in the Illustration section below.

Application: The methods may be used in any situation to answer questions about the median value of a population or median response expected under a treatment or from a process.

Computation: The aim should be to obtain an approximate confidence interval for the median, M. This can serve the purpose of providing a basis for judging the acceptability of a proposed value M_0 and for estimating the likely range of values for M.

If a statistical package is to be used, the basic requirement is to produce two pairs of confidence intervals which may be inserted into (11.2.3) to obtain an approximate confidence interval of the desired level of confidence. Alternatively, the instructions provided below may be employed for direct computation of (approximate) confidence intervals.

If n' is 13, approximate 95% confidence intervals are given by Table 11.2.1. Intervals for other levels of confidence may be constructed using (11.2.3).

Table 11.2.1. Approximate 95% Confidence Limits for the Median Based on (11.2.3)

Sample Size	Lower Limit	Upper Limit
6	$0.9y_{(1)} + 0.1y_{(2)}$	$0.1y_{(5)} + 0.9y_{(6)}$
6	$0.7y_{(1)} + 0.3y_{(2)}$	$0.3y_{(6)} + 0.7y_{(7)}$
8	$0.4y_{(1)} + 0.6y_{(2)}$	$0.6y_{(7)} + 0.4y_{(8)}$
9	$0.9y_{(2)} + 0.1y_{(3)}$	$0.1y_{(7)} + 0.9y_{(8)}$
10	$0.7y_{(2)} + 0.3y_{(3)}$	$0.3y_{(8)} + 0.7y_{(9)}$
11	$0.3y_{(2)} + 0.7y_{(3)}$	$0.7y_{(9)} + 0.3y_{(10)}$
12	$0.9y_{(3)} + 0.1y_{(4)}$	$0.1y_{(9)} + 0.9y_{(10)}$

Where n' is 13 or greater, use may be made of the Normal approximation to the Binomial to produce the limits given in (11.2.4).

$$\begin{array}{ll} \text{lower limit} & (i+1-a)y_{(i)} + (a-i)y_{(i+1)} \\ \text{upper limit} & (a-i)y_{(n-i)} + (i+1-a)y_{(n-i+1)} \end{array} \tag{11.2.4}$$

where

$$a = \frac{1}{2}(n+1) - \sqrt{n} \qquad \text{approximate 95\% interval}$$
$$= \frac{1}{2}(n+1) - \frac{1}{2}z_{\frac{1}{2}\alpha}\sqrt{n} \quad \text{approximate } 100(1-\alpha)\% \text{ interval}$$
i = largest whole number less than a.

To test the hypothesis $M = M_0$, equation (11.2.1) provides the required p-value. Provided n' exceeds 12, the Normal approximation to the Binomial may be employed and Table 7.1.1 provides the necessary computational formulae. (To use this table note that x_1 is the number of values above M_0, $\pi_0 = 1/2$ and n must be set equal to n'.) Where n' is 12 or less, computation of p from first principles is necessary (see Section 7.1.4).

General statistical packages typically offer the option of testing an hypothesis or constructing a confidence interval for M.

Illustration: The procedure is applied in Example 11.3 where the aim is to estimate the average percentage increase in property values over a ten year period. The minimum and maximum observations are 0.05 and 5.21, respectively, which are within acceptable bounds.

A point estimate of the average is provided by the sample median which is 0.98.

Analysis using the MINITAB statistical package provides the following information from which to construct a 95% confidence interval:

Achieved Confidence	Confidence interval
91.2%	0.88 to 1.30
96.4%	0.83 to 1.33

The calculations based on (11.2.3) yield the following:

$$100(1-\alpha_1) = 91.2 \qquad \text{and} \qquad 100(1-\alpha_2) = 96.4.$$

From Table 11.1.1,

$$y_{(i)} = y_{(9)} = 0.83 \qquad \text{and} \qquad y_{(n'-i+1)} = y_{(20)} = 1.33$$
$$y_{(i+1)} = y_{(10)} = 0.88 \qquad \text{and} \qquad y_{(n'-i)} = y_{(19)} = 1.30.$$

Since 95% is proportionately $d = (95-91.2)/(96.4-91.2) = 0.73$ of the distance between 91.2 and 96.4, by (11.2.3) the lower limit of the 95% confidence interval is approximately

$$(1-0.73) \times 0.88 + 0.73 \times 0.83 = 0.84$$

and the upper limit is approximately

$$(1 - 0.73) \times 1.30 + 0.73 \times 1.33 = 1.32.$$

For comparison, use of (11.2.4) and the Normal approximation to the Binomial provides the limits:

lower limit	$(9 + 1 - 9.21) \times 0.83 + (9.21 - 9) \times 0.88 = 0.84$
upper limit	$(9.21 - 9) \times 1.30 + (9 + 1 - 9.21) \times 1.33 = 1.32$

where $a = \frac{1}{2}(28 + 1) - \sqrt{28} = 9.21$; $i = 9$.

Note that this interval can simply be converted to an average yearly increase in property values by multiplication by 12 (the number of months in a year) which suggests the likely average yearly percentage increase in property values is between 10 and 16%.

However, when the statistical model is considered there is concern about the sampling assumption. The data were only collected from sales by one company in the city and there is the real possibility that the transactions of that company may not be typical of those throughout the city.

The possible bias in sample selection is potentially serious and any claim that the results of the study refer to the city at large should be accompanied by a clear statement that the findings are based on the assumption that the figures for that one agency are typical of figures for all agencies. Where the investigator feels unable to support such a claim, the wise course of action is to simply treat the statistical component as a descriptive study and publish only the sample median plus the data set. By simply presenting what was observed there can be no criticism of the findings being misleading.

Additionally, there is concern about the size of the sample compared with the total number of sales over the ten year period. The statistical model on which the estimation procedure is based assumes the sample constitutes an arbitrarily small proportion of the population. The investigator in this study could not establish the population size since the information would have required access to confidential documents held by his competitors. However, he was prepared to concede that his sample might have constituted ten percent of the total.

Where investigators are aware that a statistical assumption in a model is violated, they should seek advice from a statistician as to the best way to proceed. In this case, the violation is not likely to be serious. There are, however, many occasions in which the use of a procedure is unwise if a statistical assumption has been violated.

11.2.3 Large Sample, Distribution-Free Methods for the Mean

Model: The model is distribution-free, assuming only that there exists a frequency or probability distribution with an assumed value for the mean of M_0.

Note that if the condition of symmetry is added, then the methods have application to the median since the mean and median are equal for a symmetric distribution.

Statistic: Given a sample of size n with a mean and standard deviation of $\overline{y}$ and s, respectively, the Central Limit Theorem (Section 7.3.5) establishes that the statistic z defined by (11.2.5) has a sampling distribution which is well approximated by a standard Normal distribution provided the sample size, n, is sufficiently large.

$$z = (\overline{y} - M_0)/(s/\sqrt{n}). \tag{11.2.5}$$

If M_0 is the true value of M, the statistic z should take a value in the vicinity of zero. Increasing distance from zero reflects a decreasing level of agreement between the data and the model which includes the assumed value M_0 for the mean.

Measure of agreement: If the value of z calculated from the data is z_0, and the hypothesized mean is M_0, the p-value is equal to the probability that the magnitude of z exceeds z_0 which is defined by (11.2.6).

$$p = \Pr(|z| > |z_0|) \tag{11.2.6}$$

where z_0 is computed from (11.2.5) with $\overline{y}$ and s being the sample mean and variance, respectively.

Estimation: The sample mean provides a point estimate of M and, as shown in Section 8.3.4, approximate confidence intervals are easily constructed. Based on the statistic $\overline{y}$, the confidence intervals are defined by (11.2.7).

$$\begin{array}{rl} 95\% \text{ confidence interval} & \overline{y} - 1.96 s_m \text{ to } \overline{y} + 1.96 s_m \\ 100(1-\alpha)\% \text{ confidence interval} & \overline{y} - z_{\frac{1}{2}\alpha} s_m \text{ to } \overline{y} + z_{\frac{1}{2}\alpha} s_m \end{array} \tag{11.2.7}$$

where

$$s_m = s/\sqrt{n} \text{ is known as the } \textit{standard error of the mean.}$$

Data checking: Check that there are no values which are outside the range of possible values.

Application: The rule-of-thumb is that the approximations offered by this approach are adequate when the sample size exceeds 30. If a line diagram or histogram suggests a symmetric distribution, the methods may also be used for the median since in a symmetric distribution, the mean and median take identical values.

Computation: Given the values of the sample mean ($\overline{y}$) and the standard error of the mean (s_m), the computation of a confidence interval may be performed directly using (11.2.7).

To check the hypothesis that $M = M_0$, reference may be made to a confidence interval or direct calculations of a p-value based on (11.2.6).

Illustration: The methodology was employed in the study reported in Example 11.5. A perusal of the data in Table 1.3.2 reveals a potential error. Can you see it?

The reading in row 10, column 8 is not unreasonable given the range of values, but since the values appear to be in increasing order, it is reasonable to deduce that the reading has been incorrectly recorded. Presumably, the figure is in error rather than simply misplaced.

In general, it is dangerous to attempt to guess what the correct value is when a value appears to have been mistyped. The wiser course of action is to seek confirmation of the correct value from the original records or, when this is not possible, to treat the value as a missing value. (Reference to the original recordings of the numbers which appear in Table 1.3.2 revealed that the figure of 345 should in fact be 245 and that is the figure which was employed in the calculations reported below.)

Statistical analysis yielded a sample mean of 223.7 and a value of 3.25 for the standard error of the mean. The 95% confidence interval for the expected mean level of cholesterol is computed from (11.2.7) as 217 to 230.

If it is accepted that the sample of persons providing the data could be regarded as typical of workers in an industrial plant in which heavy metal contamination of workers occurs, then the confidence interval provides the likely range for the mean level of cholesterol in that population.

However, the researcher wished to carry the analysis further by comparing mean cholesterol levels in persons working in that environment with the mean level of persons who are not exposed to heavy metal contamination. A figure of 220 mgm% is quoted for the mean level in the non-contaminated group. This lies inside the likely range for the contamined group. Hence, the data are consistent with the claim that the two groups have the same mean level of cholesterol.

11.2.4 Small Sample Procedures for the Mean

Model: The frequency or probability distribution is presumed to be well approximated by a Normal distribution with mean M_0.

Statistic: Given a sample of size n with mean $\bar{y}$ and standard deviation s, under the model proposed, the statistic t as defined in (11.2.8) may be employed.

$$t = (\bar{y} - M_0)/(s/\sqrt{n}). \tag{11.2.8}$$

If the frequency or probability distribution of the response variable is Normal with mean M_0, then t has a sampling distribution which is a *t-distribution* with $n - 1$ degrees of freedom.

Values of t in the vicinity of zero indicate agreement with the model which proposes a mean of M_0 and increasing distance from zero indicates decreasing agreement between model and data.

Measure of agreement: A value of t calculated from the data, t_0, can be converted to a p-value by determining the probability that the magnitude of t exceeds that of t_0 as defined in (11.2.9).

$$p = \Pr(|t(n-1)| \geq |t_0|). \tag{11.2.9}$$

Estimation: The sample mean provides a point estimate of M. The derivation of confidence intervals based on the statistic t is described in Section 8.3.4 and leads to the confidence intervals for M defined by (11.2.10).

$$\begin{aligned} &95\% \text{ C.I.} \quad \bar{y} - \mathrm{t}_{0.05}(n-1)s/\sqrt{n} \text{ to } \bar{y} + \mathrm{t}_{0.05}(n-1)s/\sqrt{n} \\ &100(1-\alpha)\% \text{ C.I.} \quad \bar{y} - \mathrm{t}_{\alpha}(n-1)s/\sqrt{n} \text{ to } \bar{y} + \mathrm{t}_{\alpha}(n-1)s/\sqrt{n}. \end{aligned} \tag{11.2.10}$$

Model and data checking: The following checks should be made.

1. Check the data visually for values which are outside acceptable bounds.
2. Produce a line diagram (Section 1.3.5) and check for (i) a single value which is surprisingly large or surprisingly small; and (ii) asymmetry in the pattern which may indicate the Normality assumption is unreasonable.
3. Construct a Normal probability plot (Section 10.3.4) and check for (i) a single value which is far from the general trend line and (ii) a curved trend line which suggests the assumption of Normality is unacceptable.

If there is evidence of an error in the data this should be attended to first. If there is sound evidence for deleting or correcting the value, Steps 1 to 3 should be applied to the amended data.

Where the Normality assumption appears unacceptable, the possible courses of action depend on the nature of the departure from Normality. If the distribution appears long-tailed (shown by an S-shaped trend in the Normal probability plot) but symmetric, the sign-based procedures described in Section 11.2.2 may be employed by utilizing the fact that for a symmetric distribution, mean and median are equal.

If there is evidence of asymmetry in the distribution, a distribution-free procedure for examining the mean is required. There is no widely available technique which provides a confidence interval for the mean. In this case, it is wise to seek assistance from a consultant statistician.

Computation: A confidence interval for M can be computed using (11.2.10) with the value of $\mathrm{t}_{\alpha}(n-1)$ obtained by one of the methods described in Section 7.4.4. Alternatively, the confidence interval may be obtained from a statistical package.

To check the hypothesis that $M = M_0$, either use a statistical package or compute a value of p from (11.2.9) using the instructions in Section 7.4.4. This p-value may be used to judge the acceptability of the hypothesis $M = M_0$.

Application: The procedures can be applied when the assumption of Normality is judged to be acceptable.

Illustration: The procedure was employed to establish if the data in Example 11.2 are consistent with a mean temperature of 750°C on the tungsten surface. Model examination involved the production of a line diagram and a Normal probability plot which are shown in Figures 11.2.2 and 11.2.3, respectively. The points in the Normal probability plot show no evidence of a non-linear trend and there was nothing of concern in the line diagram. Hence, the use of the t-statistic would appear reasonable.

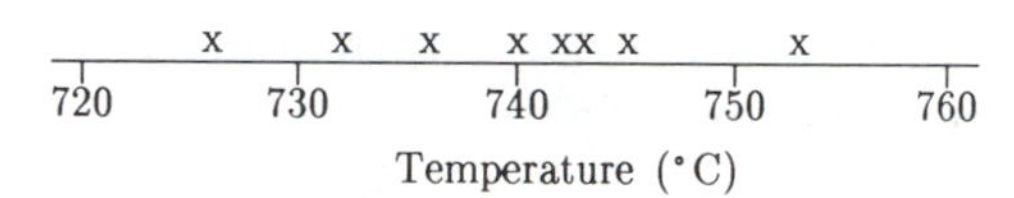

Figure 11.2.2. Line diagram for the data in Example 11.2.

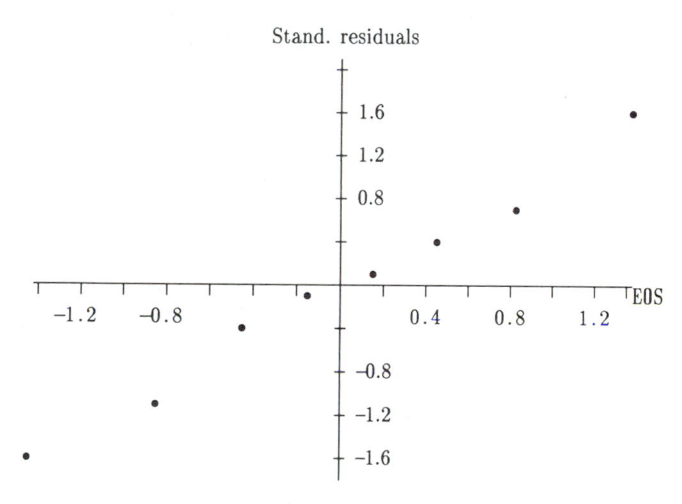

Figure 11.2.3. Normal probability plot for the data in Example 11.2.

Data examination revealed no obvious errors.

A 95% confidence interval for the mean surface temperature was computed by (11.2.10) to be 733 to 747. This suggests the actual mean temperature may have been lower than the nominal value of 750°C.

In a report on this analysis it was recommended that the confidence interval be included. Simply presenting the sample mean (740) gives no indication of the precision associated with it. This problem would not be overcome by also presenting the standard error of the mean since one needs access to t-tables and the ability to construct a confidence interval to make use of this information.

Problems

11.1 *Confidence intervals for medians — sign-based procedure*

(a) Based on the data in Table 1.3.5 which were collected in the study reported in Example 11.4, and following the procedure recommended in Section 11.2.2, compute separate 95% confidence intervals for median daily energy intake for *Girls* and *Boys.*

11.2 *Example* 3.1

The data in Table 3.2.1 were introduced in Example 3.1 and represent numbers of reported cases of a condition of the thyroid gland known as *thyrotoxicosis.* The numbers constitute 36 consecutive monthly readings.

(a) Verify the investigator's analysis that, based on the methodology of Section 11.2.2, an approximate 95% confidence interval for the median number of occurrences is 2.0 to 5.0.

(b) Plot *Occurrences* versus *Month.*

(c) Find the sample median, and hence, the proportion of observations above the median in each of the first, second and third twelve month periods.

(d) Is there evidence that the numbers of occurrences show a systematic variation with time? If so, how would this affect the practical value of the confidence interval computed in (a)?

11.3 *Confidence intervals for means*

The information recorded in Table 1.3.6 was collected by a group of students who were studying traffic flow along a road.

The road has a dual carriage way and no traffic may leave the road or stop between the two sites. One site is half a kilometer beyond a set of traffic lights and the other further along the road. There are no traffic lights between the two sites.

(a) Construct histograms based on the tables of frequencies and note the difference in patterns.

(b) Which site is closer to the traffic lights? Justify your answer.

Because of the way the students devised the survey, the same set of cars was observed at each site. Hence, the sample means for the two sets must be equal.

(c) Establish that this is so.

(d) Establish that the variances are unequal.

(e) Based on the sample data collected at each site, use (11.2.7) to compute separate 95% confidence intervals for the mean number of cars expected to travel along the road under conditions applying when the measurements were made (e.g. same time of day, day of week, etc.).

Since cars could neither leave nor enter the road between the two sites, the confidence intervals produced at (e) are estimating the same quantity and are using information on passage of the same set of cars.

(f) On the basis of the confidence intervals computed in (e), which appears to be the better site for the estimation of the mean number of cars per minute passing along the road?

11.4 *Confidence intervals for medians*

Data are provided in Table 1.3.1 for studying the effects of changes in levels of blood chemicals during a long distance run.

(a) For each of the four blood chemicals on which data are provided, *Glucose, Sodium, Potassium* and *Cholesterol,* compute the set of differences for the 22 runners, i.e. compute *difference* = *post reading* − *pre-reading* for each runner for every chemical.

(b) Using the sets of differences for the eleven *Untrained* runners, construct 95% confidence intervals for the median difference for each of the four chemicals. In which chemicals is there strong evidence that the level of the chemical has changed during the course of the race?

(c) Repeat the analysis and interpretation employed in (b) using the sets of differences for the eleven *Trained* runners. Are the findings in (b) and (c) consistent?

11.5 *Example* 11.2

Reproduce the model and data checking and the 95% confidence interval presented in Section 11.2.4 based on the data in Example 11.2.

12

Comparing Two Groups, Treatments or Processes

12.1 Forms of Comparison

One of the most common applications of statistical analysis is to assist scientists establish (i) if one population is different from another; (ii) if one treatment is better than another; or (iii) if one process has different characteristics than another. Scientists must specify what is meant by *different* or *better* and a range of possibilities is considered in Section 12.1.3. There are, however, two new features which require an introduction before proceeding to the possible forms of comparison.

12.1.1 Statistical Validity and the Comparison of Treatments

Methodology for the study of populations and experimental processes has been considered in both Chapter 9 and Chapter 11. A key element in all models introduced in those chapters has been the assumption of random selection of sample members to ensure valid application of the statistical methods.

Where an experimenter applies different treatments to subjects or units in order to judge the relative effects of the treatments, it is the method of allocating sample members between treatments rather than the selection of sample members which is critical to the *validity* of the statistical methodology. Validity of the probabilistic methods of Statistics is met by an allocation process in which sample members are independently allocated to treatments and with all sample members having the same probability of receiving a given treatment. Methods which satisfy these criteria satisfy the conditions of *random allocation.*

The above statement does not imply that the method of sample selection is unimportant. *Validity* is a necessary consideration, but it is not sufficient. Of great importance is the *sensitivity* of the experiment to the detection of treatment differences and the *scope of the conclusions.* These points are referred to in Section 12.1.2 and are examined in more depth in Chapter 13.

A word of explanation is called for about the term *treatment.* Statisticians use it in a technical sense which is much broader than common usage.

In the statistical sense, it applies to any situation in which an experimenter actively influences a subject or unit for the purpose of examining the effect of that influence. Thus, placing children in a particular learning experience is, in statistical eyes, as much a *treatment* as giving them a course of medication.

12.1.2 Matching and Adjustment Using Supplementary Information

Unrestricted random sampling from two groups or unrestricted random allocation of sample members between two treatments provides no guard against claims of unfair comparison. Random allocation of a sample containing ten males and ten females between two treatments does not protect against the possibility that one treatment receives all males and the other all females with the consequence that a so-called treatment difference may have a sexual component.

Two ways are open to avoid such an objection. The most obvious is to match the samples, either collectively or member by member. This involves ensuring that the collections of sample members assigned to the two treatments have identical characteristics in respect of one or more defined characteristics. For example, if *sex* is the defined characteristic, there might be a requirement that the proportion of females should be identical for the two collections. The other way is to adjust responses to compensate for differences in non-treatment effects.

The methods of this chapter assume no matching or adjustment of responses. The general statistical approaches employed to remove the effects of extraneous variation are discussed in Chapter 13.

12.1.3 Specific Forms of Comparison

A range of possible forms of comparison are listed and illustrated below. Included is a guide to the sections in the book where relevant methodology can be found.

1. *Comparison of proportions and success rates.* Where there are only two possible categories for the response variable, the problem can be posed as a comparison of two parameters, π_1 and π_2 which represent either population proportions in groups 1 and 2 or success rates under treatments 1 and 2. The selection and application of methods can be found in Section 12.2.

Example 12.1 provides an illustration of a study in which two groups are being compared.

2. *Looking for a shift in response.* When the response is scaled, the investigator's aim is commonly to seek to establish if there tends to be a higher response in one population than the other, one treatment than the other, or one process than the other.

The shift may be conceived as (i) a change in average response as reflected by change in the mean or median value; or (ii) a change in the distribution of responses.

As a basic rule, it is preferable to employ a method which detects differences in means or medians since these specific methods are more likely to detect a difference, should one exist.

In cases where there is no matching of samples or sample members and no adjustment for extraneous factors is required, the methodology for comparison of means is provided in Section 12.3 and that for comparison of medians is provided in Section 12.4. General distributional comparisons are considered in Chapter 16.

Example 12.2 presents a situation in which a comparison of treatments based on means was required and Example 12.3 a situation in which a comparison of two population medians was employed.

3. *Comparison of mean rates of occurrence of events.* Where outcomes are recorded as the number of occurrences of an event in time periods of fixed length or regions of fixed size, there may be interest in comparing the mean rates of occurrence under different conditions or treatments or for different groups.

In some cases, as illustrated in Example 12.4, there may be only a single reading from each source.

Where the Poisson model, as defined in Section 6.3, is applicable, a method for comparing the mean rates of occurrence of the events is provided in Section 12.3.8. Should there be reason to believe that the occurrence of one event may influence the chance of another event occurring, the Poisson model does not apply. Seek guidance from specialized literature or from a consultant statistician in such cases.

4. *Comparison of spread or variability.* When the response is scaled, there may be interest in establishing if the variability is different under the two treatments or between the two groups. Particularly in chemical and technological areas, interest is in *precision* which can be defined by one of the standard measures of variability, e.g. range, variance, or interquartile range. A method for comparing variances is provided in Chapter 19.

5. *Comparisons based on frequency tables.* Where data are summarized in the form of frequency tables, there is methodology which examines the hypothesis that the data from the two sources come from the same distribution. This situation arises when the response is categorical. Example 12.5 provides an illustration. The methodology is described in Chapter 16.

Note, however, that where there are only two possible categories for the response variable, it is generally preferable to treat the problem as one in which the aim is to compare proportions or probabilities and to use the methodology of Section 12.2.

Example 12.1 (Comparison of Different Types of Care for the Elderly). *The movement of elderly persons to a retirement home may involve either leaving those persons with much responsibility for managing their daily living or leaving them with little or no responsibility.*

A study was initiated to seek to establish the extent to which loss of responsibility might be related to retardation in physical or mental capabilities.

A retirement center offering various options in respect of responsibility was involved in the study. For a period of twelve months, every person taking up residency in the Center was asked to participate in the study. Ninety percent of those asked agreed to do so. The participants were divided into two groups, one group comprising persons responsible for their day-to-day living and the other group having all major decisions made for them. At the beginning of the study all subjects were given a test of recall ability. This was repeated at the end of the study and consequently each subject was classified as showing either (i) little or no memory deterioration or (ii) showing considerable memory deterioration. The results are recorded in Table 12.1.1.

The social scientist conducting the study was seeking to establish if removal of responsibility for day-to-day living affected the power of recall.

Table 12.1.1. Data on Memory Retention (Example 12.1)

	Group 1 responsibility maintained	Group 2 responsibility not maintained
No. with little or no deterioration	68	41
No. with much deterioration	10	24
Proportion with little or no deterioration	0.87	0.63

Example 12.2 (The Effect of Vitamin C Supplement on Ascorbate Levels in the Body). *A group of scientists working at a remote research station in a cold climate agreed to take part in a nutritional study investigating vitamin C levels in the bodies of persons living in cold climates. The* 26 *subjects were randomly allocated between the two groups so that half received vitamin C supplements and the other half formed the control group who received no supplement. Subsequently, two members of the control group were withdrawn from the study because they were found to be taking food sources which contained vitamin C supplements.*

The data recorded on the subjects comprise the level of leukocyte ascorbic acid in the blood, this being a measure of the level of vitamin C held in the body.

Based on this information, is there evidence that the prescribed vitamin C supplement affects leukocyte ascorbic acid levels in the body?

Level of Leukocyte Ascorbic Acid ($\mu g/10^8$ Cells)
(Data by courtesy of Dr. P. Gormley, Antarctic Div.
Aust. Dept. of Sci. and Technol.)

Group	
Supplement	Control
18.3	24.9
9.3	16.0
12.6	26.3
35.9	25.5
15.7	19.3
14.2	16.8
13.1	15.7
14.3	24.6
16.2	19.9
18.1	9.4
19.4	17.4
15.5	
11.7	

Example 12.3 (Energy Intake by Tasmanian School Children). *Table 1.3.5 contains data on the energy intake over a 24 hour period for a sample of 15 year old school children. The information is recorded separately for boys and girls. As one of the comparisons of interest, the investigators wished to compare median intakes for boys and girls.*

Example 12.4 (A Comparison of Accident Rates). *The Tasmanian Year Book [1984] records the following details of the number of road traffic accidents which are attributable to motorcycles.*

1982	90 *accidents*
1983	108 *accidents*

The observed rise led to a claim in the media of a "significant" increase in motor cycle related accidents from one year to the next. Is this a reasonable claim or might it be merely a consequence of chance? Might the difference be explained by the presence of more motorcyclists on the road?

Example 12.5 (Comparison of Spending Patterns). *The frequency table below classifies visitors to Tasmania according to their place of origin and the amount of money they spent. The aim is to compare the spending patterns of persons from the different groups.*

Table of Frequencies

Origin	Amount Spent		
	Low	Medium	High
Victoria	546	1294	375
Other places	408	1299	661

For visual comparison it is more informative to present the data in the form of relative frequencies in the different spending categories for each origin.

Table of Relative Frequencies

Origin	Amount Spent		
	Low	Medium	High
Victoria	0.25	0.58	0.17
Other places	0.17	0.55	0.28

The comparison in this case involves the frequency distributions for the two origins.

12.1.4 The Role of Confidence Intervals

Where the comparison is between values taken by a parameter in two groups or under two treatments, a common approach to statistical analysis is to examine the hypothesis that the values of the parameters are equal for the two groups or treatments, e.g. $\pi_1 = \pi_2$ in the case of proportions or probabilities; $M_1 = M_2$ in the comparison of means or medians.

There are substantial arguments against this restrictive approach since it may lead to one of the following two situations.

1. The possibility that acceptance of the hypothesis of no difference will be misinterpreted as evidence against the possibility of difference.

2. The possibility that rejection of the hypothesis of no difference will be interpreted as evidence of a difference of scientific significance even though the difference is not of sufficient magnitude to warrant this conclusion.

To avoid either of these situations arising, there is strong support for the presentation of a confidence interval for the difference (or ratio) in parameter values. By quoting the likely range of values for $\pi_1 - \pi_2$ or $M_1 - M_2$, scientists can (i) assess the weight to be assigned to the findings by noting the width of the interval; and (ii) draw conclusions as to the nature and relevance of the findings. These points are amply illustrated in the application of methodology in this chapter.

12.2 Comparisons Based on Success Rates and Proportions

12.2.1 The Forms of Comparison

Consider the comparison of two treatments labelled 1 and 2, for which outcomes are labelled *success* and *failure*, with π_1 and π_2 being the probabilities of success for treatments 1 and 2, respectively. The information may be summarized in the following table.

Treatment	Outcome	
	Success	Failure
1	π_1	$1-\pi_1$
2	π_2	$1-\pi_2$

Based on this table, three quantities are defined which are in common use for the purposes of comparing the success rates of the two treatments. They are:

1. the **difference** in success rates or proportions, $D = \pi_1 - \pi_2$;
2. the **relative risk** or ratio of success rates, $R = \pi_1/\pi_2$; and
3. the **odds ratio** $\omega = \dfrac{\pi_1/(1-\pi_1)}{\pi_2/(1-\pi_2)}$.

The difference, D, has application where there is interest in the numerical difference in success rates. For example, if under treatment 1 and 2 the success rates are 0.25 and 0.10, respectively, then for 100 applications of each treatment, an additional 15 successes are expected under treatment 1 than are expected under treatment 2.

Commonly in scientific situations, it is the relative risk or the odds ratio which is more informative. Thus, with success rates of 0.25 and 0.10 for treatments 1 and 2, respectively, it can be said that treatment 1 has a success rate which is two and one half times greater than that of treatment 2 ($R = 2.5$), or that the odds of success under treatment 1 are three times those under treatment 2 ($\omega = 3.0$).

The availability and ease of application of statistical methodology has influenced the choice in the past and continues to do so today. The construction of an approximate confidence interval for the difference (D) is a simple task and the formula to employ is widely available. By contrast, the construction of a confidence interval for either R or ω requires extensive computation and is not a widely available option in statistical packages.

Where feasible, the choice should be based on the most desirable form of comparison for the user. To this end, the means of constructing approximate confidence intervals for D, R and ω are all provided in this chapter.

12.2.2 Sample Size and Sensitivity

Possibly the greatest waste of experimental energies and resources is to be found in comparative studies in which sample sizes are too small to provide the sensitivity which scientists would regard as acceptable.

In this context, *sensitivity* may be defined by

(i) the probability that an estimate of the difference in success rates is within a distance δ of the true value $\pi_1 - \pi_2$; or

(ii) the probability of rejecting the hypothesis $\pi_1 - \pi_2 = 0$ when the true difference in success rates has a magnitude of D.

Sensitivity increases with increasing sample size. Hence, it is possible to define the minimum sample size required to reach a chosen level of sensitivity. Guidelines are provided in Fleiss ([1981], Chap. 3).

12.2.3 Choosing a Method

The choice is determined by two factors:

(a) the form of comparison; and

(b) whether the aim is to compare treatments, groups or processes.

The three commonly used forms of comparison are the difference, the relative risk and the odds ratio which are defined and discussed in Section 12.2.1.

Where the comparison is between two treatments, the manner of allocating sample members between treatments dictates the choice of statistical methods. The methods described in this chapter presume random allocation. Where matching is employed, reference may be made to Fleiss ([1981], Chap. 8).

When the comparison is between populations or processes, the manner of sample selection is important. The methods in this chapter are based on the assumption of random selection from each population and independence of selection between populations. Where there is matching of sample members or collections of sample members, reference may be made to Fleiss ([1981], Chap. 8).

12.2.4 Models

1. *Comparison of processes.* The presumptions are that the set of outcomes for each process are values derived under a Binomial model (Section 6.1) and that the samples are independently selected.

2. *Comparison of groups.* The presumptions are that the set of outcomes from each group are values derived under the Two-state population model (Section 6.2) and that samples are independently selected.

3. *Comparison of treatments.* The presumptions are that (i) the sample members have been randomly allocated between treatments, and (ii) that responses from sample members are independent and fall into one of two categories.

12.2.5 Data

For the application of the methods in this chapter, the data may be summarized in a table of frequencies, which has the following layout.

	Source 1	Source 2	Total
Number of successes	x_{10}	x_{20}	x_0
Number of failures	$n_1 - x_{10}$	$n_2 - x_{20}$	$n - x_0$
Total number in sample	n_1	n_2	n

From this table can be computed the sample proportions of successes, $p_{10} = x_{10}/n_1$ and $p_{20} = x_{20}/n_2$.

Where the data are provided in the form of individual responses, the reduction to a tabular form is simply a matter of counting the number of each type (success or failure) from each source. This is a task to which computers are well adapted and there will be found a simple means of constructing the table in all major packages.

12.2.6 Methods — Comparison Based on the Difference $\pi_1 - \pi_2$

Statistic: If p_1 and p_2 are the statistics denoting the probabilities of success in groups 1 and 2, respectively, then, for large sample applications, the statistic $d = p_1 - p_2$ has a sampling distribution which is well approximated by a Normal distribution with mean $\pi_1 - \pi_2$ and variance given by (12.2.1).

$$s_d^2 = p_1(1 - p_1)/n_1 + p_2(1 - p_2)/n_2. \qquad (12.2.1)$$

This large sample approximation derives from the Central Limit Theorem (Section 7.3.5). A necessary condition for this approximation to be adequate is that each of the statistics, p_1 and p_2, should have a sampling distribution which is well approximated by a Normal distribution. Following the guideline introduced in Section 9.2.4, the requirement is that the sample size exceeds n_{min} as defined in (12.2.2).

$$n_{\text{min}} = \frac{2}{3\{p_i(1 - p_i)\}^2} \qquad (12.2.2)$$

where p_i is the sample proportion in the group ($i = 1, 2$).

If either n_1 or n_2 is less than the respective values of n_{min} computed from (12.2.2), then use of the approximation based on the Normality assumption may result in misleading findings.

In practical terms, the failure of the Normality approximation has two possible causes. They are

(a) too little information due to one or both sample sizes, i.e. n_1 and n_2, being too small; or

(b) the comparison of rare events.

The statistics which might be employed given the latter possibility are discussed in Section 12.5. Theoretically, such statistics may also have application when sample sizes are small, although the low sensitivity of the statistical procedures in such circumstances makes their practical application of dubious value.

For the comparison of populations, the above approximation relies on the Binomial distribution providing a good approximation to the Hypergeometric distribution. This is the case if the population sizes are large relative to the sample sizes. Where either population size is less than twenty times the corresponding sample size, (12.2.1) and (12.2.2) require modification in the manner described in Section 9.2.5.

Measure of agreement: Where the Normal approximation is applicable and the hypothesis $\pi_1 - \pi_2 = D_0$ is correct, the most likely value for the statistic $d = p_1 - p_2$ is the value D_0, and increasing distance from D_0 reflects decreasing agreement between model and data.

If the observed difference is d_0, an approximate value for the p-value is provided by (12.2.3).

$$p = \Pr\{|z| \geq |(d_0 - D_0)/s_d|\} \tag{12.2.3}$$

where s_d is determined from (12.2.1).

Estimation: The difference in sample proportions, $p_1 - p_2$, provides the Maximum Likelihood estimator of the corresponding difference in parameter values, $\pi_1 - \pi_2$.

If the Normal approximation is applicable, an approximate confidence interval for $\pi_1 - \pi_2$ can be constructed from (8.3.3) and yields the intervals defined in (12.2.4).

$$\begin{array}{ll} 95\% \text{ confidence interval} & d - 1.96s_d \text{ to } d + 1.96s_d \\ 100(1-\alpha)\% \text{ confidence interval} & d - z_{\frac{1}{2}\alpha}s_d \text{ to } d + z_{\frac{1}{2}\alpha}s_d \end{array} \tag{12.2.4}$$

where s_d is determined from (12.2.1).

Data checking: Ensure that all responses fall into one of the two possible categories.

Computation: For the comparison of treatments or processes, ensure that sample sizes n_1 and n_2 each exceeds the minimum value defined by (12.2.2).

Where the condition is not met, refer to Section 12.5 for methodology appropriate for rare events.

If the condition is met, compute a value for s_d using (12.2.1) and construct a confidence interval for $\pi_1 - \pi_2$ using (12.2.4). The steps are illustrated below.

Illustration: The calculations are illustrated for the comparison of two groups in the study reported in Example 12.1.

From the data in Table 12.1.1, the following figures are obtained:

$$p_{10} = 0.87 \quad p_{20} = 0.63 \quad n_1 = 78 \quad n_2 = 65.$$

For Group 1, the value of n_{min} is computed from (12.2.2) to be 53 and for Group 2, the value of n_{min} is computed to be 13. In each case, the sample size exceeds the minimum for application of the Normal approximation.

The value of $d = p_1 - p_2$ is 0.24, and, using (12.2.1), the value of s_d is computed to be 0.071.

Thus, from (12.2.4) an approximate 95% confidence interval for $\pi_1 - \pi_2$ is

$$\begin{array}{ccc} 0.24 - 1.96 \times 0.071 & \text{to} & 0.24 + 1.96 \times 0.071 \\ 0.10 & \text{to} & 0.38. \end{array}$$

The fact that the confidence interval does not include the value zero indicates evidence against the hypothesis $\pi_1 = \pi_2$, i.e. against the proposition that the proportions of persons with little or no memory loss are identical under each treatment.

A useful practical interpretation of the 95% confidence interval is that somewhere between 10 and 38 more persons per hundred are likely to suffer a deterioration in memory in the group with no responsibility than in the group which retains some responsibility.

Such a finding may **not** be taken as evidence that lack of responsibility **causes** a deterioration in memory. That conclusion would rely on (i) eliminating other possible causal factors, and (ii) establishing medical grounds for expecting that memory retention is dependent on factors defined by responsibility.

12.2.7 Relative Risk and Odds Ratios: Choosing a Statistic

Statistics: By analogy with the study of differences, intuition would suggest that methodology for the examination of the relative risk, $R = \pi_1/\pi_2$, would be based on the sampling distribution of the statistic p_1/p_2. As is the case in many statistical applications, the approach suggested by intuition is not the one adopted. There are both statistical and computational reasons for this. Instead, the statistics which are in use are based on the comparison of frequencies in Tables 12.2.1 and 12.2.2.

Table 12.2.1. Table of Observed Frequencies

	Treatment A	Treatment B	Total
No. of successes	$o_{11} = x_{10}$	$o_{21} = x_{20}$	x_0
No. of failures	$o_{12} = n_1 - x_{10}$	$o_{22} = n_2 - x_{20}$	$n - x_0$
Total number	n_1	n_2	n

Table 12.2.2. Table of Expected Frequencies

	Treatment A	Treatment B
No. of successes	$e_{11} = n_1\pi_1$	$e_{21} = n_2\pi_2$
No. of failures	$e_{12} = n_1(1 - \pi_1)$	$e_{22} = n_2(1 - \pi_2)$
Total number	n_1	n_2

The statistical aim is to define a set of frequencies expected under a model in which the relative risk, $R = \pi_1/\pi_2$, is assigned a chosen value R_0, and to define a statistic which provides a measure of closeness of the observed and expected tables of frequencies.

Since the frequencies in Table 12.2.2 require specification of values for both π_1 and π_2 it is apparent that restricting the ratio π_1/π_2 to a value R_0 does not supply sufficient information to evaluate expected frequencies. There are many possible sets of expected frequencies which would meet the condition that π_1/π_2 equals R_0. Hence, there is not an obvious way of constructing a unique measure of closeness between model and data.

Statisticians have a number of options for overcoming this problem. The most desirable way is to find a statistic which takes the same value for all pairs, π_1 and π_2, which satisfy the condition $\pi_1/\pi_2 = R_0$.

That option is not available in this case and an alternative approach is adopted. It involves placing a restriction on π_1 and π_2 which permits only one possible combination for a given value of R. Statistical theory establishes the most desirable restriction as

$$n_1\pi_1 + n_2\pi_2 = x_0,$$

which, in words, states that the total number of successes expected under the model should equal the total number of successes observed in the data.

With this restriction, it is possible to define unique entries for the table of expected frequencies in terms of either the relative risk or the odds ratio. Table 12.2.3 provides the formulae when the relative risk is hypothesized to take the value R_0.

Tables 12.2.1 and 12.2.3 provide the basic elements for examining hypothesized values for the relative risk or for constructing a confidence interval for R. However, there remains the task of defining a statistic which reflects the closeness of model and data based on the information contained in the frequency tables.

Table 12.2.3. Table of Expected Frequencies Under the Hypothesis $R = R_0$ Given That $n_1\pi_1 + n_2\pi_2 = x_0$

	Treatment A	Treatment B	Total
No. of successes	$e_{11} = \frac{n_1 R_0 x_0}{n_1 R_0 + n_2}$	$e_{21} = \frac{n_2 x_0}{n_1 R_0 + n_2}$	x_0
No. of failures	$e_{12} = n_1 - e_{11}$	$e_{22} = n_2 - e_{21}$	$n - x_0$
Total number	n_1	n_2	n

The commonly used statistics for comparing model and data are (12.2.5) and (12.2.6).

$$X^2 = \sum_{i=1}^{2}\sum_{j=1}^{2}\left[\frac{(|o_{ij} - e_{ij}| - 0.5)^2}{e_{ij}}\right] \tag{12.2.5}$$

$$X_r^2 = \sum_{i=1}^{2}\sum_{j=1}^{2}\left[\frac{(o_{ij} - e_{ij})^2}{e_{ij}}\right]. \tag{12.2.6}$$

In each case, the values taken by the statistic must be non-negative and increasing values are presumed to reflect worsening agreement between the data and the model which presumes $R = R_0$. Thus, for an observed value X_0^2 the calculation of a p-value to measure agreement between model and data requires the calculation of the probability of equaling or exceeding X_0^2, i.e. $\Pr(X^2 \geq X_0^2)$ or $\Pr(X_r^2 \geq X_0^2)$.

In neither case does the statistic have a sampling distribution which lends itself to the direct calculation of this probability. Statistical theory does, however, establish that the Chi-squared distribution with one degree of freedom can provide a good approximation to the sampling distribution of either statistic in large samples.

The choice between the statistics X^2 and X_r^2 is judged by the closeness with which their sampling distributions can be approximated by a Chi-squared distribution. The general consensus among statisticians is that (12.2.5) is the preferred choice although (12.2.6) appears widely in both the literature and statistical packages.

In either case, there is yet another consideration in establishing the quality of the approximation and that is a requirement that none of the e_{ij}'s be too small. What constitutes *too small* is given different values by different statisticians, the problem being that the difference between the exact p-value and that estimated by use of the Chi-squared distribution depends in a complex manner on the configuration of the expected values and overall sample size.

A rule-of-thumb which is widely adopted requires that **no e_{ij} should be less than 3** irrespective of which statistic, X^2 or X_r^2, is employed.

12.2.8 Relative Risk and Odds Ratio: Hypothesis Testing

Hypothesis: The traditional approach to the comparison of relative risks or odds ratios is to examine the hypothesis that π_1 and π_2 are equal. The hypotheses $R = 1$ and $\omega = 1$ are equivalent and arise when $\pi_1 = \pi_2$. Hence, the same procedure may be used for both.

Under the hypothesis that $R = 1$, $\omega = 1$ or $\pi_1 = \pi_2$, Table 12.2.3 can be used to provide the expected frequencies simply by setting R_0 to be equal to 1.0. The values obtained are recorded in Table 12.2.4.

Table 12.2.4. Table of Expected Frequencies Under the Hypothesis $R = 1$ Given That $n_1\pi_1 + n_2\pi_2 = x_0$

	Treatment A	Treatment B	Total
No. of successes	$e_{11} = n_1 p_0$	$e_{21} = n_2 p_0$	x_0
No. of failures	$e_{12} = n_1(1 - p_0)$	$e_{22} = n_2(1 - p_0)$	$n - x_0$
Total number	n_1	n_2	n

where $p_0 = x_0/n$.

Computations: The computations involved in testing the hypothesis $R = 1$ or $\omega = 1$ are described below and illustrated using the information provided in Example 12.6.

The steps to be followed are listed below. Statistical packages will automatically perform all steps except Step 3.

Step 1 Construct a table of observed frequencies in the manner of Table 12.2.1.

Step 2 Construct a table of expected frequencies using the formulae in Table 12.2.4.

Step 3 Check that no expected frequency is less than 3—if there should be one or more frequencies less than 3, proceed to Section 12.5 for an alternative method.

Step 4 Calculate a value X_0^2 using (12.2.5), or, less desirably, (12.2.6).

Step 5 Convert X_0^2 to a p-value using the Chi-squared distribution with one degree of freedom to determine $\Pr\{\chi^2(1) \geq X_0^2\}$ which is an approximate value for $p_0 = \Pr(X^2 \geq X_0^2)$ or $p_0 = \Pr(X_r^2 \geq X_0^2)$.

Example 12.6 (Comparing Rates of Defective Vehicles at Two Checkpoints). *As a result of publicity over several fatal road accidents which were attributed to defective vehicles, police conducted a series of checks of vehicles in different suburbs. Data from two suburbs are recorded* *below. At each locality, vehicles passing in a thirty minute period were stopped and checked. Vehicles were identified as defective or defect-free.*

	Locality 1	Locality 2	Total
No. of defect-free vehicles	38	51	89
No. of defective vehicles	7	19	26
Total number of vehicles	45	70	115

Illustration: The results of applying the procedure to the data in Example 12.6 are presented below.

Table of Expected Frequencies
assuming $R = 1$

	Locality 1	Locality 2	Total
No. of defect-free vehicles	34.8	54.2	89
No. of defective vehicles	10.2	15.8	26
Total number of vehicles	45	70	115

Based on (12.2.5), $X_0^2 = 1.52$. From the Chi-squared distribution with one degree of freedom,

$$p_0 \cong \Pr\{\chi^2(1) \geq 1.52\} = 0.22.$$

The conclusion would be reached that the two localities may have the same expected proportion of defective vehicles on the roads since p has a value in excess of 0.05.

12.2.9 Relative Risk and Odds Ratio: Estimation

Sample estimates of R and ω are provided by replacing π_1 and π_2 by their sample estimates, p_1 and p_2, respectively.

The steps in constructing a confidence interval for R and/or ω are presented below.

To construct a $100(1-\alpha)\%$ confidence interval for R, the aim is to find all values of R_0 in Table 12.2.3 which result in (12.2.5) having a value which is less than or equal to X_α^2 where X_α^2 satisfies $\Pr(\chi^2(1) \geq X_\alpha^2) = \alpha$. This is the range of values of R_0 which would lead to a p-value in excess of α if a test of significance of the hypothesis $R = R_0$ were performed.

In the construction of confidence intervals which have been presented in previous chapters, it has been possible to find analytic expressions for the limits of confidence intervals. In this case such expressions cannot be obtained and an iterative process is required.

Both for reasons of length of calculation and the need to maintain a large number of significant figures throughout the calculations, it is not suited to hand calculators. The methodology is provided because it is not widely available in statistical packages.

The steps in the computation are presented below using the notation of Tables 12.2.1 and 12.2.3. They are illustrated using the data in Example 12.6.

Step 1 — *preliminary check on validity of Chi-squared approximation to sampling distribution of statistic.* If n, the overall sample size, is less than 40 the method should not be applied—there is simply too little information.

Set $p_0 = x_0/n$. If any one of $n_1 p_0$, $n_2 p_0$, $n_1(1-p_0)$ or $n_2(1-p_0)$ is less than 3, this may be taken as a warning that the method is unlikely to be valid. A more definitive check occurs during the course of the calculations. If the method is rejected for this reason, the only recourse is to the methodology presented in Section 12.5.

Based on Example 12.6,

> $p_0 = 89/115$ *and hence,* $n_1 p_0 = 35$, $n_2 p_0 = 54$, $n_1(1-p_0) = 10$ *and* $n_2(1-p_0) = 16$.

Thus, the preliminary check suggests the method will be valid.

Step 2 — compute

$$w = \frac{(x_{10}+0.5)(n_2 - x_{20} + 0.5)}{(x_{20}+0.5)(n_1 - x_{10} + 0.5)} \qquad \text{and}$$

$$S = \sqrt{\left[\frac{1}{x_{10}+0.5} + \frac{1}{x_{20}+0.5} + \frac{1}{n_1 - x_{10}+0.5} + \frac{1}{n_2 - x_{20}+0.5}\right]}.$$

Use these values to compute initial guesses at the lower and upper $100(1-\alpha)\%$ confidence limits of the odds ratio using the formulae

$$\omega_L(1) = \exp[\log(w) - z_{\frac{1}{2}\alpha} S] \qquad \text{and}$$
$$\omega_U(1) = \exp[\log(w) + z_{\frac{1}{2}\alpha} S].$$

(To provide the precision needed for this method, greater than usual accuracy is required. The three commonly used values are

- for a 95% confidence interval $z_{\frac{1}{2}\alpha} = 1.95996$
- for a 99% confidence interval $z_{\frac{1}{2}\alpha} = 2.57583$
- for a 90% confidence interval $z_{\frac{1}{2}\alpha} = 1.64485$.)

(*Application to Example 12.6 yields values of*

$$\omega_L(1) = 0.759276 \qquad \textit{and} \qquad \omega_U(1) = 4.975694$$

for 95% *confidence limits.*)

Step 3 — set $\omega = \omega_L(1)$ and perform the calculations listed in this step. Then repeat the calculations but this time setting $\omega = \omega_U(1)$ (i.e. the calculations must be performed separately for each limit but the method of calculation is the same in both cases).
Calculate

$$c_1 = \omega(x_0 + n_1) + n_2 - x_0$$
$$c_2 = \sqrt{[c_1^2 - 4x_0 n_1 \omega(\omega - 1)]}$$

and hence,

$$e_{11}(1) = \frac{(c_1 - c_2)}{2(\omega - 1)}$$
$$e_{12}(1) = n_1 - e_{11}(1)$$
$$e_{21}(1) = x_0 - e_{11}(1)$$
$$e_{22}(1) = n_2 - e_{21}(1).$$

If any $e_{ij}(1)$ is less than 3.0 the sample size should be regarded as too small for valid application of the test procedure. Apart from the statistical validity, the numerical process becomes unstable if any e_{ij} is near zero.
(*Application to Example 12.6 gives*

$$c_1 = 82.742978 \qquad c_2 = 98.865927$$

which yields expected frequencies

$$e_{11}(1) = 33.488448$$
$$e_{12}(1) = 11.511552$$
$$e_{21}(1) = 55.511552$$
$$e_{22}(1) = 14.488448.)$$

Step 4 — set

$$s = +1 \qquad \text{if } x_{10} > e_{11}(1)$$
$$= -1 \qquad \text{otherwise.}$$

Calculate

$$t = s\{|x_{10} - e_{11}(1)| - 0.5\}$$
$$W = \frac{1}{e_{11}(1)} + \frac{1}{e_{12}(1)} + \frac{1}{e_{21}(1)} + \frac{1}{e_{22}(1)}$$
$$F = t^2 W - (z_{\frac{1}{2}\alpha})^2$$
$$T = \frac{1}{2(\omega - 1)^2}\left[c_2 - n - \frac{\omega - 1}{c_2}[c_1(x_0 + n_1) - 2x_0 n_1(2\omega - 1)]\right]$$
$$U = \frac{1}{[e_{12}(1)]^2} + \frac{1}{[e_{21}(1)]^2} - \frac{1}{[e_{11}(1)]^2} - \frac{1}{[e_{22}(1)]^2}$$
$$V = T(t^2 U - 2Wt).$$

Adjust the value of ω by subtracting an amount equal to F/V, i.e.

$$\omega = \omega - \frac{F}{V}.$$

Calculate the ratio of the magnitude of the adjustment to the new value, i.e. $\frac{|F/V|}{\omega}$.

If this value is less than a predetermined error bound (0.001 would seem to be a suitable practical bound) then the calculations are complete. Otherwise, use the new value of ω in Step 3 and repeat the process.

(Application to Example 12.6 is shown below.)

1. *Lower limit — 3 iterations were required with intermediate values as detailed.*

	Iteration		
i	1	2	3
$e_{11}(i)$	33.488448	33.122923	33.151140
$e_{12}(i)$	11.511552	11.877077	11.848860
$e_{21}(i)$	55.511552	55.877077	55.848860
$e_{22}(i)$	14.488448	14.122923	14.151140
F/V	0.0544	-0.0041	-0.0000
$\|F/V\|\omega$	0.077	0.006	0.000
ω	0.704872	0.708923	0.708949

2. *Upper limit — 3 iterations were required with intermediate values as detailed.*

	Iteration		
i	1	2	3
$e_{11}(i)$	41.166111	41.664181	41.648087
$e_{12}(i)$	3.833889	3.335819	3.351913
$e_{21}(i)$	47.833889	47.335819	47.351913
$e_{22}(i)$	22.166111	22.664181	22.648087
F/V	-1.0044	0.0373	0.0000
$\|F/V\|/\omega$	0.167	0.006	0.000
ω	5.980128	5.942878	5.942838

Step 5(a) — *approximate confidence limits for R.* Use the values of e_{11} and e_{21} from the last iteration for each of the lower and upper limits to compute lower and upper limits for R from the formula

$$R = \frac{e_{11}/n_1}{e_{21}/n_2}.$$

(For Example 12.6, the limits are

$$R_L = \frac{(33.15)/45}{(55.85)/70} = 0.92 \quad \textit{and} \quad R_U = \frac{(41.65)/45}{(47.35)/70} = 1.37.)$$

Step 5(b) — *approximate confidence limits for* ω. Use the values of e_{11}, e_{12}, e_{21} and e_{22} from the last iteration for each of the lower and upper limits to compute lower and upper limits for ω from the formula

$$\omega = \frac{e_{11}e_{22}}{e_{21}e_{12}}.$$

(*For Example 12.6 the limits are*

$$\omega_L = \frac{(33.15)(14.15)}{(55.85)(11.85)} = 0.71 \qquad \text{and} \qquad \omega_U = \frac{(41.65)(22.65)}{(47.35)(3.35)} = 5.9.)$$

Interpretation: If the interval includes 1.0, the hypothesis of no difference, $\pi_1 = \pi_2$, is reasonable. The limits of the confidence interval may be interpreted in the manner described below for Example 12.6.

For Example 12.6, *the limits on the relative risk of* 0.92 *to* 1.4 *includes* 1.0 *as do the limits of* 0.71 *to* 5.9 *on the odds ratio. Hence, the data are consistent with the assumption that the odds of finding a non-defective car are the same at the two localities.*

However, the true odds may be almost six times as great of finding a non-defective car at locality 1 *than at locality* 2. *Since the lower limit is below one, it may be more convenient to invert the figure* (1/0.71 = 1.4) *and hence, conclude that the odds of finding a non-defective car at locality* 2 *may be up to* 1.4 *times the odds of finding a non-defective car at locality* 1.

12.3 Comparisons of Means

12.3.1 Data

The data are presumed to comprise a set of n_1 readings from one source (treatment, group or process) and a set of n_2 readings from a second source. The following notation is used to denote the sets of readings and derived statistics:

	Readings					Mean	Variance
Source 1	y_{110}	y_{120}	y_{130}	$\cdots$	y_{1n_10}	$\bar{y}_{10}$	s_{10}^2
Source 2	y_{210}	y_{220}	y_{230}	$\cdots$	y_{2n_20}	$\bar{y}_{20}$	s_{20}^2

12.3.2 Experimental Aims

The experimental questions relate to the relative values of expected means for Sources 1 and 2 which are referred to as M_1 and M_2, respectively.

Methodology is provided for (i) testing the hypothesis $M_1 = M_2$; (ii) estimating the likely size of the difference in means, i.e. $M_1 - M_2$, and (iii) estimating the ratio of means, i.e. M_1/M_2 in circumstances where this is a sensible form of comparison.

12.3.3 Statistical Models

There are three basically different models which are considered:

- models based on the Normal distribution;
- models based on the Poisson distribution;
- distribution-free models.

Those based on the Poisson distribution are identified by the nature of the response variable. They apply when the experiment produces counts of the number of occurrences of an event. Section 6.3 lists a range of situations where this model has application. Example 12.4 provides an illustration where the model might be employed for comparative purposes.

Models based on the Normal distribution are widely employed in the comparison of means. When sample sizes are small, the decision to use a model which includes the Normality assumption is based primarily on the nature of the response variable (as discussed in Section 6.5).

Distribution-free models have application when the sample sizes are large, when there is doubt as to the acceptability of the Normality assumption or when the distribution is known to be non-Normal and the appropriate distribution-based methodology is not available.

There is extensive discussion in Section 12.1 of sampling requirements for comparison of groups and processes and of treatment allocation when two treatments are to be compared. The conditions required for valid application of the methods are important and should be fully understood.

12.3.4 Choosing a Method

The methods are grouped according to the distributional requirements detailed above. Selection can be based on the following simple rules.

1. Poisson-based methodology described in Section 12.3.8 should be selected if the data comprise counts of independent events.
2. The distribution-free method described in Section 12.3.5 is available when both sample sizes, i.e. n_1 and n_2, exceed 30.
3. For smaller sample sizes, methodology based on the Normal distribution and described in Section 12.3.6 should be selected where the nature of the response variable indicates the Normal distribution model (Section 6.5) is likely to apply, i.e. if the response variable can be considered to be under the influence of many factors each of which makes a small contribution to the final response.
4. Transformation of data to a logarithmic scale is commonly required in cases where the aim is to compare ratios of means rather than

differences. Such a situation is likely to arise when (i) the response variable takes only non-negative values; (ii) the variability is expected to increase as the mean increases; and (iii) the ratio of the largest to smallest observation is of the order of ten or greater.

The method of analysis and interpretation of results is described in Section 12.3.7.

5. Distribution-free methodology should be considered when one or both of the sample sizes is small and acceptance of the Normality assumption is doubtful. A test of the hypothesis $M_1 = M_2$ can be made using an approximate randomization test (Section 7.8.2) or a confidence interval may be obtainable for $M_1 - M_2$ using the method described in Section 12.4.3.

12.3.5 A Large Sample Distribution-Free Method

Model: The model requires only an hypothesized difference D_0 for the difference in means, $M_1 - M_2$.

Statistic: The statistic employed is the difference in sample means, $d = \overline{y}_1 - \overline{y}_2$, which is to be compared with the hypothesized difference in expected values, D_0. If d takes a value close to D_0 then there is good agreement between model and data. As the distance between d and D_0 grows, so the agreement between model and data diminishes.

Provided both n_1 and n_2 are at least thirty, the Central Limit Theorem (Section 7.3.5) establishes that, for the vast majority of distributional forms encountered in practice, the sampling distribution of d is well approximated by a Normal distribution with mean $D = M_1 - M_2$ and variance, s_d^2, given by (12.3.1).

$$s_d^2 = \frac{s_1^2}{n_1} + \frac{s_2^2}{n_2}. \qquad (12.3.1)$$

Measure of agreement: The p-value for comparing the model which assumes an expected difference of D_0 with data which provides a difference in sample means of d_0 is based on the distance between the observed and expected values. The value of p is equal to the probability of obtaining a distance which is equal to or greater than the observed distance, i.e. $p = \Pr(|d - D_0| \geq |d_0 - D_0|)$. By utilizing the Central Limit theorem, an approximate value of p may be computed using (12.3.2).

$$p \cong \Pr(|z| \geq |(d_0 - D_0)/s_d|) \qquad (12.3.2)$$

where s_d is computed from (12.3.1).

Estimation: The observed difference in sample means, d_0, provides a point estimate of $M_1 - M_2$.

Approximate confidence intervals are constructed using (8.3.3) and may be computed from (12.3.3).

$$\begin{array}{llll} 95\% \text{ confidence interval} & d_0 - 1.96s_d & \text{to} & d_0 + 1.96s_d \\ 100(1-\alpha)\% \text{ confidence interval} & d_0 - z_{\frac{1}{2}\alpha}s_d & \text{to} & d_0 + z_{\frac{1}{2}\alpha}s_d \end{array} \quad (12.3.3)$$

where s_d is determined from (12.3.1).

Data checking: Check that there are no values which are outside the range of possible values and no values which are regarded as unusually large or unusually small.

Application: The rule-of-thumb is that the approach may be used if both sample sizes exceed 30.

Computation: Approximate confidence intervals can be constructed for the difference in means, $D = M_1 - M_2$, using (12.3.3).

To test the hypothesis $D = D_0$, the recommended method is to utilize a confidence interval for D and establish if D_0 is contained within the interval. Alternatively, a p-value may be determined by computing a value of p from (12.3.2) using either Normal distribution tables or a statistical computing package.

Illustration: The calculations are illustrated using data extracted from Table 12.3.1 which were obtained in Example 12.7.

Example 12.7 (Length of Labor and the Use of 'Syntocinon'). *For a period of twelve months, data were recorded on women giving birth in Tasmanian hospitals in respect of the use of 'Syntocinon,' a drug used to induce labor. Four groups of women were identified in respect to drug usage:*

Group	*Details*
1	*No syntocinon*
2	*Syntocinon within 2 hours of amniotomy**
3	*Syntocinon 2 to 6 hours after amniotomy*
4	*Syntocinon more than 6 hours after amniotomy*

**the operation to artificially rupture the membrane.*

Data were collected on the following variables:

- *Time from amniotomy to established labor;*
- *Time from amniotomy to delivery;*
- *Time from established labor to delivery.*

A summary of the data collected is given in Table 12.3.1.

The purpose of collecting the data was to compare average times between selected groups at each stage of the birth process.

Table 12.3.1. Summary Statistics from the Syntocinon Study (Example 12.7) (Data Courtesy of Prof. J. Correy, Dept. of Obstets., Univ. of Tasmania)

Group	Sample size	Amniotomy to Established labour		Amniotomy to Delivery		Established labour to Delivery	
		Mean	Variance	Mean	Variance	Mean	Variance
1	315	4.66	19.1199	9.43	32.4616	4.77	10.9990
2	301	3.65	11.7142	9.14	26.2455	5.49	14.5384
3	47	7.40	36.4200	12.49	59.6466	5.09	10.7752
4	58	14.93	40.4513	23.62	92.7659	8.69	42.7441

The calculations are illustrated for the comparison of the mean times from amniotomy to established labor for Groups 1 and 2 (M_1 and M_2).

From the information in Table 12.3.1,

$$d_0 = 4.66 - 3.65 = 1.01$$

$$s_d = \sqrt{\frac{19.1199}{315} + \frac{11.7142}{301}} = 0.32.$$

Hence, by (12.3.3), an approximate 95% confidence interval for $M_1 - M_2$ is

$$1.01 - 1.96 \times 0.32 \quad \text{to} \quad 1.01 + 1.96 \times 0.32$$
$$0.38 \quad \text{to} \quad 1.64.$$

The confidence interval suggests that the mean time from amniotomy to established labor is likely to be between 0.38 hours (about 20 minutes) and 1.64 hours (about 1 hour 40 minutes) less in the group who receive Syntocinon as compared with the group who received no Syntocinon.

The fact that the interval does not include zero is evidence against the hypothesis $M_1 = M_2$.

It is important to appreciate that the statistical analysis cannot provide evidence that the administration of Syntocinon is responsible for the difference.

12.3.6 Small Sample Methods

Model 1: The response variable is presumed to have patterns of variation which are well approximated by Normal distributions in both groups, with the distributions for Groups 1 and 2 having means M_1 and M_2, respectively.

The pattern of variation is presumed to be the same for both groups, except for possible differences in means. This implies the frequency distributions have the same (unknown) value, σ^2, for their variance.

Samples are presumed to have been independently selected from the groups.

Where the comparison is between the effects of two treatments, the assumptions differ only by requiring that sample members are randomly allocated between treatments.

Model 2: The conditions are those of model 1 except for the relaxation of the condition that the two distributions must have the same value for their variances. In this model, Groups 1 and 2 are presumed to have variances σ_1^2 and σ_2^2, respectively.

Statistics: Two statistics are considered, the statistic t_1 which applies under Model 1 and the statistic t_2 which applies under Model 2.

The statistic t_1 defined by (12.3.4) has a sampling distribution which is a t-distribution with $\nu_1 = n_1 + n_2 - 2$ degrees of freedom.

$$t_1 = \frac{(\overline{y}_1 - \overline{y}_2) - (M_1 - M_2)}{s_{1d}} \tag{12.3.4}$$

where

$$s_{1d} = \sqrt{\frac{(n_1 - 1)s_1^2 + (n_2 - 1)s_2^2}{n_1 + n_2 - 2}\left[\frac{1}{n_1} + \frac{1}{n_2}\right]}.$$

The statistic t_2 defined by (12.3.6) has a sampling distribution which is approximated by a t-distribution with degrees of freedom ν_2 where ν_2 is the nearest whole number to the value computed from (12.3.5).

$$\left[\frac{s_1^2}{n_1} + \frac{s_2^2}{n_2}\right]^2 \Bigg/ \left[\frac{(s_1^2/n_1)^2}{(n_1 - 1)} + \frac{(s_2^2/n_2)^2}{(n_2 - 1)}\right] \tag{12.3.5}$$

$$t_2 = \frac{(\overline{y}_1 - \overline{y}_2) - (M_1 - M_2)}{s_{2d}} \tag{12.3.6}$$

where

$$s_{2d} = \sqrt{\frac{s_1^2}{n_1} + \frac{s_2^2}{n_2}}.$$

Measure of agreement: Let t be either t_1 or t_2 (whichever is appropriate) and t_0 be the value of t when $M_1 - M_2$ is replaced by an hypothesized value D_0 in either (12.3.4) or (12.3.6). Then a value of zero for t_0 shows the best agreement with the hypothesized difference and increasing magnitude for t_0 represents lessening support for D_0. Hence, the p-value for comparing the data with the hypothesized difference in means, D_0, is computed from (12.3.7).

$$p = \Pr\{|t(\nu)| \geq |t_0|\}. \tag{12.3.7}$$

Whether the probability is based on the statistic defined in (12.3.4) or (12.3.6) depends on the acceptability of the assumption of equal variances in the model.

Model and data checking: The recommended procedures are

1. Check the data visually and ensure that no values are outside acceptable bounds.

2. Produce line diagrams (Section 1.3.5) for each group using the same scale for both diagrams. Check that the spread of values is approximately the same for each group and there is no value which appears surprisingly large or surprisingly small for either group.

3. Calculate the set of standardized residuals and check for any residual which exceeds 2.0.

   (The set of standardized residuals can be computed from the formula

$$r_{s_{ijo}} = (y_{ijo} - \overline{y}_{i.0})/s_{i0} \qquad \text{for } i = 1,2;\ j = 1,2,\ldots,n_i.)$$

4. Construct a line diagram based on the set of standardized residuals and look for evidence of asymmetry.

5. Produce a Normal probability plot based on standardized residuals and check that the trend line is approximately linear. (Guidelines for interpreting a Normal probability plot can be found in Section 10.3.4.)

6. Compare the variability of the data from the two sources by visually comparing the line diagrams, computing the ratio of sample variances, and, if available, applying a test of the hypothesis of equal variances.

The first consideration in analyzing the results of this model and data checking procedure is to establish if there is evidence of an error in the data. Where visual inspection or a standardized residual of large magnitude suggests a possible odd value **and** a value out of character with the general trend is apparent in the Normal probability plot, the possibility of an error in the data must be considered. It is unwise to ignore such a finding. Possible courses of action when an odd value is detected are considered in Section 10.4.3.

A non-linear trend line in the Normal probability plot and asymmetry in the line diagram based on standardized residuals indicates that a basic assumption in the model may be violated and it is wise to seek a procedure for constructing a confidence interval for the difference in means which is less sensitive to the failure of the Normality assumption than are procedures based on the t-distribution. Where the pattern of variation appears similar for the two sets of responses, the distribution-free method described in Section 12.4.3 is recommended.

If there is both evidence of asymmetry in the line diagram for the standardized residuals and evidence of unequal variances it would be wise to seek assitance from a consultant statistician. The approximate randomization test described in Section 7.8.2 is valid if the aim is merely to examine a single hypothesized value for the difference, e.g. to test the hypothesis that $M_1 = M_2$.

Model 1 should be employed unless one variance is much larger than another (rejection of the hypothesis of equal variances by a statistical test or, more simply, a ratio in excess of 5:1 may be used as a guide.) Even where there is a marked discrepancy in variances, the need to use Model 2 only arises if, additionally, (i) there is a substantial difference in sample sizes; and (ii) the larger variance is associated with the smaller sample size.

Estimation: Based on the statistic defined in (12.3.4) if variances are assumed equal, or (12.3.6) if they are not, confidence intervals for $M_1 - M_2$ can be constructed using (8.3.4). The intervals are only approximate if based on (12.3.6). The limits are defined in (12.3.8).

$$\overline{y}_1 - \overline{y}_2 - t_\alpha(\nu)s_d \qquad \text{to} \qquad \overline{y}_1 - \overline{y}_2 + t_\alpha(\nu)s_d \tag{12.3.8}$$

where $\nu = n_1 + n_2 - 2$ and s_d is replaced by s_{1d} as defined in (12.3.4) if variances are assumed equal, or ν is the nearest whole number to (12.3.5) and s_d is replaced by s_{2d} as defined in (12.3.6) if the variances are not assumed equal.

Computation: A confidence interval can be computed using (12.3.8) or by employing a statistical package.

A test of the hypothesis $M_1 - M_2 = D_0$ can be based on a confidence interval or a p-value computed from (12.3.7).

Illustration: The calculations are illustrated using Example 12.2.

Model and data checking: Visual examination of the data from Example 12.2 which is presented in Table 12.3.2(a) reveals that the fourth reading in the Supplement group is large in comparison with other values. The line diagram (Figure 12.3.1(a)) confirms this observation as does the large standardized residual (3.0) for the value.

The fact that the reading is unlikely under the assumption of Normality is demonstrated by the Normal probability plot in Figure 12.3.2(a) in which all points, with the exception of that point corresponding to the reading 35.9, lie on or close to a linear trend line.

The combined evidence suggests that, under the model which assumes Normality, the value 35.9 is an unlikely outcome with respect to the other observations which form the data set.

Consultation with the researcher led to sound reasons for discarding the response and performing the analysis on the remaining data set. The Normal probability plot and the set of standardized residuals on the reduced data set appear in Figure 12.3.2(b) and Table 12.3.2(b), respectively. The line diagram for the standardized residuals is provided in Figure 12.3.1(b).

Examining the line diagram without the value 35.9 being considered, points to less variability in the supplement group than in the control group. This is confirmed by the sample variances, 8.7 for the Supplement group versus 27.7 for the Control group. The Normal probability plot shows considerable deviation of points from a trend line but no strong evidence of a non-linear trend line. The line diagram for the standardized residuals

Table 12.3.2. Data and Summary Statistics for Example 12.2 (Data by courtesy of Dr. P. Gormley, Antarctic Div., Aust. Dept. of Science and Technol.)

(a) Complete data set

Supplement Group			Control Group		
Reading	Residual	Stand. residual	Reading	Residual	Stand. residual
18.3	1.8	0.3	24.9	5.3	1.0
9.3	−7.2	−1.1	16.0	−3.6	−0.7
12.6	−3.9	−0.6	26.3	6.7	1.3
35.9	19.4	3.0	25.5	5.9	1.1
15.7	−0.8	−0.1	19.3	−0.3	−0.1
14.2	−2.9	−0.4	16.8	−2.8	−0.5
13.1	−3.4	−0.5	15.7	−3.9	−0.7
14.3	−2.2	−0.3	24.6	5.0	0.9
16.2	−0.3	0.0	19.9	0.3	0.1
18.1	1.6	0.2	9.4	−10.2	−1.9
19.4	2.9	0.4	17.4	−2.2	−0.4
15.5	−1.0	−0.5			
11.7	−4.8	−0.7			

(b) Modified data set with reading 35.9 deleted

Supplement Group			Control Group		
Reading	Residual	Stand. residual	Reading	Residual	Stand. residual
18.3	3.4	1.2	24.9	5.3	1.0
9.3	−5.6	−1.9	16.0	−3.6	−0.7
12.6	−2.2	−0.8	26.3	6.7	1.3
*	*	*	25.5	5.9	1.1
15.7	0.8	0.3	19.3	−0.3	−0.1
14.2	−0.7	−0.2	16.8	−2.8	−0.5
13.1	−1.8	−0.6	15.7	−3.9	−0.7
14.3	−0.6	−0.2	24.6	5.0	0.9
16.2	1.3	0.5	19.9	0.3	0.1
18.1	3.2	1.1	9.4	−10.2	−1.9
19.4	4.5	1.5	17.4	−2.2	−0.4
15.5	0.6	0.2			
11.7	−3.2	−1.1			

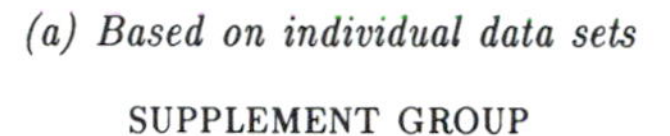

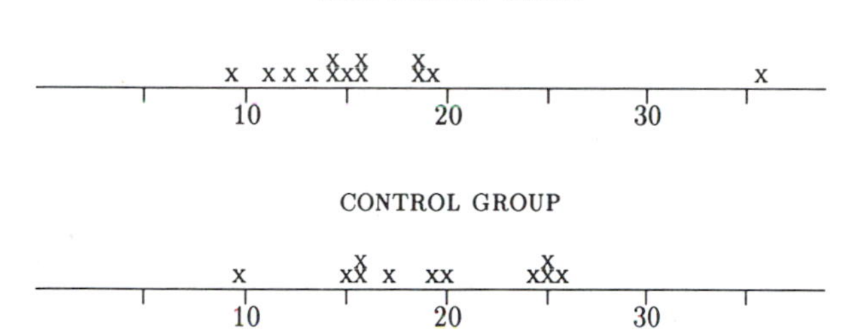

(b) Based on standardised residuals from combined data set

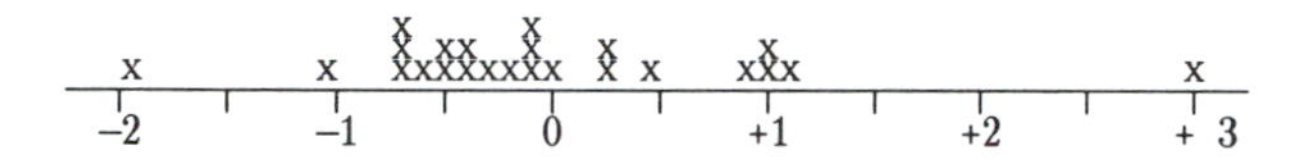

Figure 12.3.1. Line diagrams based on data in Table 12.3.2.

shows no obvious evidence of asymmetry. On balance, the assumption of Normality would seem acceptable.

Since the ratio of variances is approximately 3:1 and the sample sizes are approximately equal, there is insufficient reason to reject the assumption of equality of variances. Hence, Model 1 is assumed.

Analysis. Calculations using the data in Table 12.3.2(b) yield

$$\overline{y}_1 - \overline{y}_2 = 14.9 - 19.6 = -4.7$$

$$s_d = \sqrt{\frac{(12-1)8.7006 + (11-1)27.7856}{12+11-2}\left[\frac{1}{12} + \frac{1}{11}\right]} = 1.76.$$

From the t-distribution with 21 degrees of freedom, $t_{0.05}(21) = 2.08$. Thus, by (12.3.8), a 95% confidence interval for $M_1 - M_2$ is

$$-4.7 - 2.08 \times 1.76 \quad \text{to} \quad -4.7 + 2.08 \times 1.76$$
$$-8.4 \quad \text{to} \quad -1.0.$$

The negative sign indicates that the second group (the Control group) is likely to have a higher mean than the first group (the Supplement group).

Interpretation: The fact that the interval does not include zero is evidence against the hypothesis that the treatments are producing the same effect. The theory that persons taking a vitamin C supplement can be expected to have a higher level of the vitamin in their body is not supported by the analysis. If anything, there is some evidence for a decrease in level.

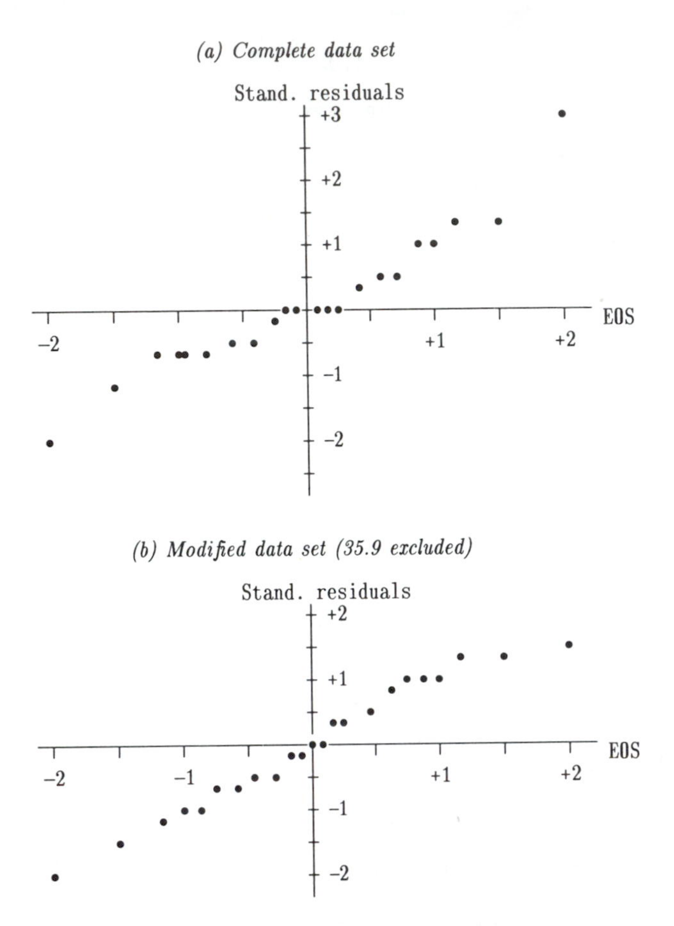

Figure 12.3.2. Normal probability plots based on data in Table 12.3.2.

This findings runs contrary to what was expected. Discrepancy between statistical findings and experimental expectation is grounds for careful appraisal of both the statistical model and analysis as well as the experimental set-up and operation. Apart from the possibility that the experimenter had the wrong hypothesis, the disagreement may be caused by (i) an incorrect choice of statistical model; (ii) an error in the data or the analysis; (iii) misinterpretation of the statistical findings; or (iv) sampling variation.

In fact, it seems that the lower reading for the supplement group may be attributable to sampling variation in this case. The data provided with this example are only one of a sequence of sets of readings which were taken at monthly intervals for seven months. This data set is the only one in which evidence of difference in means was found.

Alternative analysis: The data are also analyzed on the assumption of a model which presumes unequal variances to provide a comparison with the above results. The fact that both analyses are being presented should not be taken as a recommendation that it would be proper in real application to perform and present both analyses on the one data set.

The value obtained for s_{2d} from (12.3.6) is 1.80, and (12.3.5) yields a value of 15.4 from which it is determined that ν should be set equal to 15. The 95% confidence interval which is calculated has limits -8.5 and -0.9. This contrasts with the interval $(-8.4, -1.0)$ based on the model in which variances are assumed equal.

12.3.7 Comparisons Based on the Ratio M_1/M_2

The conditions under which it may be preferable to compare the ratio of means rather than the difference have been considered in Section 12.3.4. No new methodology is required. Data should be transformed to the logarithmic scale of measurement and either the large sample method of Section 12.3.5 employed or, if either sample size is less than thirty, the methodology of Section 12.3.6 employed (subject, of course, to the conditions being met for valid application of the method).

These methods will provide confidence limits L and U for the difference in logarithms of the means, $\log(M_1) - \log(M_2)$. The corresponding confidence limits for the ratio M_1/M_2 are $\exp(L)$ and $\exp(U)$.

The hypothesis $M_1/M_2 = R_0$ must be transformed to $\log(M_1) - \log(M_2) = \log(R_0)$ for the analysis. In particular, note that the hypothesis stating the two treatments have the same average effect, namely, $M_1/M_2 = 1$, transforms to $\log(M_1) - \log(M_2) = 0$.

The calculations are illustrated in Example 12.8.

Example 12.8 (Age and Manipulative Skills). *As part of a college course, students training to become physical education teachers were required to observe children of different ages undertake the same set of tasks to establish the skill levels which could be expected at the different ages. One task required a hockey ball to be pushed between a series of poles by a hockey stick. The time to successfully complete the task was recorded for each child.*

The results for a sample of 15 *ten year old boys and* 15 *sixteen year old boys are recorded in Table* 12.3.3.

The nature of the response variable, namely 'Time taken to complete a task,' and the large difference in average levels expected between the two groups is indicative of a situation in which the more meaningful comparison is likely to be in terms of ratios rather than differences.

The process of model and data checking is given as an exercise in Problem 12.9 *and displays (i) skewness in the line diagrams for each group, (ii) a curved trend line for Normal probability plots based on standardized residuals from each group, (iii) a large positive standardized residual in the set of standardized residuals from each group and (iv) greater variability in the data set with the larger mean.*

Model and data checking on the logarithmically transformed data suggests that the assumptions of Normality and equality of variances are reasonable and therefore the method described in Section 12.3.6 *may be applied. It yields a* 95% *confidence interval for* $\log(M_1) - \log(M_2)$ *of* -1.53 *to* -0.88. *Hence, the* 95% *confidence interval for* M_1/M_2 *is* $\exp(-1.53)$ *to* $\exp(-0.88)$ *which is* 0.217 *to* 0.415.

For purposes of presentation, this might more conveniently be expressed as an interval for the ratio M_2/M_1, *in which case it becomes* $1/0.415 = 2.4$ *to* $1/0.217 = 4.6$. *It can be concluded that the sixteen year old boys are likely to be between about two and one half and four and one half times as fast as the ten year old boys in completing the task.*

Table 12.3.3. Data and Derived Information from Example 12.8

Time to complete task (in secs.)		Log data		Standardised residuals		Standardised residuals (log. scale)	
Age 16	10	16	10	16	10	16	10
9	37	2.20	3.61	−0.70	−0.36	−0.73	−0.24
14	45	2.64	3.81	0.08	−0.01	0.30	0.20
11	41	2.40	3.71	−0.39	−0.19	−0.26	−0.01
14	87	2.64	4.47	0.08	1.82	0.30	1.70
9	53	2.20	3.97	−0.70	0.33	−0.73	0.57
18	27	2.89	3.30	0.71	−0.79	0.88	−0.96
6	105	1.79	4.65	−1.17	2.60	−1.67	2.13
8	46	2.08	3.83	−0.86	0.03	−1.00	0.25
30	27	3.40	3.30	2.60	−0.80	2.07	−0.96
8	35	2.08	3.56	−0.86	−0.45	−1.00	−0.37
10	38	2.30	3.64	−0.54	−0.32	−0.48	−0.18
12	54	2.48	3.99	−0.23	0.38	−0.06	0.62
16	19	2.77	2.94	0.40	−1.15	0.61	−1.76
23	36	3.14	3.58	1.50	−0.41	1.46	−0.30
14	30	2.64	3.40	0.08	−0.67	0.30	−0.72

12.3.8 Comparisons of Mean Rates of Occurrences of Events

Model: The Poisson model (Section 6.3) is presumed to apply for each source with the mean rates of occurrences of events being M_1 and M_2 for Sources 1 and 2, respectively.

Data: The data available from each source constitute one or more observations, with each observation being the number of occurrences of an event over a fixed period of time or a fixed region of space. The following notation is used.

	Reading				Mean	Total
Source 1	x_{110}	x_{120}	$\cdots$	x_{1n_10}	$\overline{x}_{10}$	X_{10}
Source 2	x_{210}	x_{220}	$\cdots$	x_{2n_20}	$\overline{x}_{20}$	X_{20}

(Note: it is essential that each observation is based on the same length of time or same-sized region of space for application of the methodology in this section. If this condition is not met, seek assistance from a consultant statistician.)

Statistic: The statistic employed is the difference in the sample means, i.e. $d = \overline{x}_1 - \overline{x}_2$. While the exact sampling distribution of d is not easy to express or simple to apply for computational purposes, the Central Limit Theorem establishes that the large sample distribution of d can be well approximated by a Normal distribution with mean $D = M_1 - M_2$ and variance $\overline{x}_1/n_1 + \overline{x}_2/n_2$. In practice, the Normal approximation is adequate provided the total number of occurrences, $X_1 + X_2$, is at least five.

The outcomes become less likely under the model as the distance between D and the sample mean difference, d, increases.

Measure of agreement: If the hypothesized difference in means is D_0 and the observed difference in mean number of occurrences is d_0, an approximate p-value between model and data is defined by (12.3.9).

$$p = \Pr\{|z| \geq |d - D_0|/s_d\} \tag{12.3.9}$$

where $s_d = \sqrt{\overline{x}_1/n_1 + \overline{x}_2/n_2}$.

Model and data checking: There is only elementary data checking required. Obviously, counts must be whole numbers which are not negative. In general, model checking is non-statistical, depending on establishing that the experimental procedure meets the requirements of the Poisson model. In particular, the occurrences of events must be independent.

Estimation: A point estimate of $M_1 - M_2$ is provided by $\overline{x}_1 - \overline{x}_2$.

Approximate confidence intervals for $M_1 - M_2$ can be based on (8.3.3) and yield the formulae presented in (12.3.10).

$$\begin{array}{ll} 95\% \text{ C.I.} & \overline{x}_1 - \overline{x}_2 - 1.96s_d \text{ to } \overline{x}_1 - \overline{x}_2 + 1.96s_d \\ 100(1-\alpha)\% \text{ C.I.} & \overline{x}_1 - \overline{x}_2 - z_{\frac{1}{2}\alpha}s_d \text{ to } \overline{x}_1 - \overline{x}_2 + z_{\frac{1}{2}\alpha}s_d \end{array} \tag{12.3.10}$$

where $s_d = \sqrt{\overline{x}_1/n_1 + \overline{x}_2/n_2}$.

Computation: Approximate confidence intervals for the difference in means, $M_1 - M_2$, can be computed using (12.3.10) provided the total of all counts from both sources is at least five.

If the total number of events $(X_1 + X_2)$ is less than five there is no point in seeking an alternative procedure since the sensitivity of the procedure would be too low to have practical value.

A test of the hypothesis $M_1 - M_2 = D_0$ can be based on a confidence interval or a p-value can be computed directly using (12.3.9).

Illustration: The calculations are illustrated using the data in Example 12.4. This example represents the special case in which there is only one observation for each source, i.e. $n_1 = n_2 = 1$, $\overline{x}_1 = 90$ and $\overline{x}_2 = 108$.

An approximate 95% confidence interval for $M_1 - M_2$ is computed from (12.3.10) to be

$$\begin{array}{lcl} 90 - 108 - 1.96\sqrt{\{90 + 108\}} & \text{to} & 90 - 108 + 1.96\sqrt{\{90 + 108\}} \\ -46 & \text{to} & +10. \end{array}$$

With reference to the claim that there had been a *significant increase* in motorcycle related accidents from 1982 to 1983, it is apparent that such a claim could not be based on the statistical findings since they fail to dispel the possibility that there is no increase and that there might indeed be a decrease in the expected rate of motorcycle accidents.

Of considerable practical importance is the width of the confidence interval. It is clearly far wider than desirable if the findings are to be of practical value. The width is determined by the variance of the difference in sample means, which, from a practical viewpoint, can only be reduced by increasing the size of one or both samples, n_1 and n_2.

Yet, in the case of Example 12.4, it is impossible to obtain more than one observation for the year. This will always be the case in respect of annual accident rates or crime rates and suggests that claims of increased accident rates or increased crime rates from one year to the next must be viewed with some suspicion unless there is a statistically significantly difference or the basis of the claim extends beyond the simple set of yearly figures.

12.4 Comparisons of Medians

12.4.1 Data

The data are presumed to comprise a set of n_1 readings from one source (treatment, group or process) and a set of n_2 readings from a second source. The following notation is used to denote the sets of readings:

	Readings			
Source 1	y_{110}	y_{120}	$\cdots$	y_{1n_10}
Source 2	y_{210}	y_{220}	$\cdots$	y_{2n_20}

Readings Ordered from Smallest to Largest
Within Sources

Source 1	$y_{1(1)0}$	$y_{1(2)0}$	$\cdots$	$y_{1(n_1)0}$
Source 2	$y_{2(1)0}$	$y_{2(2)0}$	$\cdots$	$y_{2(n_2)0}$

12.4.2 Choosing a Method

The methodology most widely recommended is the rank-based method described in Section 12.4.3. It is based on a model which assumes the distributions for two sources have the same shape but may differ in the values of the medians. It is the recommended methodology unless the model checking described in Section 12.4.3 provides evidence of differences in the patterns of variation. If so, the sign-based methodology described in Section 12.4.4 is available. That methodology is based on a model which is free of any distributional assumptions.

12.4.3 Methods Based on Ranks

Model: Given that M_1 and M_2 are the medians of the distributions for Sources 1 and 2, respectively, then $M_1 - M_2$ is assumed to take the value D_0. No assumption is made about the mathematical forms of the distributions although the two distributions are assumed to have the same shape.

Note that, while means and medians may differ within distributions, the fact that the distributions have the same shape implies that the difference between means will be equal to the difference between medians and hence the procedure may be used for examining the difference in means.

Statistic: Consider the model which includes the assumption that the two medians are equal, i.e. $M_1 = M_2$ or $D_0 = 0$.

Suppose the data from the two sources are combined and ordered from smallest to largest, and a number attached to each observation which is its rank in the ordered data. This is illustrated in Table 12.4.1 for two data sets.

Let S_1 be the sum of the ranks which correspond to the n_1 observations from Source 1 and S_2 be the sum of ranks which correspond to the n_2 observations from Source 2. Then $s_1 = S_1/n_1$ and $s_2 = S_2/n_2$ are the means of the ranks from the respective sources.

Intuitively, it would be reasonable to use the statistic $d = s_1 - s_2$ as the basis for establishing if the hypothesis $M_1 = M_2$ is reasonable since increasing magnitude for d implies increasing evidence against the hypothesis that the two medians are equal. In fact, because the sum of the ranks is constant, i.e. $S_1 + S_2 = \frac{1}{2}(n_1 + n_2)(n_1 + n_2 + 1)$, any one of a range of statistics, including s_1, s_2, S_1 and S_2, would be equally suited.

If the model includes the assumption $M_1 - M_2 = D_0$, where D_0 is not zero, the statistical development is unchanged. The modification required is in respect of the data. To each response from Source 2 it is necessary to add an amount D_0, i.e. to set $y'_{2(j)} = y_{2(j)} + D_0$ for $j = 1, 2, \ldots, n_2$, and to employ the value $y'_{2(j)}$ in place of $y_{2(j)}$ in the determination of ranks and rank-based statistics. In effect, this transformation converts the problem into one in which the model to be examined included the hypothesis $M_1 - M'_2 = 0$ where M'_2 is the median of the distribution which produced the values $y'_{2(j)}$ $(j = 1, 2, \ldots, n_2)$.

Table 12.4.1. Data Sets and Associated Rank Information

Data Set 1 Source		Data Set 2 Source	
1	2	1	2
15	10	10	26
25	26	15	38
33	40	25	40
38		33	

Combined Data Set 1			Combined Data Set 2		
Source	Data	Rank	Source	Data	Rank
2	10	1	1	10	1
1	15	2	1	15	2
1	25	3	1	25	3
2	26	4	2	26	4
1	33	5	1	33	5
1	38	6	2	38	6
2	40	7	2	40	7

Combined Data Set 1: $S_1 = 16$, $S_2 = 12$; $s_1 = 4.0$, $s_2 = 4.0$

Combined Data Set 2: $S_1 = 11$, $S_2 = 17$; $s_1 = 2.75$, $s_2 = 5.67$

Measure of agreement: For the model which includes $M_1 = M_2$, i.e. $D = 0$, values of $d = s_1 - s_2$ of increasing magnitude reflect worsening agreement between model and data. Hence, if the observed difference in mean ranks is d_0, the p-value is determined from (12.4.1).

$$p_0 = \Pr(|d| \geq |d_0|) \tag{12.4.1}$$

where the probability is based on the sampling distribution of d.

(If the model includes the assumption $M_1 - M_2 = D_0$, then the modification to the data described in the **Statistics** section above should be employed before the assignment of ranks and the computation of d_0.)

Model and data checking: Check the data to ensure there are no values which are surprisingly large or surprisingly small.

Construct line diagrams for data from each source separately using the same scale for both. Check for evidence that the spread of values, and, more generally, the pattern of variation is similar in the two plots. Where the patterns are markedly dissimilar, the method described in Section 12.4.4 is to be preferred.

Estimation: The difference in sample medians provides the best point estimate of the difference in medians of the distributions.

There is no simply analytic expression by which to define confidence limits for $D = M_1 - M_2$. The construction of a confidence interval for D from first principles which has a level of confidence of exactly $1 - \alpha$ would require the determination of upper and lower limits, $D_L(\alpha)$ and $D_U(\alpha)$ such that addition of either $D_L(\alpha)$ or $D_U(\alpha)$ to each response from Source 2 would result in (12.4.1) yielding a p-value of exactly α and the addition of any value D_0 which lay between these limits would yield a p-value of at least α.

Since rank statistics can only take a finite number of values, it follows from Section 8.3.6 that such limits could be found for only a limited number of values of α. The approach generally adopted is to find that confidence interval which has the smallest level of confidence equal to or greater than $1 - \alpha$. (Alternatively, using the approach described in Section 8.3.6, an approximate confidence interval with the desired level of confidence can be constructed for any chosen level of confidence by use of linear interpolation.)

Application: The methodology is widely used for the comparison of medians. Where distributions are asymmetric but of similar shape it is useful for the comparison of means.

There is also scope for application where the response variable is not directly measurable but it is possible to identify the interval in which the response lies. In effect, there is a measurable categorical variable which has categories corresponding to the intervals on the underlying scale. Example 12.9 provides an illustration.

Example 12.9 (Drug Comparisons in Laboratory Trials). *Twenty rats were assigned between a drug treatment and a control, ten to each group. The method of allocation is discussed in Example* 13.2. *Each rat was subsequently assigned to one of ten categories which reflect differing levels of activity, with the categories being identified by the numbers* $1, 2, \ldots, 10$ *such that one equates with least activity and ten with greatest activity. The classifications obtained for the twenty rats are recorded in Table* 12.4.2.

Table 12.4.2. Level of Activity of Rats (Example 12.9)

	Rat*									
	1	2	3	4	5	6	7	8	9	10
Treatment										
Control	3	7	2	5	8	6	4	2	4	5
Drug	8	6	7	3	3	6	5	6	9	4

**note that the numbering of rats is performed independently in the two groups, i.e. Rat 1 in the Control group has no connection with Rat 1 in the Drug group.*

Table 12.4.3. Conversion of Ordered Categorical Data to Ranks (Based on Data in Example 12.9)

Group	Observed Category	Rank	Averaged* Rank
C	2	1	1.5
C	2	2	1.5
C	3	3	4
D	3	4	4
D	3	5	4
C	4	6	7
C	4	7	7
D	4	8	7
C	5	9	10
C	5	10	10
D	5	11	10
C	6	12	13.5
D	6	13	13.5
D	6	14	13.5
D	6	15	13.5
C	7	16	16.5
D	7	17	16.5
C	8	18	18.5
D	8	19	18.5
D	9	20	20

**where two more more observations have the same value the ranks which would have been assigned to them are averaged and it is the averaged ranks which are used in the statistical analysis.*

Since the numbers are merely codes for the ten categories, it would not be meaningful to treat them as scaled values. However, they do reflect an ordering of responses since, for example, category 3 *is above category* 2 *but below category* 4. *Hence, either the rank-based or sign-based procedures may be applied. The conversion of the numerical categories to ranks to permit the application of the rank-based procedures of this section is illustrated in Table* 12.4.3.

Computation: There is both a graphical and a non-graphical algorithm for the computation of confidence intervals. Both methods are tedious, although straightforward, and are not reproduced here. The interested reader will find the steps described in Conover ([1980], Section 5.4).

Where the required level of confidence is stated to be $1-\alpha$, the methods provide a confidence interval with level of confidence which is guaranteed to be at least $1-\alpha$. If the difference between the confidence level obtained and that required is substantial, the interpolation method described in Section 8.3.6 may be employed to produce an approximate confidence interval with the desired level of confidence.

The procedure is ideally suited to a computer and the option to construct a confidence interval for the difference in medians will be found in most general statistical packages—commonly associated with the name *Mann–Whitney* in honor of the two statisticians who popularized the use of the rank statistic on which it is based.

The test of the hypothesis $D = D_0$ can be based on a confidence interval. However, it may also be applied in a direct form which utilizes the p-value defined in (12.4.1). In this form it is commonly referred to as the *Mann–Whitney U-test* when testing the hypothesis $D = 0$. There is not a simple formula for the computation of the p-value, although in large samples an approximation based on the Normal distribution is available. (See, for example, Conover ([1980], Section 5.3).) As with the determination of a confidence interval, the construction of a p-value is best done by computer.

Illustration: Application of the MINITAB statistical package to the data in Table 12.4.2 with the aim of producing an approximate 95% confidence interval yields the interval (−3.01, 0.99) with a level of confidence 95.5% for the difference in medians, $M_1 - M_2$.

Since the interval includes zero, the data are consistent with the assumption that the median level is identical under Treatment and Control conditions.

A closer approximation to the nominal 95% level of confidence could be obtained by additionally determining that the 94.6% confidence interval is (−2.99, 0.99) and using linear interpolation in the manner described in Section 8.3.6.

12.4.4 Methods Based on Signs

Model: The model includes the assumption that the difference in medians, $M_1 - M_2$, is equal to D_0.

There are no distributional assumptions.

Statistic: The construction of the statistic is firstly described for the case where $D_0 = 0$, i.e. the two medians are presumed equal.

Let $\tilde{y}$ be the median of the combined sample from both sources. Consider the following frequency table which, for simplicity of explanation, assumes the combined sample size, n, is even and there are no responses equal to $\tilde{y}$.

	Source 1	Source 2	Total
Number below $\tilde{y}$	f_{1-}	f_{2-}	$\frac{1}{2}n$
Number above $\tilde{y}$	f_{1+}	f_{2+}	$\frac{1}{2}n$
Total	n_1	n_2	n

Note that the row sums are fixed as are the column sums. Hence, stating any one frequency in the body of the table fixes the other frequencies. Consequently, any one of the frequencies in the body of the table may be used as the statistic on which a test of significance is based. For the purposes of the discussion, suppose the chosen statistic is f_{1-}.

The sampling distribution of the statistic f_{1-} is a member of a well known family of distributions referred to collectively as the *Hypergeometric distribution* which is defined by (6.2.1).

The construction of a sampling distribution for f_{1-} has been considered for the situation in which the combined sample size is even and there are no values equal to the median of the combined sample. Violation of either condition requires only minor change to the development and does not alter the form of sampling distribution.

A simple variation to the construction of the statistic also caters for the case where the assumed difference in median values, D_0, is possibly non-zero. It is simply a matter of adding the value D_0 to each of the responses from Source 2 before combining the data from the two Sources for the construction of the combined median and the table of frequencies.

Measure of agreement: Under the assumption that $M_1 = M_2$, the most likely outcomes are those for which $f_{1-} = \frac{1}{2}n_1$. The sampling distribution of f_{1-} is symmetric and values decrease towards either extreme as the distance between f_{1-} and $\frac{1}{2}n_1$ increase. Thus, if the observed number of observations from Source 1 which lie below the median is f_0, the p-value is defined by $p = \Pr(|f_{1-} - \frac{1}{2}n_1| \geq |f_0 - \frac{1}{2}n_1|)$ and can be evaluated using

(12.4.2).

$$p = 2\sum_{i=0}^{f_0} \pi(i) \qquad \text{if } f_0 \leq \tfrac{1}{2}n_1$$
$$= 2\sum_{i=f_0}^{\frac{1}{2}n_1} \pi(i) \qquad \text{otherwise,} \tag{12.4.2}$$

where

$$\pi(i) = \frac{\binom{n_1}{i}\binom{n_2}{\frac{1}{2}n - i}}{\binom{n}{\frac{1}{2}n}}$$

and sources are numbered such that n_1 is less than or equal to n_2.

Direct calculation is tedious and best left to a computer. The Central Limit Theorem (Section 7.3.5) may be invoked when the sample size is large and establishes that the sampling distribution of f_{1-} tends towards a Normal distribution with mean M_f and variance σ_f^2 defined by (12.4.3).

$$M_f = \tfrac{1}{2}n_1$$
$$\sigma_f^2 = \frac{n_1 n_2}{4(n_1 + n_2)}. \tag{12.4.3}$$

Thus, in large samples, if $M_1 = M_2$ and f_{1-} has a value f_0, an approximate value for p is defined by (12.4.4).

$$p = \Pr(|z| \geq |(f_0 - M_f)/\sigma_f|)] \tag{12.4.4}$$

where M_f and σ_f are defined by (12.4.3).

Estimation: The difference in sample medians provides a point estimate of the difference in the medians of the two distributions.

The construction of a confidence interval for $D = M_1 - M_2$ uses order statistics in an analogous manner to that described in Section 8.3.6 for the one-sample case. In practice, the procedure is only feasible using the large sample approximation to the sampling distribution of the statistic f_{1-} defined in the Measure of agreement section above. Based on this large sample approximation, an expression for an approximate confidence interval for $M_1 - M_2$ is provided by (12.4.5) which is obtained from Pratt [1964].

The direct application of (12.4.5) is only possible for those values of α for which $\frac{1}{2}n_1 \pm d_\alpha$ and $\frac{1}{2}n_2 \pm d_\alpha$ are whole numbers. By the use of linear interpolation in the manner described in Section 8.3.6, approximate limits for any level of α may be obtained. The method of computing these limits is given in the Computation section below.

$$\Pr(y_{1(\frac{1}{2}n_1-d_\alpha)} - y_{2(\frac{1}{2}n_2+d_\alpha)} < M_1 - M_2 < y_{1(\frac{1}{2}n_1+d_\alpha)} - y_{2(\frac{1}{2}n_2-d_\alpha)}) = 1-\alpha \tag{12.4.5}$$

where $d_\alpha = z_{\frac{1}{2}\alpha}\sigma_f$ and σ_f is the square root of the variance defined in (12.4.3).

Computation: The formulae employed rely on a large sample approximation and should not be applied if either sample size, n_1 or n_2, is less than ten.

Based on (12.4.5), the construction of an approximate confidence interval for the difference in medians, $M_1 - M_2$, is by the following simple steps.

Step 1 — for an approximate $100(1-\alpha)\%$ confidence interval, compute a value of d_α, from (12.4.6).

$$d_\alpha = \frac{1}{2} z_{\frac{1}{2}\alpha} \sqrt{\frac{n_1 n_2}{n_1 + n_2}}. \tag{12.4.6}$$

Step 2 — compute values for each cell in the following tables.

Data from Source 1

b	a^*	$y_{1(a-1)}$	$y_{1(a)}$	$(a-b)y_{1(a-1)} + (b-a+1)y_{1(a)}$
$\frac{1}{2}n_1 + d_\alpha$				U_1
$\frac{1}{2}n_1 - d_\alpha$				L_1

*smallest whole number greater than b

Data from Source 2

b	a^*	$y_{2(a-1)}$	$y_{2(a)}$	$(a-b)y_{2(a-1)} + (b-a+1)y_{2(a)}$
$\frac{1}{2}n_2 + d_\alpha$				L_2
$\frac{1}{2}n_2 - d_\alpha$				U_2

*smallest whole number greater than b

Step 3 — compute the lower and upper limits of an approximate $100(1-\alpha)\%$ confidence interval for $M_1 - M_2$ as $L_1 - L_2$ and $U_1 - U_2$, respectively.

Illustration: The calculations are illustrated using data in Table 1.3.5 with the aim of establishing the likely size of the difference in median intakes of girls (M_1) and boys (M_2). A 95% confidence interval for $M_1 - M_2$ is sought.

Step 1

$$d_{0.05} = \frac{1}{2}(1.96)\sqrt{\frac{110 \times 108}{110 + 108}} = 7.23.$$

Step 2 — compute values for each cell in the following tables.

Data from Girls' Intakes

b	a	$y_{1(a-1)}$	$y_{1(a)}$	$(a-b)y_{1(a-1)} + (b-a+1)y_{1(a)}$
62.23	63	911	912	911.2
47.77	48	782	784	783.5

Data from Boys' Intakes

b	a	$y_{2(a-1)}$	$y_{2(a)}$	$(a-b)y_{2(a-1)} + (b-a+1)y_{2(1)}$
61.23	62	1235	1281	1245.6
46.77	47	1118	1118	1118.0

Step 3 — The approximate 95% confidence interval for $M_1 - M_2$ ranges from a lower limit of $783.5 - 1245.6 = -462.1$ to an upper limit of $911.2 - 1118.0 = -206.8$.

The negative sign indicates that the median value for the Group labelled 2 (Boys) exceeds that for the Group labelled 1 (Girls).

The 95% confidence interval for difference in medians suggests that fifteen year old boys in the population sampled have a daily energy intake which is likely to be between 207 and 462 megajoules above that for girls from the same population.

12.5 A Consideration of Rare Events

12.5.1 Methodological Limitations

Methodology for the comparison of proportions or probabilities which appears in Section 12.2 has required either the use of the Normal approximation to the Binomial in the study of differences or a Chi-squared approximation in the study of relative risks or odds ratios. When either π_1 or π_2 is near zero or one, these approximations will not be valid even when sample sizes are large.

While the experimenter's question remains unchanged in these circumstances, it is necessary to vary the statistical approach and to seek alternative methods which can be applied without restrictions on sample size.

Two methods are introduced—in Section 12.5.2, an all-purpose method is described which may be used for hypothesis testing for any of the three methods of comparison; in Section 12.5.3, a method for constructing a confidence interval for the relative risk is described.

Despite the fact that sample sizes may be large, neither method is particularly sensitive to detecting variations in the success rates between the sources.

12.5.2 An Exact Test Based on Frequencies

The development of a test of the hypothesis $\pi_1 = \pi_2$ is based on the tables of frequencies in Table 12.5.1.

Table 12.5.1. Tables of Frequencies Employed in the Construction of a Test of the Hypothesis $\pi_1 = \pi_2$

(a) observed frequencies

	Treatment A	Treatment B	Total
No. of Successes	$o_{11} = x_{10}$	$o_{21} = x_{20}$	x_0
No. of Failures	$o_{12} = n_1 - x_{10}$	$o_{22} = n_2 - x_{20}$	$n - x_0$
Total Number	n_1	n_2	n

(b) expected frequencies

	Treatment A	Treatment B	Total
No. of Successes	$e_{11} = n_1\pi_1$	$e_{21} = n_2\pi_2$	
No. of Failures	$e_{12} = n_1(1 - \pi_1)$	$e_{22} = n_2(1 - \pi_2)$	
Total Number	n_1	n_2	n

(c) expected frequencies given $\pi_1 = \pi_2$ and $n_1\pi_1 + n_2\pi_2 = x_0$

	Treatment A	Treatment B	Total
No. of Successes	$e_{11} = n_1 p_0$	$e_{21} = n_2 p_0$	x_0
No. of Failures	$e_{12} = n_1(1 - p_0)$	$e_{22} = n_2(1 - p_0)$	$n - x_0$
Total Number	n_1	n_2	n

where $p_0 = x_0/n$.

For reasons advanced in Section 12.2.7, it is necessary to impose a condition on the construction of expected frequencies in addition to the requirement that $\pi_1 = \pi_2$. The condition suggested by statistical theory is that *observed number of successes equals expected number of successes.* This in turn implies that the observed number of failures equals the expected number of failures. Hence, the expected and observed tables of frequencies are constrained to have the same row and column totals. With the addition of the hypothesized condition $\pi_1 = \pi_2$, the table of expected frequencies is uniquely defined by Table 12.5.1(c).

Having established the table of expected frequencies, a test of significance may be constructed by (i) defining all possible outcomes which have the same marginal totals as those of the frequency tables in Table 12.5.1, (ii) defining an ordering of the possible outcomes from that most consistent with the model to that outcome least consistent with the model; and (iii) computing a p-value as the probability of obtaining an outcome which is at or beyond the observed outcome in the ordering.

In fact, the ordering relies only on the numbers of successes in the two treatments and, in the case of rare events, the number of possible tables is typically small, numbering only one more than the number of successes. The possible tables are listed in Table 12.5.2, and Table 12.5.4 provides an illustration based on the data in Table 12.5.3.

Table 12.5.2. Possible Data Sets if Marginal Totals Must Equal Those in the Observed Data Set*

	Treatment A	Treatment B
Successes	0	x_0
Failures	n_1	$n_2 - x_0$

	Treatment A	Treatment B
Successes	1	$x_0 - 1$
Failures	$n_1 - 1$	$n_2 - x_0 + 1$

	Treatment A	Treatment B
Successes	2	$x_0 - 2$
Failures	$n_1 - 2$	$n_2 - x_0 + 2$

to

	Treatment A	Treatment B
Successes	x_0	0
Failures	$n_1 - x_0$	n_2

*The list of possible tables is based on the assumptions that $n_1 > x_0$ and $n_2 > x_0$, conditions which are reasonable where x_0 is the number of rare events.

Table 12.5.3. Comparison of the Incidence of SIDS in Huon and Esperance Local Government Areas (Data supplied by Dr. N. Newman, Royal Hobart Hospital, Tasmania)

(a) table of observed frequencies

	Region: Huon	Region: Esperance	Total
Number of SIDS deaths	$o_{11} = 1$	$o_{21} = 2$	$x_0 = 3$
Number not dying of SIDS	$o_{12} = 530$	$o_{22} = 371$	$n - x_0 = 901$
Total Number	$n_1 = 531$	$n_2 = 373$	$n = 904$
Proportion dying of SIDS	0.0019	0.0054	0.0033

Table 12.5.3 (*cont.*)

(b) table of expected frequencies assuming $e_{11} + e_{22} = x_0$ and $\pi_1 = \pi_2$

	Region		
	Huon	Esperance	Total
Number of SIDS deaths	$e_{11} = 1.76$	$e_{21} = 1.24$	$x_0 = 3.00$
Number not dying of SIDS	$e_{12} = 529.24$	$e_{22} = 371.76$	$n - x_0 = 901.00$
Total number	$n_1 = 531.00$	$n_2 = 373.00$	$n = 904.00$
Proportion dying of SIDS	0.0019	0.0054	0.0033

Under the model which contains the hypothesis $\pi_1 = \pi_2$, the probability of exactly i successes in a sample of n_1 under treatment 1 and $x-i$ successes in a sample of n_2 from treatment 2 is determined from (12.5.1) for the case where $n_1 \geq i$ and $n_2 \geq x - i$. (This is not a practical restriction since any meaningful examination of rare events will require that the two sample sizes are each larger than the total number of occurrences of the rare event in the sample.)

$$\Pr(o_{11} = i) = \frac{\binom{n_1}{i}\binom{n_2}{x-i}}{\binom{n}{x}} \quad \text{for} \quad i = 0, 1, \ldots, x^* \tag{12.5.1}$$

*provided $n_1 \geq i$, $n_2 \geq x - i$.

The set of probabilities define an ordering of the data sets, based on the statistic *number of successes for Treatment A* (i), from those most consistent with the hypothesis $\pi_1 = \pi_2$ to those least consistent with the hypothesis. Since there are $x+1$ possible values for i, the ordering produces the sequence $i_1, i_2, \ldots, i_{x+1}$ for the possible values of i. If the observed data set has i_s successes under treatment 1, the p-value for the test of significance is given by (12.5.2).

$$p_0 = \sum_{k=s}^{x+1} \Pr(o_{11} = i_k) \tag{12.5.2}$$

where $\Pr(o_{11} = i_k)$ is determined from (12.5.1).

Table 12.5.4 illustrates the construction of data sets and presents the probabilities computed from (12.5.1) and p-values calculated from (12.5.2). There are four possible outcomes. Since $x = 3$, the ordering is

$$i_1 = 2 \qquad i_2 = 1 \qquad i_3 = 3 \qquad i_4 = 0.$$

The observed data set is Set number 2, i.e. $i_s = 1$. Hence, by (12.5.2),

$$
\begin{aligned}
p &= \sum_{k=2}^{4} \Pr(o_{11} = i_k) \\
&= \Pr(o_{11} = 1) + \Pr(o_{11} = 3) + \Pr(o_{11} = 0) \\
&= 0.30 + 0.20 + 0.07 \\
&= 0.57.
\end{aligned}
$$

By the usual intepretation of a p-value, the figure of 0.57 is consistent with the hypothesis $\pi_1 = \pi_2$, i.e. that the incidence of SIDS is the same in the two local government areas.

Table 12.5.4. Comparison of the Incidence of SIDS in Huon and Esperance Local Government Areas (Based on information in Table 12.5.3)

(a) possible data sets*

Set Number

Region 1		Region 2		Region 3		Region 4	
H	E	H	E	H	E	H	E
$i=0$	3	$i=1$	2	$i=2$	1	$i=3$	0
531	370	530	371	529	372	528	373

*Data sets constrained to have marginal totals which are identical with those for the table of observed frequencies.

(b) probabilities

Set	1	2	3	4
i	0	1	2	3
$\Pr(o_{11} = i)$	0.07	0.30	0.43	0.20

(c) p-values

Set	1	2	3	4
p	0.07	0.57	1.00	0.27

The usual warning about the need for caution in interpretation of the result of a test of significance in the absence of a confidence interval applies. It has particular relevance in this case because of the low power of the test.

12.5.3 Confidence Interval for the Relative Risk

Suppose that two treatments have success rates π_1 and π_2 and the following data have been collected on subjects which have been randomly allocated between the treatments.

	Treatment A	Treatment B	Total
Number of successes	x_{10}	x_{20}	x_0
Total number	n_1	n_2	n

Statistic: If the probabilities of success are small and the sample sizes are large, the sampling distributions of the numbers of successes under each treatment are well approximated by Poisson distributions with means $n_1\pi_1$ and $n_2\pi_2$, respectively.

Statistical theory establishes that for two independent Poisson statistics, z_1 and z_2, with parameters $n_1\pi_1$ and $n_2\pi_2$, respectively, the sampling distribution of z_1, conditional on the sum $z_1 + z_2$ being fixed at the value $x_1 + x_2$, is Binomial with n equal to $x_1 + x_2$ and probability of success, P, determined from (12.5.3).

$$P = n_1\pi_1/(n_1\pi_1 + n_2\pi_2). \tag{12.5.3}$$

Estimation: Using the methodology of Chapter 9, a confidence interval for P can be obtained which yields values P_L and P_U for lower and upper limits, respectively.

These can be converted to confidence limits for the relative risk $R = \pi_1/\pi_2$ using the relationship $R = \frac{n_2}{n_1}\frac{P}{1-P}$.

Application and computations: The application is warranted only when the method presented in Section 12.2.9 for constructing a confidence interval for R is not applicable.

The calculations may be performed by following the steps listed below and on the assumption that x is less than $n - x$.

Step 1 — Check that at least one of n_1x/n and n_2x/n is less than 3.0. If both are 3.0 or larger, the methodology of Section 12.2.9 should be considered since it will provide, in general, the shorter confidence interval for any chosen level of confidence.

Step 2 — Check that the conditions for valid application of the Poisson approximation to the Binomial distribution are valid by using the guidelines that (i) both n_1 and n_2 are in excess of fifty; and (ii) both x_1/n and x_2/n are less than 0.1. (Failure of these conditions to be met indicates that at least one of the sample sizes is too small for a meaningful confidence interval to be constructed.)

Step 3 — Compute confidence limits P_L and P_U using the appropriate methodology of Section 9.2.4 with the value for n being set equal to $x_1 + x_2$ and the values of x_1 and x_2 being the numbers of successes under treatments A and B, respectively.

Step 4 — Transform the limits P_L and P_U to corresponding limits R_L and R_U for the relative risk using the relationship $R = \frac{n_2}{n_1} \frac{P}{1-P}$.

Illustration: The method is applied below to the following data.

	Group A	Group B	Total
Number of SIDS deaths	8	0	8
Total number	1056	547	1603

Step 1 — since n_2x/n is less than 3.0 the methodology of Section 12.2.9 is of doubtful validity.

Step 2 — both n_1 and n_2 are in excess of 50 and x_1/n_1 and x_2/n_2 are less than 0.1. Hence, the Poisson approximation to the Binomial is reasonable.

Step 3 — the value of n is taken to be the sum of the number of SIDS deaths, namely $8 + 0 = 8$, and x_1 is the number of successes in Group A, namely 8. Based on these values, 95% confidence limits for P are $P_L = 0.64$ and $P_U = 1.00$.

Step 4 — the limits for R are

$$R_L = \frac{547}{1056} \times \frac{0.64}{1 - 0.64} = 0.9 \qquad R_U = \frac{547}{1056} \times \frac{1}{1 - 1} = \infty.$$

Since the ratios includes unity, the possibility that the rate of SIDS deaths is the same for both groups cannot be discounted.

Problems

12.1 *Comparison of proportions — difference in expected proportions*

An examination of tutorial records for 245 students who completed a first year course in pure mathematics provides the information in the following table.

Number of Tutorials Attended[1]	Performance in Subject Pass	Fail
more than 20	137	34
20 or fewer	38	36

Number of Assignments Submitted[2]	Performance in Subject Pass	Fail
more than 15	142	22
15 or fewer	23	58

[1]total number of tutorials for course is 26.

[2]total number of assignments for course is 22.

Following the procedure recommended in the 'Computation' section of Section 12.2.6,

(a) compute 95% confidence limits for the difference in pass rates between students who attended more than 20 tutorials and students who attended 20 or fewer tutorials.

(b) Compute 95% confidence limits for the difference in pass rates between students who completed more than 15 assignments and students who completed 15 or fewer assignments.

Would the course teacher be warranted in using the results of these analyses to advise students in future classes that a student's chance of success in the subject are increased by attending more than 20 tutorials or by completing more than 15 assignments?

12.2 *Comparisons of proportions — ratio of expected proportions*

A large hotel and resort company was concerned at the turnover rate in unskilled staff. The proportion of staff leaving within two months of appointment averaged 42% and, with a uniform labor management policy throughout the organization, was relatively constant from region to region.

The company's operation were subsequently split into two regions, a northern region and a southern region, and the regional management given complete control over the criteria for selecting employees and for placing them in jobs within their individual regions. Note, however, that wages paid and formal conditions of employment were necessarily the same for both regions.

After the new policies were in place in a region, the next 100 employees placed in unskilled positions formed a sample for a study. For each of these employees it was noted if the person resigned within two months of taking the job. The results of the study are recorded below.

		Region A	Region B
Stayed more than 2 months	Yes	49	38
	No	51	62

Visual examination of the results establishes that Region A has a better retention rate (i.e. proportion of staff staying more than two months) than Region B in the sample data.

(a) Either follow the steps given in Section 12.2.8 to test the hypothesis that the retention rates are the same for both regions, or follow the steps in Section 12.2.9 to produce a 95% confidence interval for the ratio of retention rates.

Are there grounds for confidently predicting that the retention rate is higher in Region A than in Region B?

(b) Compare the observed rate in Region A with the previous rate (42% stayed longer than two months). Has the management of Region A strong grounds for claiming that the retention rate under their new policy is more effective than under the old policy? (Construct a confidence interval for the retention rate in Region A or test the hypothesis that the retention rate is 42%—use the methodology of Chapter 9.)

12.3 *Confidence interval for an odds ratio*

The method of packaging fruit in cartons plays an important role in preventing damage to the fruit in transit between the point of packing and the market. In general, better quality packaging equates with more expensive packaging.

Two types of packaging were compared for the transport of apricots. The type labelled 1 is known to be less effective than the type labelled 2. However, Type 1 is cheaper than Type 2. The question of interest is the comparative odds of a carton of fruit reaching its destination without damage to the fruit under the different types of packaging.

To provide information which might help answer this question, ten suppliers of export quality fruit were each asked to pack 30 cartons of fruit using 15 cartons of each type.

Assessment was made at the point of marketing for each carton with a classification of either *damaged* or *not damaged.* The results are recorded below.

		Type 1	Type 2
State of Carton	Damaged	35	9
	Not damaged	115	141

Obtain a 95% confidence interval for the ratio of the odds of a carton being damaged under the different types.

12.4 *Confidence intervals for differences in means — large samples*

Based on the data provided in Table 12.3.1, and using (12.3.3), compute 95% confidence intervals for the differences in mean times from Established labor to Delivery for

(i) Groups 1 and 2;

(ii) Groups 3 and 4.

Note the effects of sample size and variance on the relative widths of the two intervals.

12.5 *Comparison of means — large samples*

Persons who have undergone orthopedic surgery typically must visit the surgeon at least once for a check-up after they have been discharged from the hospital. Data are recorded below for the number of post-operative visits by patients of two surgeons. The data were collected from all patients who entered the hospital as *public* patients in a twelve month period (i.e. persons who did not request the surgeon of their choice and who were assigned the specialist who was on roster at the time of their admission).

		Number of visits 0	1	2	3	4	5	6	Total
Freq.	Surgeon X	41	11	52	93	12	2	2	213
	Surgeon Y	60	25	69	115	6	1	3	279

Is there evidence that the mean number of post-operative visits differs between the two surgeons? (Compute a 95% confidence interval for the difference in means using (12.3.3).)

(*Note that there are* 213 *observations for surgeon* X *and* 279 *observations for surgeon* Y. *If a computer is to be used for the analysis, the data to be entered for surgeon* X *comprises* 41 *zeros,* 11 *ones, etc. and for surgeon* Y *comprises* 60 *zeros,* 25 *ones, etc. There will most likely be a shorthand method for entering grouped data of this type.*

To perform the calculations directly, the following formulae may be used for computing the mean and variance:

$$\bar{y} = \sum_{i=1}^{k} f_i m_i / n \qquad \text{and} \qquad s^2 = \sum_{i=1}^{k} f_i (m_i - \bar{y})^2 / (n-1)$$

where f_i is the number of observations with value m_i $(i = 1, 2, \ldots, k)$ n is the sample size, i.e. $n = \sum_{i=1}^{k} f_i$.)

12.6 *Comparison of means when Normality can be assumed*

In an ecological study of the effects of discharge from an industrial plant into a bay, data were collected on plant species from three regions of the bay, one region on the western side of the bay near the point of discharge, a second region on the same side of the bay but well removed from the point of discharge and a third region on the opposite side of the bay.

The belief is that Regions 2 and 3 are areas of the bay which are unaffected by any discharge from the industry.

At each site a species of seaweed was growing and the data presented below record the lengths of individual plants which formed the sample.

Length of Seaweed (cm)

Region 1 94 90 78 95 73 77 95 78 98 95 90 79 88 109
Region 2 89 75 65 82 73 87 76 72 83 58 89 78
Region 3 60 53 55 62 74 71 53 75 68 83 67 71 67 51

Following the model and data checking procedures in Section 12.3.6,

(a) construct line diagrams for data from each region using the same scale for each diagram and visually compare the pattern for the three data sets; and

(b) construct separate Normal probability plots for data from the different regions.

Does the assumption of Normality seem reasonable in all cases?

(c) Compute variances for all data sets.

(d) Based on the guidelines in Section 12.3.6, decide whether the assumption of equal variances is reasonable and, if so, employ (12.3.8) to construct 95% confidence intervals for (i) the difference in mean length in Regions 1 and 3; and (ii) the difference in mean length in Regions 2 and 3.

What conclusions can be drawn as to the effect of the industrial pollutant on the mean length of seaweed in the bay?

12.7 *Confidence interval for differences in means — non-Normal distributions*

The following data were obtained from an experiment to compare the emission rates from wood heaters using two different species of wood.

Emission Rate (g/hour)

Pine	5.18	10.23	6.02	3.88	3.85	4.34	2.84
billets	2.64	11.82	9.24	8.73	13.48	11.37	
Eucalyptus	2.93	3.14	2.23	2.58	1.08	11.79	0.94
billets	2.73	1.66	3.00	2.07	1.03	1.61	3.85
	2.48	4.52	3.70	4.62	10.91	9.35	13.92

The experimenter wished to compare the mean emission rates for the two species.

(a) Construct line diagrams for the two data sets using the same scale for both.

(b) Note the skewness in the distributions which suggest that the assumption of Normality is unreasonable.

(c) Follow the recommendation in Section 12.3.6 and employ the method described in Section 12.4.3 to construct a confidence interval for the difference in mean emission rates.

Is there strong evidence that one species has a lower mean emission rate than the other?

12.8 *Comparing mean rates of occurrence*

A microbiologist investigating methods of storing cheese, was examining different protective coverings for use once the cheese had been exposed to the air. He was comparing a thin plastic covering with a treated cloth covering.

The experimental technique he adopted was to cut twenty equal sized pieces of cheese from a block of cheese and randomly assign the twenty pieces between the two forms of protective covering such that each covering was assigned ten pieces.

After storing the covered pieces for ten days at constant temperature, the number of mold colonies formed on each piece of cheese was recorded.

The data from the experiment are recorded below

Number of Colonies

Plastic covering	0	8	9	6	5	12	10	15	7	7
Cloth covering	8	4	10	4	3	9	6	9	7	3

Assuming the colonies form independently of one another, use the method based on the Poisson model (Section 12.3.8) to determine if there is evidence of a difference in the mean number of colonies expected under the two forms of covering.

12.9 *Estimating a ratio of means — logarithmic transformation*

(a) Using the data in Table 12.3.3 (which derives from Example 12.8), follow the model and data checking steps given in Section 12.3.6 to establish that the assumption of Normality is unreasonable and there is evidence of a greater variability with the larger mean.

(b) Transform the data to a logarithmic scale and apply the model and data checking procedures given in Section 12.3.6 to the transformed data.

(c) Compute a 95% confidence interval for the difference in means on the logarithmic scale and convert the limits to the original scale of measurement using the transformation given in Section 12.3.7. (This provides a 95% confidence interval for the ratio of mean time to complete the set task for 16 year old and 10 year old boys.) Check that your answer agrees with that provided in Section 12.3.7.

12.10 *Comparison of medians*

Data are provided in Table 1.3.1 for four variables—glucose, sodium, potassium and cholesterol.

(a) For each variable construct the set of differences between pre- and post-race readings for all of the 22 runners for whom data are recorded.

(b) Analyze each variable separately, using the method given in Section 12.4.4, to establish if there is evidence that the size of the change in median level of the blood chemical during the race varies between *Untrained* and *Trained* runners.

12.11 *Comparison of medians and preparation of a report*

Data are recorded in Table 1.2.1 for two measures of the level of pain experienced by women during childbirth, (i) *Pain rating index* which is a scaled variable recording the womens' level of perceived pain, and (ii) *Pain reaction* which is an ordered categorical variable with five possible categories, labelled $1, 2, \ldots, 5$. This is the assessment of level of pain as perceived by the nursing staff.

(a) Compare the median value of *Pain rating index* for the *Physiotherapy trained* group and the *Others* group. (Compute a 95% confidence interval for the difference in medians.)

(b) Use the methodology of Section 12.4.3 to compare the median values of *Pain reaction* for the *Physiotherapy* trained group and the *Others* group.

(c) Compare the median time in labor for the *Physiotherapy trained* group and the *Others* group.

(d) Using whatever summary statistics and graphical description you consider appropriate to supplement the findings in (a), (b) and (c), present a short report on your findings suitable for a non-statistical audience.

13

Observational Studies, Surveys and Designed Experiments

13.1 Observational Study or Designed Experiment?

There are important reasons for distinguishing studies in which the experimenter allocates treatments between subjects after the sample has been selected, from those studies in which the investigator merely records responses from sample members. Studies in which the experimenter applies treatment to the subjects or units and has control over the allocation of the treatments are termed *designed experiments.* The others are termed *observational studies.*

In many cases, the form of study is decided by the nature of the comparison. A medical researcher may be in the position to decide which subjects will receive the new drug treatment and which subjects will receive the standard. An educator who wishes to compare academic performance between males and females cannot allocate a sex to each subject in the investigation!

Statistically speaking, the differences between the two types of studies are profound and may affect the role of sample selection, the method of statistical analysis and the nature and scope of the conclusions which can be drawn. The important distinctions are presented in the following subsections.

13.1.1 Sample Selection and Validity of Statistical Conclusions

When methods of Inferential Statistics are employed, there are two important considerations in respect of the statistical conclusions reached. The first relates to the *validity* of the method used and the second relates to the *scope* of the statistical conclusions.

In observational studies which involve the comparison of groups, the manner in which samples are selected affects the validity of the statistical conclusions. As a general rule, statistical validity of methods of Inferential Statistics requires that (i) random selection, or its equivalent, be employed from each group, and (ii) responses obtained from members of one group are independent of the responses obtained from any other group.

In designed experiments, validity is dependent solely on the method of allocation of subjects or experimental units to treatments. Thus, it is possible to pass a statistical judgment on the relative effects of two or more treatments without any knowledge of the method of sample selection. Although there may be connections between different sample members, the fact that sample members are independently assigned between treatments negates the effects of these connections from the viewpoint of ensuring validity of the resulting methods of statistical analysis.

13.1.2 Sample Selection and the Scope of Statistical Conclusions

The declaration that statistical evidence in a designed experiment provides strong evidence for a difference in treatment effects is only part of the information which the person conducting the experiment requires. There is also a need to know the scope of the conclusions.

For example, a medical researcher who correctly applies methodology from Chapter 12 to provide evidence of a difference in two treatments when the subjects are deliberately chosen to be of the same age, sex and from the same background cannot presume the *statistical* conclusions have application in the population at large, an agricultural scientist who finds evidence from valid statistical analysis that one variety of a crop out yields another based on an experiment at a single agricultural research station cannot claim the *statistical* findings have validity across a farming region.

In designed experiments, the scope of the conclusion is decided separately from the validity of the statistical analysis and depends on the method of sample selection. In the simplest situation, that in which the sample members are randomly selected, the statistical conclusions relate to the population or process sampled.

There may be deliberate non-random selection of sample members to achieve a pre-treatment uniformity in sample members in order to improve the chance of detecting treatment differences should they exist. In this case, it is for the experimenter to produce a non-statistical argument to extend the conclusions beyond the immediate sample. For example, a drug experiment may establish the worth of a drug in a collection of young males. The extrapolation of this finding to other sections of the community requires either further experimentation or medical argument.

13.1.3 Observational Studies — Difficulties of Interpretation

Scientists rarely undertake comparative studies simply to determine if two or more groups are different. Generally, the comparison is made in the context of attempting to establish why the groups differ or to measure the extent of the difference. Consider the problem described in Example 13.1.

Example 13.1 (Are Boys Better Than Girls at Mathematics?). *Surveys in many countries in the world document a* **belief** *that boys are better than girls at mathematics. There is evidence from some observational studies establishing better performances at secondary school level by groups of males and there is evidence from tertiary levels that the ratio of females to males in mathematics courses is commonly much lower than in arts courses.*

Even if one accepts the evidence that there is a tendency for boys to perform better in mathematics courses than girls, how does one establish if this difference is inherently sexual or if it is the result of the environment in which the children live and work?

This last question characterizes the dilemma which arises in any observational study where the aim is to determine if the difference established by statistical analysis is related to the factor which is presented as distinguishing the groups.

In Example 13.1, there are two obstacles standing in the way of formulating a study which separates the effects of *Sex* from the effects of other contributory factors. The most obvious problem is the identification of other possible contributory factors and the elimination of their contributions. Many factors have been suggested, but there always remains the argument that the removal of effects of known or suspected contributors does not eliminate the possibility of undetected factors being the cause of the claimed sex difference. Investigators are in the impossible position of never being able to substantiate the claim that *all* important contributors apart from the sex factor have been eliminated.

Even if widespread agreement could be obtained that all important contributory factors had been identified, there remains the practical difficulty of selecting samples which ensure the elimination of all identified contributors other than the sex factor.

A related problem concerns the **advisability** of eliminating the effects of all other contributory factors. This is well illustrated by research undertaken by Dr. Jane Watson at The University of Tasmania. In a study of factors which influence performance in tertiary mathematics, Dr. Watson established that examination performance is associated with *test anxiety*—more anxious students tend to have worse performances. Additionally, Dr. Watson found that girls tended to exhibit more *test anxiety* than boys. Should the effects of *test anxiety* be removed before comparing the performances of the sexes or is a higher level of *test anxiety* something which is inherently female? If the latter is true, then it would be misleading to eliminate the effects of *test anxiety* before the sex comparison is made.

Why do the same concerns not arise in designed experiments? After all, there is no requirement that the groups of sample members to be assigned to different treatments must be identical in all respects.

The answer lies in the nature of the allocation process. While it is true that in any **one** experiment, treatment differences may be confounded with a specific non-treatment effect, the nature of the process of *random* allocation of subjects between treatments ensures that in repetitions of the experiment, the same confounding becomes increasingly less probable. Thus, independent experiments which reach the same findings provide increasing support for the difference being related to differing treatment effects rather than to non-treatment effects.

A particularly valuable consequence lies in the fact that the experimenter is relieved of the need to identify the set of non-treatment factors which might influence response. In this context, adhering to random allocation provides an insurance policy against the type of serious blunder which is described in Example 13.2.

Example 13.2 (Allocating Rats in a Laboratory Experiment). *Statistical consultants are often visited by scientists when the results from an experiment do not match with expectations and there is concern that the statistical analysis may be incorrect. This was the case in respect of the laboratory experiment which produced the unexpected results described in Example* 12.9.

The unexpected finding reported in that example was the apparent failure of the sedative to decrease the level of activity in the treated rats.

In the experiment, twenty rats available for an experiment were placed in a holding cage. The rats were individually taken from the cage and either given a piece of food containing a sedative or given a piece of food containing no sedative. Each rat was then assessed for level of activity and awarded a score on a zero to ten scale. The analysis and findings are reported in Example 12.9.

In fact, the experiment was repeated because the experimental results were inconsistent with expectations. In both experiments, the sedative failed to show the desired effect. Yet the chemical structure of the sedative suggested it should have worked. The failure was a serious setback to the biochemist who was undertaking the research.

The clue to the flaw in the experiment came in discussion with the laboratory assistant who had the task of applying the treatments to the animals. Contrary to instructions, the method employed in allocating rats to treatments was to give the untreated food to the first ten rats pulled from the holding cage and the treated food to the remaining ten rats. Not surprisingly the first ten rats pulled from the cage were the more timid rats and the rats given the sedative were the more active rats. The effects of the sedative were negated by the natural variation in activity between the animals.

With hindsight, it may seem reasonable to have taken account of the fact that the rats would show a natural variation in level of activity and to attempt to remove the effects of this source of variation in the design of the experiment or in the statistical analysis. The problem with this way of thinking is that many possible sources of extraneous variation are present in any experimental situation and it is neither realistic nor practical to expect that each can be identified and its effect removed. Hence, there is need for an approach which remains valid even when undetected contributory factors are present.

13.2 Experimental and Treatment Designs

13.2.1 The Control of Extraneous Variation

Through the judicious selection of experimental units or subjects and allocation of those units or subjects between treatments, it is possible to improve the sensitivity of statistical analysis to treatment differences without reducing the scope of the statistical conclusions.

There is a substantial area of Statistics concerned with the specification of designs which allows separation of treatment and non-treatment effects. The breadth and complexity of designs has necessitated a formal means of describing the selection of sample members and their allocation to treatments. These aspects of the design are referred to respectively as the *experimental design* and the *treatment design*. The basic features of these components are considered in the following sections.

13.2.2 Experimental Designs

The function of experimental designs is to permit comparison of treatment effects free of confusion with other factors which may seriously influence response.

Experimental designs had their origins in field studies. In the simplest design, a uniform area of land is available for growing plants and the units selected for the experiment are formed by subdividing the land into *plots,* commonly by imposing a rectangular grid on the field as illustrated in Experimental Design I.

Experimental Design I

→

1	5	9	13
2	6	10	14
3	7	11	15
4	8	12	16

The sixteen plots formed in Experimental design I are available for allocation between treatments. If there is no identified factor contributing to variation in response between the plots, there is no reason for restricting treatment comparisons in any way. Hence, the usual method of treatment allocation given Experimental design I would be by random allocation of units to treatments following the method outlined in Appendix 13.2.

Such a design would not be appropriate in the following situation. Consider a site at which there is an area of land of the same size as that for which Experimental design I could be employed, but with a moisture gradient across the land. The effect of moisture variation would be expected to result in higher crop growth at the wet end than the dry end. In this case, uniformity in growth is only found within areas of the same moisture level. Hence, the wise course of action would be to limit comparisons to plots which are subject to the same level of moisture. This aim can be met by adopting the two-stage process which is illustrated in Experimental Design II.

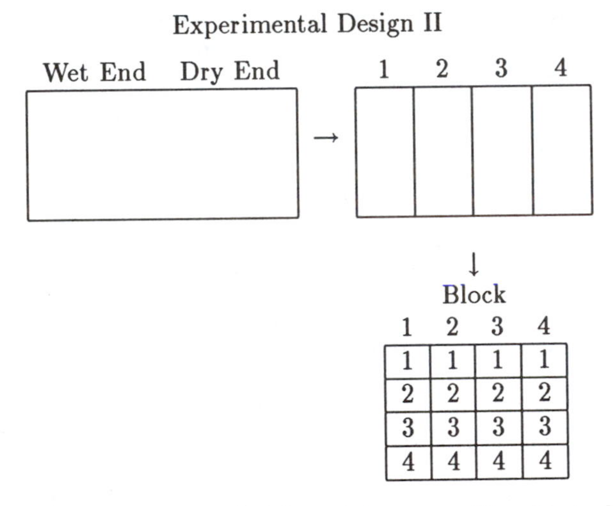

The first stage requires the division of the area of land into *blocks* which are represented in the diagram by vertical strips and which have the property that, within a block, the level of moisture is presumed constant. The second stage is to subdivide the blocks into plots.

The significance of the division into blocks lies in the fact that the effect of variations in moisture level is confined to *between block* variation while *within block* variation is free of the effect of the moisture factor.

Where Experimental design II is appropriate, sensible treatment comparisons would take place only *within blocks* since these comparisons avoid confusion between treatment and moisture effects.

The essential distinction between the two designs lies in the fact that in Experimental design I there is only one unit size and that is represented by the *plot*, whereas in Experimental design II there are two unit sizes—the *blocks* and the *plots*.

Designs of greater complexity than that defined by Experimental design II are commonly employed in field trials, but all are constructed with the aim of separating the effects of non-treatment factors from the effects of treatment factors.

Experimental designs have an equally important role in scientific studies involving people and in laboratory experiments as they do in studies of plants and field trials. Consider the situation in which selection of individuals is to be made for a psychology experiment in which the first task is to divide the population into four age groups, with subsequent selection of persons from within each group. From the design viewpoint, this is equivalent to Experimental design II with the *blocks* corresponding to the age groups and the *plots* corresponding to the persons selected for the experiment from within each age group.

Another common situation arises when each person corresponds to a *block*. This occurs when a response is to be obtained from each sample member on more than one occasion. Then the *plots* correspond to the times at which responses are to be collected. The particular advantage of this design lies in the fact that the treatment comparisons can be made free of person-to-person variation.

13.2.3 Strata in Experimental Designs

A valuable description of designs identifies one or more unit sizes or *strata*. According to this description, Experimental Design I would be a *one-stratum design* with the single unit size being identified with the *plots*. Experimental Design II would be a *two-strata design* with *blocks* being the larger units and *plots* the smaller units.

Example 12.2 illustrates a one-stratum design in which the 26 persons used in the experiment constitute the units to be allocated between the two treatments.

Examples 13.3 and 13.4 offer illustrations of experimental designs which are two-strata designs while Example 13.5 illustrates a three-strata design.

Example 13.3 (The Effect of Hormone on Growth Rate of Corn). *It commonly happens in glass house experiments that growing conditions are not constant across the glass house. Thus, like-treated plants may grow at different rates, perhaps in the manner illustrated below.*

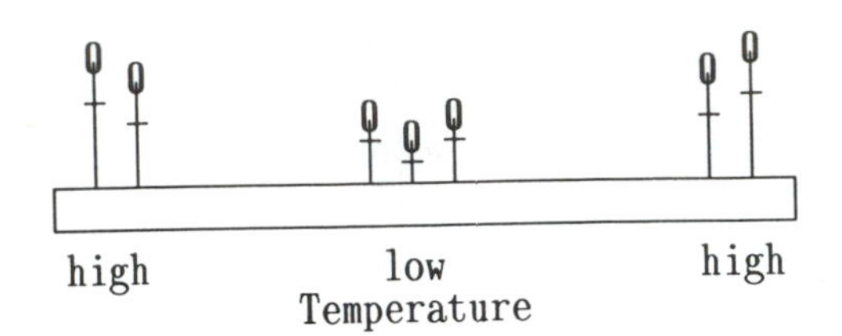

An experimenter faced with this situation was required to lay out a series of twenty pots in the glass house for the comparison of two treatments such that ten pots are assigned to a hormone treatment and ten are untreated, serving as the 'controls.' A tray was available for holding the pots in the following configuration.

	Position									
	1	2	3	4	5	6	7	8	9	10
Row 1	x	x	x	x	x	x	x	x	x	x
Row 2	x	x	x	x	x	x	x	x	x	x

← Temperature variation →

The Experimental Design was constructed in the following steps.

1. *Ten positions were selected on the table on which the pots were to be placed such that within each position there was a constant temperature environment.*

Position									
1	2	3	4	5	6	7	8	9	10

← Temperature variation →

2. *Two pots were selected and placed in each position*

	Position									
	1	2	3	4	5	6	7	8	9	10
Row 1	x	x	x	x	x	x	x	x	x	x
Row 2	x	x	x	x	x	x	x	x	x	x

← Temperature variation →

The two strata are clearly identifiable. There is a **position** *stratum and a* **pots within position** *stratum.*

The two pots in each position were randomly allocated between the two treatments, thereby ensuring that at each position both the hormone treatment (H) *and the 'no hormone' treatment* (N) *were represented. The layout which resulted is illustrated below.*

Position	1	2	3	4	5	6	7	8	9	10
Row 1	N	N	H	N	N	N	H	N	N	H
Row 2	H	H	N	H	H	H	N	H	H	N

← Temperature variation →

The data obtained from the experiment are presented in Table 13.2.1. *As is customary, the rows of the table reflect the treatment applied rather than the location on the table.*

Analysis of the data is considered in Section 13.3.

Table 13.2.1. Weights of Corn Cobs (g) (Data from Example 13.3)

Position	1	2	3	4	5	6	7	8	9	10
Treatment										
Hormone	600	570	610	510	250	320	300	450	710	550
Control	600	630	640	470	260	310	310	490	770	620

Example 13.4 (Comparison of Methods of Culturing from Blood Samples). *Two different methods of culturing bacteria present in blood samples were under comparison. In the standard method, known as the* PHA-PHA *method, the same process was used at each stage. A simpler and cheaper method identified as the* PHA-LK *method which employed a different second stage was under investigation.*

The measured response was the number of bacterial colonies grown from a standard amount of blood. The relative quality of one method compared with another is determined by the comparative numbers of colonies which could be produced from equal quantities of blood which have the same level of infection by the bacteria.

The experiment was conducted in three steps.

1. *The selection of* 11 *subjects with a sample of blood taken from each sample member.*
2. *From each blood sample four portions were selected for culturing.*
3. *The four portions from each blood sample were randomly assigned between the two methods such that each method received two portions.*

A two-strata design was employed in this study. The first stratum is the **subject** *stratum and the units at this level are the samples of blood taken from the different subjects. The second stratum is the* **portion within subject** *stratum and the units are the portions of blood which come from the sample drawn from each subject.*

As noted above, it is the portions of blood from within a subject which were randomly allocated between the two treatments.

The use of a design in which both methods were applied to the blood of the one subject, allowed the comparison of methods within subjects. This was particularly important in this sitation since different subjects had different levels of infection of the organism which was being cultured.

The data which were obtained in the experiment are presented in Table 13.2.2. Analysis of the data can be found in Section 13.3.

Table 13.2.2. Numbers of Colonies Formed Under Different Culturing Methods (Data from Example 13.4)

Subject	Method PHA:PHA Duplicate		Method PHA:LK Duplicate	
	1	2	1	2
134A	4	36	8	6
134B	23	10	39	40
141	59	40	10	160
142	61	90	70	78
147	110	89	190	470
148	98	190	290	370
150	280	260	160	250
163	240	370	910	400
164	350	420	500	700
165	310	510	270	380
170	120	740	480	620

Example 13.5 (The Role of Nutrients and pH on the Uptake of Magnesium in Plants). *An agricultural scientist wishing to study the effects of calcium and pH in nutrient solution on the uptake of magnesium in three plant species, employed an experimental design which was constructed in the following manner.*

Step 1 — *two growth chambers were selected for the experiment.*
Step 2 — *within each growth chamber, four trays were placed.*
Step 3 — *within each tray, three pots were placed.*

The steps are illustrated in Figure 13.2.1. *In the figure, the three strata are clearly evident—the* **chamber** *stratum, the* **trays within chambers** *stratum and the* **plots within trays within chambers** *stratum.*

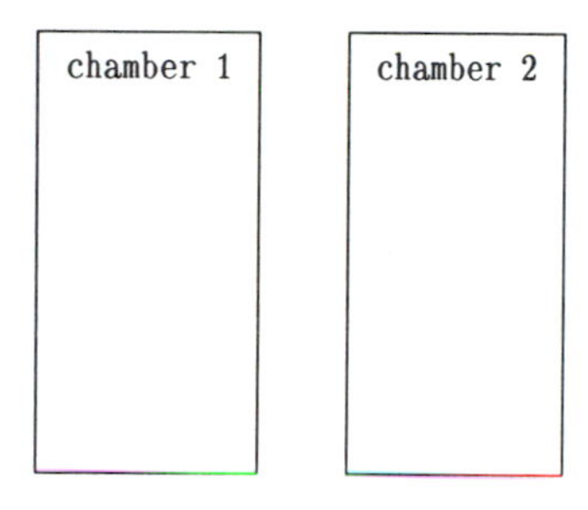

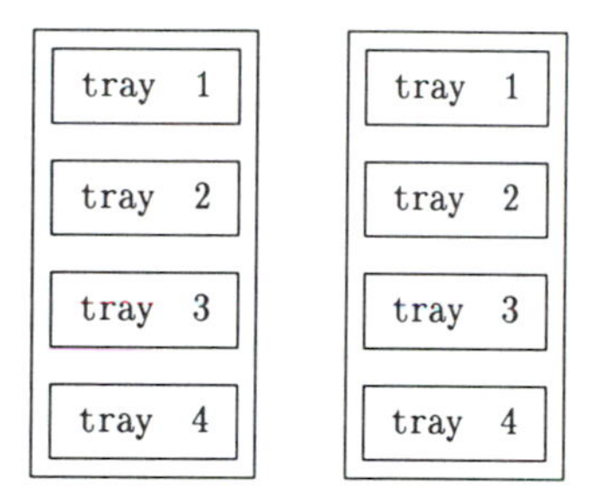

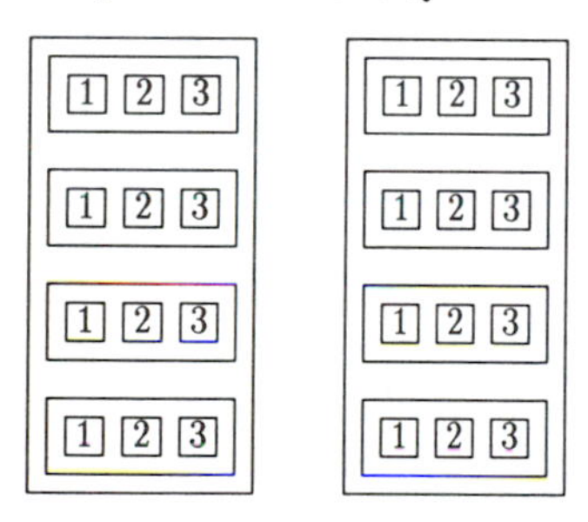

Figure 13.2.1. Schematic portrayal of the experimental design for Example 13.5.

*Each of the trays was allocated a nutrient solution with a chosen level of calcium ('high' or 'normal') and a chosen level of pH (*4.0 *or* 4.6*) subject to the condition that the four combinations of calcium and pH are all represented in both chambers. The four trays in each chamber were randomly allocated between the four combinations of calcium and pH to provide the allocation which is shown in Figure* 13.2.2.

Step 1 – allocation of combinations of calcium and pH between trays in each chamber

Calcium level is either high (H) or normal (N)
pH level is either 4.0 or 4.6

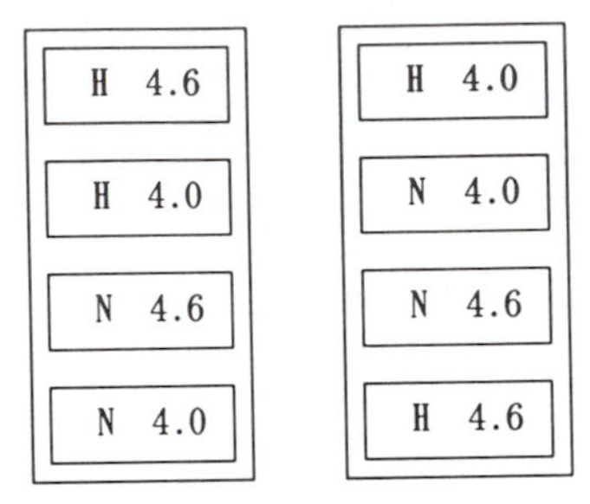

Step 2 – allocation of plant species between pots within each tray

Plant species are Kikuyu (K), lettuce (L) and cabbage (C)

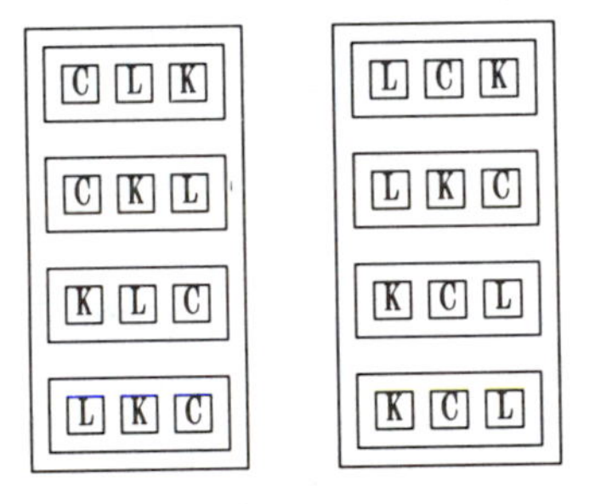

Figure 13.2.2. Schematic portrayal of the treatment design for Example 13.5.

Three plant species were used in the experiment, 'Kikuyu,' 'cabbage' and 'lettuce.' The three pots within each tray were randomly allocated between the three plant species. The actual randomization is shown in Figure 13.2.2.

Example 13.5 illustrates both the complexity which may arise in the construction of designed experiments and the simplicity of description of designs if they are broken down into component parts.

13.2.4 Treatment Designs

The description of experimental designs makes no reference to the manner in which experimental units are to be allocated between treatments. This is contained within the statement of the *treatment design.*

In the simplest treatment designs, it is only at one stratum level that units are allocated between treatments. This is the case in both Example 12.2 and 13.4. In Example 12.2, the persons taking place in the study were randomly allocated between two groups—one group who were given a vitamin supplement and the other or 'control' group whose members received no vitamin supplement. In Example 13.4, persons used in the study were the units in the **subject** stratum. However, there was no allocation of subjects between treatments in this experiment. The allocation of treatments was between the four portions of blood taken from each subject. The random allocation took place *within* subjects with the restriction that two portions of blood from each subject were allocated to the PHA-PHA method and two portions to the PHA-LK method.

The treatment design employed in Example 13.5 introduces two new elements. Note that the allocation of units to treatments occurred at more than one stratum level. Trays within chambers were allocated between combinations of calcium and pH while pots within trays were allocated between plant species. Additionally, there is a structure in the treatments. The four treatments defined at the tray stratum level are distinguished by the combinations of levels of two *factors*—calcium and pH. This is referred to as a *factorial arrangement* of treatments. With this design it is possible to (i) separate the effects of calcium and pH on magnesium uptake, and (ii) to examine the influence of pH on the effect which changing levels of calcium has on magnesium uptake.

13.3 Paired Comparisons

When comparing two treatments, the sensitivity of the comparison is reduced if there is large variability in the set of responses due to non-treatment sources. The practice of using uniform experimental material or like subjects to reduce the level of non-treatment variation has the disadvantage of reducing the scope of the statistical findings. There is also the practical difficulty that a sample composed of like members may be difficult or unacceptably expensive to obtain.

An alternative approach is to employ a two-strata experimental design in which the major source of non-treatment variability is contained in variation between units of the first stratum and treatment comparisons take place between units at the second stratum level. In this design, uniformity is required only in pairs of sample members. This approach is employed in Example 13.3.

13.3.1 The Nature and Application of 'Pairing'

Obtaining pairs of like sample members does not necessarily involve searching through experimental material for two units which are similar. Other possibilities include

- placing pairs of units in a like environment as was done in Example 13.3 where pairs of pots were placed in the same temperature conditions in a glass house;
- subdividing experimental material to form like units as was done with a blood sample in Example 13.4; and
- making two measurements on the one unit, one measurement after applying treatment A and the other measurement after applying treatment B, with the decision as to which treatment is applied first being determined by random allocation.

13.3.2 The Experimental Aim

Where the response variable is a scaled variable, the purpose of this study is presumed to be the examination of the difference in average response between the two treatments, $D = M_1 - M_2$ or, less commonly, the ratio $R = M_1/M_2$. If the study relates to proportions or success rates, the aim may be to study differences, ratios or odds ratios. Methodology is provided in this chapter for comparisons of means or medians. When the comparison is based on proportions or probabilities, methodology can be found, for example, in Fleiss ([1981], Chap. 8).

13.3.3 Data

The data are in the form of a sequence of n pairs of responses and associated differences, which are denoted in the following manner.

Pair	1	2	$\cdots$	n
Treatment 1	y_{11}	y_{12}	$\cdots$	y_{1n}
Treatment 2	y_{21}	y_{22}	$\cdots$	y_{2n}
Difference	$d_1 = y_{11} - y_{21}$	$d_2 = y_{12} - y_{22}$	$\cdots$	$d_n = y_{1n} - y_{2n}$

13.3.4 Statistical Models

The description and examination of statistical models involves the introduction of a *component equation* and the concept of *additivity*. Component equations were introduced in Section 10.2.3 and additivity is discussed in Section 10.3.8. An understanding of those sections is presumed in the following discussion.

Variation between the observed responses in a paired comparison study is presumed to be attributable to three sources. They are

- treatment differences;

- changes in conditions between pairs; and
- differences between two members of the same pair which are not related to the different treatments applied.

For example, in Example 13.3, some variation in responses between pots at the same temperature can be attributed to the fact that a plant in one pot had a hormone treatment while its partner did not, some variation can be associated with the positions of the pairs of pots in the glass house (which includes the effect of temperature variations across the glass house) and the remaining variation between pots in the same position is due to unidentified causes.

The common presumption in models employed in paired comparisons is that the contributions of the different factors to the responses are combined in the manner portrayed in (13.3.1), i.e. the components may be added to give the response.

$$y_{ij} = M + T_i + P_j + e_{ij} \qquad \text{for } i = 1, 2;\ j = 1, 2, \ldots, n \tag{13.3.1}$$

where y_{ij} is the response from the member in the jth pair which received the ith treatment; M is the overall mean (expected value); T_i is the adjustment for the fact that the unit received the ith treatment; P_j is the adjustment for the fact that the unit is in the jth pair; and e_{ij} is the contribution of unidentified effects to the response of the unit which received the ith treatment and is a member of the jth pair.

Equation (13.3.1) is termed an *additive* component equation because it is assumed that the observed response is formed by the addition of the components.

The purpose of using this two-strata design is to eliminate the effects of between-pair variability, i.e. the contributions of the effect P_j in (13.3.1). This aim can be met if the analysis is based on the set of differences, $d_1, d_2, \ldots, d_n$, where $d_i = y_{1j} - y_{2j}$ for $j = 1, 2, \ldots, n$, since (13.3.1) then reduces to the simpler form (13.3.2) which contains no contribution from the elements P_j $(j = 1, 2, \ldots, n)$.

$$d_j = D + e'_j \qquad \text{for } j = 1, 2, \ldots, n \tag{13.3.2}$$

where

$$\begin{aligned} d_j &= y_{1j} - y_{2j}; \\ D &= \{M + T_1\} - \{M + T_2\} = T_1 - T_2; \\ e'_j &= e_{1j} - e_{2j}. \end{aligned}$$

The quantity D which appears in (13.3.2) is the expected difference in average response between units having treatment 1 and treatment 2 and is the quantity which is of primary interest to the investigator. In practice it may represent either the difference in mean response or difference in median response.

Validity of statistical methodology rests on the assumption that within each pair the two units are randomly allocated between the two treatments.

13.3.5 Preliminary Model Checking

Preliminary model checking involves an examination of the additivity assumption. As explained in Section 10.3.8, there are two types of non-additivity, one associated with the assumption that the two identified components have an additive contribution, i.e. $T_i + P_j$, and the other being the assumed additivity of the identified and unidentified components, i.e. $(T_i + P_j) + e_{ij}$.

From a practical point of view, the additivity of the identified components demands that the expected difference in treatment effects, $D = T_1 - T_2$, is identical under the different conditions which apply between pairs. Thus, in Example 13.3, it is presumed that the average difference in weight of corn cobs between plants treated with the hormone and those which are untreated is the same over all temperature conditions experienced in the glass house.

The additivity of the identified and unidentified components influences the form of comparison which is relevant. For reasons given in Section 10.3.8, the only case of non-additivity which is considered involves a multiplicative model and the comparison of ratios of treatment effects.

Using the notation introduced in (13.3.1) the multiplicative model has a component equation represented by (13.3.3).

$$y_{ij} = MT_iP_je_{ij} \qquad \text{for } i = 1, 2;\ j = 1, 2, \ldots, n. \qquad (13.3.3)$$

To eliminate the contribution of the component P_j it is necessary to consider the ratio of paired responses as defined in (13.3.4).

$$r_j = Rf_j \qquad \text{for } j = 1, 2, \ldots n \qquad (13.3.4)$$

where

$$r_j = y_{1j}/y_{2j};$$
$$R = \{MT_1\}/\{MT_2\} = T_1/T_2;$$
$$f_j = e_{1j}/e_{2j}.$$

For purposes of statistical analysis, it is convenient to represent (13.3.4) in the additive form (13.3.5).

$$\log(r_j) = \log(R) + \log(f_j) \qquad \text{for } j = 1, 2, \ldots, n. \qquad (13.3.5)$$

Expressed in the additive form of (13.3.5), it is possible to use the same methodology as applies when (13.3.2) is assumed. The methods can be employed directly to examine hypotheses about $\log(R)$ or to construct a confidence interval for $\log(R)$. By employing the exponential transformation, it is possible to construct a confidence interval for the ratio $R = T_1/T_2$.

13.3.6 Choosing and Applying a Method

There are two basic considerations when choosing a method—the parameter to be used in the comparison (mean or median) and the nature of the comparison (comparing differences or ratios). The basis for deciding between the mean and the median is discussed in Section 11.1. Whether the ratio or the difference is the appropriate form of comparison is based on the procedure described below.

The steps to be employed in model and data checking are presented below and are illustrated using the data in Table 13.2.1 which were obtained in the experiment reported in Example 13.3.

Step 1 — *Preliminary data checking*

A check should be made on the measured responses to ensure that no response has been wrongly recorded and no observation is outside acceptable bounds.

Step 2 — *Preliminary model checking*

Checking for constancy in the difference. $D = T_1 - T_2$. Plot d_j versus either y_{1j} or y_{2j} for $j = 1, 2, \ldots, n$. There should be no evidence of a trend in the points.

The plot presented in Figure 13.3.1 *displays clear evidence of a decreasing trend line. Examination of the data in Table* 13.2.1 *from which the plot is constructed provides the explanation. The difference in the effects of the hormone and the no-hormone treatments is greater in conditions of higher temperature.*

Where the difference $D = T_1 - T_2$ is not constant within the range of environments of interest, any statistical conclusions concerning its value is dependent on the particular conditions applying to the sample members. It is unwise to make general pronouncements on the average value of D in such cases. More fruitful may be an exploration of the manner in which the difference in treatment effect varies with the change in environmental conditions.

Checking for additivity in identified and unidentified components

1. Plot the unsigned difference, $|d_j| = |y_{1j} - y_{2j}|$, versus the total, $t_j = y_{1j} + y_{2j}$ $(j = 1, 2, \ldots, n)$. Look for evidence of increasing variability in values of $|d_j|$ with increasing values of t_j as evidence that a multiplicative model may be required.

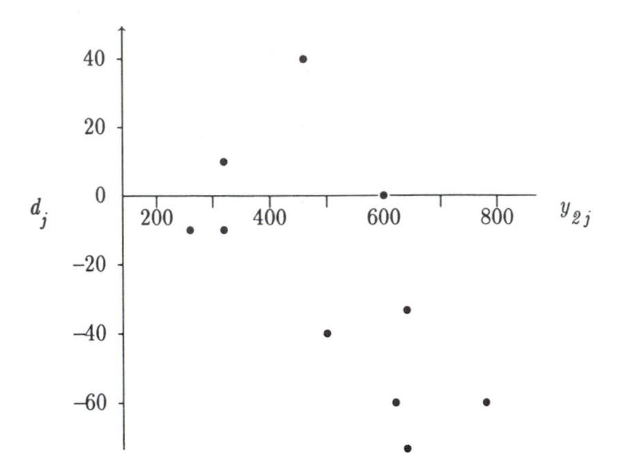

Figure 13.3.1. Model checking — plot of difference in weight (d_j) vs. weight (y_{2j}).

2. Check that the data contain no negative values. Look for evidence of great variability in the data. Does the largest value exceed the smallest value by an order of magnitude of at least ten? If these checks are answered in the affirmative, this can be taken as evidence of a multiplicative model and a study of ratios might be more informative.

The application of these procedures is illustrated using the data contained in Table 13.3.1 *which were obtained in the experiment reported in Example* 13.4. *It is apparent from the counts in Table* 13.3.1(a) *that there is great variation in responses. Figure* 13.3.2(a) *provides the evidence that the variability in differences does increase as the total increases.*

Hence, there is evidence favoring a multiplicative model rather than an additive model to represent the set-up in Example 13.4.

Where the multiplicative model is suggested from this preliminary model checking, the data should be converted to a logarithmic scale and the steps in model checking described above should be repeated. If there remains doubt about either additivity assumption, advice should be sought from a consultant statistician.

The plot based on the log-transformed data in Table 13.3.1(b) *appears in Figure* 13.3.2(b). *The transformation appears to have stabilized the variance.*

Futher examination of the model and data occurs when the method of analysis is selected and any problem concerned with individual readings or distributional assumptions can be considered at that point.

Table 13.3.1. Information Derived from the Data in Table 13.2.2

(a) Untransformed data

Subject	PHA:PHA	PHA:LK	Difference	Total
	y_{1j}	y_{2j}	$\|d_j\|$	t_j
134A	40	14	26	54
134B	33	79	46	112
141	99	170	71	269
142	151	148	3	299
147	199	660	461	859
148	288	660	372	948
150	540	410	130	950
163	610	1310	700	1920
164	770	1200	430	1970
165	820	650	170	1470
170	860	1100	240	1960

(b) log transformed data

Subject	PHA:PHA	PHA:LK	Difference	Total
	$\log(y_{1j})$	$\log(y_{2j})$	$\|d'_j\|$ [1]	t'_j [2]
134A	3.689	2.639	1.050	6.328
134B	3.497	4.369	0.873	7.866
141	4.595	5.136	0.541	9.731
142	5.017	4.997	0.020	10.014
147	5.293	6.492	1.199	11.786
148	5.663	6.492	0.829	12.155
150	6.292	6.016	0.275	12.308
163	6.413	7.178	0.764	13.591
164	6.646	7.090	0.444	13.736
165	6.709	6.477	0.232	13.186
170	6.757	7.003	0.246	13.760

Notes:
[1] $|d'_j| = |\log(y_{1j}) - \log(y_{2j})|$
[2] $t'_j = \log(y_{1j}) + \log(y_{2j})$.

Step 3 — *Choosing a method*

(a) *Additive model.* If the additivity of the component equation in (13.3.1) has been established, the statistical analysis may be based on the set of differences and the statistical problem is essentially the same as that considered in Chapter 11, with the observations being the differences $d_1, d_2, \ldots, d_n$ and the parameter of interest being $D =$

$T_1 - T_2$. An hypothesis concerning the value of D may be examined or a confidence interval constructed for D. The choice of methods and the computational procedures are described in Section 11.2.

(b) *Multiplicative model.* When a multiplicative model is suggested from the checks made at step 2, the data should be converted to a logarithmic form, as illustrated in Table 13.3.1. Once in logarithmic form, the analysis can proceed using the methods described in Section 11.2. Note that an hypothesis $T_1/T_2 = R_0$ must be converted to logarithmic form and becomes $\log(R) = \log(R_0)$. Confidence limits for $\log(R)$ obtained as L' and U' must be converted to confidence limits for $R = T_1/T_2$ by application of the exponential function. Thus, the limits for T_1/T_2 are $\exp(L')$ and $\exp(U')$. An illustration is provided below.

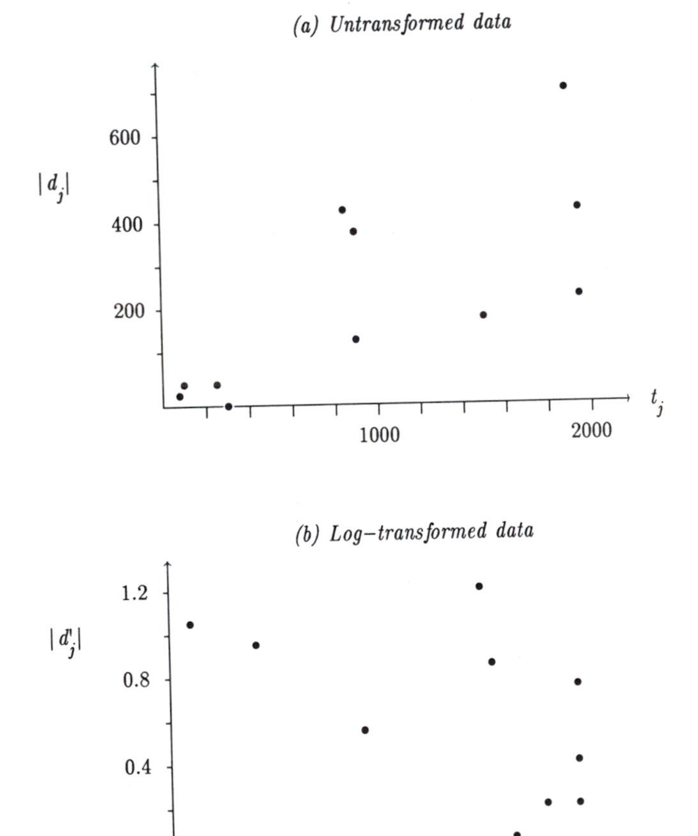

Figure 13.3.2. Plots of unsigned differences versus totals (based on data in Table 13.3.2).

Illustration. Preliminary model checking using data in Table 13.3.1 (which were obtained in the experiment reported in Example 13.4) suggests that a logarithmic transformation is required to produce an additive model. Hence, the analysis is based on the set of differences $d'_j = \log(y_{1j}) - \log(y_{2j})$ in Table 13.3.1(b).

The aim is to provide a confidence interval for the ratio of the mean number of colonies produced under the $PHA : PHA$ method and the $PHA : LK$ method. Reference to Section 11.2.1 suggests the use of the procedure described in Section 11.2.4 which is based on the t-statistic. Model checking suggests the assumption of Normality is reasonable for the differences computed from the logarithmic-transformed data and consequently (11.2.10) is employed to compute a 95% confidence interval for $\log(R)$—with $\bar{y}$ and s_m being obtained from the set of differences. The arithmetic mean and the standard deviation of the set of differences $d'_1, d'_2, \ldots, d'_{11}$ contained in Table 13.3.1(b) are −0.3017 and 0.6521, respectively. From the t-distribution the tabulated value for $t_{0.05}(10)$ is 2.23. Hence, by (11.2.10), the limits for $\log(R)$ are −0.740 and 0.136.

The 95% confidence limits for T_1/T_2 are consequently $\exp(-0.740) = 0.48$ and $\exp(0.136) = 1.1$. Thus, the ratio of the mean number of colonies produced by the $PHA : PHA$ method to mean number produced by the $PHA : LK$ method is expected to lie between 0.48 and 1.1. In practical terms, this suggests that the $PHA : PHA$ method is, on average, unlikely to produce many more colonies than the $PHA : LK$ method. On the other hand, it is possible that it produces only about half as many colonies as the alternative method. Given that the $PHA : LK$ method is the cheaper of the two, there is strong support for its use.

13.3.7 'Before and After' Studies

There are experimental situations which have the appearance of designed experiments and paired comparisons even though they are observational studies. These are studies in which two responses are obtained from each experimental unit, one before application of a treatment and one after application of the treatment, with the aim of assessing the effect of the treatment. They are not designed experiments because the experimenter has no control over the order of application of the treatment. Example 13.6 provides an illustration.

Example 13.6 (Changes in Blood Glucose After Exercise). *Data presented in Table* 1.3.1 *were collected from blood samples taken from* 22 *runners before a race and after the race. While the race might be seen as the application of a 'treatment' and the pre-race reading as the 'control,' there is no possibility of randomly allocating the treatment and the control between the two times at which the subjects were sampled—the 'control' must always precede the 'treatment.' This runs contrary to the conditions* *which apply for a designed experiment.*

Provided the sample has been randomly selected, the method of analysis presented for paired comparisons applies, although there are limitations on the nature of the conclusions reached. Because the 'before' and 'after' treatments are not randomly allocated, there is the usual problem which arises with observational studies concerning the cause of the difference. Evidence of a difference between the responses at the two times cannot necessarily be attributed to the application of the treatment.

13.4 Surveys

The process of sampling a population to estimate numerical characteristics is termed a *survey.* It is a form of observational study which has assumed considerable importance with the increasing investigations in the area of social science and, through a non-scientific role, as a source of speculation for the media.

For reasons outlined below, there are features of surveys which make them a rather specialized area of Statistics.

In most scientific applications of Statistics, the major source of error in statistical conclusions is a consequence of incomplete knowledge arising from the fact that the data come from a sample rather than the whole population. This leads to the existence of *sampling variability* or *chance variability.* Furthermore, since the extent of this source of error can generally be estimated, it is possible in the statistical conclusions to provide a measure of reliability of the results.

In the analysis of survey data, it is more commonly non-statistical considerations which are the major sources of error—poorly worded questions, bias in question presentation, false information supplied by the respondent, etc. Such sources of error may not be recognized, let alone quantified in the presentation of conclusions.

Constructing and implementing surveys requires considerable expertise. Scientists who do not have this expertise are strongly advised to seek assistance from persons or organizations who specialize in this field. The information presented in this section is primarily aimed at explaining the complexity and difficulties of conducting surveys rather than equipping the reader with the knowledge to conduct high quality surveys.

13.4.1 Stages in Conducting a Survey

As noted in the introduction, there is the possibility of major errors arising from diverse sources. If the investigator is to make the most effective use of the resources available, it is essential to carefully examine all aspects of the design and application of surveys at the planning level. The various stages which must be considered are outlined below.

Stage 1 — *Defining the questions to be asked and the objectives of the survey.* Clear, precise, and unambiguous questions must be the goal of a person who is constructing a survey. If the questions in a survey require interpretation by others, there is the possibility of misinterpretation or bias.

Stage 2 — *Defining the population and population unit.* Ideally, the investigator would like a list of the population members which is current at the time the population is to be sampled. Additionally, for a mobile population, the location of each population member at the time of sampling is desired. These ideal conditions are found only in specialized situations. More commonly, a list of population members, if it is available, is out of date. At a more fundamental level, there may be difficulties in defining the population to be sampled. There may, for example, be controversy over the groups to be included in the population or over the boundaries of the region within which population members are to be found. The population unit may be in dispute. For example, what constitutes a *family* in an economic survey?

Stage 3 — *Listing the available resources.* A wide variety of resources is likely to be required for the operation of the survey and the analysis of the results. These range from physical requirements such as lists of population members or maps of regions to human resources. Money will probably be required for planning, implementation and analysis.

Stage 4 — *Choosing a survey method.* The primary aim when considering the method is to minimize bias and maximize precision within the constraints imposed by limited resources. The role of survey designs in this regard is considered in Section 13.4.5.

Stage 5 — *Application.* Particularly where a questionnaire or interview is involved, a pilot survey is generally desirable to establish potential difficulties either with the form of questioning or the method of collection. Great care must be taken to ensure that the data collectors are equipped to follow precisely their instructions.

Stage 6 — *Data storage and analysis.* Commonly there is a large amount of data collected in a survey and the first task is generally to store these data on a computer. There is need to ensure that the necessary computer programs are available to input and summarize the data and to perform the calculations required for any statistical analysis which might be required.

Stage 7 — *Interpretation and presentation of results.* The manner in which the results of a survey are presented is at least as important as any other part of the survey since a failure to convey the results accurately and concisely will negate the most meticulous care and attention in the earlier stages.

To use the resources to minimize error and maximize precision demands that each and every one of these stages be considered **when the survey is being planned.** More detailed discussion of this information is contained in the following sections.

13.4.2 'Tools' Needed in a Survey

The 'tools' needed to conduct a survey are summarized below.

Lists, maps, etc. The population to be sampled must be defined. This may involve finding lists of population members, maps identifying positions of population members, etc. Check that such information is accurate and complete.

Survey designs. The choice of a design usually requires the involvement of an expert. Arbitrarily selecting sample members is fraught with danger. Unrestricted random selection is usually unsatisfactory because it is both expensive and fails to provide adequate precision. Basic alternatives are presented in Section 13.4.5. The final selection of a design is dependent on cost considerations and known characteristics of the population being sampled.

Questionnaire and interview information. In a survey the information to be collected is contained in a series of questions which may be in a questionnaire or in a list of questions to be asked by the data collectors. Forming the set of questions is an extremely difficult task since they must be interpreted by either the respondent or the data collector. Where possible, use a set of questions which have been designed by an expert and have been well tested. If feasible, conduct a pilot survey to judge if the questions are being properly interpreted.

Computing facilities. Commonly, there is a need to store the information collected in a computer. Obviously there must be a computer available in this case. Additionally, there must be the necessary programs to accept the data. Usually there must also be programs which can summarize and analyze the data. There must be sufficient time available on the computer and sufficient money to pay for the computer time if a payment is required. Before entering the data into a computer, ensure the computer has the capacity to meet all requirements or, alternatively, ensure there are facilities for transferring the data to a more powerful or larger computer at some later stage.

13.4.3 Human Resources Needed in a Survey

The human resources which are required fall into three major areas. They are

- expertise in survey design;

- data collectors and data processors; and
- expertise in the processing, analysis and interpretation of results.

Expertise in survey design. Expertise comes at several levels. A commercial organization or the Bureau of Statistics may provide advice or conduct the survey. Fellow researchers who have expertise in designing surveys of the same or similar type may provide a blueprint from which to work. In some respect, most scientists require expert assistance in survey design if they are to avoid the possibility of serious errors and/or inefficient designs.

Data collectors and data processors. The physical acts of collecting information and entering that information into a computer or other storage medium requires some human input. If the data collection involves interviews, a minimum level of expertise is required. This may necessitate training.

Expertise in analysis and interpretation. Quite distinct from the ability to input and manipulate data, there may be a requirement for statistical expertise. In part, this relates to finding and applying suitable computer programs in statistics and in part to the interpretation of results of the analysis.

13.4.4 Potential Errors

Errors may arise at three stages—the design stage, during data collection and during processing and analysis. Additionally, there is sampling variability which adds to uncertainty in the findings.

1. *Errors at the design stage*

   (a) *Errors in sampling methods.* A potentially serious error occurs when the sampling method leads to bias in the results. This occurs when one subgroup in a population is unwittingly over- or under-represented. It also arises when there is deliberate over- or under-representation of one or more subgroups but an incorrect adjustment for this fact is made.

   (b) *Errors in factual information.* The possibility of inaccurate or incomplete lists or maps has been noted above.

   (c) *Errors in questionnaires or instructions for data collectors.* There are innumerable sources of error in questionnaires. Some of the more common are listed below.

      (i) Motive behind the question: most persons who conduct surveys have expectations about the results. It is easy to allow those expectations to unconsciously influence the nature of the questions in a manner which leads the respondents to provide the answer which the investigator expects. Experiments have established that apparently insubstantial changes in question composition and structure can result in dramatic variations in responses.

(ii) Bias in question content: questions can be worded in a way which make a respondent feel that certain answers are unacceptable or degrading. For example, the question

"To which health fund do you belong?"

may discourage the answer "none."

(iii) Lack of specificity: the question

"How many persons live in this household?"

may be interpreted to mean *on this day, including persons who used to live here, including persons who are sometimes here, who will be living here in the future,* etc.

Depending on the purpose of the question there may be serious underestimation, overestimation or lack of precision in the estimate produced from these data.

(iv) Length and order of questionnaire. In many studies there is a conflict in the preparation of the questionnaire between (i) maintaining an adequate number of questions to provide accurate and complete information, and (ii) avoiding a questionnaire which is so lengthy that respondents lose interest and either begin to answer without due care or fail to complete the questionnaire. Furthermore, it is known that respondents pay most attention to questions at the beginning and the end.

(v) Recall questions. Although a respondent may be anxious to provide complete and accurate information, asking questions which rely on memory may lead to serious error. The two factors which determine recall ability are time since the event and the importance of the event. For example, in a survey seeking information about leisure activities, the question was asked

"Did you go to the cinema in the past two week?"

From earlier information collected it was established that 38% of people surveyed failed to recall a visit to the cinema while 24% thought they had been in the past two weeks when in fact they were recalling an earlier visit.

(vi) Embarrassing questions. Considerable skill is required to obtain information to which the respondent does not wish to publicly admit. Even the experts have difficulty here. The Australian Bureau of Statistics, which has a world-class reputation in data collection, sought details of alcohol consumption in the Australian community. If one were to believe the results of their survey, only 50% of the alcohol purchased is actually consumed!

Apart from errors deriving from the questions in the questionnaire itself, errors may arise because the respondents fail to correctly understand what is required of them. The same problem can arise from instructions to interviewers. If an interviewer misunderstands instructions as to the manner in which the interview should take place, serious distortions in the findings may result. There is a need to seek a compromise between making instructions so long that the person reading them is overwhelmed by the detail, or so short that they are incomplete or ambiguous. This is an area which is as important as the construction of the questions in the survey.

2. *Errors at the collection stage*

(a) Missing information: Lack of response to all or part of the questions in a survey may result from

(i) failure to reach the respondent;

(ii) refusal to the respondent to provide information; and

(iii) inability of the respondent to provide information.

There is a large body of research which suggests that results from the group who do respond and the group who do not respond may differ markedly, with the consequence that a serious bias is introduced into the findings. Non-response is a potentially serious problem since it may cause the findings of the whole survey to be unacceptable. It is essential the problem is addressed at the design stage.

(b) Respondent errors: As noted above, there is the possibility of a respondent deliberately giving false answers because of the embarrassing nature of the question. In similar vein, deliberately false information may be given to raise the social status of the respondent. Commonly, people alter their occupation to give an impression of being on what they perceive as a higher social level or raise the level of their income. Older people may deliberately lower their age.

Many errors are, however, unintentional. The problem of recall has a number of facets. Rare events are more easily recalled and placed in time than are common events; more important events are more accurately placed in time than are trivial or commonplace events; the length of the reference period is critical, and inexperienced investigators would normally be astounded at the shortness of the period of accurate recall.

People will forget that they have had a one month holiday within the past six months!

Particular care must be taken if respondents are being repeatedly surveyed. They are likely to attempt to give answers which are consistent with those given previously. Equally, if the respondents live or work together, they are likely to attempt to give information which they feel matches with the information being sought from their partner.

(c) Interviewer errors: Care must be taken to avoid influence from an interviewer on the respondent questions. It is clearly established, for example, that an interviewer who is clearly of a minority race will attract more sympathetic responses about racial harmony than will an interviewer from the majority race.

Hopefully, by this point, it will be obvious that there are a large number and diversity in types of pitfalls which may arise in the construction and implementation of a survey. Except for surveys which follow a tried and true method of design and collection, it is highly desirable to

(i) have the questions and instructions critically assessed by colleagues; and

(ii) test the method and questions in a pilot survey in the circumstances in which the actual survey is to be employed.

The energy, cost and time consumed in these operations is generally a small price to pay for the detection of flaws which may result in serious distortion or failure of the survey.

If statistical analysis is required, there are potential errors in the incorrect application or interpretation of the findings. These are not pursued at this point since they appear in other chapters in the book.

13.4.5 Basic Sampling Procedures

Given that the investigator can meet the practical requirements discussed in the previous sections, the method of sample selection is the key to reliable and accurate findings. Rarely does unrestricted random sampling (Section 2.1) provide an adequate basis. It is generally impractical to implement, too expensive and fails to ensure the precision required. The basic alternatives are introduced below and the introduction is followed by a comparison of methods. (For a more detailed coverage see Cochran [1977].)

The key factors in selecting a design are *variability* and *cost.* Where a population can be subdivided into sub-populations such that there is large variation between sub-populations relative to variation within sub-populations, there is scope for greatly increasing precision over that deriving from methods based on unrestricted random sampling. Where there is the possibility of employing a scheme which reduces unit collection costs, there is the possibility of sampling more units for the same cost and hence, increasing precision.

The primary sampling schemes are described below.

Stratified random sampling. The population is partitioned into sub-populations which are termed *strata.* Random sampling then takes place within each stratum. There is scope for (i) unequal sampling of strata to reflect different unit collection costs or different levels of variability; and (ii) further subdivision to define third and higher strata.

Systematic sampling. The method of systematic sampling was introduced in Section 2.1.2.

Multi-stage sampling. Suppose, as in Stratified random sampling, the population is partitioned into sub-populations. However, instead of randomly sampling every sub-population, only a random selection of sub-populations are sampled. This is *two-stage random sampling.* The way is open to extend this approach by partitioning each sub-population into sub-sub-populations and introducing a third selection stage.

Cluster sampling. In multi-stage sampling, the presumption is that at the final selection stage, a random sample of population members is selected from within each sub-unit. Cluster sampling represents an alternative in which all sample members within the sub-unit are selected at this final stage. For example, the final sub-unit might be the household and all members in the household would be included in the sample.

13.4.6 Comparison of Sampling Methods

The effectiveness of partitioning into strata for either stratified random sampling or multi-stage sampling depends on the ability of the designer to construct strata which have large between-stratum variability relative to within-stratum variability. In very simple terms,

$$\begin{array}{c}\text{VARIABILITY FROM}\\ \text{SIMPLE RANDOM SAMPLING}\end{array} = \begin{array}{c}\text{BETWEEN STRATUM}\\ \text{VARIABILITY}\end{array} + \begin{array}{c}\text{WITHIN STRATUM}\\ \text{VARIABILITY}\end{array}$$

In stratified random sampling only the within-stratum variability contributes to the variance of the estimator. Presuming that the designer has sufficient knowledge to partition the population to obtain large stratum-to-stratum variation, stratified random sampling will always provide more precise estimators than will unrestricted random sampling.

Two-stage sampling, or more generally, multi-stage sampling, provides the same type of advantage as stratified random sampling but with more flexibility. The disadvantage lies in the added complexity of defining and implementing two sampling schemes, one for the strata and one for the units within each stratum.

Cluster sampling is effective under the opposite conditions to those described above. For cluster sampling to be effective, there should be minimum between stratum variation and maximum within stratum variation. The primary reason for selecting Cluster sampling is its reduced unit collection cost when the units to be sampled are located in close proximity. The disadvantages derive from the possible lack of independence of responses from units in the cluster. The basic problem which exists is the difficulty of obtaining a reliable estimate of precision for estimators derived under Cluster sampling. For this reason, it is a technique which is rarely used alone. However, it does have an important role in composite schemes.

The disadvantage of Systematic sampling have been discussed in Section 2.1.3. Generally, it is wiser to use one or a combination of the other methods defined in Section 13.4.5.

For schemes which involve sampling strata, an important consideration is the number of population members to sample from each sub-population. The simplest choice is to select the same number from each sub-population. Where the number of population members in each stratum is known, a better choice is to make the proportion of sample members selected from each sub-population as close as possible to the proportion of population members contained within that sub-population. Two other factors come into play. They are (i) the relative magnitudes of variances within sub-populations; and (ii) the relative unit costs of collecting information from within sub-populations. The basic rules for increasing precision are to make sampling more intensive as (i) within sub-population variability increases; and (ii) the unit cost of collection decreases.

13.4.7 Using Supplementary Information

There exists the possibility of using information collected on variables which are related to the quantity or quantities under investigation to improve the precision of estimators. Three commonly used approaches are described and discussed below.

Ratio estimators. Consider the following two examples:

1. Building contractors are asked to provide information on the value of building work they are undertaking in order to permit the government to estimate the value of building activity currently being undertaken in the country. The builders are notorious for underestimating the value of their work (perhaps fearing the information may fall into the hands of the taxation office). There is, however, scope for adjusting the stated *value* of work in progress based on the *level* of work being undertaken. This is a variable for which a figure can be obtained without reference to the builder. Experience suggests that a builder who underestimates his level of activity by, for example, 20%, will also underestimate his value by 20%. Hence, the Bureau of Statistics

can obtain a less biased estimate of the value of building activity by increasing claimed values of activity in proportion to the ratio of actual to claimed level of activity.

2. Australia conducts a census every five years. This provides detailed information for small regions. In the intervening years it may feasible to obtain information on changes in a quantity of interest over a large region, e.g. a state, but not for a small region. If there is reason to believe that the trend in the larger region will be reflected in the smaller region, there is scope for adjusting the figure in the small region proportionately.

In these cases there is the presumption that the change in the quantity of interest is proportional to changes in a supplementary variable for which there is information available. The process is termed *ratio estimation.* The essential feature of ratio estimators is that the estimator (y_x) is presumed to be related to the supplementary variable (x) according to the equation $y_x = Rx$ where R has a value which is known or can be estimated.

Regression estimators. The idea introduced with ratio estimators may be generalized by lifting the restriction that the estimator must take the value zero when the supplementary variable takes the value zero. The simplest extension assumes a linear relation, i.e. $y_x = a + bx$. A more illuminating way to present this equation is in the form $y_x = y + bx$ where y is the estimator which does not use information contained in x and bx is the adjustment which is made as a result of having knowledge of the supplementary variable x. The name *regression estimator* derives from the statistical technique of regression analysis (Chapter 18).

Double sampling. The basic idea in regression analysis may be carried further and applied in situations in which information is available only on the supplementary variable. To do this, a two-stage sampling scheme is employed. Hence, the name *double sampling.* Consider the following example.

The volume of usable timber in a tree may be determined by chopping the tree down, running the tree through a saw mill and measuring the amount of timber obtained. Alternatively, it may be estimated cheaply and in a non-destructive fashion from simple measurements made on the tree.

The ideal situation is to take a sample in which the exact method and the approximate method are used to provide estimates, y and x, respectively, for the volume of the timber in the forest from which the trees were taken. From these data a relationship can be established between y and x. Usually, this relationship is assumed linear, i.e. $y = a + bx$.

A second sample of trees is identified in the forest and for each tree in this sample, an estimate based on the non-destructive method is obtained. This information is used to adjust the estimate obtained from the first sample.

The circumstances in which double sampling has application are those in which there are two variables available on which to base the estimation, one of which has a high unit cost and the other of which has a low unit cost but is less reliable. The relative costs and levels of precision of the estimators based on the two variables dictate if double sampling is worth using and if so, the number of units to be employed in each sample.

Appendix 1. Steps in Randomly Selecting a Sample from a Population

Two methods are described below for generating a sample which is based on the process of random selection. Method 1 is the most commonly needed method and is more efficient if the sample size is small relative to the population size.

Method 1

Step 1 — assign each of the N population members a different number from the set $1, 2, \ldots, N$.

Step 2 — to obtain a sample of size n, generate n pseudo-random numbers in the range 0 to 1, multiply each number by N and add one. Discard the fractional part of each number and call the sequence of whole numbers which remain $i_1, i_2, \ldots, i_n$.

If any integers are identical, generate further pseudo-random numbers and construct more integers until there are n distinct integers $i_1, i_2, \ldots, i_n$. Choose population members numbered $i_1, i_2, \ldots, i_n$ to form the sample. (*In practice, it is generally more convenient to generate twice as many numbers as required and then select the first distinct n numbers in the sequence.*)

Illustration. Suppose the aim is to generate a sample of size 3 from a population comprising 10 members. The 3 pseudo-random numbers generated at Step 2 are 0.835 0.177 0.226 which become 9.35 2.17 3.26 after multiplication by 10 and addition of 1. Thus, $i_1 = 9$ $i_2 = 2$ $i_3 = 3$.

Hence, the sample is formed of population members 2, 3 and 9.

Method 2

Step 1 — assign each of the N population members a different number from the set $1, 2, \ldots, N$.

Step 2 — to obtain a sample of size n, use a statistical package to generate N distinct pseudo-random numbers. Call this sequence of numbers $p_1, p_2, \ldots, p_N$.

Step 3 — Determine the ranks of these numbers when ordered from smallest to largest and denote the ranks by $r_1, r_2, \ldots, r_N$. (Thus, r_1 is the rank of p_1, r_2 is the rank of p_2, etc.)

Step 4 — form the sample from population members numbered $r_1, r_2, \ldots, r_n$. (*Note that the only numbers which need be displayed are the set obtained at Step* 4.)

Illustration. Suppose the aim is to generate a sample of size 3 from a population comprising 10 members.

The 10 pseudo-random numbers generated at Step 2, namely $p_1, p_2, \ldots, p_n$ are

0.835 0.117 0.226 0.039 0.487 0.666 0.731 0.705 0.519 0.324

When these numbers are ranked from smallest to largest the respective ranks are

10 2 3 1 5 7 9 8 6 4

Thus, $r_1 = 10$, $r_2 = 2$ and $r_3 = 3$.

Hence, the sample is formed of population members numbered 10, 2 and 3.

Appendix 2. Steps in Random Allocation of Units Between Treatments

Random allocation. Consider the allocation of n sample members between t treatments such that n_1 members are to be allocated to treatment 1, n_2 members to treatment 2,..., and n_t members to treatment t where $n = n_1 + n_2 + \cdots + n_t$.

Step 1. Number the sample members from 1 to n.

Step 2. Generate n distinct pseudo-random numbers using a statistical package. Call the numbers generated $p_1, p_2, \ldots, p_n$.

Step 3. Rank the numbers generated at Step 2 from smallest to largest. Let r_1 be the rank of p_1, r_2 be the rank of p_2, etc. Associate the rank r_1 with the sample member numbered 1, r_2 with the sample member numbered 2, etc.

Step 4. Assign the sample members with ranks 1 to n_1 to treatment 1, those with ranks $n_1 + 1$ to $n_1 + n_2$ to treatment 2, etc.

Illustration. Suppose there are 6 sample members to be assigned to three treatments such that each treatment is to receive two sample members.

The 6 pseudo-random numbers generated by Step 2, namely $p_1, p_2, \ldots, p_n$ are

0.899 0.410 0.801 0.672 0.299 0.943

These yield the ranks

$$5 \quad 2 \quad 4 \quad 3 \quad 1 \quad 6$$

Thus, sample members numbered 5 and 2 would be assigned to treatment 1, sample members numbered 4 and 3 would be assigned to treatment 2 while sample members numbered 1 and 6 would be assigned to treatment 3.

Problems

13.1 *Example* 13.3

Follow Steps 1 and 2 in Section 13.3.6 and reproduce Figure 13.3.1.

13.2 *Example* 13.4

Reproduce the analysis for Example 13.4 which is presented in Section 13.3.6.

13.3 *Analysis of data in Table* 1.3.1

(a) For each of the variables *glucose, sodium, potassium and cholesterol* in Table 1.3.1 apply Steps 1 and 2 in Section 13.3.6 to the data for the *untrained* group of runners and report your findings. Where appropriate, proceed to Step 3 and establish which, if any, of the variables show evidence of difference in median level between *pre-race* and *post-race* readings.

(b) Repeat the analysis in (a) based on data for the *trained* group of runners.

13.4 *A study of drug compliance*

For various reasons, elderly persons are inclined to forget to take prescribed medicine. Different strategies have been examined in an attempt to minimize the problem. Data presented below come from ongoing research in a collaborative study involving ten institutions. At each institution, 100 subjects were employed and they were randomly allocated between two groups—the 'red' group and the 'white' group—with 50 subjects per group. (The names for the groups come from the different colors of the tablets which were prescribed.) The groups differ in the strategies adopted to maximize compliance with instructions. The data recorded below comprise the number of subjects in each group of 50 who were considered to have complied with instructions in respect of medication. The aim is to establish if there is a higher average compliance in one group than the other.

To allow for variations in the type of persons in the different institutions, it is wise to compare the groups within each institution.

Apply the steps in Section 13.3.6 to establish if there is evidence of mean difference in compliance rates between the groups.

(Note that the problem could be expressed as a comparison of success rates and methodology associated with the Binomial model applied. The design structure makes the relevant Binomial methodology less easily accessible. The above approach relies on the Normal approximation to the Binomial and is limited in application to situations in which (i) each observation is based on the same number of trials, and (ii) the number of trials is sufficiently large for the Normal approximation to have application—as judged by (9.2.1).)

Number of Compliant Subjects in Total of 50

Institution	1	2	3	4	5	6	7	8	9	10
Red group	32	27	48	33	32	26	32	28	24	33
White group	30	19	37	26	27	18	31	25	23	33

14

Comparing More Than Two Treatments or Groups

14.1 Approaches to Analysis

14.1.1 New Dimensions

Statistical methods applicable for the comparison of three or more treatments or groups must cater for complexities not encountered in the methods of previous chapters. The complexities are of two forms

- experimental and treatment designs (introduced in Chapter 13); and
- a broader range of questions which may be posed in respect of treatment comparisons.

A consequence is that scientists cannot rely on a 'cookbook' approach to the selection of statistical methods. There is a need to understand the basis of the statistical methodology, the rules for definition of designs and the translation of experimental questions into statistical hypotheses if the correct interpretation of analyses is to be achieved. Computers and statistical packages become an essential part of the process of statistical analysis to cope with the greater computational demands.

An expansion of the range of statistical models and methods has also been demanded to cater for the increasingly more intricate and more detailed scientific experiments in today's world. Whereas earlier scientific experiments may have sought evidence of gross differences between treatments, there is a tendency nowadays to construct experiments which seek evidence of smaller differences and which presume multi-factorial rather than single factor effects on response. A by-product of increasing scientific sophistication is the need for statistical models to provide more accurate representations of the experimental set-ups.

In Chapters 14 to 16 an attempt is made to present the basis and some applications of the wide ranging methodology available to scientists for the comparison of treatment effects. Chapter 14 is devoted to a presentation of the bases of the methods, with Section 14.1 giving an informal introduction and Sections 14.2 and 14.3 providing the formal definitions of the models and methods. Chapters 15 and 16 illustrate the application of the methods.

14.1.2 Scope of the Applications

1. *The study of average response.* In the absence of treatment structure, statistical analysis may be called upon to

(i) test the hypothesis that all treatments produce the same average response versus the alternative that at least two treatments have different effects;

(ii) make pairwise comparisons of selected pairs of treatments; and

(iii) make pairwise comparisons of all pairs of treatments.

If there is a factorial arrangement of treatments, the function of statistical analysis may be to

(i) contrast response under different levels of a factor; and

(ii) seek evidence of interaction between factors and examine the nature of any interaction which is detected.

Table 14.1.1. Data from Example 13.5

Mg. concentration in roots

Treatment	Calcium	pH	Species	Chamber 1	Chamber 2
1	Low	4.0	Kikuyu	3507	3221
2	Low	4.0	Cabbage	833	769
3	Low	4.0	Lettuce	1154	1077
4	Low	4.6	Kikuyu	3595	3443
5	Low	4.6	Cabbage	1399	1215
6	Low	4.6	Lettuce	1594	1539
7	High	4.0	Kikuyu	2804	2809
8	High	4.0	Cabbage	603	446
9	High	4.0	Lettuce	460	372
10	High	4.6	Kikuyu	2502	2993
11	High	4.6	Cabbage	1052	723
12	High	4.6	Lettuce	648	245

(Table 14.1.1 provides an illustration of a factorial arrangement of treatments in which the twelve treatments are defined by combinations of three factors—*calcium, pH* and *species.* Statistical analysis may be used to establish the effect of variations in calcium level on the uptake of magnesium in plants and the manner in which the influence of calcium is affected by changes in pH. By including the three plant species in the one experiment, it is possible to establish if the effects of calcium and pH on magnesium uptake are consistent from one species to another.)

Questions requiring the examination of treatment contrasts may also arise in the absence of a factorial arrangement of treatments as is illustrated in Example 14.1.

Where treatments or groups are distinguished by levels of a scaled variable (as when treatments comprise different levels of a fertilizer or different intensities of a stimulus) consideration should be given to the possibility that the experimental aim concerns the relation between the response and the scaled variable rather than the comparison of treatments. If a stimulus is applied at levels 0, 10, 20, 40 and 60, the experimenter must decide if the particular levels selected have an intrinsic interest or, alternatively, if they represent convenient points along a scale and have been selected to permit a study of the manner in which the response changes with increasing level of intensity. If the latter is true, there is methodology in Section 18.7 which is more effective.

Example 14.1 (Environment and Learning: A Study with Rats). *Eighty rats from a laboratory stock were made available for an experiment. The rats were randomly allocated between four groups with twenty rats per group. The groups were distinguished in the following manner.*

1. *Rats in the first group* (G1) *were used in experiments with mazes with a food reward each time the maze was correctly negotiated.*

2. *The same procedure was used with the second group* (G2) *but with no food reward for successfully completing the maze.*

3. *The third and fourth groups* (G3 *and* G4) *had no contact with mazes. They were distinguished by their accomodation. Group* 3 *had 'open-plan' accommodation in which all animans were in a single area. Group* 4 *lived in a 'house-style' accommodation in which the cage was divided into rooms with small interconnecting doorways.*

At the end of one week in a chosen environment, each rat was placed at the entrance of a maze with the same path as those used in the treatment of Groups 1 *and* 2. *Each rat was required to make one successful journey through the maze and was given a food reward. The rat was then returned to the entrance and the time to make a second successful passage through the maze was recorded.*

Table 14.1.2 *presents a series of questions which provided the motivation for the study and the statistical hypotheses deriving from these questions.*

Table 14.1.2. Questions and Hypotheses from Example 14.1

(In the hypotheses below, T_i is the expected time for rats in Group i ($i = 1, 2, 3, 4$) to successfully complete a maze.)

Question	Hypothesis
Is it reasonable to assume that, on average, all groups take the same time to complete the maze?	$T_1 = T_2 = T_3 = T_4$
Does group 1 appear to differ from the other groups?	$T_1 = \frac{T_2+T_3+T_4}{3}$
Does the average time depend on previous contact with mazes?	$\frac{T_1+T_2}{2} = \frac{T_3+T_4}{2}$
Does the average time depend on previous contact with mazes and the nature of the contact?	$T_1 = T_2 = \frac{T_3+T_4}{2}$
Is there evidence of difference in average time taken between the groups who are not given a food reward?	$T_2 = T_3 = T_4$
Is there evidence of difference in average time between groups who had no no contact with mazes?	$T_3 = T_4$

2. *The study of proportions and frequency distributions.* In general terms, the statistical task relates to the comparison of frequency distributions. In the particular case where the response variable has only two categories, each treatment or group can be identified by a single parameter which might be the expected proportion of successes, the expected success rate or the expected odds of success. Hence, the differences between the treatments can be quantified by expressing the difference in terms of proportions, success rates or the odds ratio.

Example 14.2 (Monitoring Heavy Metal Pollution Using a Plant Indicator). *In a research program to monitor metal pollution in rivers, the central component of the investigation was a plant which appeared to change color from green in unpolluted water to brown in polluted water. An experiment was devised to explore the use of the plant as an indicator of pollution.*

The experiment involved the use of 450 *plants which were randomly allocated between nine treatments arranged in a factorial arrangement. The factor 'mercury' which identifies the levels of mercury pollution in the water is under examination. The effect of varying mercury levels is examined in water of different salinity levels. Thus, 'salinity' is the second factor.*

Table 14.1.3. Data from Example 14.2

(a) Based on two-category response

Table of Frequencies

Treatment	1	2	3	4	5	6	7	8	9
Salinity level[1]	F	F	F	B	B	B	S	S	S
Mercury level[2]	Z	L	H	Z	L	H	Z	L	H
Colour									
green	40	36	12	49	33	12	43	39	22
brown	10	14	38	1	17	38	7	11	28

(b) Based on four-category response

Table of Frequencies

Treatment	1	2	3	4	5	6	7	8	9
Salinity level[1]	F	F	F	B	B	B	S	S	S
Mercury level[2]	Z	L	H	Z	L	H	Z	L	H
Colour									
green	40	36	12	49	33	12	43	39	22
yellow-green	9	7	14	1	13	21	6	8	20
yellow-brown	0	5	16	0	1	13	1	3	7
brown	1	2	8	0	3	4	0	0	1

Notes.

[1]Salinity levels are fresh (F), brackish (B) and salty (S).

[2]Mercury levels are zero (Z), low (L) and high (H).

Response is categorical and arises from the classification of the plants used in the study. Two response variables are considered, one in which there are only two categories, 'green' and 'brown,' and a second in which the 'brown' category is subdivided into three categories labelled 'yellow-green,' 'yellow-brown' and 'brown.' Data for each variable are summarized in Table 14.1.3.

If a four-category classification is employed, as in Table 14.1.3(b), *the general task is to establish whether the frequency distribution varies between the treatments. A refinement is possible by noting that the categories are ordered from the green end of the spectrum to the brown end. With this added information, there is scope for seeking evidence of a shift in frequencies from one end of the spectrum to the other.*

14.1.3 Comparison of Means — Analysis of Variance

Table 14.1.4(a) contains data collected on cotton yields under three different treatments. The data are presented visually in the following line diagrams.

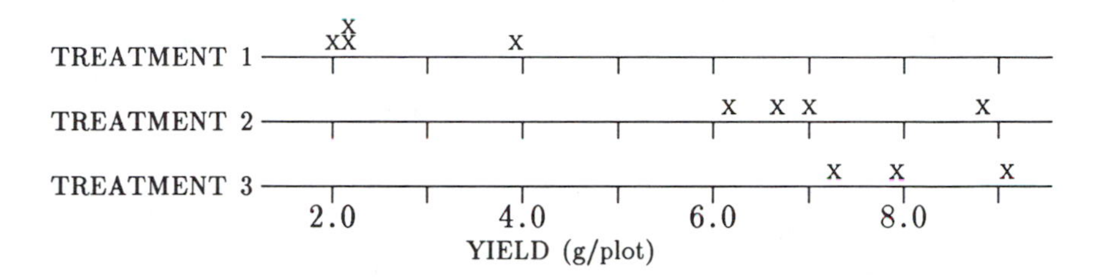

Table 14.1.4. Data and Fitted Values Used in Treatment Comparisons

(a) Scaled responses

Treatment	Yield			
1	3.86	2.17	2.18	1.95
2	8.84	6.23	7.05	6.55
3	7.28	9.11	7.85	

(b) Fitted values based on a model which assumes no treatment differences

Treatment	Yield			
1	5.73	5.73	5.73	5.73
2	5.73	5.73	5.73	5.73
3	5.73	5.73	5.73	

(c) Fitted values based on a model which assumes treatment differences

Treatment	Yield			
1	2.54	2.54	2.54	2.54
2	7.17	7.17	7.17	7.17
3	8.08	8.08	8.08	

On the assumptions that (i) the plots used in the experiment were randomly allocated between the treatments and (ii) there are no identified sources of variation other than treatment effects, visual examination of the line diagrams suggests that the expected yield is lower under treatment 1 than under treatment 2. The comparison of yields under treatments 2 and 3 is less clear cut. While yields under treatment 2 have a lower sample mean than those under treatment 3, perhaps the difference in means could be explained as sampling variation.

The decision that treatment 1 is likely to produce lower average yields than treatment 2 is based on the fact that the difference in treatment means (the **between** treatment variation) is too large to reasonably be explained as sampling variation (the **within** treatment variation). *Analysis of variance* is a statistical method which quantifies this comparison.

The basis of the approach lies in the partitioning of the variation in the combined sample into two components—variation *between* treatments which reflects variability in treatment means and variation *within* treatments which reflects variability between responses which received the same treatment. The primary statistic which is used to partition variability is the *sum of squares.* Equation (14.1.1) provides the decomposition of the sum of squares for the complete sample into component parts.

$$\underset{\text{sum of squares}}{\text{Total}} = \underset{\text{sum of squares}}{\text{Between treatment}} + \underset{\text{sum of squares}}{\text{Within treatment}}$$

$$\sum_{i=1}^{3}\sum_{j=1}^{n_i}(y_{ij}-\overline{y}_{..})^2 = \sum_{i=1}^{3}\sum_{j=1}^{n_i}(\overline{y}_{i.}-\overline{y}_{..})^2 + \sum_{i=1}^{3}\sum_{j=1}^{n_i}(y_{ij}-\overline{y}_{i.})^2 \quad (14.1.1)$$

where

y_{ij} is the yield in the jth plot ($j = 1, 2, \ldots, n_i$) for treatment i ($i = 1, 2, 3$) with $n_1 = n_2 = 4$ and $n_3 = 3$;

$\overline{y}_{i.}$ is the (sample) mean yield for the ith treatment ($i = 1, 2, 3$); and

$\overline{y}_{..}$ is the (sample) mean yield for all plots.

The sums of squares relate to variances in the following manner. If $n = n_1 + n_2 + n_3$, and the sample variances for treatments 1, 2 and 3 are s_1^2, s_2^2 and s_3^2, respectively, then

1. the overall sample variance is

$$s_t^2 = (\text{Total sum of squares})/(n-1);$$

2. the average within treatment variance is

$$\begin{aligned} s_w^2 &= [(n_1-1)s_1^2 + (n_2-1)s_2^2 + (n_3-1)s_3^2]/(n_1-1+n_2-1+n_3-1) \\ &= (\text{Within treatment sum of squares})/(n-3); \text{ and} \end{aligned}$$

3. the between treatment variance based on the three treatment means is

$$s_b^2 = (\text{Between treatment sum of squares})/(3-1).$$

If the three treatments have the same expected yield, these three variances estimate the same quantity, namely the variability due to unidentified sources which is commonly termed the *random variation*. However, if the expected yield differs between treatments, both s_t^2 and s_b^2 will provide inflated estimates of random variation since they include positive components reflecting treatment differences. It is conventional in Analysis of variance to use the statistic $F = s_b^2/s_w^2$ to judge the acceptability of the hypothesis that expected yields are identical for all treatments. Increasing values of F equate with lessening support for the hypothesis and hence, provide the basis of a statistical test of the hypothesis that all treatments have the same effect.

While the above view of the Analysis of variance technique is useful in that it relates directly to the intuitive reasoning one might employ in deciding if expected responses vary between treatments, it is limiting in that it fails to point to the means of generalizing the approach to encompass the wider range of experimental situations which must be considered. An alternative view is offered below which meets this need.

The fundamental task when analyzing the data in Table 14.1.4(a) is to fit two models to the data. In the first model, the assumption is made that all treatments have the same effect. The second model varies from the first only by the removal of the assumption that all treatments have the same effect. The statistical task is to establish if the second model provides a significantly better fit than the first. If it does, evidence of treatment differences exists.

The primary statistical task is to provide sets of predicted or *fitted* values based on the two models for comparison with the set of observed values. Tables 14.1.4(b) and (c) provide sets of fitted values. The model underlying the values in Table 14.1.4(b) is based on the hypothesis that all treatments produce the same average response, while the values in Table 14.1.4(c) arise under the more general model which permits treatment differences. While the basis for constructing fitted values has not been discussed at this stage (see Section 14.3), the values in Tables 14.1.4(b) and (c) are intuitively reasonable. In the first of these tables, all fitted values are presumed equal since there is no identified source of variation in the responses. To make the set of fitted values comparable with the set of observed values in Table 14.1.4(a), the condition imposed is that the overall means should agree. Hence, all the fitted values in Table 14.1.4(b) are set equal to the sample mean. In constructing the fitted values in Table 14.1.4(c), allowance is made for the possibility that the treatments may have different effects. However, for plots receiving the same treatment, the predicted responses are identical and are set equal to the treatment means.

The sums of squares in the identity (14.1.1) can all be expressed in terms of the three sets of figures in Table 14.1.4, namely,

- the data $\{y_{ij0}\}$;

- estimates under the model which assumes no treatment differences $\{\hat{y}^0_{ij0}\}$; and
- estimates under the model which allows for treatment differences $\{\hat{y}^1_{ij0}\}$.

It is apparant that

- the Total sum of squares measures the agreement between $\{y_{ij0}\}$ and $\{\hat{y}^0_{ij0}\}$, i.e., compares the data with the model which presumes no treatment differences;
- the Within treatment sum of squares measures the agreement between $\{y_{ij0}\}$ and $\{\hat{y}^1_{ij0}\}$, i.e., compares the data with the model which allows for the possibility of treatment differences; and
- the Between treatment sum of squares measures the agreement between $\{\hat{y}^0_{ij0}\}$ and $\{\hat{y}^1_{ij0}\}$, i.e., compares the model which presumes no treatment differences with the model which allows for the possibility of treatment differences.

The appeal to intuition for the construction of sets of fitted values and the *ad hoc* choice of the sum of squares statistic for the comparison of sets of observed and fitted values, will be shown to have a sound statistical basis in Section 14.3.

In the example, the only identified source of variation is the treatment effect. There is scope for the introduction of non-treatment effects through the use of experimental designs and for the introduction of structure in the treatments through the use of treatment designs. Yet, the basic approach introduced above is unchanging—fit a model which assumes no treatment differences, fit a second model which allows for the possibility of treatment differences and establish if the improvement in fit by relaxing the restriction on treatment effects is sufficiently large to warrant rejection of the hypothesis that all treatments have the same effect. The process, titled *Analysis of variance,* is formally described in Section 14.3.4.

14.1.4 Comparison of Proportions and Frequency Distributions — Analysis of Deviance

Table 14.1.5(a) contains data in the form of a frequency table. Three treatments, representing levels of mercury pollution in water, distinguish the rows of the table, while the columns are distinguished by categories of the response variable. The experimental aim is to establish if the expected proportion of brown plants varies with changing levels of pollution.

The standard statistical approach to analysis is to construct a set of values which represent the frequencies expected on the assumption that the expected proportion of brown plants is the same for all treatments. These predicted values, which are termed the *fitted values*, are constructed from the following argument. In each row there are 50 counts. In the complete sample the proportion of brown plants is 62/150. If each row is to have the same proportion of brown plants and row totals are to be maintained at their observed numbers, then the number of brown plants predicted for each row is $50 \times 62/150$. By the same argument, the predicted number of green plants in each column is $50 \times 88/150$. The set of fitted values produced by this argument are presented in Table 14.1.5(b).

Table 14.1.5. Observed and Fitted Values of Green and Brown Plants
(Data extracted from Table 14.1.3)

(a) Table of observed values

	Category		
Mercury Level	green	brown	Total
zero	40	10	50
low	36	14	50
high	12	38	50
total	88	62	150

(b) Table of fitted values

	Category		
Mercury Level	green	brown	total
zero	29.3	20.7	50
low	29.3	20.7	50
high	29.3	20.7	50
total[1]	87.9	62.1	150

Note.
[1] In theory, the column totals for observed and expected frequences should agree. The difference is round-off error.

The model which includes the assumption that the expected proportion of brown plants is unaffected by the level of mercury pollution becomes less likely as the sets of observed and fitted values become more dissimilar. Two statistics are in common use for comparing the set of observed frequencies and the set of fitted values—the *deviance* defined by (14.1.2) and the *Chi-squared statistic* defined by (14.1.3).

$$D = 2\sum_{i=1}^{3}\sum_{j=1}^{2} y_{ij} \log[y_{ij}/\hat{y}_{ij}] \qquad (14.1.2)$$

$$X^2 = \sum_{i=1}^{3} \sum_{j=1}^{2} [(y_{ij} - \hat{y}_{ij})^2 / \hat{y}_{ij}] \tag{14.1.3}$$

where y_{ij} is the observed frequency in the jth category ($j = 1, 2$) at the ith mercury level ($i = 1, 2, 3$); and $\hat{y}_{ij}$ is the predicted frequency (*fitted value*) in the jth category at the ith mercury level under the assumption of no effect from varying levels of mercury pollution.

In mathematical terms, X^2 is an approximation to D, being the first term in the logarithmic expansion of D. Both share the property that a value of zero implies perfect agreement between the data and the model. Increasing values of either statistic equates with lessening support for the hypothesis of no effect of changes in mercury levels.

The approach described above has broad application and remains essentially unchanged when the complexity of the design is increased. It applies without alteration if the number of categories of the response variable is increased beyond two. There is scope for using the technique for the comparison between two models in addition to the comparison of a model with the data. The methodology is formally developed as the *Analysis of deviance* in Section 14.3.3.

14.1.5 Component Equations

Whether the data set comprises a table of scaled responses or a table of frequencies, each number can be viewed as an expression of the combined effect of all of the factors which influence the response. The manner in which the effects are postulated to combine to form the response dictates the core of the statistical model and, consequently, the method of statistical analysis. Following the pattern established in Chapter 10, the expression of the decomposition of the response into component parts is provided by *component equations.* For simplicity, it is convenient to divide the overall component equation into two parts. The first describes the separation of response into contributions from identified and unidentified sources and has the simple structure portrayed in (14.1.4).

$$\text{RESPONSE} = \begin{matrix}\text{IDENTIFIED}\\ \text{CONTRIBUTION}\end{matrix} + \begin{matrix}\text{UNIDENTIFIED}\\ \text{CONTRIBUTION}\end{matrix}. \tag{14.1.4}$$

The contribution from the unidentified sources is often termed the *random error.* The value of the effect is unpredictable and varies between sample members. It is a consequence of the process of *random* selection of sample members and/or the *random* allocation of sample members between treatments. The only information about random effects comes from the nature of the probability distributions which generate them.

Unlike the random error, the identified component is presumed to be expressible through a mathematical equation in terms of the factors which are identified as contributing to the response. This secondary component equation can take a wide variety of forms and is used to identify the different classes of statistical models which are commonly required. Two examples are presented below, the first commonly having application in the study of mean responses and the second in the study of proportions. These examples precede the general development of statistical models in Section 14.2.

1. *Linear equations.* Consider a set of scaled responses, $\{y_{ij0}\}$, in which the subscript i identifies the treatment applied ($i = 1, 2, \ldots, g$) and j distinguishes responses arising under the same treatment ($j = 1, 2, \ldots, n_i$). The identified contribution to the response y_{ij0} is conventionally denoted by $E(y_{ij})$ to reflect the fact that this contribution is the deterministic or non-random component. The simplest component equation presumes there are no treatment differences and no other factors identified as contributing to the responses—all variation between responses is presumed to be random variation (sampling variation). The component equation which describes this situation is defined by (14.1.5).

$$E(y_{ij}) = M \qquad \text{for } i = 1, 2, \ldots, g;\ j = 1, 2, \ldots, n_i. \tag{14.1.5}$$

The parameter M is the population mean or long term average response. This equation underlies the fitted values in Table 14.1.4(b).

If allowance is made for possible treatment differences, a second component must be added to the equation which adjusts the mean upwards if the treatment produces above average yield or downwards if the treatment produces below average yield. The component equation may be expressed by (14.1.6).

$$E(y_{ij}) = M + T_i \qquad \text{for } i = 1, 2, \ldots, g;\ j = 1, 2, \ldots, n_i \tag{14.1.6}$$

where M is the expected yield over all treatments; and T_i is the adjustment for the fact that the ith treatment has an expected yield which may differ from the overall average.

In (14.1.6), the expected values are constant for units receiving the same treatment (i.e., for constant i) but may vary between units which receive different treatments. This equation underlies the fitted values in Table 14.1.4(c).

The equation must be expanded when there are two identified sources of variation, as in the data of Table 14.1.6. In that table, variation between rows is associated with time of sampling (i.e., pre-race or post-race) while variation between columns is associated with differences between runners. This suggests (14.1.7) as an expression of the composition of the identified contribution, $E(y_{ij})$.

$$E(y_{ij}) = M + T_i + R_j \qquad \text{for } i = 1, 2;\ j = 1, 2, \ldots, 11 \tag{14.1.7}$$

where

M is the overall average level of glucose;

T_i is the deviation from the average for the ith time of sampling; and

R_j is the deviation from the average for the jth runner.

Table 14.1.6. Data Extracted from Table 1.3.1

	Blood Glucose Level										
Runner	1	2	3	4	5	6	7	8	9	10	11
Source											
Pre-race	67	46	67	66	70	60	72	76	72	53	66
Post-race	100	73	89	95	76	108	77	88	104	108	84

Equations (14.1.6) and (14.1.7) have in common the assumption that the components may be expressed in terms of parameters which are added together to give the expected response. Such equations are termed *linear equations* and are components of *Linear models* which are defined in Section 14.2.3.

2. *Multiplicative equations.* Consider a table of frequencies, $\{y_{ij0}\}$, where i identifies the treatments ($i = 1, 2, \ldots, g$) and j identifies the response category ($j = 1, 2, \ldots, c$). The contribution to the frequency y_{ij0} from identified sources, $E(y_{ij})$, is commonly referred to as the *expected frequency.* The simplest expression for $E(y_{ij})$ arises when all treatments are postulated to have the same frequency distribution. This assumption implies that (14.1.8) is the appropriate component equation.

$$E(y_{ij}) = n_i \pi_j \qquad \text{for } i = 1, 2, \ldots, g;\ j = 1, 2, \ldots, c \tag{14.1.8}$$

where n_i is the sample size of group i; and π_j is the proportion of population members in response category j. Equation (14.1.8) is termed a *multiplicative* equation since the components are multiplied rather than added.

If the requirement is relaxed that all treatments must have the same frequency distribution, the parameter π_j in (14.1.8) represents the expected proportion of population members in category j. For the ith treatment, the expected proportion of population members in the jth category might vary from the average, and this possibility may be allowed for by the inclusion of an adjustment, γ_{ij}, as portrayed in (14.1.9).

$$E(y_{ij}) = n_i \pi_j \gamma_{ij} \qquad i = 1, 2, \ldots, g;\ j = 1, 2, \ldots, c. \tag{14.1.9}$$

It is apparent that (14.1.8) is the restricted form of (14.1.9) in which all parameters in the set $\{\gamma_{ij}\}$ are equal to one. Hence, a test of the hypothesis that all treatments have the same expected frequency distributions can be expressed statistically by the hypothesis $\gamma_{ij} = 1$ for all combinations of i and j.

By adopting the strategy that contributions are portrayed as deviations from an average, representation of more complex design structures is straightforward and parallels the expansion illustrated for linear equations in (14.1.7). Statisticians have carried the similarity further by adopting as standard, the representation of multiplicative equations on a logarithmic scale. For example, (14.1.9) would be represented as

$$\log[E(y_{ij})] = n'_i + \pi'_i + \gamma'_{ij}$$

where

$$n'_i = \log[n_i], \quad \pi'_i = \log[\pi_j] \quad \text{and} \quad \gamma'_{ij} = \log[\gamma_{ij}].$$

The resulting equation is a linear equation, which, since it is on a logarithmic scale, is termed a *log-linear equation.* Statistical theory establishes that the common form of representation can be carried over into a common methodology and opens the way for a more general structure for models and methodology based on *Generalized linear models* and *Analysis of deviance.* These models and the associated methodology are introduced in Sections 14.2 and 14.3, respectively.

14.2 Statistical Models

14.2.1 General Assumptions

The definition of a statistical model is in terms of (i) an hypothesis, (ii) the method of sample selection and treatment allocation, and (iii) statistical assumptions which include the forms of the component equations and distributional assumptions.

Hypothesis. Among the parameters in the component equation are those which define treatment effects. Hypotheses which arise from experimenters' questions are based on these parameters. There are numerous forms of experimental questions which can be expressed in terms of parameters representing treatment effects. These are identified in Section 14.1.2.

Methods of sampling and treatment allocation. When sampling from groups, the most common assumptions are that (i) sample members are selected independently from the different groups and (ii) random sampling is employed within groups.

Violation of the independence condition is rare and is not considered further. More complex selection procedures which arise from multi-stage selection were introduced in Chapter 13 and their implications for statistical analysis are considered in Chapter 15.

When comparing treatments, the method of analysis to be employed is determined by the experimental and treatment designs (see Chapter

13). The variations in analysis which associate with different designs are considered in Chapters 15 and 16.

Distributional assumptions. While there are distribution-free models available for the comparison of median responses, they are of limited application and discussion of distribution-free methodology is only briefly considered in Section 14.5. Far more common is the use of distribution-based models. Those distributions which are commonly found to have application are members of a group of distributions which are referred to as the *Exponential family* of distributions. The important members of this family are:

1. the Normal distribution for scaled responses (Section 6.5) which is extensively employed when comparing means between groups;

2. the Poisson distribution (Section 6.3) which has application when the response is in the form of counts representing the number of occurrences of independent events;

3. the Binomial distribution (Section 6.1) used in the study of proportions or success rates;

4. the Multinomial distribution (Section 6.4) when the response variable is unordered categorical;

5. the Logistic distribution (Section 6.6) when there is an underlying or *latent* variable and the observed response is ordered categorical; and

6. the Gamma distribution (Section 6.3) when the response is in the form of the time between successive occurrences of independent events.

The circumstances in which each distribution may have application are outlined in the referenced sections.

Component equations. The statistical model must define the response variable as a function of its component parts. As explained in Section 14.1.5, there are two parts to this expression. The first describes the combination of the identified and unidentified components. The second describes the manner in which the identified component is expressed in terms of the treatment and non-treatment contributions.

The presumption in the statistical models described in this chapter is that identified and unidentified components combine in an additive manner, as defined by (14.1.4). Note that there is an important situation in which this condition is met only after transformation of the data to a logarithmic scale of measurement (see Section 10.3.8).

The manner in which the identified component is expressed in terms of the treatment and non-treatment contributions is determined by (i) the nature of the experimental aim, (ii) the nature of the response variable;

and (iii) the design of the experiment. This is an aspect of the model which is discussed in detail in the following sections, and subsequently, in Chapter 15 when experimental and treatment designs are considered.

14.2.2 Selecting a Model

From the viewpoint of scientists, there are five statistical aspects to be considered when designing an experiment for the purpose of treatment comparison. Two are general considerations—computing facilities and statistical expertise. Before the widespread availability of statistical packages, computing facilities were commonly the limiting factor in choice of design. This is generally not the case today. Statistical expertise will always be a consideration and reference is made throughout the book to situations in which assistance may be required from professional statisticians. The three remaining considerations are

(i) the nature of the treatment comparison, i.e. whether it is based on differences, ratios, odds ratios, etc.;

(ii) the nature of the random or unexplained component of the response, in particular, the assumed distributional form; and

(iii) the experimental and treatment designs employed in the experiment.

The roles of the first two of these considerations are discussed in this section and the role of experimental and treatment designs is discussed in Chapter 15.

Three classes of statistical models are introduced which have wide application in scientific experimentation. These models are shown to be particular cases of what is termed the *Generalized linear model* and the scope of the more general model is explained. All models fit the structure for statistical models presented in Section 14.2.1.

14.2.3 Linear Models

Where the response variable is continuous, the most common situation encountered in practice is that in which

(i) the pattern in the random or unexplained variation is well approximated by a Normal distribution; and

(ii) the treatment comparison is based on differences.

These assumptions are often extended to discrete scaled variables, although other options may be preferable (see, particularly, the Log-linear model below). For experiments in which complex experimental designs are presumed necessary, the assumptions may also have application when the response is ordered, categorical (e.g. when responses are scored using 0 to 10 grading).

A *linear* model includes the presumption that the identified component, $E(y_{ij})$, is the sum of the individual contributions. Formally, this may be expressed by (14.2.1).

$$E(y_{ij}) = c_{1ij}P_1 + c_{2ij}P_2 + \cdots + c_{kij}P_k \tag{14.2.1}$$

where $c_{mij}P_i$ is the contribution from the mth identified source ($m = 1, 2, \ldots, k$) to the value recorded for the jth unit which received the ith treatment ($i = 1, 2, \ldots, g$; $j = 1, 2, \ldots, n_i$) and comprises a known constant c_{mij} and a parameter P_i whose value may be known.

In practice, the equation can usually be expressed in a much simpler form. Typically, the constants $\{c_{mij}\}$ take either the value zero or one. Equations (14.1.6) and (14.1.7) provide simple illustrations of linear equations. In (14.1.6) there are the three treatment effects, T_1, T_2, and T_3, in addition to the overall mean M. Thus, the full representation of the linear equation would be

$$E(y_{ij}) = c_{1ij}M + c_{2ij}T_1 + c_{3ij}T_2 + c_{4ij}T_3.$$

However, as is apparent from (14.1.6), only the parameters M and T_1 make a contribution to plots which received treatment 1. This requirement is met by setting the values of the constants preceding T_2 and T_3 equal to zero for plots in which treatment 1 is applied, i.e. when i takes the value one. By adopting this approach for each treatment in turn, and setting all the constants which are non-zero to one, (14.1.6) results. For computational purposes it is sometimes necessary to represent the equation in the form of (14.2.1), i.e. to formally assign values to the set of constants $\{c_{mij}\}$. This is known as the *regression* formulation of the design and is discussed in Section 18.7.2.

The connection between *linear* equations and the study of *differences* in treatment effects is easily established. Consider the linear equation employed in (14.1.7), i.e. $E(y_{ij}) = M + T_i + R_j$. In this equation, the effects T_1 and T_2 are treatment effects while the effects R_j ($j = 1, 2, \ldots, 11$) are non-treatment effects. The difference, $E(y_{1j}) - E(y_{2j}) = T_1 - T_2$ for $j = 1, 2, \ldots, 11$, is a treatment contrast free of other effects. The possibility of separating treatment and non-treatment effects by taking differences between expected responses is a characteristic of linear equations.

Unless otherwise stated, it is usual to assume in Linear models that

(i) the probability distribution of the response variable y_{ij}, and hence, the unexplained component, e_{ij}, is a Normal distribution;

(ii) the parameters are defined in such a way that the expected value of the random component (e_{ij}) is zero;

(iii) the mechanism generating the random components is unchanging over the course of or locality of the experiment, a fact which is represented in the model by an assumption that the e_{ij}'s are identically distributed; and

(iv) the values of the random components are mutually independent, a fact which is represented in the model by the assumption that the e_{ij}'s are independently distributed.

In designed experiments, random allocation of subjects or experimental units between treatments is assumed and this has a counterpart in observational studies where random selection of subjects or experimental units is presumed. These conditions are implicit in the assumption that the e_{ij}'s are independently distributed and underlie the commonly used methodology which is applied in the comparison of treatments based on Linear models.

Possibilities exist for including more than one random component in the linear equation. One area in which this occurs is in multi-strata, designed experiments which are discussed in Chapter 15. Another area is in the partitioning of variation between components and is discussed in Chapter 19.

14.2.4 Log-Linear Models

Where the response is categorical, either binary in the study of proportions or multi-state in the study of frequency distributions, the reason for assuming a multiplicative relation between $E(y_{ij})$ and the individual effects is given in Section 14.1.4. The general form of a multiplicative relationship is defined by (14.2.2).

$$E(y_{ij}) = [P_1]^{c_{1ij}}[P_2]^{c_{2ij}} \cdots [P_k]^{c_{kij}} \tag{14.2.2}$$

where $[P_m]^{c_{mij}}$ is the contribution from the mth identified source ($m = 1, 2, \ldots, k$) to the value recorded in the jth category ($j = 1, 2, \ldots, c$) for the ith treatment ($i = 1, 2, \ldots, g$) and comprises a known constant c_{mij} and a parameter P_i whose value may be unknown. Equation (14.2.2) has much simpler expressions in specific applications. For example, (14.1.8) is a representation of (14.2.2) for the situation in which there are no treatment differences.

As noted in Section 14.1.5, it is common practice to represent (14.2.2) in the *log-linear* form of (14.2.3). Note that the ratio of treatment effects on the original scale is transformed into a difference in treatment effects on the logarithmic scale, i.e. $P'_r - P'_s = \log[P_r/P_s]$. Since it is generally the parameters on the original scale which are of interest to scientists, analyses based on multiplicative equations or log-linear equations pertain to the study of ratios of effects rather than their differences.

$$\log[E(y_{ij})] = c_{1ij}P'_1 + c_{2ij}P'_2 + \cdots + c_{kij}P'_k \tag{14.2.3}$$

where $P'_1, P'_2, \ldots, P'_k$ are parameters with $P'_m = \log[P_m]$ for $m = 1, 2, \ldots, k$, $\{c_{mij}\}$ is the set of constants as defined in (14.2.2).

It is important to realize that the change of scale for the purposes of representing the equation does not affect the purpose of the analysis, namely to compare frequency distributions for different treatments. Estimates of those parameters in the multiplicative model which are of interest to experimenters are recoverable from the analysis based on the Log-linear model. The introduction of the Log-linear form is motivated by statistical considerations, primarily the opportunity to provide a unified basis for statistical analysis.

In the current context, Log-linear models are being considered for the comparison of frequency distributions between treatments or groups. The Binomial model (Section 6.1) or Multinomial model (Section 6.4) is presumed to apply for each treatment or group and responses are presumed to be obtained independently from different groups.

Two other applications of Log-linear models are noted. They arise

(i) where means are being compared and the conditions for the Poisson model apply (Section 6.3); and

(ii) where there is an examination of the relationship between two or more categorical variables (see Section 17.3).

14.2.5 Logistic Models

Where the response variable has only two categories, perhaps labelled *success* and *failure,* there may be interest in comparing the odds of success under different treatments. Since there are only two categories, the expected frequency in category j for treatment i, $E(y_{ij})$, can be expressed in terms of the probability of success under treatment i, π_i, by (14.2.4).

$$\begin{aligned} E(y_{ij}) &= n_i\pi_i && \text{for } j=1 \\ &= n_i(1-\pi_i) && \text{for } j=2 \end{aligned} \tag{14.2.4}$$

where n_i is the sample size for treatment i ($i = 1,2,\ldots,g$). The odds of success under treatment i is $\omega_i = \pi_i/(1-\pi_i)$. By analogy with linear and log-linear equations, the study of the ratio of odds may be based on component equations which are linear on the *logit* scale, as defined by (14.2.5).

$$\log(\omega_i) = c_{1i}P_1 + c_{2i}P_2 + \cdots + c_{ki}P_k \tag{14.2.5}$$

where $c_{mi}P_m$ is the contribution from the mth identified source ($m = 1,2,\ldots,k$) to the observed odds of success under the ith treatment ($i = 1,2,\ldots,g$) and comprises a constant c_{mi} and a parameter P_m.

Equation (14.2.5) can be derived from the Logistic model described in Section 6.6. The basic assumption is the existence of a scaled response variable, u, such that a response above a fixed threshold, u_T, is regarded as a *success* while a value at or below the threshold is regarded as a *failure.* Hence, π_i is the probability of a response above u_T from a unit to which the ith treatment has been applied. If u is expressed as a linear function of the parameters $P_1, P_2, \ldots, P_k$, (6.6.4) establishes that the logarithm of the odds of success has the component equation defined in (14.2.5).

14.2.6 Generalized Linear Models

The Linear, Log-linear and Logistic models are special cases of a broader class of models in which the relation between $E(y_{ij})$ and the parameters $P_1, P_2, \ldots, P_k$ is defined by (14.2.6).

$$\varphi[E(y_{ij})] = c_{1ij}P_1 + c_{2ij}P_2 + \cdots + c_{kij}P_k \qquad (14.2.6)$$

where $c_{mij}P_m$ is the contribution from the mth identified source ($m = 1, 2, \ldots, k$) to the jth unit receiving treatment i ($i = 1, 2, \ldots, g$; $j = 1, 2, \ldots, n_i$) and comprises a known constant c_{mij} and a parameter P_m whose value may be unknown; and φ is the *link function* which defines the transformation linking the scale on which observed values are measured with the scale on which the identified contribution, $E(y_{ij})$, can be expressed as a linear function of the parameters.

Linear equations are an example of Generalized linear equations in which no transformation is required, i.e. $\varphi[E(y_{ij})] = E(y_{ij})$; Log-linear equations arise when a logarithmic transformation is required, i.e. $\varphi[E(y_{ij})] = \log[E(y_{ij})]$ while Logistic equations arise when a *logit* transformation is required, i.e. $\varphi[\pi_i] = \log[\pi_i/(1-\pi_i)]$ where the connection between π_i and $E(y_{ij})$ is provided by (14.2.4).

The distributional forms which may be employed in the Generalized linear model are restricted to the Exponential family, the important members of which are described in Section 14.2.1.

Table 14.2.1. Forms of Comparison in Common Use

Quantity	Comparison		Scale for Additivity
Proportion	Difference	$\pi_i - \pi_j$	π
	Ratio	π_i/π_j	$\log(\pi)$
	Odds Ratio	$\frac{\pi_i/(1-\pi_i)}{\pi_j/(1-\pi_j)}$	$\log[\pi/(1-\pi)]$
Mean	Difference	$M_i - M_j$	M
	Ratio	M_i/M_j	$\log(M)$

The practical application of Generalized Linear Models is immense. They offer scope for statistical analysis based on realistic models over many scientific disciplines and in many different situations. They offer a freedom in the choice of the form of comparison of parameters through the choice of the link function. Table 14.2.1 presents forms of comparison which are commonly required and the corresponding link functions.

The general development of methodology based on Generalized Linear Models is too extensive to present in this introductory book. McCullagh and Nelder [1983] provide a definitive coverage of the topic for the reader with a background in mathematical Statistics.

14.3 Statistical Methods

14.3.1 Fitted Values, Residuals and Parameter Estimation

The task of comparing one of the statistical models defined in Section 14.2 with data requires the determination of the set of values predicted by the model, namely the set of *fitted values* $\{\hat{y}_{ij0}\}$. The fitted value $\hat{y}_{ij0}$ is an estimate of the identified contribution to the response y_{ij0}, namely $E(y_{ij})$. The difference between the observed value, y_{ij0}, and the fitted value, $\hat{y}_{ij0}$, denoted by $\hat{e}_{ij0} = y_{ij0} - \hat{y}_{ij0}$, is termed the *residual.* The set of residuals $\{\hat{e}_{ij0}\}$ provides estimates of the contributions to the set of responses $\{y_{ij0}\}$ from unidentified factors and, as such, provides information about the nature and size of variability attributable to unidentified sources. Fitted values and residuals have an important role in model and data checking and in model construction.

Computation of fitted values and residuals. The Linear, Log-linear and Generalized linear equations, (14.2.1), (14.2.3) and (14.2.6) are expressed as functions of parameters $P_1, P_2, \ldots, P_k$. Fitted values are constructed by replacing the parameters in these equations by sample estimates. For the Generalized linear model, the relation between the estimator of $E(y_{ij})$, $\hat{y}_{ij}$, and the estimators of the parameters is defined by (14.3.1).

$$\hat{y}_{ij} = \varphi^{-1}(c_{1ij}\hat{P}_1 + c_{2ij}\hat{P}_2 + \cdots + c_{kij}\hat{P}_k)$$
$$\text{for } i = 1, 2, \ldots, g; \;\; j = 1, 2, \ldots, n \qquad (14.3.1)$$

where φ^{-1} is the inverse function of the link function φ; $\hat{P}_1, \hat{P}_2, \ldots, \hat{P}_k$ are estimators of $P_1, P_2, \ldots, P_k$, respectively.

The application of (14.3.1) for linear and log-linear equations is defined by (14.3.2) and (14.3.3), respectively.

$$\hat{y}_{ij} = c_{1ij}\hat{P}_1 + c_{2ij}\hat{P}_2 + \cdots + c_{kij}\hat{P}_k \text{ for } i = 1, 2, \ldots, g; \;\; j = 1, 2, \ldots, n \qquad (14.3.2)$$

$$\hat{y}_{ij} = \exp(c_{1ij}\hat{P}_1 + c_{2ij}\hat{P}_2 + \cdots + c_{kij}\hat{P}_k) \text{ for } i = 1, 2, \ldots, g; \; j = 1, 2, \ldots, n. \tag{14.3.3}$$

Parameter estimation. Two methods of parameter estimation are widely used. They are

(i) the distribution-based *Method of maximum likelihood* which is defined in Section 8.4; and

(ii) the distribution-free method known as *Least squares estimation* in which the values assigned to the parameters are those which minimize

$$\sum_{i=1}^{g}\sum_{j=1}^{n}[y_{ij0} - E(y_{ij})]^2.$$

As a general rule, the Method of Maximum Likelihood is preferred since it uses information about the assumed form of the distribution and it possesses statistical properties which guarantee its estimators have desirable statistical properties. Note, however, that for an important class of models—Linear models which include the assumption of Normality—the two methods provide the same estimates. These are the estimates which minimize R in (14.3.4).

$$R = \sum_{i=1}^{g}\sum_{j=1}^{n}\left[y_{ij0} - \{c_{1ij}P_1 + c_{2ij}P_2 + \cdots + c_{kij}P_k\}\right]^2. \tag{14.3.4}$$

Degrees of freedom. If n denotes the number of entries in the data set $\{y_{ij0}\}$ and p is the number of parameters in the component equation which are independently estimated in the formation of the fitted values, then d, as defined in (14.3.5) is termed the *degrees of freedom* associated with the data set and the model which generates the set of fitted values.

$$d = n - p. \tag{14.3.5}$$

While p may be equal to the number of parameters appearing in the component equation, it often happens that the number of parameters which can be independently estimated is less than p. This would be the case, for example, in a frequency table where the expected proportions $\pi_1, \pi_2, \ldots, \pi_g$ are required to sum to one. In this case $\pi_g = 1 - (\pi_1 + \pi_2 + \cdots + \pi_{g-1})$. Hence, there are only $g - 1$ rather than g proportions to be independently estimated. Another commonly occurring situation is that in which the treatment effects, $T_1, T_2, \ldots, T_g$, which reflect deviations from a mean, are required to sum to zero. Again, the degrees of freedom are reduced by one.

14.3.2 Comparing Models and Data

Having defined a set of observed values based on the data and a set of fitted values derived under conditions defined in a model, there remains the task of choosing a statistic to measure the agreement between the set of observed values and the set of fitted values and converting the value of the statistic into a p-value.

Choice of statistic. Two statistics are widely employed to serve this role. They are the *Deviance* (D) which is a Likelihood ratio statistic (Section 8.4), and the *Pearson Chi-squared statistic* (X^2) which is a distribution-free statistic.

The Pearson Chi-squared statistic has the general form defined in (14.3.6)

$$X^2 = \sum_{i=1}^{g}\sum_{j=1}^{n}\left[\frac{(y_{ij} - \hat{y}_{ij})^2}{V(\hat{y}_{ij})}\right] \tag{14.3.6}$$

where $V(\hat{y}_{ij})$ is the estimated variance of the statistic $\hat{y}_{ij}$. The Deviance has the following forms.

1. For the Normal distribution, the Deviance is based on the sums of squares of differences between the observed and estimated responses as defined in (14.3.7).

$$D = \sum_{i=1}^{g}\sum_{j=1}^{n}(y_{ij} - \hat{y}_{ij})^2/\sigma^2 \tag{14.3.7}$$

   where σ^2 is the variance of the Normal distribution.

2. For the Poisson, Binomial and Multinomial distributions, the Deviance is defined by (14.3.8).

$$D = 2\sum_{i=1}^{g}\sum_{j=1}^{n} y_{ij}\log[y_{ij}/\hat{y}_{ij}]. \tag{14.3.8}$$

3. For the Gamma distribution, the Deviance is defined by (14.3.9).

$$D = 2\sum_{i=1}^{g}\sum_{j=1}^{n}[-\log(y_{ij}/\hat{y}_{ij})]. \tag{14.3.9}$$

Since the Deviance uses information about the distributional form and the Pearson Chi-squared statistic does not, in theory it is the Deviance which should be the preferred statistic when there is confidence about the correctness of the distributional assumptions. However, other factors intervene. In particular, there is a strong historical preference for the Pearson

Chi-squared statistic. With the widespread and growing availability of computational facilities for analysis based on the Deviance statistic plus additivity properties introduced below, it can be expected that this statistic will become the more commonly used statistic in future. For that reason, illustrations of methods in this book are based on the Deviance statistic.

Note, however, that in practical terms, the two statistics produce values which are generally similar. They agree to the extent that a zero value for either statistic provides the best agreement between model and data and arises when every observed value agrees with its corresponding fitted value. Increasing values of either statistics imply worsening agreement with the model. Under a range of important practical situations, both the Deviance and the Chi-squared statistic have sampling distributions which are Chi-squared distributions under the assumption of Normality and which are well approximated by Chi-squared distributions under other important models.

The manner in which the Deviance statistic is employed has a standard form known as the *Analysis of deviance.* It is described in Section 14.3.3 and its application is illustrated in Chapter 16. A modification of the method is required when the Normal distribution is assumed and the value of the variance parameter σ^2 is unknown. The modified form is known as *Analysis of variance.* It is introduced in its basic form in Section 14.3.4 and is expanded upon and illustrated in Chapter 15.

14.3.3 Analysis of Deviance

Notation. The Deviance is determined from knowledge of a data set $\{y_{ij0}\}$ and a model M. To reflect this fact in the following discussion, the Deviance statistic is denoted by $D(y, M)$. Where models M_1 and M_2 are fitted to the same data set, the two Deviance statistics defined may therefore be identified as $D(y, M_1)$ and $D(y, M_2)$, respectively.

Where the aim of the analysis is to seek evidence of treatment differences, the two models which are commonly fitted are

1. the model which includes the assumption that there are no differences in treatment effects and is denoted by M_0; and

2. the least restrictive model which includes no hypotheses about treatment effects and is denoted by M_*.

There may be intermediate models fitted which place partial restrictions on treatment effects. These will be represented by $M_1, M_2, \ldots$ where increasing numerical value of the subscript implies lessening restrictions on treatment effects. For example, if T_1, T_2, T_3, and T_4 represent the treatment effects, then

M_0 includes the hypothesis H_0: $T_1 = T_2 = T_3 = T_4$;
M_1 may include the hypothesis H_1: $T_1 = T_2 = T_3$;
M_2 may include the hypothesis H_2: $T_1 = T_2$; and
M_* places no restrictions on the treatment effects.

Model comparisons and additivity. The application of methodology based on the Deviance statistic requires a means of comparing the fits of two models as well as the fit of a model with data. The statistic $D(M_1, M_0)$ as defined by (14.3.10) provides the basis for a comparison of models M_0 and M_1 (subject to the restriction that M_1 is a generalization of M_0).

$$D(M_1, M_0) = D(y, M_0) - D(y, M_1). \tag{14.3.10}$$

The Deviance statistic has the property that $D(M_1, M_0)$ could be computed directly from an expression for the Deviance statistic $D(y, M)$ by inserting the set of fitted values under M_1 into the expression in place of the set of observed values and the set of fitted values under M_0 into the expression in place of M.

The *additivity* property can be extended to three (or more) models as portrayed in (14.3.11).

$$D(y, M_0) = D(M_1, M_0) + D(M_2, M_1) + D(y, M_2). \tag{14.3.11}$$

If there is lack of agreement between the data and the model proposing no differences in treatment effects, it is possible to provide a decomposition of that lack of fit into meaningful parts. For example, suppose there are four treatments with effects T_1, T_2, T_3 and T_4, and

- model M_0 includes the hypothesis H_0: $T_1 = T_2 = T_3 = T_4$;
- model M_1 includes the hypothesis H_1: $T_1 = T_2 = T_3$;
- model M_2 includes the hypothesis H_2: $T_1 = T_2$.

Then

(i) a comparison of M_0 with the data is a test of the hypothesis H_0 against the alternative that at least one pair of treatment effects are different;

(ii) a comparison of M_0 with M_1 is a test of the hypothesis H_0 against the alternative that treatment 4 has a different effect to at least one of treatments 1, 2 and 3;

(iii) a comparison of M_1 with M_2 is a test of the hypothesis H_1 against the alternative that treatment 3 has a different effect from at least one of treatments 1 and 2; and

(iv) a comparison of M_2 with the data is a test of the hypothesis H_2 against the alternative that treatments 1 and 2 have different effects.

Sampling distributions. If the models fitted are $M_0, M_1, M_2, \ldots$ and M_a $(a = 0, 1, 2, \ldots)$ includes the correct model, then $D(y, M_a)$ has a sampling distribution which is a Chi-squared distribution with degrees of freedom d_a determined from (14.3.5) when the underlying distribution is Normal. For other distributions in the *Exponential family* (which includes the Binomial, Multinomial and Poisson) the Chi-squared distribution with d_a degrees of freedom provides only an approximation to the sampling distribution of the Deviance.

Furthermore, if M_a is a correct model, then any model which contains fewer restrictions is also a correct model and the above statements apply to those models also. Thus, if M_0 is a correct model then so too are M_1, M_2, etc. More generally, if M_a is a correct model, then not only does $D(y, M_a)$ have a sampling distribution which is (approximately) Chi-squared with degrees of freedom d_a but

(i) $D(y, M_{a+1})$ has a distribution which is (approximately) Chi-squared with degrees of freedom d_{a+1}; and

(ii) $D(M_{a+1}, M_a)$ has a distribution which is (approximately) Chi-squared with degrees of freedom $d_a - d_{a+1}$.

The presumption is always made that the model which imposes no restrictions on the treatment effects, M_*, is a correct model. Since M_* places no restrictions on the treatment effects, evidence against this model suggests incorrect assumptions in the sampling assumptions, the form of the component equation or the form of distribution assumed.

Table 14.3.1. The Analysis of Deviance Table

Source of Variation	D.F.[1]	Deviance	p[2]
Not explained by M_0	d_0	$D(y, M_0)$	p_0
Not explained by M_1	d_1	$D(y, M_1)$	p_1
Gain from M_1 over M_0	$d_0 - d_1$	$D(y, M_0) - D(y, M_1)$	p_{01}
Not explained by M_2	d_2	$D(y, M_2)$	p_2
Gain from M_2 over M_0	$d_0 - d_2$	$D(y, M_0) - D(y, M_2)$	p_{02}
Gain from M_2 over M_1	$d_1 - d_2$	$D(y, M_1) - D(y, M_2)$	p_{12}

Notes.

[1] D.F. is the degrees of freedom.

[2] p is the probability of exceeding the observed value of the deviance statistic.

Analysis of deviance table. To provide a concise form of presentation of analyses when fitting a sequence of models using the Deviance statistic, it is usual to collect the results in an *Analysis of deviance table* which has the layout displayed in Table 14.3.1 when models M_0, M_1 and M_2 are compared with the data. Illustrations of the application of Analysis of deviance are provided in Chapter 16.

14.3.4 Analysis of Variance

When the distributional form is Normal, the Deviance statistic, defined by (14.3.7) includes a parameter σ^2 which is the variance of the Normal distribution. In most experimental situations the value of σ^2 is unknown and hence, values of the Deviance can only be computed as multiples of σ^2. In such cases, the Analysis of Deviance approach as described in Section 14.3.3 cannot provide p-values. A commonly used alternative is to use statistics which are ratios of Deviances. In this case, the parameter σ^2, which appears in both numerator and denominator, is canceled out. The procedure employed is termed *Analysis of variance* and is informally introduced in Section 14.1.3. The basis of the technique is explained below for Linear models. The application of the methodology in the comparison of treatment effects is described in Chapter 15. It is also employed in Chapter 18 in regression analysis and in Chapter 19 for variance component analysis.

The universal aim in Analysis of variance is to construct a statistic as a ratio in which the denominator is unaffected by hypotheses made in respect of treatment effects while the numerator becomes increasingly larger as the treatment effects show increasing differences. The development of such a statistic is considered below.

Sums of squares. The basic statistic employed in Analysis of variance is the quantity $SS(y, M)$ defined in (14.3.12) which is known as the *Residual sum of squares* under the model M (for the good reason that it is the sum of squares of the residuals under model M). It is equal to $\sigma^2 D(y, M)$ where $D(y, M)$ is the Deviance statistic defined by (14.3.7). Equations (14.3.13) and (14.3.14) follow directly from the additivity property of the Deviance statistic defined in (14.3.10).

$$SS(y, M) = \sum_{i=1}^{g} \sum_{j=1}^{n} (y_{ij} - \hat{y}_{ij})^2 \tag{14.3.12}$$

where y_{ij} is the statistic providing the response y_{ij0}; and $\hat{y}_{ij}$ is the Maximum Likelihood estimator of $E(y_{ij})$ under the model M.

$$SS(y, M_a) = SS(M_{a+1}, M_a) + SS(y, M_{a+1}) \tag{14.3.13}$$

$$SS(M_{a+1}, M_a) = \sum_{i=1}^{g} \sum_{j=1}^{n} (\hat{y}_{ij}^{a+1} - \hat{y}_{ij}^{a})^2 \tag{14.3.14}$$

where $\hat{y}_{ij}^{a}$ and $\hat{y}_{ij}^{a+1}$ are Maximum Likelihood estimators of $E(y_{ij})$ under models M_a and M_{a+1}, respectively.

Mean square. The statistics $MS(y, M)$ and $MS(M_{a+1}, M_a)$ which are defined by (14.3.15) are known as *mean squares.*

$$\begin{aligned} MS(y, M) &= SS(y, M)/d \\ MS(M_{a+1}, M_a) &= SS(M_{a+1}, M_a)/(d_a - d_{a+1}) \end{aligned} \tag{14.3.15}$$

where d, d_a and d_{a+1} are degrees of freedom based on models M, M_a and M_{a+1}, respectively. Under the assumption of a Linear model which includes the distributional assumption of Normality and independence of responses, the Mean square statistics have the following properties.

Property 1. When M_a $(a = 0, 1, 2, \ldots)$ is a correct model, the expected value of the statistic $MS(y, M_a)$ is σ^2.

(Since M_* is always presumed to be a correct model, $MS(y, M_*)$ has an expected value of σ^2 irrespective of any hypotheses about treatment effects.)

Property 2. When the additional restriction in M_a relative to M_{a+1} is correct, the expected value of $MS(M_{a+1}, M_a)$ is σ^2.

Property 3. When the additional restriction in M_a relative to M_{a+1} is false, the expected value of $MS(M_{a+1}, M_a)$ exceeds σ^2.

Property 4. The statistic $MS(M_{a+1}, M_a)$ for $i = 0, 1, 2, \ldots$ is distributed independently of $MS(y, M_*)$.

F-statistic. Consider then the statistic F defined in (14.3.16) which is known as the *F-statistic* (in honor of a distinguished statistician, Sir Ronald Fisher). If the additional restriction in M_a relative to M_{a+1} is correct, then, by property 2, $MS(M_a, M_{a+1})$ has an expectation of σ^2. Hence, F is a ratio of two statistics with the same expectation, namely σ^2. If the additional restriction imposed in M_a relative to M_{a+1} is false, then, by Property 3, the numerator has a higher expectation than the denominator. Therefore, it is reasonable to regard increasing values of F as providing lessening support for the additional restrictions in M_a.

$$F = MS(M_{a+1}, M_a)/MS(y, M_*). \tag{14.3.16}$$

F-test. By invoking property 4 of the Mean square statistic, the statistic defined by (14.3.16) has an $F([d_a - d_{a+1}], d_*)$ distribution (see Section 7.6) when the additional restriction imposed in M_a relative to M_{a+1} is true. Given a value F_0 for the statistic F, a p-value can then be constructed as

$$p_0 = \Pr(F([d_a - d_{a+1}], d_*) \geq F_0).$$

If the p-value is below 0.05 (0.01) there is weak (strong) evidence that the additional restriction imposed in M_a relative to M_{a+1} is incorrect.

Table 14.3.2. The Basic Layout of the Analysis of Variance Table

Source of Variation	D.F.	S.S.	M.S.	F	p
Gain from M_{a+1} over M_a	$d = d_a - d_{a+1}$	$SS = SS_a - SS_{a+1}$	$MS = \frac{SS}{d}$	$F = \frac{MS}{MS_*}$	p_0
Not explained by M_*	d_*	SS_*	$MS_* = \frac{SS_*}{d_*}$		

Legend.
D.F. – Degrees of freedom
S.S. – Sum of squares
M.S. – Mean square
F – F-statistic
p – p-value
SS_a and SS_{a+1} are values of $SS(y, M_a)$ and $SS(y, M_{a+1})$
SS and SS_* are values of $SS(M_{a+1}, M_a)$ and $SS(y, M_*)$
MS and MS_* are values of $MS(M_{a+1}, M_a)$ and $MS(y, M_*)$

Analysis of variance table. The computations which arise in the application of *Analysis of variance* are typically summarized in an *Analysis of variance table.* The basic layout of the table is displayed in Table 14.3.2.

The general theory and widespread application of the technique of *Analysis of variance* came much later than various specialized applications of the technique. The consequence is that there are traditional forms of presentation of Analysis of variance tables which do not obviously relate to the essential structure of Table 14.3.2. These are discussed in Chapter 15.

14.4 Practical Considerations

14.4.1 Experimental Aims

Section 14.1.2 includes a consideration of the nature of the questions for which the methodology in this chapter has relevance. In this section, there is a brief consideration of some of the practical aspects of establishing experimental aims.

The scientific questions which provide the motivation for the statistical analysis must obviously be decided by the scientists who are undertaking the experimentation. However, the following broad guidelines may assist scientists in making the correct selection and application of statistical methodology.

1. *Predefined comparisons.* If the stated pre-experimental aim identifies one or more comparisons which have particular scientific interest, then those comparisons should be examined individually and separate conclusions drawn in respect of each specific question.

2. *Comparisons when there is no obvious group or treatment structure.* When there are three or more groups and the aim is to seek evidence of differences between them, with no pre-experimental theories as to which pairs might be different, the most widely employed approach is to firstly examine the hypothesis that all groups are identical in respect of the characteristic of interest. Only if this hypothesis is rejected is a more detailed examination made to establish where the differences are likely to lie. This approach offers protection against the chance that spurious differences will be found as a consequence of making an arbitrary number of comparisons.

3. *Factorial arrangement of treatments.* If there is a factorial arrangement of treatments, the primary analysis examines the effects of the individual factors and their interactions. Subsequent analysis may be used to determine which factor levels have different effects and to explore the nature of the interaction between the factors.

14.4.2 Selecting a Method

The process of selecting a method involves the identification of the form of comparison. The possibilities and the relevant approaches to statistical analysis are outlined below.

1. *Comparison of proportions and success rates.* If the expected proportions or success rates are denoted by $\pi_1, \pi_2, \ldots, \pi_g$, the simplest task is to examine the hypothesis that all π_i's are equal and, if this hypothesis is unacceptable, to determine which groups appear to differ. The methodology to do this is provided in Sections 16.2.7 and 16.2.8.

As explained and illustrated in Section 16.1, it is common for more detailed experimental questions to arise because of structure in the group or treatment set. Relevant methodology is discussed and illustrated in Sections 16.2.9, 16.2.10 and 16.2.11.

Note that the methodology of Chapter 9 should be employed in the case where there is a stated standard or normal rate, and the aim is to determine if individual groups or treatments show differences from the stated rate. In that case, separate analyses may be performed for each group or treatment.

If the requirement is for one or more pairwise comparisons, then the methodology of Section 12.2 is appropriate.

Two special cases which are of some practical importance, but which are not covered in this book, are the following.

(a) *Supplementary information.* There may arise occasions in which the aim is to compare expected proportions or success rates in different groups or treatments but there is variation from an additional source which must first be removed.

For example, in an industrial or commercial situation the expected proportion of times a mistake is made in completing a task may be compared under different work practices. However, the employees being used in the study have different levels or experience in performing the task. If the level of experience can be quantified, perhaps by recording the period of time each person has been doing the task, there is potential for including this information in the statistical model and adjusting responses to eliminate this source of variation.

If the comparison is based on odds ratios a supplementary variable may be included directly in the component equation when a Logistic model is assumed. If the comparison is based on ratios, the possibilities for including this information are limited. In either case, assistance should be sought from a consultant statistician both in respect of the analysis and the interpretation of the data.

(b) *Underlying continuous variable.* There is discussion in Section 6.6 of situations in which there is a continuous variable under investigation, although the response which is recorded only provides information as to whether the variable takes a value above or below a threshold. Even though measurement does not take place on the underlying scale, it is possible to estimate the relative distances between the mean values for different groups or treatments on the underlying scales. This requires the assumption of a distributional form for the latent variable and the use of methodology based on Logistic models. A consultant statistician should have the necessary facilities to undertake this analysis.

2. *Comparison of means.* Interest usually lies in the comparison of mean differences. There is a vast methodology developed for this purpose under the name of *Analysis of variance,* which is introduced in Chapter 15.

The development of methods of statistical inference based on the Analysis of variance rely on the assumption of Normality as the underlying distribution for the unidentified component in the component equation. Where this assumption is unacceptable, an alternative with broad application and adequate power is provided by an approximate permutation test as described in Section 7.8.2.

A variation of the conditions in which Analysis of variance may be employed arises in the study of ratios of means when a multiplicative model applies. The basic conditions under which this model applies are discussed in Section 10.3.8. The experimental conditions under which the model might be expected to apply are presented in Section 15.3.4.

Through the use of Generalized linear models, other forms of comparison of means may be possible. Seek assistance from a consultant statistician if there is reason to base comparisons of means on alternative forms.

3. *Comparison of median rates.* The examination of the hypothesis that all groups have the same median response or that the median effects of different treatments are identical is considered in Section 14.5.

4. *Comparison of variability.* Methods for comparing variances are presented in Chapter 19.

5. *Comparisons based on frequency tables.* When data are summarized in the form of frequency tables, there is methodology presented in Chapter 16 for comparing distributional forms between groups or treatments.

14.5 Comparisons Based on Medians

14.5.1 Applications and Models

There is not a unified and complete methodology available for the comparison of medians to rival Analysis of variance plus the methods of pairwise comparisons which are available for the comparison of means. The methods discussed in this section apply to studies in which the only identified source of variation are group or treatment differences. For methods which have application with more complex experimental designs consult, for example, Hollander and Wolfe [1973].

Apart from the possibility that the investigator wishes to study medians rather than means, important reasons for using methodology of this section are

(i) doubts over the correct distributional assumption; and

(ii) data which identify a category of response rather than a scaled measurement, such as is illustrated in Example 14.3.

Two methods for examining the hypothesis of equality of medians are the sign-based method introduced in Section 14.5.2 and the rank-based method introduced in Section 14.5.3. Both methods can be considered as extensions of the two-group methods described in Section 12.4 and are constructed from the concepts introduced in Section 7.7.

Example 14.3 (Effects of Alcohol on Coordination). *An experiment was conducted in which* 57 *volunteers were used in a study of the effect of alcohol on driving ability. One of the tests performed involved a computer-based assessment of the ability of each subject to coordinate brain and hand activity after the application of a 'treatment.'*

The three treatments employed in the study comprise

1. *consumption of* 5 *glasses of low alcohol beer in the period of one hour, where the beer was known to the recipient to be low alcohol beer;*

2. *consumption of* 5 *glasses of low alcohol beer in the period of one hour but with the recipient being told that the beer was high alcohol beer;*

3. *consumption of* 5 *glasses of high alcohol beer in the period of one hour where the beer was known to the recipient to be high alcohol beer.*

The 57 subjects were randomly allocated between the three treatments with the restriction that the treatments should be equally represented.

Coordination ability was scored on a 1 *to* 10 *scale. From a statistical point of view, the observed variable can be considered as an ordered categorical variable which is providing information on an underlying scaled variable and it is the median values of the scaled variable under the three treatments which are to be compared.*

The scores obtained in the test are presented in Table 14.5.1.

Table 14.5.1. Results from the Coordination Test Described in Example 14.3

	Treatment					
	High alcohol		False high alcohol		Low alcohol	
Subject within Treatment	Score	Rank[1]	Score	Rank[1]	Score	Score[1]
1	4	20.0	5	30.0	4	20.0
2	2	3.5	7	50.5	8	56.0
3	6	40.5	4	20.0	6	40.5
4	5	30.0	8	56.0	4	20.0
5	5	30.0	7	50.5	7	50.5
6	7	50.5	5	30.0	6	40.5
7	3	10.0	3	10.0	6	40.5
8	4	20.0	6	40.5	6	40.5
9	3	10.0	6	40.5	4	20.0
10	3	10.0	3	10.0	8	56.0
11	2	3.5	5	30.0	6	40.5
12	6	40.5	7	50.5	7	50.5
13	5	30.0	4	20.0	3	10.0
14	4	20.0	5	30.0	7	50.5
15	2	3.5	6	40.5	5	30.0
16	1	1.0	6	40.5	5	30.0
17	2	3.5	4	20.0	3	10.0
18	3	10.0	4	20.0	7	50.5
19	3	10.0	4	20.0	6	40.5

Note.
[1]averaged ranks are constructed in the manner illustrated in Table 12.4.3

14.5.2 Testing Equality of Medians: A Sign-Based Test

In this test a scaled response variable is reduced to a binary variable by coding all responses below the sample median for the combined sample to category 1 and all responses above the median to category 2. The resultant values may then be combined in a table of observed frequencies.

On the assumption that all groups have the same expected median response, the expected proportions of observations above and below the sample median should be identical for all groups. On this premise, a table of *fitted values* can be constructed which split the total number of counts for each group equally between categories 1 and 2, i.e., if the number of observations above and below the median for treatment i ($i = 1, 2, \ldots, g$) are n_{+i} and n_{-i}, respectively, the corresponding fitted values are identically $\frac{1}{2}(n_{+i} + n_{-i})$.

Based on these observed and fitted frequency tables, the methodology of Sections 16.2.7 and 16.2.8 may be employed to establish if there is evidence against the hypothesis that all groups have the same median, and, if so, where the differences are likely to lie. Note that the Deviance statistic employed to test the hypothesis that the median response is identical for all treatments has $g - 1$ degrees of freedom.

The application of the methodology is illustrated using the data in Table 14.5.1. The sample median for the 57 scores is 5. The observed and fitted frequencies and standardized residuals based on the methodology described in Sections 16.2.7 and 16.2.8 are presented in Table 14.5.2.

Table 14.5.2. Observed and Fitted Frequencies Plus Residuals Obtained from Analysis of Data in Table 14.5.1

Treatment	High alcohol			False high alcohol			Low alcohol		
	obs.[1]	res.[3]	fit.[2]	obs.[1]	res.[3]	fit.[2]	obs.[1]	res.[3]	fit.[2]
Number below 5	13	+1.77	8.0	7	−0.18	7.5	5	−1.20	8.5
Number above 5	3	−1.77	8.0	8	+0.18	7.5	12	+1.20	8.5
Total	16		16.0	15		15.0	17		17.0

Notes.
[1]observed frequencies
[2]fitted frequencies under the model which assumes medians for all treatments are equal
[3]standardized residuals computed using (16.2.9)

Following the steps outlined in Section 16.2.7, the Deviance is computed using (16.2.5) to be 9.8. By reference to the Chi-squared distribution with 2 degrees of freedom, an approximate p-value is computed as $p = \Pr\{\chi^2(2) \geq 9.8\}$ is less than 0.01 which provides some evidence against the hypothesis that all medians are equal. Reference to standardized residuals in the manner suggested in Section 16.2.8 indicates that the *low alcohol* group is likely to have above average coordination and the *high alcohol* group is likely to have below average coordination. The *false high* group is intermediate.

14.5.3 Testing Equality of Medians: A Rank-Based Test

Model. The presumption is that all groups have the same distribution, although the form of the distribution need not be specified. Random sampling is presumed, with independent selection from each group. In the case of treatment comparisons, the treatments are all presumed to produce the same effect and sample members are presumed to be randomly allocated between treatments.

Data. The observed data set is presumed to comprise a set of responses $\{y_{ij0}\}$ from a scaled variable or an ordered categorical variable. In the latter case, there is presumed to be an underlying scaled variable and it is the median of the underlying variable which is being compared between groups. For the purpose of applying the rank-based procedure it is necessary to transform the data in the following manner.

> Order the set of observed responses $\{y_{ij0}\}$ from smallest to largest and replace the ordered values by their ranks to provide a set of ranks $\{r_{ij0}\}$. (Where two or more responses are equal, replace each with the average of their ranks, computed in the manner illustrated in Table 12.4.3.)

Statistic. The statistic H, defined by (14.5.1), has a sampling distribution which is well approximated in large samples by a Chi-squared distribution with $g - 1$ degrees of freedom provided all groups have the same distribution. Furthermore, where the hypothesis is true, the statistic z_i defined by (14.5.2), has a sampling distribution which is approximately $N(0, 1)$.

$$H = \frac{12}{n(n+1)} \left[\sum_{i=1}^{g} n_i (\overline{r}_{i\cdot} - \overline{r}_{\cdot\cdot})^2 \right] \tag{14.5.1}$$

where

$\overline{r}_{i\cdot}$ is the average rank in the ith group $(i = 1, 2, \ldots, g)$;

$\overline{r}_{\cdot\cdot} = \frac{1}{2}(n+1)$ is the average of all ranks;

n_i is the number of observations for group i; and

$n = n_1 + n_2 + \cdots + n_g$ is the total number of observations.

$$z_i = [\bar{r}_{i.} - \bar{r}_{..}]/\sqrt{(n+1)(n/n_i - 1)/12} \qquad \text{for } i = 1, 2, \ldots, g. \qquad (14.5.2)$$

Measure of agreement. The statistic H takes a value of zero when all treatments have the same average rank. Increasing values of H equate with decreasing support for the hypothesis that all groups have the same median (and hence, the same distribution). Given a value H_0 computed using (14.5.1), an approximate p-value can be computed from (14.5.3).

$$p_0 \cong \Pr\{\chi^2(g-1) \geq H_0\}. \qquad (14.5.3)$$

More detailed examination. If p_0 is less than 0.05, there is evidence of differences in medians between groups. Examination of the set of values computed from (14.5.2) assists in determining where the difference lie. Increasing magnitude of z_i equates with increasing evidence that treatment i has a median which differs from the median over all treatments. The sign of z_i indicates whether the median is below or above the average.

Computations. A value of H_0 may be computed directly from (14.5.1). For most practical purposes, (14.5.3) provides an adequate approximation to the p-value. It is not recommended if every treatment has fewer than five readings.

Illustration. Based on the set of ranks provided in Table 14.5.1, the value of H_0 is 12.7 and, by (14.5.3),

$$p = \Pr\{\chi^2(2) \geq 12.7\} < 0.01.$$

Hence, there is strong evidence for claiming there are differences in the medians. The mean ranks and corresponding z-values, computed from (14.5.2), are presented below.

Group	Median	Mean Rank	z_i
High alcohol	3.0	18.2	−3.46
False high alcohol	5.0	32.1	0.99
Low alcohol	6.0	36.7	2.47

It is apparent that the *high alcohol* group has a median which is significantly below average while the *low alcohol* group has a median which is significantly above average. Perhaps surprisingly, those persons who thought they were drinking high alcohol beer, but were actually drinking low alcohol beer, have a median which is above the value for the group who knew they were drinking low alcohol beer. Whether this difference is simply sampling variation is not clear. Another study might clarify the situation.

Problems

14.1 *Example* 14.3

Reproduce the analyses made on data in Example 14.3 using both the sign-based test (Section 14.5.2) and the rank-based test (Section 14.5.3).

14.2 *Comparing pollution levels between sites*

Data are presented in the following tables for levels of *dust* and *zinc* taken from 12 collectors at each of twelve different sites.

(a) Use the sign-based method (Section 14.5.2) to establish if there is evidence of differences in median levels of dust between the sites, and if so, which sites show the largest differences from the overall average.

(b) Use the rank-based method (Section 14.5.3) to establish if there is evidence of differences in median levels of zinc between the sites, and if so, which sites show significant differences from the overall average.

Site	Dust Levels											
1	9	10	7	72	10	96	*	12	20	3	5	4
2	24	15	14	9	12	53	41	5	6	18	58	4
3	8	8	11	11	11	85	30	14	10	5	5	10
4	19	14	14	21	17	62	68	29	9	14	9	10
5	46	4	12	13	36	36	10	22	24	6	7	14
6	63	42	*	12	11	49	24	18	9	8	7	9
7	21	31	*	62	16	66	8	13	9	8	9	40
8	22	78	*	46	22	16	2	24	25	5	1	1
9	43	8	46	29	40	84	43	11	13	8	43	9
10	30	14	8	25	45	*	6	2	2	36	4	6
11	13	9	37	61	15	42	38	15	3	5	5	4
12	17	9	23	9	83	22	6	6	8	8	5	21

*Note: the symbol * denotes a missing value.*

Site	Zinc Levels											
1	45	4	67	10	9	122	*	0	0	40	0	0
2	12	833	6	28	15	1	25	55	829	0	27	1262
3	49	63	69	338	18	9	20	145	26	254	78	10
4	55	74	59	322	20	13	167	31	103	16	10	35
5	62	4	7	17	13	6	99	36	19	16	38	0
6	0	7	171	178	7	40	21	43	8	*	0	0
7	12	0	3	15	15	2	40	407	*	79	224	*
8	6	13	*	557	7	4	46	0	8	491	103	1369
9	0	7	*	16	0	19	453	0	0	158	0	0
10	371	161	408	70	551	13	446	13	8	151	214	8
11	665	0	*	4	89	9	*	13	8	13	0	*
12	0	101	2	6	41	2	21	14	0	14	0	10

14.3 *Analysis of variance — for the more mathematically inclined*

There is a fundamental identity of Analysis of variance based on the statistical model with the component equation

$$y_{ij} = M + T_i + e_{ij} \qquad \text{for } i = 1, 2, \ldots, g;\ j = 1, 2, \ldots, n$$

where M is a constant; $T_1, T_2, \ldots, T_g$ are constants which sum to zero; e_{ij} is distributed $N(0, \sigma^2)$ for $i = 1, 2, \ldots, g$; $j = 1, 2, \ldots, n$; and $\{e_{ij}\}$ is a set of independent variables.

Use the fact that $E(e_{ij}) = 0$, $\text{var}(e_{ij}) = E(e_{ij}^2) = \sigma^2$, and by (5.4.20), $E(e_{ij}, e_{km}) = 0$ for $(i,j) \neq (k,m)$, plus relevant properties from Section 5.4 to prove that

(i) $E(y_{ij}) = M + T_i$; (ii) $E(\bar{y}_{i.}) = M + T_i$; (iii) $E(\bar{y}_{..}) = M$;

(iv) $E(WSS) = E\left[\sum_{i=1}^{g}\sum_{j=1}^{n}(y_{ij} - \bar{y}_{i.})^2\right] = (n-g)\sigma^2$; and

(v) $E(BSS) = E\left[\sum_{i=1}^{g} n(\bar{y}_{i.} - \bar{y}_{..})^2\right] = n\sum_{i=1}^{g} T_i^2 + (g-1)\sigma^2$.

Result (iv) establishes that the *Residual mean square, $RMS = WSS/(n-g)$*, is an unbiased estimator of σ^2 irrespective of any condition imposed on the treatment effects $T_1, T_2, \ldots, T_g$, whereas the *Between treatment mean square, $BMS = BSS/(g-1)$*, is an unbiased estimator of σ^2 only when the treatment effects are identically zero. If there are treatment differences, the expectation of BMS exceeds σ^2 by an amount $n\sum_{i=1}^{g} T_i^2/(g-1)$. Hence, the larger the treatment differences, the greater is the expectation of BMS. These results justify the use of the F-statistic, $F = BMS/RMS$, to test the hypothesis $T_1 = T_2 = \cdots = T_g$. (For those who would like to extend these results, Chapter 15 contains examples of more complex model equations. The same rules and methods for constructing expectations of mean squares apply.)

15

Comparing Mean Response When There Are Three or More Treatments

15.1 Experimental and Statistical Aims

From the experimenter's point of view, the primary function of this chapter is to provide methods which extend the comparison of treatment or group means beyond two treatments or groups. The statistical approach must not only accommodate this aim, but must do so in the most effective way possible. This leads to the need for methodology which incorporates the effect of employing experimental designs (which are used to control the level of non-treatment or extraneous variation). Much of the content of this chapter is devoted to rules for correctly selecting and applying methodology for designed experiments. The chapter has the following format:

- Section 15.2 is devoted to identification and description of designs;
- Section 15.3 describes the methods of model and data checking;
- Sections 15.4 to 15.6 document the construction of statistical methods for application with common designs;
- Section 15.7 describes the analysis and interpretation when there is a factorial arrangement of treatments; and
- Section 15.8 provides the methodology for comparison of treatments in pairs and in groups.

15.2 Defining and Choosing Designs

The definition of a design involves the identification of the *experimental* and *treatment* designs. The introduction to these different aspects of design which is given in Section 13.2 is presumed in the discussion which follows.

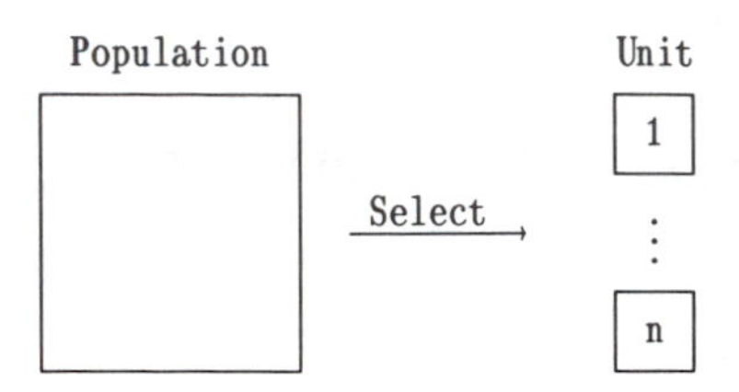

Figure 15.2.1. Experimental design structure for a one-stratum design.

15.2.1 Experimental Designs

Experimental designs are distinguished by the method of selecting units from which responses are to be obtained. The two basic ways in which selection takes place are

- spatial selection, in which the units are physical entities in space—for example, members of a population, plots of land formed by subdividing an area of land, portions of liquid obtained by sampling liquid in a container; and
- temporal selection in which the same spatial unit is repeatedly sampled, e.g. sampling a subject after each new treatment has been applied.

One-stratum designs. The simplest experimental designs involve only one stage of selection and are called *one-stratum designs.* The selection process may be represented in the manner of Figure 15.2.1. Example 15.1 provides an illustration of a one-stratum design in which the units are human subjects, while Example 15.2 provides an illustration in which the units are pots containing trees.

Example 15.1 (Self Concept Enhancement). *A group of* 24 *subjects with impaired reading ability were randomly allocated between three treatments with each treatment being allocated eight subjects. The treatments employed were* (i) *a standard remedial reading treatment for the complete therapy session;* (ii) *a standard remedial reading treatment for one half of the session and self concept enhancement for the remainder; and* (iii) *no treatment* (*i.e. the control group*). *After the session, each subject was given a test of self concept and the scores obtained are recorded in Table* 15.2.1. *The aim of the experiment is to establish if there are differences between the treatments and, if so, to determine where those differences might lie.*

Table 15.2.1. Scores for Self Concept (Example 15.1) (Data by courtesy of Dr. J. Davidson, Psychology Dept., Univ. of Tasmania)

Remedial Reading	Remedial reading + self concept enhancement	Control
49	41	50
67	67	43
52	44	66
52	73	43
66	57	53
47	61	49
47	72	49
48	68	53

Example 15.2 (Tolerance of Pinus radiata to Waterlogging). *The effect of moisture deficiency on the growth of pine trees (Pinus radiata) is relatively well known, but the extent to which P. radiata is affected by waterlogging has not been well researched. An experiment was devised to compare the effects of different levels of moisture content in a range of soils.*

The experiment utilized thirty two pots, with each pot containing one young tree. Sixteen treatments were defined as combinations of four levels of moisture level and four soil types. The thirty-two pots were randomly allocated between the sixteen treatments with two pots receiving each treatment.

At the end of two years, the trees were removed from the pots and the root weights determined. The results are recorded in Table 15.2.2. *The agricultural scientist who undertook the experiment was interested in the extent to which changing the moisture level in the soil affects root growth and how varying the soil type might influence the effect of changing moisture level.*

Table 15.2.2. Data on Root Growth from Example 15.2

Weight of roots in two year old trees

Moisture Level	Soil Type: Heavy		Med. heavy		Med. light		Sandy	
Waterlogged	1026	899	2400	1765	8673	4465	820	574
Very wet	3626	7775	4186	2626	1371	2040	2967	2091
Wet	3371	6286	4089	1391	2719	2033	9842	4894
Optimal	5964	10285	6365	13466	8094	15514	5616	10918

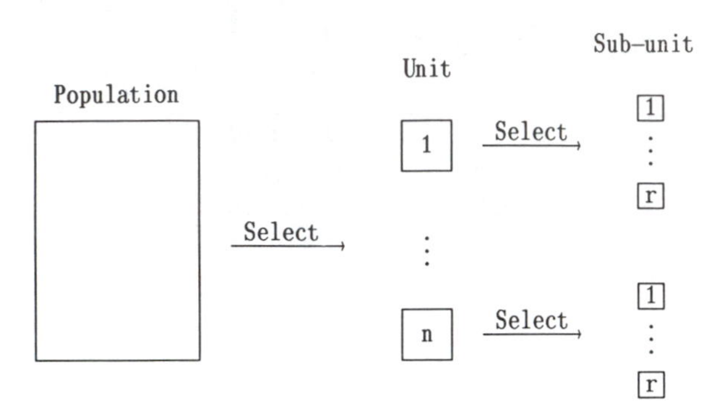

Figure 15.2.2. Experimental design structure for a two-strata design with equal numbers of sub-units per unit.

Experimental design equation for a one-stratum design. If there is a response for variable y on each of the n units, the set of responses obtained can be denoted by $y_{10}, y_{20}, \ldots, y_{n0}$. Each response is presented as the sum of the two components in (15.2.1).

$$y_{i0} = M + U_{i0} \qquad \text{for } i = 1, 2, \ldots, n \tag{15.2.1}$$

where M is expected (mean) response (e.g. the population mean); and U_{i0} is the deviation of the response for the ith unit from the expected response. For purposes of statistical development, it is convenient to regard y_{i0} and U_{i0} as the values of statistics y_i and U_i, and to represent the component equation as (15.2.2)

$$y_i = M + U_i \qquad \text{for } i = 1, 2, \ldots, n. \tag{15.2.2}$$

Two-strata designs. A *two-strata design* arises when there are two stages to the selection, as is portrayed in Figure 15.2.2. The design is extensively employed in the following forms.

1. *Randomized complete block designs.* In agricultural experiments, the classical two-strata design arises when an area of land is divided into strips or *blocks* and subsequently each block is divided into *plots* such that the number of plots within a block is a multiple of the number of treatments to be compared. Within each block, the plots are randomly allocated between treatments such that every treatment is equally represented in each block. The blocks are the units of the **first** stratum and the plots are the units of the **second** stratum. Example 15.3 provides an illustration of a Randomized complete block design (also known as an RCB design).

The Randomized complete block design appears in many forms. Example 13.4 is another illustration which, from a scientific viewpoint, may appear quite different to Example 15.3. The common features are the two stages of selection with the allocation of second stage units (*plots*) between the treatments.

2. *Repeated measures designs.* In the field of Psychology, a common form of experimentation is based on the selection of a group of human subjects and the use of those subjects in successive sessions for the comparison of treatments. Thus, the units at the first stage of selection are the human subjects and these form the units of the **first** stratum. The units at the second stage of selection are the times of sampling for each subject and these form the units of the **second** stratum. The fact that each person is repeatedly sampled leads to the name *Repeated measures design.* Example 15.4 provides an illustration.

Example 15.3 (Effect of Seeding Rate and Fertilizer on Pasture Yields). *Data in Table* 15.2.3 *come from an agricultural experiment laid out as a Randomized complete block design. The experiment comprised the selection of an area of ground which was divided into three strips (blocks) such that within each strip, conditions of growth were similar. Each strip of ground was then divided into twelve equal-sized plots.*

Table 15.2.3. Pasture Yields Under Different Treatments (Data by courtesy of Dr. J. Yates, Faculty of Agric. Sci., Univ. of Tasmania)

Dry matter/plot (g)

Treatment			Block I	Block II	Block III
sr1	p0	n0	23.1	27.3	22.1
sr1	p0	n1	59.0	68.1	39.9
sr1	p1	n0	30.2	53.2	20.5
sr1	p1	n1	67.6	85.8	49.5
sr1	p2	n0	28.4	55.9	17.8
sr1	p2	n1	48.1	64.4	14.6
s2	p0	n0	32.4	79.1	23.9
sr2	p0	n1	72.9	81.7	61.0
sr2	p1	n0	40.8	38.0	57.8
sr2	p1	n1	93.8	121.2	74.8
sr2	p2	n0	58.9	66.8	25.1
sr2	p2	n1	55.1	84.1	26.9

Treatments.

sr1, sr2 — low and high seeding rates respectively
p0, p1, p2 — Nil, 312.5kg, 625kg superphosphate per hectare
n0, n1 — Nil, 15kg sulphate of ammonia per hectare

The twelve plots in each block were randomly allocated between twelve treatments which were formed of combinations of seeding rate, level of nitrogen fertilizer and level of phosphorus fertilizer.

The experimental aim is to establish the role of the three treatment factors —seeding rate, nitrogen fertilizer and phosphorus fertilizer in promoting growth.

Example 15.4 (Perception and Instruction). *Does the viewing of aggressive behavior in news items influence the attitude of viewers to the acceptability of aggression?*

Thirty-six students took part in a study in which they each spent three sessions viewing videos. At each session a different perspective on aggression was the dominant theme. The three themes represented were:

Theme A — aggression is a desirable characteristic.

Theme B — aggression is an undesirable characteristic.

Theme C — aggression is an undesirable but understandable characteristic.

For every subject, the order in which the theme material was presented was decided by a process of random allocation.

At the end of a session, each subject was asked a series of questions on current items of news which pertained to aggressive acts. The set of answers was converted into a score on a zero to ten scale which measured the subject's attitude to aggression. The results for all subjects are contained in Table 15.2.4.

Experimental design equations for two-strata designs. There is not a conventional presentation of component equations to reflect the strata in multi-strata experimental designs, a fact which contributes to the widespread misuse of statistical analysis for designed experiments. The premise adopted in this book is that component equations should uniquely identify designs and these, in turn, should uniquely identify the correct form of the Analysis of variance technique to be used in the statistical analysis. To this end, (15.2.3) is offered as a description of the experimental design for a two-strata design in which n primary units are selected and, from within each primary unit, r sub-units are selected.

$$y_{ij} = M + U_i + \{S_{i,j}\} \qquad \text{for } i = 1, 2, \ldots, n; \ j = 1, 2, \ldots, r \qquad (15.2.3)$$

where

y_{ij} is the response from the jth sub-unit in the ith unit;

M is the expected (mean) response;

U_i is the deviation of the mean response for the ith unit from the overall mean; and

Table 15.2.4. Measuring Attitudes to Aggression

"Aggression Scores"
(Higher values imply more positive view of aggression)

	Theme				Theme		
Subject	A	B	C	Subject	A	B	C
1	4	6	5	19	3	5	5
2	4	5	5	20	3	6	5
3	3	6	7	21	6	7	7
4	4	4	4	22	2	4	4
5	4	5	6	23	1	4	1
6	5	7	5	24	5	6	10
7	4	6	6	25	3	5	8
8	5	6	8	26	5	5	5
9	7	6	9	27	4	7	5
10	5	6	8	28	8	7	8
11	4	8	6	29	3	1	3
12	3	4	4	30	4	6	4
13	4	5	4	31	4	3	7
14	5	6	7	32	4	6	6
15	7	8	8	33	5	4	5
16	8	9	9	34	5	6	8
17	5	7	5	35	3	4	5
18	6	8	9	36	7	7	6

$S_{i,j}$ is the deviation of the response in the jth sub-unit from the mean in the ith unit.

(Note that the equation assumes equal numbers of sub-units per unit. More complex designs may require r to vary between units.)

Representing the subscripts for the sub-unit effect, S, in the form i, j (rather than the form ij) identifies the fact that the values taken by j only have relevance within a level of i. In other words, the sub-unit labelled 1 in the first unit has no relation to the sub-unit labelled 1 in any other unit. This is expression of the *hierarchical* nature of the design. (*Non-hierarchical designs* are discussed in Section 15.6.5.)

Three-strata designs. If there were a third stratum, a second set of parentheses would be introduced, in the manner portrayed in (15.2.4)

$$y_{ijk} = M + U_i + \{S_{i,j} + \{R_{i,j,k}\}\} \quad \text{for } i = 1, 2, \ldots n;$$
$$j = 1, 2, \ldots, r; \;\; k = 1, 2, \ldots, s. \qquad (15.2.4)$$

Equation (15.2.4) has application in Example 15.7 and Example 13.5. In Example 15.7, the strata comprise *blocks* (U_i), *plots within blocks* ($S_{i,j}$) and *sub-plots within plots within blocks* ($R_{i,j,k}$). The construction of the experimental design used in Example 13.5, as portrayed in Figure 13.2.1, identifies the three strata effects—*chambers* (U_i), *trays within chambers* ($S_{i,j}$) and *pots within trays within chambers* ($R_{i,j,k}$).

15.2.2 Treatment Designs

Experimental designs are concerned with the manner of selecting units to be used in an experiment. They make no reference to the treatments which may be applied or to the manner in which units are allocated between treatments. That is the province of *treatment designs.*

The function of treatment design equations is the partitioning of between unit variation within a stratum into treatment effects and non-treatment effects. Thus, in Example 15.1, where the experimental design equation may be expressed by (15.2.5), a portion of the variation is the set $\{S_i\}$ may be attributable to the different treatments applied to the subjects and a portion to non-treatment contributions. This partitioning of the within stratum variation can be represented in a component equation, the *treatment design equation,* for the stratum. For Example 15.1, the treatment design equation is represented by (15.2.6).

$$y_i = M + S_i \qquad \text{for } i = 1, 2, \ldots, 24 \tag{15.2.5}$$

where y_i denotes the score for the ith subject; and S_i is the difference between the expected score for the ith subject and the mean over all treatments.

$$S_i = T_j + e_{jk} \qquad \text{for } j = 1, 2, 3;\ k = 1, 2, \ldots, 8 \tag{15.2.6}$$

where T_j is the amount by which the mean effect of the jth treatment differs from the overall mean; and e_{jk} is the effect which represents the contribution from unidentified (non-treatment) factors which causes the response for the kth subject who receives the jth treatment to differ from the mean value for the jth treatment.

Subscripts: note that the subscript i which identifies the subjects in (15.2.5) *is replaced by the pair of subscripts j, k in* (15.2.6). *Unique identification of a subject can be made by quoting either the value of i or the pair of values* (j, k).

Factorial arrangement of treatments. Where the treatments are distinguished by two or more factors, there is scope for providing analyses which pertain to the factors rather than individual treatments. This requires the expression of the treatment design in terms of the factors. Thus, in Example 15.2, the sixteen treatments can be defined in terms of two factors—*Soil type* (S) and *Moisture level* (W)—in order to allow consideration of the following questions:

1. *Does root growth vary with soil type?*
2. *Does root growth vary with moisture level?*
3. *Does variation of root growth with moisture level depend on the soil type?*

To answer these questions, the treatment design equation is presented in the form of (15.2.7).

$$P_i = S_m + W_n + (S.W)_{mn} + e_{mnp} \quad \text{for } m = 1,2,3,4;\ n = 1,2,3,4;\ p = 1,2 \tag{15.2.7}$$

where

P_i is the contribution from the ith pot ($i = 1, 2, \ldots, 32$);

S_m is the amount by which the mean response in the mth soil type differs from the overall mean;

W_n is the amount by which the mean response in the nth moisture level conditions differs from the overall mean;

$(S.W)_{mn}$ is the amount by which the combined contribution from soil type and moisture level differs from the sum of their individual contributions, namely $S_m + W_n$ (i.e. $(S.W)_{mn}$ is the *interaction* effect); and

e_{mnp} is the combined contribution from all non-treatment contributors.

Representing treatment designs. Units may be allocated to treatments at any stratum level. The description of the experiment should clearly identify the stratum or strata at which allocation takes place and whether or not there is treatment structure. The effect of a treatment design at any stratum level can be displayed by representing the corresponding element in the experimental design component as the sum of one or more treatment effects plus a component for non-treatment effects. The latter is commonly referred to (rather misleadingly) as the *error term* for that stratum.

The structure of treatment designs for a collection of commonly employed designs is shown below and is illustrated by examples which appear in this chapter.

One stratum design

Experimental design equation: $y_i = M + U_i$

Treatment design equation:

- without treatment structure

$$U_i = T_j + e_{jk}$$

Illustration: Example 15.1

- with factorial arrangement of treatments (2 factors)

$$U_i = A_j + B_k + (A.B)_{jk} + e_{jkm}$$

Illustration: Example 15.2

Two strata designs

Experimental design equation: $y_{ij} = M + U_i + \{S_{i,j}\}$

Treatment design equation:

- treatments applied only in the second stratum (no treatment structure)

$$S_{i,j} = T_k + e_{i,k}$$

Illustration: Example 15.4

- treatments applied only in the second stratum (factorial arrangement)

$$S_{i,j} = A_k + B_m + (A.B)_{km} + e_{i,km}$$

Illustration: Example 15.3 (3 factors)

- treatments applied only in the second stratum (with repetition)

$$S_{i,j} = T_k + (U.T)_{ik} + e_{i,km}$$

Note: The interaction term $(U.T)_{ik}$ allows for the possibility that treatment differences may not be the same for all units or subjects.

Illustration: Example 15.5

- treatments applied only in first stratum (no treatment structure)

$$U_i = T_k + e_{km}$$

Illustration: Example 15.6

- treatments applied in both strata (factorial arrangement)

$$U_i = A_k + e_{km}$$
$$S_{i,j} = B_n + (A.B)_{kn} + e'_{km,n}$$

Notes:

1. *The interaction term appears in the lower stratum.*
2. *There are two terms representing unexplained variation—*e_{km} *representing unexplained variation between units in the first stratum and* $e'_{km,n}$ *representing unexplained variation between subunits in the second stratum.*

Three strata designs. Experimental design equation:

$$y_{ijk} = M + U_i + \{S_{i,j} + \{P_{i,j,k}\}\} \quad \text{for } i = 1, 2, \ldots, n;$$

$$j = 1, 2, \ldots, r; \; k = 1, 2, \ldots, s.$$

There are various treatment designs which may be employed with three strata designs. Example 15.7 provides an illustration of a design in which each plot in a Randomized complete block design is subdivided to form sub-plots and treatments are applied at both the plot and sub-plot levels. The fact that a plot has been subdivided has led to the title *split-plot design.* There are two treatment design equations required, one at the plot level and one at the sub-plot level. These are displayed in (15.2.8) on the assumption that the r plots per unit are allocated between r levels of factor A and s sub-plots per plot are allocated between the s levels of factor B.

$$\begin{aligned} S_{i,j} &= A_m + e_{i,m} && \text{for } m = 1, 2, \ldots, r; \; i = 1, 2, \ldots, n \\ P_{i,j,k} &= B_n + (A.B)_{mn} + e'_{i,m,n} && \text{for } n = 1, 2, \ldots, s. \end{aligned} \tag{15.2.8}$$

Note the introduction of two unexplained or residual components, $e_{i,m}$ which represents contributions from non-treatment factors which vary between plots within blocks, and $e'_{i,m,n}$ which represents contributions from non-treatment factors that vary between sub-plots within plots. At each stratum in which treatments are applied, there is necessarily a non-treatment or *residual* component.

A more complex treatment design based on the same experiment design is provided by Example 13.5 where there is a factorial arrangement at the second stratum plus a factor allocated within the third stratum. The design is displayed in Figures 13.2.1 and 13.2.2.

Example 15.5 (Evaluation of Fitness). *The scientific evaluation of fitness to assist athletes prepare more effectively for events is now well established. In this study, three athletes were 'wired' to record, among other things, the level of exertion they required in the performance of ten different exercises.*

For each athlete, a maximum desirable exertion level was established and during a single exercise session the athlete was electronically monitored one hundred times to establish the number of times the exertion level was above the pre-determined maximum desirable level. It is this variable for which data are recorded in Table 15.2.5.

Each athlete took part in the monitoring program for a period of one month. In that time, 30 *readings were obtained from the athlete, three from each of ten exercises. The order in which the exercises were monitored was decided by a random process. (The random order in which exercises were assessed not only served to validate the statistical analysis, but also to ensure that neither the athlete nor his coach was aware which exercise in the set was being monitored.) The results are recorded in Table* 15.2.5.

Table 15.2.5. Data on Exertion Levels for Example 15.5

Number of times exertion level exceeded max. desirable level in 100 samplings

Exercise	Athlete								
	CT			LT			PW		
Basic	50	66	54	3	16	17	18	18	15
EA1	64	68	64	62	49	21	31	24	23
EA2	41	38	27	20	24	27	28	28	16
EA3	64	71	59	28	42	36	32	27	21
EM1	36	57	47	27	52	33	15	22	16
RA1	66	74	74	48	47	38	34	13	50
RA2	85	64	71	22	41	49	30	21	30
RA3	38	52	79	42	43	46	20	11	43
TB1	77	63	88	29	25	31	48	20	35
TB2	77	60	62	30	31	23	56	10	19

Example 15.6 (Activity of Children at Play in Different Environments). *At a pre-school which is used for research into child behavior, the reactions of children confronted with a new game were observed in nine distinct environments. For the experiment, eighteen children were used. These children were randomly allocated between the nine environments subject to the condition that there should be* 2 *children examined in every environment.*

All children were given the same game to play. This game required them to move between two boxes containing items for the game. Observers counted the number of times each child moved from one box to another before the child tired of the game. All children played the game at least twice in the observation period, and counts are recorded in Table 15.2.6 *for the first two times the game was played by each child.*

In statistical terms, a second stratum is introduced to the design by the repeated observations on individual children although no treatments were applied at the second stratum since each child remained in a constant environment.

Table 15.2.6. Data Collected in Example 15.6

Envir.	Child	Scores[1]		Envir.	Child	Scores[1]	
1	1	21	15	5	10	9	5
1	2	1	13	6	11	1	11
2	3	1	2	6	12	16	15
2	4	3	4	7	13	11	10
3	5	11	8	7	14	13	10
3	6	8	8	8	15	5	2
4	7	1	4	8	16	6	10
4	8	16	7	9	17	10	8
5	9	14	14	9	18	6	7

Note.
[1]Number of trips between two boxes.

Example 15.7 (Comparison of Lucerne Varieties at Different Cutting Dates). *The data in Table* 15.2.7 *were obtained from an experiment in which a field was divided into six strips (blocks) with each strip being divided into three equal-sized plots. Within each block the three plots were randomly allocated between three varieties of lucerne such that every variety was sown in one plot. Each plot was divided into four sections. Four dates of cutting were defined and, for each plot, the four sections were randomly allocated between the cutting dates. At each cutting date, the chosen section in each plot was harvested and the lucerne in the section then left to regrow until the following year.*

Table 15.2.7 Yields of Lucerne Varieties (Example 15.7)
(Data by courtesy of Dr. J. Yates, Faculty of Agric. Sci., Univ. of Tasmania)

		Block					
Variety	Date	1	2	3	4	5	6
Ladak	A	2.17	1.88	1.62	2.34	1.58	1.66
	B	1.58	1.26	1.22	1.59	1.25	0.94
	C	2.29	1.60	1.67	1.91	1.39	1.12
	D	2.23	2.01	1.82	2.10	1.66	1.10
Cossack	A	2.33	2.01	1.70	1.78	1.42	1.35
	B	1.38	1.30	1.85	1.09	1.13	1.06
	C	1.86	1.70	1.81	1.54	1.67	0.88
	D	2.27	1.81	2.01	1.40	1.31	1.06
Ranger	A	1.75	1.95	2.13	1.78	1.31	1.30
	B	1.52	1.47	1.80	1.37	1.01	1.31
	C	1.55	1.61	1.82	1.56	1.23	1.13
	D	1.56	1.72	1.99	1.55	1.51	1.33

The experimenter chose this design to permit an examination of the effects of changes in cutting data on the following year's yield. By including different varieties, it is possible to establish if the effect of changing cutting date is consistent between varieties.

15.2.3 Model Equations

While it is easier to establish the correct design by considering the experimental and treatment designs separately, it is usual to combine the experimental and treatment design equations for purposes of presentation. The combined equation is termed the *model equation* in this book. Model equatons for some widely used designs are presented below. The notation introduced with experimental and treatment design equations in Section 15.2.1 and 15.2.2 is retained.

One-stratum design

- no treatment structure $y_{jk} = M + T_j + e_{jk}$
- two factors $y_{jkm} = M + A_j + B_k + (A.B)_{jk} + e_{jkm}$

Two-stratum design

- treatments applied only in the second stratum, no treatment structure

  (i) no replication of treatments within strata

$$y_{ik} = M + U_i + \{T_k + e_{i,k}\}$$

  (ii) replication of treatments within strata

$$y_{ikm} = M + U_i + \{T_k + (U.T)_{ik} + e_{i,km}\}$$

Comment: In the first of these designs, the number of sub-units within each unit is identical with the number of treatments. Thus, each treatment is represented once, and only once, within each unit. A consequence is that any variations in treatment differences between units, as would be represented by the term $(U.T)_{ij}$ are indistinguishable from the effect of non-treatment factors, $e_{i,k}$, a fact which is made apparent by the two terms sharing the same pair of subscripts. Hence, the design only has application when it is reasonable to assume that expected treatment differences are identical for all units. There is limited scope for assessing the acceptability of this assumption (see Section 15.3) and for making allowance in the model for some forms of interaction between U and T.

- treatments applied only in second stratum, two factors

(i) no replication of factor combinations within strata

$$y_{ikm} = M + U_i + \{A_k + B_m + (A.B)_{km} + e_{i,km}\}$$

(ii) replication of factor combinations within strata

$$y_{ikmn} = M + U_i + \{A_k + B_m + (A.B)_{km} + (U.A)_{ik} + (U.B)_{im} + (U.A.B)_{ikm} + e_{i,kmn}\}.$$

– treatments applied only in first stratum, no treatment structure

$$y_{jkm} = M + T_k + e_{km} + \{S_{km,j}\}.$$

Comment: This design is commonly employed when there is sub-sampling to improve precision. For example, a person's response to a treatment may be repeatedly measured to allow a more accurate estimate of the mean response, chemical analysis may be done in duplicate or triplicate to provide a better estimate of the level of a chemical compound in a sample, or several leaves may be taken from a plant and each leaf analyzed separately.

– treatments applied in both strata, two factors

no replication of the factor in the second stratum

$$y_{kmn} = M + A_k + e_{km} + \{B_n + (A.B)_{kn} + e'_{km,n}\}.$$

Comment: At each stratum level there is need for a term which incorporates the non-treatment contributions at that stratum level.

15.2.4 The Use of Supplementary Variables

As explained in Chapter 13, a primary function of experimental design is to separate non-treatment and treatment contributions to the response variable in order to improve the precision with which treatment comparisons can be made. Employing within-strata comparisons, as is done with Randomized block designs and Repeated measures designs, is one way in which this may be achieved. Examples 15.8 and 15.9 introduce situations for which a different approach is required.

Example 15.8 (*continuation of Example* 15.1) **Self Concept Enhancement).** *A study is reported in Example* 15.1 *in which subjects with reading impairment receive different treatments prior to a test for self concept. The aim of the study is to establish if the treatments have different effects on self concept. A potential weakness of the study, as it is presented in Example* 15.1, *is a failure to take into account the differing levels of self concept between the subjects prior to the treatments.*

In fact, in the study as performed, each subject was given a pre-treatment self concept test in addition to the post-treatment test. Scores for both tests are recorded in Table 15.2.8. *There is obvious merit in adjusting post-treatment scores to allow for pre-treatment variation in self concept in order to remove one source of non-treatment variation in the responses before making treatment comparisons. The simplest way to make the adjustment is to subtract the pre-treatment score from the post-treatment score, i.e. to base the analysis on the differences between the two scores for the* 24 *subjects in the sample.*

Table 15.2.8. Scores for Self Concept (Example 15.8) (Data by courtesy of Dr. J. Davidson, Dept. of Psychology, Univ. of Tasmania)

Remedial Reading Control		Remedial Reading + Self Concept Enhancement			
Pre-test score	Post-test score	Pre-test score	Post-test score	Pre-test score	Post-test score
44	49	31	41	58	50
62	67	60	67	44	43
39	52	36	44	65	66
50	52	57	73	45	43
57	66	57	57	50	53
57	47	51	61	43	49
45	47	50	72	45	49
38	48	51	68	52	53

Example 15.9 (Work Practice and Productivity). *There is increasing evidence that productivity in repetitive tasks may be improved by varying the tasks performed by each worker. The current study is one of a series of experiments designed to examine this proposition.*

Sixty workers volunteered to take part in the study. They were randomly split into four groups which were distinguished in the following way.

Form of Production	Time Spent on Job	
	One whole shift	Two half shifts
Serial	Group 1	Group 2
Parallel	Group 3	Group 4

The number of items produced by each worker in the equivalent of one shift is recorded in Table 15.2.9. *However, it is known that much of the variation in the numbers recorded in Table* 15.2.9 *is due to differing levels of experience of the workers. In general, the longer a worker has been employed on the production line the more items he or she can produce in a shift. A measure of this experience is provided by the number of days a worker has spent on the production line prior to the commencement of the study. This information is also recorded in Table* 15.2.9.

From the experimenter's point of view, the additional information does not alter the aim of the study. However, the statistical model and method of analysis may be varied to accommodate the extra knowledge.

Table 15.2.9. Data from the Production Line Study (Example 15.9)

Form of Production							
Serial				Parallel			
Time Spent on Job				Time Spent on Job			
Full		Half		Full		Half	
P[1]	D[2]	P	D	P	D	P	D
70	13	69	28	66	41	64	32
79	37	75	38	56	12	68	26
90	45	79	45	65	38	70	33
65	15	71	29	70	38	58	22
69	20	62	19	61	28	58	18
72	24	61	11	57	19	79	45
79	29	67	18	58	14	61	26
75	35	73	40	62	20	60	16
76	39	74	34	65	25	60	20
76	30	70	33	75	45	64	23
65	21	66	28	70	47	67	37
72	29	66	21	63	31	73	36
75	20	74	29	72	42	69	32
78	33	61	22	62	24	67	36
82	40	62	23	69	38	50	16

Notes.
[1]P is the number of items produced
[2]D is the number of days experience on the production line.

Covariates. In Examples 15.8 and 15.9, information about identified non-treatment information is contained in scaled variables—*pre-treatment score* in Example 15.8 and *number of days experience* (D) in Example 15.9. Scaled variables which provide supplementary information of use in explaining a portion of the non-treatment variation in the response variable are referred to as *covariates.*

The use of covariates in the model equation for a design requires the specification of a mathematical relationship between the response variable and the covariate.

Most commonly, a linear relation between the response variable (y) and the covariate (x) is assumed. For a one-stratum design the model equation is

– in the absence of a covariate

$$y_{ij} = M + T_i + e_{ij} \qquad \text{for } i = 1, 2, \ldots, t;\ j = 1, 2, \ldots, n; \text{ and}$$

– in the presence of a covariate

$$y_{ij} = M + T_i + \beta(x_{ij} - \overline{x}_{..}) + e'_{ij} \qquad \text{for } i = 1, 2, \ldots, t;\ j = 1, 2, \ldots, n$$

where $\overline{x}_{..}$ is the mean value of the covariate over all treatments; and β is a constant (usually of unknown value).

The function of the covariate is more readily appreciated if the equation is written in the form

$$y_{ij} - \beta(x_{ij} - \overline{x}_{..}) = M + T_i + e'_{ij}.$$

In this form, the left hand side of the equation is the response adjusted to the value which would have arisen if the covariate had taken the value $\overline{x}_{..}$ rather than x_{ij}. The inclusion of the covariate term in the model equation has the function of artificially creating a sample in which all sample members have the same value for the covariate, namely $\overline{x}_{..}$. Before the introduction of the covariate, the unexplained variation is measured by the variability in the set $\{e_{ij}\}$. The introduction of the covariate implies that e_{ij} can be represented as the sum of two components, i.e.

$$e_{ij} = \beta(x_{ij} - \overline{x}_{..}) + e'_{ij}.$$

The unexplained variability is now limited to the variation in the set $\{e'_{ij}\}$. If there is a strong relationship between the response variable and the covariate, a large reduction in unexplained variation can be achieved. This has the important implication that sensitivity of treatment comparisons is greatly increased.

The above discussion is based on the assumptions that

(i) the relation between the response variable and the covariate is linear; and

(ii) the value of β is constant over all treatments.

The acceptability of these assumptions must be checked and methods for examining the assumptions are presented in Section 15.3.3.

If there is more than one stratum, an additional covariate term must be included in the model equation for each stratum above the stratum in which measurements are made on the covariate. Thus, if the model equation without the inclusion of the covariate is

$$y_{ijkm} = M + A_i + e_{ij} + \{B_k + (A.B)_{ik} + e'_{ij,km}\}$$

and the covariate provides the set of measurements $\{x_{ijkm}\}$, the model equation which includes the covariate terms would be

$$\begin{aligned} y_{ijkm} = M + A_i + \beta_1(\overline{x}_{ij..} - \overline{x}_{....}) + f_{ij} \\ + \{B_k + (A.B)_{ik} + \beta_2(x_{ijkm} - \overline{x}_{....}) + f'_{ij,km}\} \end{aligned}$$

where

$$\begin{aligned} e_{ij} &= \beta_1(\overline{x}_{ij..} - \overline{x}_{....}) + f_{ij} \\ e'_{ij,km} &= \beta_2(x_{ijkm} - \overline{x}_{....}) + f'_{ij,km}. \end{aligned}$$

Note that the values of β_1 and β_2 are unrelated and the existence of a strong relationship between the response variable and the covariate in one stratum does not imply there will be a strong relationship in another stratum.

Pre- and Post-treatment measurements. Where the covariate is the pre-treatment measurement on the response variable, as in Example 15.8, it is common to base the analysis on the difference between pre- and post-treatment responses, i.e. $d_{ij} = y_{ij} - x_{ij}$. This is equivalent to setting β equal to one in the model equation and implies that, for like-treated subjects, the expected difference in the post-treatment score will equal the expected difference in the pre-treatment score.

15.2.5 Fixed and Random Effects

The method of selecting units at each stratum level influences the scope of the statistical findings and may affect the form of statistical analysis.

Sample selection. There are two possible forms of selection. If the principle of random selection is employed, as defined in Chapter 2, the corresponding term in the experimental design equation is called a *random effect.* Where this condition is not met, the term is called a *fixed effect.* In practice, the distinction is usually clear cut. Thus,

- where subjects are selected from a population, the nature of the selection procedure can be examined to establish whether it is random or non-random;
- where the units are formed by repeatedly sampling in time, e.g. repeatedly measuring the response from the one subject, the decision depends on the manner in which the non-treatment component of the response is presumed to be varying with time. If there is a systematic change with time, e.g. a cyclical trend or long term trend, the stratum effect is best considered to be a fixed effect, whereas, if it is considered to be merely random variation, i.e. 'noise', it can be thought of as a random effect; and

– in agricultural field trials in which an area of land at a particular geographical location is divided into blocks and plots, the components representing block and plot effects are fixed effects.

Treatment allocation. In designed experiments, where there is random allocation of units between treatments, the residual term in the treatment design equation for that stratum is a random effect.

Note that treatment *selection* as distinct from treatment *allocation* is presumed to be non-random within the context of the experimental aims considered in this chapter. The presumption is that each treatment has been specifically chosen because it is to be compared with the other treatments being applied. Hence, treatment effects (including treatment factors and their interactions) are presumed to be fixed effects. There are circumstances in which treatments are randomly selected. These are discussed in Chapter 19.

Model equations. Random effects are identified by introduction of a different typeface in subsequent model equations. A consequence is that two model equations which include the same set of terms can distinguish the different forms of sample selection. For example, (15.2.9) and (15.2.10) both represent two strata designs with treatments (randomly) applied in the second stratum. However, (15.2.9) represents a situation in which the primary units were randomly selected while (15.2.10) represents a situation in which they were non-randomly selected. The representation of the residual component, $\varepsilon_{i,jk}$, as a random component indicates random allocation of sub-units between treatments.

$$\mathcal{y}_{ijk} = M + \mathcal{U}_i + \{T_j + (\mathcal{U}.T)_{ij} + \varepsilon_{i,jk}\} \tag{15.2.9}$$

$$\mathcal{y}_{ijk} = M + U_i + \{T_j + (U.T)_{ij} + \varepsilon_{i,jk}\}. \tag{15.2.10}$$

The scope of statistical conclusions. From the experimenter's point of view, the distinction between fixed and random sample selection affects the scope of the statistical inference. Where random selection of units has been employed, the statistical conclusions in respect of treatment comparisons apply to the population from which the sample was selected. Where non-random selection is employed, the sample represents the population to which the statistical conclusions apply. In the latter case, the scientist who claims the results have application beyond the subjects or units which comprise the sample, must rely on non-statistical arguments that the treatment differences found in the sample apply in the more general population. Thus, the agricultural scientist who conducts an experiment on plots at one research station, or the medical researcher who uses volunteers from a minority group in the community, cannot rely on statistical argument to extend the scope of their conclusions beyond the particular area of land or the particular sample of people employed in the study. (This is not to say the conclusions may not have broader application, merely that the claim of a broader application cannot rest on statistical grounds.)

Effect on statistical analysis. The distinction between fixed and random effects is also important in determining the form of the statistical analysis for certain designs. The designs which are affected are those multi-strata designs in which there is allocation of treatments between units at the second or lower strata with each treatment being applied more than once to each subject or unit. (Winer [1971] provides an informative exposition.)

Experimental and statistical 'populations.' When relating the distinction between fixed and random effects to an experimental situation, care must be taken to distinguish the population of subjects in which the experimenter has an interest and the population being defined in the statistical context. Where there has been random selection of a sample of subjects from the experimenter's population then (i) the two populations are identical; (ii) the statistical conclusions may be applied to the experimenter's population; and (iii) the statistical analysis should be based on the assumption that the unit effect is a random effect. Where there has been non-random selection from the experimenter's population, (i) the statistical population becomes the sample; (ii) the statistical conclusions apply only to the sample (and hence, require non-statistical justification to extend them to the experimenter's population); and (iii) the statistical analysis should be based on the assumption that the unit effect is a fixed effect.

15.2.6 Observational Studies and Non-Random Treatment Allocation

A distinction is made between *observational studies* and *designed experiments* in Section 13.1. Both may be concerned with treatment comparison, but in the former case, the allocation of treatments is outside the scope of the experimenter. Some examples will make the point clear.

1. The factor *sex* was quoted in Chapter 13 as a prime example. Generally, it is not possible to randomly allocate a sex to each sample member.

2. In Repeated measures studies, the effect of order or time of application may be of interest. Clearly *time 1* must come before *time 2*, *time 2* before *time 3*, etc.

3. There may be circumstances in which the researcher is not in control of the allocation of treatments. For example, in a glass house study in which *temperature* is a factor, there may be one glass house which is permanently the *high temperature* glass house and another which is permanently the *low temperature* glass house.

4. In repeated measures studies, a treatment may involve an irreversible change to a subject, as when a laboratory animal has an organ removed or a nerve cut. The measurement of response in the intact animal must precede the measurement of post-operative response.

There are, potentially, three points of concern when treatment allocation is non-random. They are

(i) lack of independence when the same pair of treatments are necessarily always applied to neighboring units;

(ii) a carry-over effect in which the application of one treatment persistently influences response to subsequent treatments; and

(iii) uncertainty as to whether evidence of differences between groups which receive different treatments is a treatment effect or is due to other factors.

Lack of independence arises because neighboring units are more likely to be subjected to similar environmental, i.e. non-treatment, effects. A carry-over effect may lead to bias in estimation of treatment differences. Advice should be sought from a consultant statistician on design and analysis of experiments where either or these possibilities exists.

Experimental design offers some scope for attempting to isolate treatment and non-treatment effects when the treatments are non-randomly applied. If, for example, the aim is to compare performances of males and females where there may be an *age* effect, consider the use of a design which allows comparisons within *age* groups or which uses *age* as a covariate.

Experimental and treatment designs. To correctly identify experimental and treatment design equations for observational studies, a simple strategy is to regard the grouping factor as defining a stratum in the experimental design equation. Thus, if the aim is to compare the performance of males and females in a test and a sample of ten females and ten males are obtained by random selection from each group, the appropriate experimental equation would be

$$y_{ij} = M + G_i + \{\mathcal{B}_{i,j}\} \qquad i = 1,2;\ j = 1,2,\ldots,10$$

where y_{ij} is the response of person j from group i; $M + G_1$ and $M + G_2$ are average responses from female and male groups, respectively; $\mathcal{B}_{i,j}$ is the deviation of the response for person j in group i from the average for group i.

While the two groups are distinguished by the fact that all members of one group are male and all members of the other group are female, there is no guarantee that other factors may not also distinguish the groups.

Thus, if the researcher wishes to consider *sex* as a treatment effect to be investigated, the appropriate treatment design equation would be

$$G_i = S_k + r_k \qquad \text{for } k = 1, 2$$

where S_1 and S_2 are adjustments for sex differences; and r_1 and r_2 are contributions to G_1 and G_2 from other sources.

The fact that S and r have the same subscript is an admission that the two contributions are inseparable. While there is a statistical test to establish if the hypothesis $G_1 = G_2$ is acceptable, there is not a statistical test of the hypothesis $S_1 = S_2$. An experimenter who assumes that rejection of the hypothesis $G_1 = G_2$ implies the rejection of $S_1 = S_2$ must provide non-statistical argument that the difference is unlikely to be due to non-sexual causes.

Observational factors such as *sex* may appear in a design in addition to factors to which units have been randomly allocated. Thus, in the above example, if the ten persons in each group were randomly allocated between two treatments with five persons assigned to each treatment, the treatment design equation at the lower stratum would be

$$\mathcal{B}_{i,j} = T_m + (G.T)_{im} + \varepsilon'_{i,mn} \qquad \text{for } i = 1, 2;\ m = 1, 2;\ n = 1, 2, \ldots, 5.$$

For the same reasons as are given above in respect of the interpretation of G, the interaction between G and T can be interpreted as an interaction between *sex* and *treatment* only where non-statistical argument can support the claim that the group differences are only attributable to sexual differences.

15.2.7 Statistical Models

The presumption in this chapter is that Linear models, as defined in Section 14.2.3, are assumed. The specific features of the models as they apply here, are stated below.

1. *Model equation.* The observed response is presumed to be represented as the **sum** of treatment effects, unit effects, contributions from covariates and unexplained components as described in Sections 15.2.1 to 15.2.6.

2. *Hypotheses.* In the absence of treatment structure, the primary hypothesis included in a model is that all treatment effects are identical, i.e. $T_1 = T_2 = \cdots = T_t$. Variations on this basic hypothesis are considered in Section 15.8.2. Where there is a factorial arrangement of treatments, the hypotheses included in the model are presumed to apply to the effects of the individual factors and their interactions.

3. *Distributional assumptions.* Terms in the model equation which are random variables (random effects) are presumed to be Normally and independently distributed. (The independence condition is determined by the nature of unit selection and treatment allocation.)

4. *Sample selection and treatment allocation.* The manner in which units at each stratum are selected must be stated (either explicitly or implicitly in the statement of the distributional assumptions). The method of treatment allocation at each stratum level must be identified as random or non-random.

15.3 Model and Data Checking

15.3.1 Fitted Values, Residuals and Standardized Residuals

Corresponding to the set of responses which form the data, there can be constructed sets of fitted values, residuals and standardized residuals which provide the basic information for model and data checking. The method of constructing fitted values and residuals is discussed in Section 14.3.1 for the Linear model, although, in practice, these sets of values are generally available as part of the Analysis of variance output from a statistical package. Based on these three sets of values, the checking of data and model assumptions may be undertaken for all forms of design.

15.3.2 Basic Model and Data Checking

1. *Normality.* A Normal probability plot can be constructed using the set of standardized residuals. A linear trend line indicates acceptability of the Normality assumption. (More extensive discussion of the construction and interpretation of the plot is provided in Section 10.3.4.)

Evidence of non-Normality may indicate the need for transformation of the set of responses to a new scale of measurement on which the Normality assumption is reasonable (see Section 15.3.4). Alternatively, a distribution-free method may be required. This possibility is discussed in Section 14.5.

2. *Equality of variance.* The plot of fitted values versus residuals provides a simple visual check on the acceptability of the assumption that the pattern of variation is the same for all treatments (i.e. the assumption of equality of variance). Check that the spread of the residuals does not vary with increasing fitted values. The common form of departure is seen as increased variability in the residuals with increasing fitted values and arises because treatments which produce higher responses also produce greater variability. Typically, this pattern is associated with evidence of non-Normality. It suggests the need for a transformation or a variation in the model (see Section 15.3.4).

3. *Incorrect model equation.* If the correct model equation has been employed there should be no trend in the plot of fitted values versus residuals. In Randomized complete block designs and Repeated measures designs

in which there is no repetition of the treatments in the lower stratum, the usual practice is to assume there is not block x treatment or subject x treatment interaction. Particularly if there are large variation between blocks or between subjects, such an assumption may be unacceptable and may be detected in the plot of fitted values versus residuals by the presence of a curved trend line. Apart from this simple visual check, there are tests for specific forms of interaction which can be applied. The different forms which the interaction may take make general guidelines for detecting and countering this problem difficult to formulate. If there is evidence of non-additivity in the model equation the wisest course of action is to seek assistance from a consultant statistician.

4. *Checking for errors in the data.* An error in the data is indicated by (i) a standardized residual with a magnitude in excess of 2.0 **plus** (ii) a point in the Normal probability plot which is inconsistent with the trend line. If a response is presumed to be in error, model and data checking must recommence on the set of fitted values, residuals and standardized residuals formed from analysis of the data set in which the odd value is treated as a missing value.

> *Note: identification of an error in the data can be difficult in designed experiments because of the restrictions on residuals which requires them to sum to zero within each level of a treatment or factor. If there are only two replications of a treatment or two levels of a design factor, the set of residuals will comprise pairs of numbers with the same magnitude but of opposite sign. In this situation neither the residuals nor the Normal probability plot will identify which member of the pair of residuals is more likely to be in error.*

15.3.3 Checks When There Is a Covariate

> *Note: the following discussion is based on the assumption there is only one covariate. If there is more than one covariate, fit the model containing all covariates and apply the suggested procedures to each covariate in turn.*

There are two dimensions to the procedure of checking the contribution from a covariate. One relates to the nature of the relationship between the response variable and the covariate while the other relates to the advisability of employing the covariate. These are considered in turn.

1. *Relation between response variable and covariate.* The presumption is that the contribution from the covariate is (i) linear and (ii) the line has the same slope for all treatments. These conditions are implicit in the use of the expression $\beta(x_{ij} - \overline{x}_{..})$ for the contribution from the covariate.

Some statistical packages provide a test of the hypothesis that the slope is constant over all treatments. Make use of the test if it is available. A simple alternative is described below which exploits the fact that a non-linear relation or unequal slopes will result in patterns in the signs of the residuals.

Step 1. *Order the values of the covariate from smallest to largest and place the residuals in the same order.*

Step 2. *List the residuals as they are ordered at Step 1 separately for each treatment, i.e.*

$$\begin{array}{ccccc} \textit{Treatment} & & & & \\ 1 & r'_{11} & r'_{12} & \cdots & r'_{1n_1} \\ 2 & r'_{21} & r'_{22} & \cdots & r'_{2n_2} \\ \vdots & & & & \\ t & r'_{t1} & r'_{t2} & \cdots & r'_{tn_t}. \end{array}$$

Step 3. *Examine each row for evidence of patterning in the signs of the residuals. In particular, look within each treatment*

– *for evidence of a pattern of the type*

$$- - - + + + + + - - \quad or \quad + + + - - - - - - + +$$

as evidence of a non-linear relation between the response variable and the covariate;

– *for evidence of a pattern of the type*

$$- - - - + + + + \quad or \quad + + + + - - - -$$

as evidence of unequal slopes.

Illustration. The results are illustrated in Table 15.3.1 using the data from Example 15.8. Examination of the signs of the residuals, when ordered by values of the pre-treatment score within each treatment, reveals no evidence of pattern. Hence, the assumption of linearity and the assumption that the slopes of the line are constant over all treatments appears to be reasonable.

If there is clear evidence of non-linearity or unequal slopes, the justification for the covariate adjustment is lost. The failure of this assumption carries the implication that treatment differences vary with changing values of the covariate. Where the problem is one of a non-linear relation, it may be possible to change the scale of measurement for the covariate to achieve linearity. Where the slopes are unequal, the treatment comparisons cannot be divorced from the value of the covariate. A sensible approach in this situation is to convert the values of the covariate into levels of a factor—*low* and *high*, or *low*, *medium* and *high*, etc.—and to include the factor as part of the treatment design. The interaction between the treatment and covariate factors may then be examined in the course of the analysis. If this is not feasible, consider the possibility of performing separate analyses on responses for low and high levels of the covariate.

Table 15.3.1. Information Used to Examine the Covariate Assumptions Based on Data in Table 15.2.8

Residuals* ordered by values of pre-treatment scores within treatments

Treatment								
1	5.5	8.5	0.5	−2.5	−2.5	4.5	−14.5	0.5
2	−1.3	−3.3	10.8	−1.3	5.8	4.8	−11.3	−4.3
3	5.5	−1.5	−2.5	3.5	2.5	0.5	−8.5	0.5

*residuals based on model equation

$$y_{ij} - x_{ij} = M + T_i + \varepsilon_{ij}$$

2. *Advisability of using the covariate — extrapolation of responses.*

Compute the overall sample mean for the covariate.

Construct separate line diagrams (Section 1.3.5) *for the covariate for every treatment but using the same scale for all diagrams.*

Note if there is an obvious difference in the means and the spread of values of the covariate for the different treatments. Check particularly, if the range of values for all treatments encompasses the overall sample mean.

If there is marked variation in the average values and ranges of values taken by the covariate for the different treatments, the advisability of using the covariate adjustment must be questioned. As explained in Section 15.2.4, the function of the covariate term in the model is to adjust the observed responses to those values which would be expected if all sample members had the same value of the covariate. Since the adjustments rely on fitting relationships between the response variable and the covariate within each treatment, extrapolation of the line outside the range in which it is constructed implies an unverifiable assumption that the form of the relation is unchanging outside the observed range.

3. *Covariates affected by treatments.* A covariate should not be included in the model equation if the values taken by the covariate are affected by the treatments which are being compared. To do so can drastically alter the nature of the problem which is being considered. That this is so can be demonstrated by reference to Figure 15.3.1. This displays the uptake of a nutrient by plants, as a function of water uptake, with each point labelled by the treatment applied–full shading of plants, half shading of plants and

no shading of plants. The results of the experiment are clearly visible in Figure 15.3.1. Both nutrient uptake (response variable) and water uptake (covariate) are increased by reducing the amount of shading of the plants. What would be the effect of adjusting nutrient uptake by including water uptake as a covariate in the model equation for nutrient uptake? The function of the covariate term is to adjust the set of nutrient uptakes to the values which would be expected if the plants all had the same water uptake (which is traditionally taken to be the mean value of the covariate in the sample). Hence, the effect would be to adjust the nutrient uptakes downward for unshaded plants and upward for completely shaded plants. After adjustment, statistical analysis would reveal no differences in nutrient uptake. Does an adjustment which forces shaded plants to take up as much water as unshaded plants make sense? There may be reason for considering the ratio of nutrient uptake to water uptake as a measure of the concentration of the nutrient uptake. However, that is a different question which requires a different model.

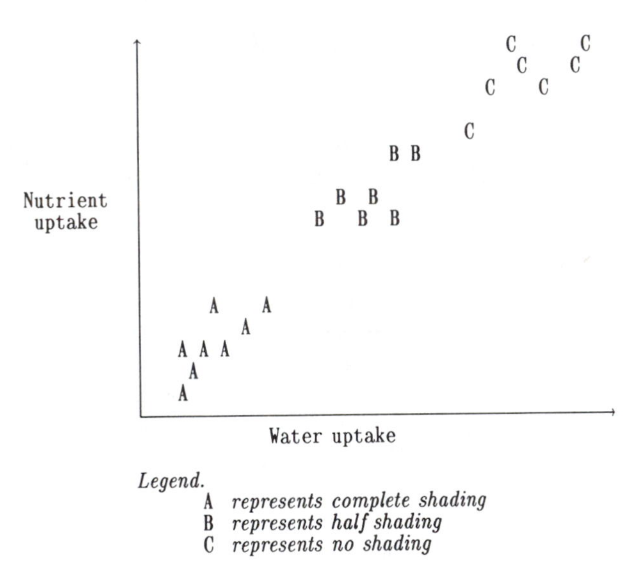

Figure 15.3.1. Nutrient uptake versus water uptake in plants.

15.3.4 TRANSFORMATIONS

The model proposed in Section 15.2.7 which underlies the methods introduced in this chapter, has three important requirements:

- additivity of explained and unexplained components;
- the assumption of Normality for random components; and

– the assumption that the variance of random components is constant between treatments.

There are some readily identifiable situations in which one or more of these conditions will not be met. Two options exist in such cases. The first is to vary the method of analysis to suit the altered conditions in the model. The second is to alter the data by transformation to another scale of measurement on which the conditions are (approximately) met. In pre-computer days, the second option commonly offered the only practical approach. Unfortunately, transformations which are introduced to satisfy statistical, as distinct from scientific, conditions restrict the interpretation of the findings. On the positive side, they offer a broader scope of applications and, for many experimenters, a familiarity not provided by alternative methods.

The principal alternative to transformation of the data is provided by the methodology based on the Generalized linear model and, in particular, on the Log-linear and Logistic models defined in Sections 14.2.4 and 14.2.5, respectively. These provide scientists with scope for constructing models and methods of treatment comparison which are more closely related to their needs. However, the methodology is generally only available for one-stratum designs and requires a significantly higher level of expertise to apply and interpret.

It is not possible to offer detailed guidance as to the optimal choice for a given situation. The decision should be based on the following (often conflicting) considerations.

1. For multi-strata designs it is wise to employ the transformation approach.

2. Where there is interest in quantitative comparisons of treatment effects, as distinct from simply testing for evidence of treatment differences, the use of Generalized linear models is recommended.

3. Where there is a factorial arrangement of treatments, use of Generalized linear models is recommended since transformation to a new scale of measurement alters the interpretation of additivity and interaction.

4. Use of transformations is recommended for experimenters who have a low level of statistical expertise both on grounds of simplicity of application and interpretation and because of the wider availability of support.

Situations in which a transformation may be required. Three situations are described below in which the model conditions defined in Section 15.2.7 are unlikely to be met. In each case, a transformation is suggested. Where there is potential for using an alternative model, the model to be employed is stated.

1. *Multiplicative models and logarithmic transformations.* Where the responses are non-negative and there is large variation in the responses, typified by the data in Table 15.3.2 which arises in Example 15.10, the Linear model proposed in Section 15.2.7 will not provide an adequate description. This fact can be established by examining the Normal probability plot and the plot of fitted values versus residuals.

An alternative exists which satisfies both statistical and experimental demands. It involves the transformation of the set of responses, $\{y_{ij0}\}$, to a logarithmic scale of measurement by conversion of y_{ij0} to $\log\{y_{ij0} + 1\}$. (The addition of one to the response allows for the possibility of a zero reading.) The function of the transformation has been explained in Sections 10.3.8 where it is noted that the logarithmic transformations allows the parameters from the original scale of measurement to be retained. However, the analysis is based on the ratios of treatment means rather than the differences between treatment means.

Table 15.3.2. Amount of Combustible Material Per Plot (Example 15.10)

	Method of Clearing			
Site	Burning	Slashing & clearing	Chemical & burning	Bradley
Construction site	110	102	2193	18
	4915	184	34	93
	493	314	43	86
	221	1626	289	259
Rugged ground	113	416	60	71
	122	53	46	22
	326	40	52	49
	192	70	332	32
Sloping ground	274	336	170	36
	87	615	38	105
	102	36	11	18
	453	37	8	124

Example 15.10 (Methods for Clearing Vegetation). *Many exotic plants introduced into Australia to provide beauty in home gardens have spread into neighboring bush land and have overrun the native vegetation. The prodigious growth of the exotics has provided a serious fire hazard and studies have been undertaken to establish the most effective method of clearing the land. One study involved the selection of equal-sized areas of land at each of three sites with each area of land being divided into 16 equal-sized plots. The plots were randomly allocated between four methods of clearing with four plots assigned to each method. The data, which are*

recorded in Table 15.3.2, comprise the amount of combustible material on the plots twelve months after clearing.

Application. The steps to be employed when a logarithmic transformation is required are listed below.

Step 1 — Transform each response, y_{ij0}, to $\log\{y_{ij0} + 1\}$.

Step 2 — Construct a Normal probability plot and a plot of fitted values versus residuals based on the transformed data and determine if the transformation has been effective, i.e. check that the trend line in the probability plot is linear and the variability in residuals is constant over the range of fitted values. If the transformation has not been successful, seek guidance from a consultant statistician.

Step 3 — Continue the analysis on the transformed scale.

Step 4 — Where there is evidence of treatment differences, retransform the treatment means to the original scale of measurement, i.e. if $\overline{y}'_{i.}$ is the mean on the logarithmic scale, $\overline{y}_{i.} = \exp(\overline{y}'_{i.}) - 1$ is the retransformed mean.

> *Note:* (i) *the retransformed mean is generally regarded as a better measure of the treatment effect than is the arithmetic mean computed on the original scale of measurement because it is less affected by individual large readings which characterize this type of data; and* (ii) *the standard deviation (or standard error) on the logarithmic scale cannot be retransformed to provide a standard deviation on the original scale—remember that the variance (and hence the standard deviation) is not constant on the original scale of measurement.*

2. *Counts of independent events.* The response variable is the number of counts of independent events occurring in a fixed period of time or a fixed geographical or spatial region (see Section 6.3 for examples). In this case the Poisson distribution is likely to apply. If a Linear model based on a Normal distribution is presumed, and the largest count is at least five times the value of the smallest non-zero count, model and data checking is likely to reveal (i) a curved trend line for the Normal probability plot with the largest standardized residual having a value in excess of two; and (ii) increasing variability in residuals with increasing fitted values to reflect the fact that the variance of a Poisson distribution is proportional to the mean. Possible approaches to analysis include

(i) transformation of the set of responses, $\{y_{ij0}\}$, by employing the transformation $\sqrt{y_{ij0}}$ or the transformation $\sqrt{y_{ij0} + \frac{1}{2}}$ if there are zero values in the data; or

(ii) employment of a Log-linear model based on a Poisson distribution (see, for example, McCullagh and Nelder [1983] for details of its application).

If the square root transformation is employed, there is scope for testing the hypothesis that all treatment effects are the same and, if this hypothesis is rejected, for establishing which pairs of treatments are likely to be different. From the viewpoint of statistical analysis, once the data have been transformed, the processes of model and data checking, hypothesis testing and interpretation apply as they would to untransformed data. When presenting results of the analysis, clearly state the transformation which has been employed. If tables of means are required in the original scale of measurement these should be reported in addition to the means on the transformed scale. Since the transformation stabilizes variance, it is reasonable to present a single value for the standard deviation on the transformed scale. It is not possible to provide a single value for the standard deviation on the original scale since variance is expected to increase in proportion to the mean.

3. *Proportions and the arcsine (angular) transformation.* When the responses comprise the number of successes, $\{x_{ij0}\}$, in a fixed number of trials (n), i.e. the Binomial model (Section 6.1) applies and the variable x_{ij} has a Binomial distribution. If π_i is the probability of success under treatment i, the mean of x_{ij} is $n\pi_i$ and the variance is $n\pi_i(1-\pi_i)$.

The comparison of treatments may be defined in terms of the mean number of successes in n trials or the expected proportion of successes in n trials.

When the value of π_i is close to zero or one, the assumptions of Normality is inadequate and where π_i exhibits wide variation, the assumption of equality of variance is unacceptable. Statistical theory establishes that, after a transformation in which the proportion of successes, $p_{ij} = x_{ij}/n$ is transformed to the value of the angle whose sine is the square root of p_{ij}, the conditions of Normality and equality of variance are reasonable. The transformation is known as the *arcsine* or *angular* transformation and is typically available as a standard option in statistical packages.

The alternative to transforming the data is to employ either the Log-linear or Logistic model. Application of the Log-linear model is described in Chapter 16. The Log-linear model should be employed if interest lies in the estimation of ratios of success rates while the Logistic model should be employed if interest lies in the estimation of the ratios of odds of success.

The steps in analysis when the arcsine transformation is to be applied are stated below. They are based on the assumption that the number of trials, n, is constant for all sample members. Where this requirement is not met, analysis based on the transformed data requires the introduction of a weighting variable since the arcsine transformation does not stabilize the variance. A simpler alternative may be to employ either the Log-linear or

Logistic model since these impose no conditions on the relative magnitudes on n.

Step 1 — Compute the average proportion of successes for each treatment. If all means lie in the range 0.3 to 0.7, no transformation is required since there is little variation in variances and the Normal distribution provides an adequate approximation to the Binomial. In this case, the methodology may be applied directly to the set of observed proportions.

Otherwise proceed to Step 2.

Step 2 — If n is less than 50, the following adjustment should be made before transformation:

- any proportion which is zero should be changed to $1/4n$
- any proportion which is one should be changed to $1 - 1/4n$.

Step 3 — Transform each proportion, p_{ij0}, to $\arcsin(\sqrt{p_{ij0}})$.

Step 4 — Analyze the transformed data based on the appropriate model equation (Section 15.2) using the Analysis of variance technique (Sections 15.4 to 15.7).

If there is evidence of treatment differences, pairwise comparisons may be based on the procedure described in Section 15.8.1.

15.4 Analysis of Variance and One-Stratum Designs

The technique of Analysis of variance, introduced in Sections 14.1.3 and 14.3.4, provides the basis for statistical comparisons of treatment effects. Its application is described in this section for one-stratum designs in which there is no treatment structure. The application of the technique to more complex designs in later sections is merely an extension of this basic analysis.

15.4.1 Statistical Models and Data

The design is a one-stratum design with no treatment structure. The model equation for the most general model assumed is given by (15.4.1). Other components of the model are the standard assumptions defined in Section 15.2.7.

$$y_{ij} = M + T_i + \varepsilon_{ij} \qquad \text{for } i = 1, 2, \ldots, t;\ j = 1, 2, \ldots, n \tag{15.4.1}$$

where

y_{ij} is the response from the jth unit receiving the ith treatment;

M is the overall mean;

T_i is the difference between the mean for the ith treatment and the overall mean; and

ε_{ij} is the contribution from unidentified sources to the response of the jth unit receiving the ith treatment.

If there are no treatment differences, $T_1 = T_2 = \cdots = T_t = 0$ and the appropriate model equation is given by (15.4.2).

$$y_{ij} = M + \varepsilon_{ij} \qquad \text{for } i = 1, 2, \ldots, t;\ j = 1, 2, \ldots, n. \tag{15.4.2}$$

The data are provided by a set of responses $\{y_{ij0}\}$. The set of differences in Table 15.4.1(a) provides an illustration.

Table 15.4.1. Observed and Fitted Values Based on Data and Models Employed in Example 15.8

Post-treatment score — pre-treatment score

(a) Observed differences

Treatment								
A	5	5	13	2	9	−10	2	10
B	10	7	8	16	0	10	22	17
C	−8	−1	1	−2	3	6	4	1

(b) Fitted values[1] under model equation $y_{ij} = M + \varepsilon_{ij}$

Treatment								
A	5.42	5.42	5.42	5.42	5.42	5.42	5.42	5.42
B	5.42	5.42	5.42	5.42	5.42	5.42	5.42	5.42
C	5.42	5.42	5.42	5.42	5.42	5.42	5.42	5.42

(c) Fitted values[2] under model equation $y_{ij} = M + T_i + \varepsilon_{ij}$

Treatment								
A	4.5	4.5	4.5	4.5	4.5	4.5	4.5	4.5
B	11.25	11.25	11.25	11.25	11.25	11.25	11.25	11.25
C	0.5	0.5	0.5	0.5	0.5	0.5	0.5	0.5

Notes.
[1]computed as overall sample mean
[2]computed as sample means within treatments.

15.4.2 Statistical Aims and Methods

The statistical task is to fit the two models distinguished by model equations (15.4.1) and (15.4.2) to the data and to establish if the model which allows for treatment differences provides a significantly better fit to the data. If it does, there is evidence of treatment differences.

Fitted values. The process of fitting the models to the data is described in Section 14.3.1 and involves the construction of two sets of fitted values,

(i) the set $\{\hat{y}^0_{ij0}\}$ based on the model equation (15.4.2) under the assumption of no treatment differences; and

(ii) the set $\{\hat{y}^1_{ij0}\}$ based on the model equation (15.4.1) under the assumption that there may be treatment differences.

Since (15.4.2) includes no identified factor to explain variation between responses, the fitted values in the set $\{\hat{y}^0_{ij0}\}$ must all have the same value. Statistical theory (see Section 14.3.1) establishes that value to be the sample estimate of M, namely $\overline{y}_{..0}$, i.e. $\hat{y}^0_{ij0} = \overline{y}_{..0}$ for all pairs (i, j). Table 15.4.1(b) provides the set of fitted values under the model based on (15.4.2). If the model equation is provided by (15.4.1), treatment differences are identified but there is no explained variation between responses taken from units which experienced the same treatment. The value of $\hat{y}^1_{ij0}$ is $\overline{y}_{i.0}$ for all pairs (i, j). This is the mean response for treatment i. Table 15.4.1(c) provides the fitted values for the data in Table 15.4.1(a) under the assumption that there may be treatment differences.

Comparing models and data. Having determined the two sets of fitted values, it only remains to determine if the fit under the more general model is significantly better than the fit of the restricted model to the data. The statistical process is based on the *Sum of squares* statistics defined by (14.3.12) which, for the one-stratum model, are given by (15.4.3) and (15.4.4).

$$TSS = \sum_{i=1}^{t}\sum_{j=1}^{n}(y_{ij} - \overline{y}_{..})^2. \tag{15.4.3}$$

The statistic TSS, which is the residual sum of squares under the model which assumes no treatment differences, is traditionally termed the *Total sum of squares* since it represents all the variation in the data set.

$$WSS = \sum_{i=1}^{t}\sum_{j=1}^{n}(y_{ij} - \overline{y}_{i.})^2. \tag{15.4.4}$$

The statistic WSS, which is the residual sum of squares under the model which allows for the possibility of treatment differences, is termed the *Within treatment sum of squares.*

The difference between these sum of squares, $TSS - WSS$, can be defined by (15.4.5) and is the *Between treatment sum of squares.* It represents the gain from including a term in the model equation which allows for treatment differences.

$$BSS = \sum_{i=1}^{t}\sum_{j=1}^{n}(\bar{y}_{i.} - \bar{y}_{..})^2 = \sum_{i=1}^{t} n(\bar{y}_{i.} - \bar{y}_{..})^2. \quad (15.4.5)$$

Sampling variability. If there are no differences in treatment effects, i.e. $T_1 = T_2 = \cdots = T_t$, the Between treatment sum of squares represents only sampling or *random* variation. If there are differences in treatment effects, the Between treatment sum of squares is the sum of two positive terms—one representing sampling variation and the other representing differences in treatment effects. Which of these situations applies cannot be determined by considering the Between treatment sum of squares alone. It is necessary to have an independent measure of the magnitude of sampling variability against which the Between treatment sum of squares can be measured.

The sampling or random variability is measured by the variance of the unexplained component (ε_{ij}) in the model equation and is traditionally denoted by σ^2 in the statistical model. The *Residual mean square,* which is represented by (15.4.6) for a one-stratum design, is an unbiased estimator of σ^2 irrespective of whether or not there are differences in treatment effects (see (14.3.15) and accompanying properties).

$$RMS = WSS/d_r = \sum_{i=1}^{t}\sum_{j=1}^{n}(y_{ij} - \bar{y}_{i.})^2/t(n-1) \quad (15.4.6)$$

where WSS is the Within treatment sum of squares defined in (15.4.4); and $d_r = t(n-1)$ is the *Residual degrees of freedom.*

Application of (14.3.15) and the properties which follow, establishes that the *Between treatment mean square,* BMS, which is equal to $BSS/(t-1)$, also provides an unbiased estimate of σ^2 if there are no treatment differences. However, in the event that there are treatment differences, the Between treatment mean square has an expected mean square in excess of σ^2. Hence, as the Between treatment mean square becomes increasingly greater than the Residual mean square, there is growing evidence to support the claim of differences in treatment effects.

F-statistic. The statistic $F = BMS/RMS$, which is a representation of (14.3.16), provides the formal basis of a test of the hypothesis $H{:}T_1 = T_2 = \cdots = T_t = 0$. Increasing values of F reflect lessening support for the hypothesis. The procedure is formalized in Section 15.4.3 under the name *Analysis of variance.*

15.4.3 The Analysis of Variance Table

The information required to construct the F-statistic is traditionally summarized in tabular form. For a one-stratum design with no treatment structure, Table 15.4.2 displays the form of the Analysis of variance table based on (15.4.1). Note that each row of the table corresponds to a term in the model equation which identifies variation in the data, i.e.

- the *Between treatments* row corresponds to the effect T_i;
- the *Within treatments* (or *Residual*) row corresponds to ε_{ij}; and
- the *Total* row corresponds to the response y_{ij}.

The p-value is defined and discussed in Section 15.4.4.

Table 15.4.2. Analysis of Variance Table for a One-Stratum Design in Which There Is No Treatment Structure

(a) General layout

Source of Variation	D.F.	S.S.	M.S.	F	p
Between treatments	$t-1$	BSS	$BMS = \frac{BSS}{t-1}$	$F_0 = \frac{BMS}{RMS}$	p_0
Within treatments (Residual)	$d_r = t(n-1)$	WSS	$RMS = \frac{WSS}{d_r}$		
Total	$n-1$	TSS			

(b) Based on analysis of data in Table 15.4.1

Source of Variation	D.F.	S.S.	M.S.	F	p
Between treatments	2	472.3	236.2	6.2	<0.01
Within treatments (Residual)	21	805.5	38.36		
Total	23	1277.8			

Legend.

D.F. — is degrees of freedom
S.S. — is sum of squares
M.S. — is mean square
F — is F-statistic
p — is p-value.

15.4.4 Interpreting the Analysis of Variance Table

Statistic. The statistic F in the Analysis of variance table has the property that increasing values reflect lessening support for the hypothesis that all treatment effects are the same. If the hypothesis $T_1 = T_2 = \cdots T_t$ is correct, the sampling distribution of F is an $F(t-1, d_r)$ distribution, where d_r is the Residual degrees of freedom from the Analysis of variance table.

Measure of agreement. If F_0 is the value of F computed from the data, the p-value for comparing model and data is $p_0 = \Pr\{F(t-1, d_r) > F_0\}$. Using the conventional interpetation,

- a value of p_0 in excess of 0.05 is consistent with the hypothesis that all treatment effects are the same;
- a value of p_0 less than 0.01 provides strong evidence that at least one pair of treatments are having different effects; and
- a value of p_0 between 0.05 and 0.01 provides weak evidence against the hypothesis that all treatment effects are the same.

Note that rejection of the hypothesis that all treatments have the same effect is generally only the first step in analysis. Subsequently, there is a need to establish which pairs of treatments appear to be producing different effects, a task considered in Section 15.8.1.

Computations. Statistical packages are widely available for the production of the Analysis of variance table and the computation of the p-value. Alternatively, formulae for computing the Total sums of squares, the Residual (Within treatments) sum of squares and the Between treatment sum of squares are provided in (15.4.3), (15.4.4) and (15.4.5). The formulae in Table 15.4.2 (a) may then be employed to compute the values required for the completion of the Analysis of variance table. The p-value can be obtained by the use of one of the methods described in Section 7.6.4.

Illustration. Table 15.4.2(b) contains the computations from the analysis of data in Table 15.4.1(a). Note that the p-value is less than 0.01 indicating there is strong evidence for rejecting the hypothesis that the treatment effects are the same. The matter of establishing which pairs of treatments are having different effects may be undertaken using the methods defined in Section 15.8.1.

15.5 Two-Strata Designs

In this section two strata designs are introduced in which

- there is no treatment structure;

- the application of treatments occurs in one stratum only;
- the number of sub-units is the same for all units; and
- every treatment is applied to the same number of units or sub-units.

Designs with factorial arrangements of treatments are considered in Section 15.7. Other treatment structure is considered in Section 15.8.2. More complex experimental designs are considered in Section 15.6.

The Analysis of variance procedure which is employed is a simple extension of that described in Section 15.4 for one-stratum designs.

15.5.1 Designs with Sub-Sampling

A simple two-stratum design is that in which the treatments are applied in the first stratum and the second stratum arises from sub-sampling or repeatedly sampling the units to which the treatments are applied. The model equation is defined by (15.5.1).

$$y_{ijk} = M + T_i + \varepsilon_{ij} + \{\varepsilon'_{ij,k}\} \quad \text{for } i = 1,2,\ldots,t; \quad j = 1,2,\ldots,r; \; k = 1,2,\ldots,n. \tag{15.5.1}$$

where ε and ε' are the non-treatment components in the unit and sub-unit strata, respectively.

Example 15.6 provides an illustration in which 18 children provide the primary units to be allocated among nine treatments (environments). The sub-sampling arises because each child is sampled twice.

The Analysis of variance table for this model is presented in Table 15.5.1(a). Note the following properties:

(i) the table is split into two sections corresponding to the two strata defined in the model equation;

(ii) the rows in the table correspond to the terms in the model equation which identify the sources of variation; and

(iii) the first stratum has the same model equation and the same analysis of variance structure as the one-stratum design introduced in Section 15.4.

The term ε_{ij} is a statistic which represents the contributions from factors which vary **between** units but are constant within units. The variance of ε_{ij}, denoted by σ_u^2, is a measure of the variability contributed by these factors to the variability in the response variability. (In Example 15.6, this is represented by the variation between children who are in the same environment.) The term $\varepsilon'_{ij,k}$ represents the contributions from factors which vary between sub-units **within** the same unit. The variance of $\varepsilon'_{ij,k}$ is denoted by σ^2. (In Example 15.6, this is represented by the variation between

Table 15.5.1. Analysis of Variance Table for a Two-Strata Design with Subsampling

(a) General formulation

Source of Variation	D.F.	S.S.	M.S.	F	
Between units stratum					
Between treatments	$t-1$	BSS	$BMS = \frac{BSS}{t-1}$	$F_1 = \frac{BMS}{s_b^2}$	p
Within treatments (Residual)	$t(r-1)$	RSS	$s_b^2 = \frac{RSS}{t(r-1)}$	$F_2 = \frac{s_b^2}{s^2}$	p
Between subunits within units stratum					
Between subunits	$tr(n-1)$	SSS	$s^2 = \frac{SSS}{tr(n-1)}$		
Total	$trn-1$	TSS			

$$BSS = rn\sum_{i=1}^{t}(\bar{y}_{i..} - \bar{y}_{...})^2 \qquad RSS = n\sum_{i=1}^{t}\sum_{j=1}^{r}(\bar{y}_{ij.} - \bar{y}_{i..})^2$$

$$TSS = \sum_{i=1}^{t}\sum_{j=1}^{r}\sum_{k=1}^{n}(y_{ijk} - \bar{y}_{...})^2 \quad SSS = \sum_{i=1}^{t}\sum_{j=1}^{r}\sum_{k=1}^{n}(y_{ijk} - \bar{y}_{ij.})^2$$

$p_{01} = \Pr\{F(t-1, t[r-1]) \geq F_{01}\}$ where F_{01} is the computed value of F_1
$p_{02} = \Pr\{F(t[r01], tr[n-1]) \geq F_{02}\}$ where F_{02} is the computed value of F_2

(b) Expected values for mean squares

Source of Variation	D.F.	M.S.	E(M.S.)
Between units stratum			
Between treatments	$t-1$	$BMS = \frac{BSS}{t-1}$	$\sigma^2 + n\sigma_u^2 + nr\sum_{i=1}^{t} T_i^2/(t-1)$
Within treatments (Residual)	$t(r-1)$	$s_b^2 = \frac{RSS}{t(r-1)}$	$\sigma^2 + n\sigma_u^2$
Between subunits within units stratum			
Between subunits	$tr(n-1)$	$s^2 = \frac{SSS}{tr(n-1)}$	σ^2

repeated samplings for the one child.) Table 15.5.1(b) records the expected values of mean squares. Note that:

(i) the ratio $F_1 = BMS/s_b^2$ has the same expectation for numerator and denominator if there are no treatment differences, i.e. $T_1 = T_2 = \cdots = T_t = 0$, whereas the numerator has the greater expectation if there are treatment differences. Hence, F_1 can be sensibly employed to test the hypothesis of no treatment differences by presuming that larger values of F_1 provide decreasing support for the hypothesis of no treatment differences.

(ii) The ratio $F_2 = s_b^2/s^2$ provides a means of testing the hypothesis $\sigma_u^2 = 0$. More usefully, the statistic $s_u^2 = (s_b^2 - s^2)/n$ provides an estimate of σ_u^2 which can be compared with the estimate of σ^2 provided by s^2. In Section 19.2.7, there is a consideration of the application of this information in the design of experiments.

Computation. Formulae are given in Table 15.5.1(a) for the statistics employed in the construction of the Analysis of variance table. Section 7.6.4 provides the means of converting the F-values F_{01} and F_{02} into p-values p_{01} and p_{02}, respectively.

Illustration. The Analysis of variance table for Example 15.6 is presented in Table 15.5.2. The test of the hypothesis that the children respond in the same way under all environments (treatments) is based on the value of p_{01} which is greater than 0.1. This result provides insufficient evidence to reject the hypothesis that average scores (number of movements between boxes) varies between environments.

Table 15.5.2. Analysis of Variance Table for Example 15.6

Source of Variation	D.F.	S.S.	M.S.	F	p
Between children stratum					
Between environments	8	311	38.87	0.9	>0.1
Within environments (Residual)	9	375	41.67	3.4	0.01
Between times within children stratum					
Between times	18	219	12.17		

15.5.2 Randomized Complete Block Designs

Consider a design in which each of r primary units (blocks) is partitioned into t secondary units (plots) and the t plots within each block are randomly allocated between t different treatments. Thus, each treatment appears once, and only once, in each block. The design is called a randomized complete block design. The model equation is given in (15.5.2).

$$y_{ij} = M + B_i + \{T_j + \varepsilon_{i,j}\} \qquad \text{for } i = 1, 2, \ldots, r; \; j = 1, 2, \ldots, t. \quad (15.5.2)$$

Table 15.5.3. Analysis of Variance Table for a Randomized Complete Block Design

(a) General formulation

Source of Variation	D.F.	S.S.	M.S.	F	p
Between blocks stratum					
Between blocks	$r-1$	BSS	$BMS=\frac{BSS}{r-1}$	$F_1=\frac{BMS}{s^2}$	p_{01}
Between plots within blocks stratum					
Between treatments	$t-1$	TSS	$TMS=\frac{TSS}{t-1}$	$F_2=\frac{TMS}{s^2}$	p_{02}
Residual	$(r-1)(t-1)$	RSS	$s^2=\frac{RSS}{(r-1)(t-1)}$		

$$BSS = t\sum_{i=1}^{r}(\bar{y}_{i.} - \bar{y}_{..})^2 \qquad TSS = r\sum_{j=1}^{t}(\bar{y}_{.j} - \bar{y}_{..})^2$$

$$RSS = \sum_{i=1}^{r}\sum_{j=1}^{t}(y_{ij} - \bar{y}_{i.} - \bar{y}_{.j} + \bar{y}_{..})^2$$

$p_{01} = \Pr\{F(r-1, [r-1][t-1]) \geq F_{01}\}$ where F_{01} is the computed value of F_1

$p_{02} = \Pr\{F(t-1, [r-1][t-1]) \geq F_{02}\}$ where F_{02} is the computed value of F_2

(b) Expected values of mean squares

Source of Variation	D.F.	M.S.	Expected Mean Square
Between blocks stratum			
Between blocks	$r-1$	$BMS=\frac{BSS}{r-1}$	$\sigma^2+t\sum_{i=1}^{r} B_i^2/(r-1)$
Between plots within blocks stratum			
Between treatments	$t-1$	$TMS=\frac{TSS}{t-1}$	$\sigma^2+r\sum_{j=1}^{t} T_j^2/(r-1)$
Residual	$(r-1)(t-1)$	$s^2=\frac{RSS}{(r-1)(t-1)}$	σ^2

Example 15.3 provides an example of a randomized complete block design with 12 treatments. (For the present, the factorial nature of the treatments is ignored.)

The general form of the Analysis of variance table is presented in Table 15.5.3(a) and is illustrated in Table 15.5.4 using the data in Example 15.3.

Table 15.5.4. Analysis of Variance Table for Example 15.3

Source of Variation	D.F.	S.S.	M.S.	F	p
Between block stratum					
Between blocks	2	6414	3207.3	$F_1 = 22$	<0.01
Between plots within blocks stratum					
Between treatments	11	12683	1153.1	$F_2 = 7.9$	<0.01
Residual	22	3220	146.4		
Total	35	22318			

Since the second stratum of the Analysis of variance table has the same structure as that for the one-stratum design described in Section 15.4.3, a test of the hypothesis may be based on the statistic F_2 and the p-value p_{02} following the guidelines offered in Section 15.4.4. The table of expected values of mean squares provided in Table 15.5.3(b) provides the justification for using the statistic F_2. If there is evidence of treatment differences, the methodology of Section 15.8.1 may be employed to identify the likely nature of the differences.

In general, the block effect is a fixed effect. Subject to the limitations on interpretation as explained in Section 15.2.5, the statistic F_1 and associated p-value, p_{01}, may be used to establish if there is evidence of differences between blocks. This information may be useful in determining if the design has been effective in removing a significant source of non-treatment variation.

15.5.3 Repeated Measures Designs

Designs are considered in which a sample of subjects is selected for a study and every subject has at least one measured response to each of the treatments which are being compared. Example 15.4 is a study in which there is only one measurement for each treatment per subject. Example 15.5 is a study in which there are three measurements made for each treatment per subject.

1. *One application of each treatment per subject.* The model equation is defined by (15.5.3) and has a different representation depending on whether

the subject effect is fixed (S_i) or random ($\mathcal{S}_i$).

$$\begin{aligned} y_{ij} &= M + S_i + \{T_j + \varepsilon_{i,j}\} \\ y_{ij} &= M + \mathcal{S}_i + \{T_j + \varepsilon_{i,j}\} \end{aligned} \quad \text{for } i = 1,2,\ldots,r;\ j = 1,2,\ldots,t. \qquad (15.5.3)$$

Illustration. Example 15.4.

Where (15.5.3) is the model equation, the method of analysis is the same as that described in Section 15.5.2 for a Randomized complete block design, i.e. the Analysis of variance table is provided by Table 15.5.3(a). An illustration for a repeated measures experiment is provided in Table 15.5.5. The distinction between a fixed or random subject effect has no bearing on the test of the hypothesis that all treatments have the same effect. It does, however, affect the scope of the conclusion for the reasons given in Section 15.2.5. Additionally, when the subjects are randomly selected, there is scope for comparing the relative magnitudes of between subject and within subject variability, a topic considered in Section 19.2.7.

Table 15.5.5. Analysis of Variance Table for Example 15.4 Based on (15.5.3)

Source of Variation	D.F.	S.S.	M.S.	F	p
Between subjects stratum					
Between subjects	35	230.07	6.574	$F_1 = 5.8$	<0.01
Between times within subjects stratum					
Between themes	2	46.46	23.23	$F_2 = 20$	<0.01
Residual	70	79.54	1.136		

2. *n applications of each treatment per subject.* If each subject has t treatments randomly applied on nt occasions such that each treatment is applied n times, the model equation is defined by (15.5.4). There are two representations depending on whether the subject effect is fixed or random.

$$\begin{aligned} y_{ij} &= M + S_i + \{T_j + (S.T)_{ij} + \varepsilon_{i,jk}\} \\ y_{ij} &= M + \mathcal{S}_i + \{T_j + (\mathcal{S}.T)_{ij} + \varepsilon_{i,jk}\} \end{aligned} \qquad (15.5.4)$$

$$\text{for } i = 1,2,\ldots,r;\ j = 1,2,\ldots,t;\ k = 1,2,\ldots n.$$

Illustration. Example 15.5.

The layout for the Analysis of variance table based on (15.5.4) is given in Table 15.5.6 and illustrated in Table 15.5.7 (on the assumption the subject effect is a random effect).

Table 15.5.6. Analysis of Variance Table for Repeated Measures Design with Repetition

Source of Variation	D.F.	S.S.	M.S.	F	p
Between subjects stratum					
Between subjects	$r-1$	BSS	$BMS=\frac{BSS}{r-1}$	$F_1=\frac{BMS}{s^2}$	p_{01}
Between times within subjects stratum					
Between treatments	$t-1$	TSS	$TMS=\frac{TSS}{t-1}$	F_2 *	p_{02}
Subjects × treatments	$(r-1)(t-1)$	ISS	IMS = $\frac{ISS}{(r-1)(t-1)}$	$F_3=\frac{IMS}{s^2}$	p_{03}
Residual	$rt(n-1)$	RSS	$s^2=\frac{RSS}{rt(n-1)}$		

*— if the subject effect is fixed, $F_2 = TMS/s^2$ and
$p_{02} = \Pr\{F(t-1, rt[n-1]) \geq F_{02}\}$ where F_{02} is the computed value of F_2

— if the subject effect is random, $F_2 = TMS/IMS$ and
$p_{02} = \Pr\{F(t-1, [r-1][t-1]) \geq F_{02}\}$ where F_{02} is the computed value of F_2

$$BSS = tn\sum_{i=1}^{r}(\overline{y}_{i..} - \overline{y}_{...})^2 \qquad TSS = rn\sum_{j=1}^{t}(\overline{y}_{.j.} - \overline{y}_{...})^2$$

$$ISS = n\sum_{i=1}^{r}\sum_{j=1}^{t}(\overline{y}_{ij.} - \overline{y}_{i...} - \overline{y}_{.j.} + \overline{y}_{...})^2 \quad RSS = \sum_{i=1}^{r}\sum_{j=1}^{t}\sum_{k=1}^{n}(y_{ijk} - \overline{y}_{ij.})^2$$

$p_{01} = \Pr\{F(r-1, rt(n-1)) \geq F_{01}\}$ where F_{01} is the computed value of F_1
$p_{03} = \Pr\{F([r-1][t-1], rt[n-1]) \geq F_{03}\}$ where F_{03} is the computed value of F_2

Fixed or random subject effects. When each subject has treatments repeatedly applied, the form of the statistical analysis is influenced by the nature of the subject effect. If it is a random effect, the test of the hypothesis that treatment means are equal is based on the statistic F_2 defined as TMS/IMS in Table 15.5.6, whereas, if the subject effect is fixed, it is based on the statistic F_2 defined as TMS/s^2 in Table 15.5.6. The reason for the two forms of the F-statistic lies in the different meanings for the treatment × subject interaction effect. (See Winer [1971] for amplification.) The table of expected values of mean squares presented in Table 15.5.8 suggests why the particular forms of F_2 have been selected.

Table 15.5.7. Analysis of Variance Table for Example 15.5

Source of Variation	D.F.	S.S.	M.S.	F	p
Between subjects stratum					
Between subjects	2	20838	10419	$F_1 = 54$	<0.01
Between times within subjects stratum					
Between exercises	9	4635	515	$F_2 = 2.7^*$	0.04
Subjects × exercises	18	3467	192.6	$F_3 = 1.6$	0.09
Residual	60	7251	120.8		

$^*F_2$ is computed on the assumption that the subjects were randomly selected.

Table 15.5.8. Expected Mean Squares for a Repeated Measures Design with Repetition of Treatments Within Subjects

		Expected Mean Square	
Source of Variation	D.F.	Subject effect fixed	Subject effect random
Between subjects stratum			
Between subjects	$r-1$	$\sigma^2+tn\sum_{i=1}^{r} S_i^2/(r-1)$	$\sigma^2+tn\sigma_u^2$
Between times within subjects stratum			
Between treatments	$t-1$	$\sigma^2+rn\sum_{j=1}^{t} T_j^2/(t-1)$	$\sigma^2+n\sigma_I^2$
Subjects × treatment	$(r-1)(t-1)$	$\sigma^2+\frac{\sum_{i=1}^{r}\sum_{j=1}^{t}(S.T)^2}{(r-1)(t-1)}$	$\sigma^2+n\sigma_I^2$
Residual	$rt(n-1)$	σ^2	σ^2

Interpreting the Analysis of variance table.

1. *Is there evidence of treatment differences?*

   The answer to this question may be based on p_{02} in Table 15.5.6 using the interpretation provided in Section 15.4.4.

2. *Is there evidence of large subject-to-subject variability?*

   In general, the repeated measures design is chosen because there is an expectation of large subject-to-subject variation. Formally, the value of p_{01} in Table 15.5.6 may be used to test whether there is evidence of between subject variability. If the subjects were randomly selected, a potentially more useful approach is to compare the magnitudes of between and within subject variability using the methods described in Section 19.2.7.

3. *Is there evidence that treatment differences vary between subjects?*

   If p_{03} in Table 15.5.6 has a value below 0.01 this may be taken as strong evidence that the relative effects of the treatments vary from one subject to another. If the subjects were randomly selected, this result would imply that the answer to question 1 relates to average treatment differences over all population members. If the subjects were non-randomly selected, this results suggests that the same conclusions may not be reproduced with another set of subjects. The importance of a significant subject × treatment interaction depends on the nature of the interaction, a topic considered in Section 15.7.1.

Illustration. The analysis of the data collected in Example 15.5 (on the assumption that the athletes (subjects) were randomly selected) is summarized in Table 15.5.7. It provides weak evidence that average exertion levels vary between exercises since p_{02} is less than 0.05 but not less than 0.01. The fact that p_{03} is greater than 0.05 provides support for the hypothesis that differences in exertion levels between exercises are consistent over all athletes.

There is strong evidence of variability between athletes in exertion level as established by p_{01} having a value less than 0.01.

15.6 More Complex Designs

Complexity may be introduced into the design in a number of ways. Among the more important are

- the use of one or more covariates;
- the addition of more strata;
- unequal representation of treatments within strata;
- unequal representation of treatments within blocks or subjects; and
- the use of non-hierarchical designs.

These are considered in turn.

15.6.1 Analysis of Covariance

If one or more covariates have been employed in the design, this fact is reflected by the presence of one or more covariate terms in the model equation. The term *Analysis of covariance* is employed to describe the application of the Analysis of variance procedure in such cases. Within each stratum in the model in which a covariate term appears, there is a row in the Analysis of variance table which can be used to establish whether the inclusion of the covariate(s) has been useful in reducing the non-treatment component of the variation in the response variable.

The computation of sums of squares for inclusion in the Analysis of variance table is more complex when there is a covariate present and the relevant computational formulae are not provided in this book. Those statistical packages which provide a comprehensive Analysis of variance package generally also offer the option of covariance analysis.

Table 15.6.1 provides an illustration of an Analysis of variance table for a one-stratum design which includes a covariate.

Table 15.6.1. Analysis of Covariance Based on Data in Table 15.2.9

(Ignoring the factorial arrangement of the treatments)

Source of Variation	D.F.	S.S.	M.S.	F	p
Between treatments	3	1155.25	385.08	38	<0.01
Covariate	1	1646.49	1646.49	163	<0.01
Residual	55	555.24	10.10		

Interpretation of an Analysis of covariate table. A test of the hypothesis that all treatments are equal is based on the p-value in the *Between treatments* row of the table. Interpretation of the p-value follows the same rules as apply in the absence of a covariate (see Section 15.4.4).

If there is a single covariate, the p-value in the *Covariate* row of the table formally tests the hypothesis that coefficient β in the term $\beta(x_{ij} - \overline{x}_{..})$ is zero. Where the hypothesis $\beta = 0$ is acceptable, the conclusion may be drawn that the covariate term is making no useful contribution and the analysis should be repeated without the covariate term included in the model equation. Some packages provide more specific measures of the usefulness of the covariate and reference should be made to accompanying manuals to establish the manner in which these measures should be interpreted.

Where two or more covariates are employed in the model equation, their joint contribution is usually represented in a single row of the Analysis of covariance table. In this case the p-value can be used to establish if at least one of the covariates is making a useful contribution. There is the possibility of testing whether one or more of the covariates can be removed without loss of precision in the treatment comparison. A statistician could offer advise on how this can be effected.

15.6.2 Designs with Three or More Strata

The manner of describing designs introduced in Section 15.2 ensures that addition of further strata does not increase the difficulty of construction of Analysis of variance tables. Just as the equations describing the designs are constructed sequentially—adding one term at a time—so the Analysis of variance tables are constructed from the model equation by adding one row at a time. With the exception of the values of the F-statistics, the presumption is that the numerical components of the Analysis of variance table can be obtained from a reputable statistical package. Experience suggests that the correct denominators for the F-statistics may not always be forthcoming. However, these may be simply established from the model equation by following the steps given in Section 15.6.6.

15.6.3 Unequal Representation of Treatments Within Strata

The effect of not representing treatments equally within strata is to produce what is termed an *unbalanced design.* There are two important consequences,

- treatment effects are not all estimated with the same precision; and
- in all but the simplest designs, the effects of the different factor appearing in the model equation cannot be completely separated—in statistical jargon they are said to be *confounded.*

From a computational viewpoint, a more complex computing algorithm is required for the computation of sums of squares when there is confounding of two or more effects in the model equation. Not all statistical packages which provide the means of constructing Analysis of variance tables will handle unbalanced designs. Others which will handle unbalanced designs require the user to define the model in what is termed a *regression format* using *indicator variables* (see Section 18.7.2).

Interpretation is made more difficult by the presence of confounding since it becomes impossible to establish if an apparent treatment difference is a real difference in treatment effects or is due to the inseparability of the treatment effect of interest from another effect.

Unequal representation of treatments within a stratum is sometimes done by design and sometimes is outside the control of the experimenter. An example of the former arises when a standard treatment is to be compared with a batch of new treatments. Since the standard is to be used in all comparisons while each of the new treatment is to be used only in comparison with the standard, there are strong grounds for having greater replication of the standard.

The analysis and interpretation of unbalanced designs is not discussed in this book. Winer [1971] provides a comprehensive coverage with an emphasis on studies in the social sciences and Cochran and Cox [1957] provides a coverage oriented towards the biological sciences.

15.6.4 Unequal Representation of Treatments Within Blocks or Subjects

In a multi-strata design, it is possible for the treatments to be applied in the second or in a lower stratum in a manner in which every treatment is represented the same number of times in the stratum, but with treatments not being equally represented in all of the units or subjects, which define the stratum. Consider the following arrangement of four treatments, labelled A, B, C, D in a repeated measures design in which each subject had three treatments applied.

Subject	1	2	3	4	5	6	7	8
Treatments	A	A	A	B	A	A	A	B
applied	B	B	C	C	B	B	C	C
	C	D	D	D	C	D	D	D

Each treatment appears six times in the *Within subjects* stratum, yet each subject only receives three of the four treatments.

Where one or more subjects or blocks is deliberately missing one or more treatments, the design is described as an *incomplete block design.* If every **pair** of treatments appears in the same block an identical number of times, the design is described as a *balanced design.* In general, *balanced* incomplete block designs are relatively straightforward to analyze and their results are easy to interpret. *Unbalanced* incomplete block designs present the same problems of analysis and interpretation as arise for unbalanced complete block designs (Section 15.6.3).

15.6.5 Non-Hierarchical Designs

To this point in the book, all designs considered have been *hierarchical designs* since each new stratum formed in a design arises by sampling or partitioning units in an existing stratum. Thus, *blocks* are subdivided to form *plots* and *plots* are subdivided to form *sub-plots; subjects* are repeatedly sampled in time in repeated measures designs. Example 15.11 provides an example of a design in which the units are overlapping rather than nested. Such a design is termed a *non-hierarchical design.*

Example 15.11 (Fish Tasting Trials). *Nine staff members in a food research laboratory took part in a comparative study of different methods for storing fish. The study involved attendance at two tasting sessions when each taster was given six pieces of fish to taste. The six pieces of fish were taken from fish subjected to three different storage methods—freezing, chemical treatment and storage in a high* CO_2 *environment. Each taster was* *given the treated fish in a random order with the restriction that each treatment must be represented twice.*

At first sight, this design might appear to be a simple repeated measures design since there are subjects identified and each subject is repeatedly sampled. The feature which distinguishes this design from hierarchical, repeated measures designs is the identification of sessions as effects in their own right. This introduces a third unit to the design—the *sessions* unit. The times of sampling are definable not only within subjects, but also within sessions. For every subject, the first three samples were tasted within session 1 and this was the same session for every taster. The *tasters* and *sessions* effects are not hierarchical since the tasters unit crosses both sessions and the sessions unit encompasses all tasters.

In general, the analysis of non-hierarchical designs is complex and there is not necessarily a unique F-statistic defined to test hypotheses of interest. Be wary of statistical packages which purport to give F-tests for non-hierarchical designs. If a non-hierarchical design appears to be required, seek the assistance of a consultant statistician.

15.6.6 Rules for Constructing F-Ratios

For hierarchical designs, the following steps may be used to construct the correct F-ratios for the examination of effects represented in a model equation for a balanced design.

Step 1 *For any row in the Analysis of variance table, note the corresponding term in the model equation and its subscript(s).*

Step 2 *Use the mean square for the row identified at Step 1 as the numerator of the F-statistic.*

Step 3 *IF the term includes a fixed effect, seek from the model equation all* **random** *components which have the same subscript(s) as the term identified at Step 1. Choose the term which has the smallest number of subscripts.*

ELSE choose the residual mean square from the stratum below that which contains the term found at Step 1.

Step 4 *Use the mean square from the row in the Analysis of variance table which associates with the term identified at Step 3 as the denominator of the F-statistic. Use the statistic to compute a value F_0.*

Step 5 *Use the degrees of freedom for the numerator (d_1) and denominator (d_2) to define the F-distribution which provides the p-value from the relation $p_0 = \Pr\{F(d_1, d_2) \geq F_0\}$.*

Illustrations.

1. *One-stratum design.* Model equation

$$y_{ij} = M + T_i + \varepsilon_{ij} \qquad \text{for } i = 1, 2, \ldots, t;\ j = 1, 2, \ldots, n$$

Analysis of variance table — see Table 15.4.2.

To obtain the correct F-ratio to test the hypothesis that all treatment effects are identical, proceed according to the following steps:

1. Note that the *Between treatment* row in the Analysis of variance table corresponds to the term T_i. Hence, at Step 2, the *Between treatment mean square* is selected as the numerator of the F-ratio.

2. Since the term T_i is a fixed effect, terms are sought in the model equation which are random effects and have the subscript i as one subscript. The term ε_{ij}, which corresponds to the *Residual* row in the Analysis of variance table is the only random term which includes the subscript i. Hence, at Step 4, the F-statistic for the Between treatment row is formed as

$$F = \text{Between treatment mean square } / \text{ Residual mean square.}$$

2. *Repeated measures design with a random subject effect and repetition of treatments within subjects.* Model equation

$$y_{ijk} = M + \mathcal{S}_i + \{T_j + (\mathcal{S}.T)_{ij} + \varepsilon_{i,jk}\} \quad \text{for } i = 1, 2, \ldots, r;$$

$$j = 1, 2, \ldots, t;\ k = 1, 2, \ldots, n.$$

Analysis of variance table — see Table 15.5.6.

(a) The *Between treatment* row corresponds to the term T_j. The terms $(\mathcal{S}.T)_{ij}$ and $\varepsilon_{i,jk}$ both have the subscript j. However, the interaction term, $(\mathcal{S}.T)_{ij}$, has fewer subscripts and hence, the *Subject × treatment* row provides the denominator for the F-statistic.

(b) The *Between subject* row corresponds to the term $\mathcal{S}_i$ in the model equation. Since this term does not include a fixed effect, the denominator for the F-statistic is the Residual mean square from the second stratum.

(c) The *Subject × treatment* row corresponds to the term $(\mathcal{S}.T)_{ij}$ which includes a fixed effect, T_i. It has two subscripts, i and j.

The *Residual* row in the second stratum corresponds to the term $\varepsilon_{i,jk}$ which is the only random term with the subscripts i and j. Hence, the Residual mean square provides the denominator for the *Subject × treatment* row.

15.7 Factorial Arrangement of Treatments

15.7.1 Factors and Interactions

Suppose that a set of treatments are formed from the combinations of the different levels of two factors, A and B. There may be sound experimental reasons for defining the treatments in terms of the two factors. In particular, the experimenter has the possibility of seeking answers to the questions:

1. *Is there evidence that changing the level of factor A affects average response?*

2. *Is there evidence that changing the level of factor B affects average response?*

3. *Is there evidence that the effect of changes in the level of factor A is dependent on the level of factor B?*

To be in a position to provide a statistical analysis which examines these questions, the treatment contribution to the response, which is represented by the single term T_j in previous sections, must be decomposed into three components which relate to these three questions. The components are

(i) the contribution from factor A at level k, A_k $(k = 1, 2, \ldots, a)$;

(ii) the contribution from factor B at level m, B_m $(m = 1, 2, \ldots, b)$; and

(iii) the *interaction* between A and B which provides the contribution $(A.B)_{km}$ when factor A is at level k and factor B is at level m.

The factor combinations define $t = ab$ treatments and the relation between the treatment effect T_j $(j = 1, 2, \ldots, t)$ and the factorial components is defined by (15.7.1).

$$T_j = A_k + B_m + (A.B)_{km}. \qquad (15.7.1)$$

Note that the subscript j is replaced by the pair of subscripts k, m. If there are 2 levels of factor A and 3 levels of factor B, the connection could be expressed in the following way.

j	1	2	3	4	5	6
k	1	1	1	2	2	2
m	1	2	3	1	2	3

The interaction effect. The effect of changing the level of factor A from level r to level s when factor B is at level m is $[A_r + (A.B)_{rm}] - [A_s + (A.B)_{sm}]$. If there is no interaction between A and B, this reduces to $A_r - A_s$, i.e. the difference is constant over all levels of B. In the presence of an interaction, the effect of varying the level of factor A is seen to be dependent on the level of factor B.

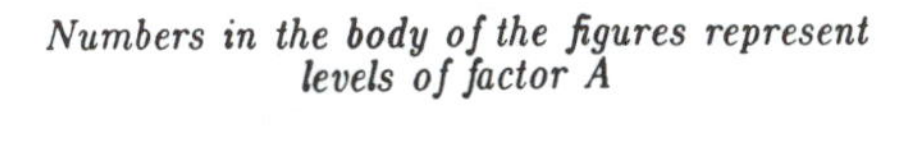

Numbers in the body of the figures represent
levels of factor A

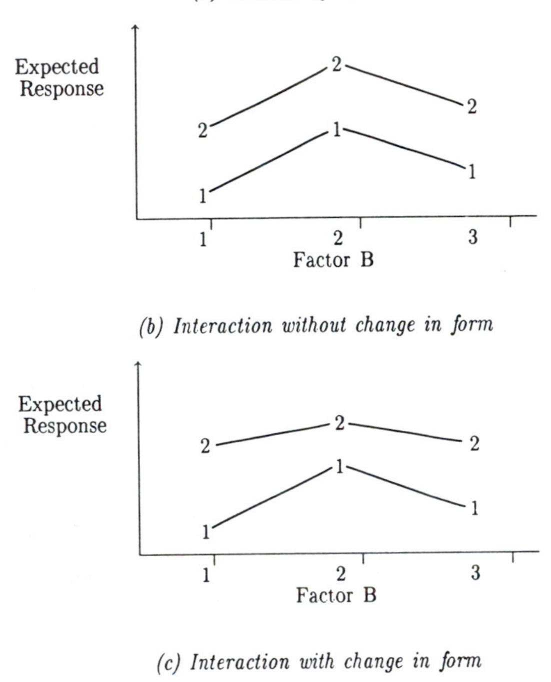

(c) Interaction with change in form

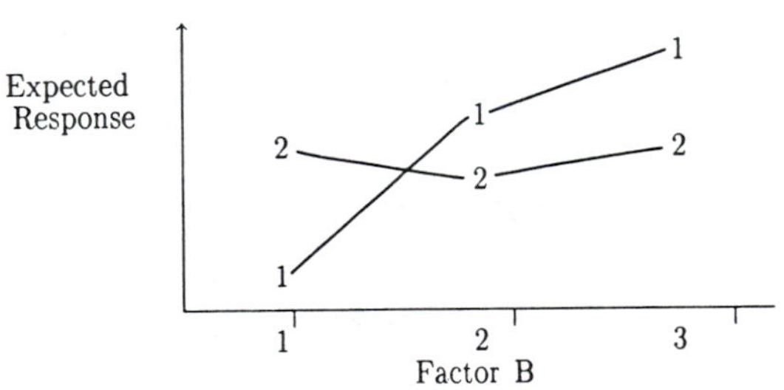

Figure 15.7.1. The nature of an interaction between two factors.

The practical role of an interaction effect can be readily visualized in a graphical form. Figure 15.7.1 presents three patterns of practical importance.

- In Figure 15.7.1(a), there is no interaction between factors A and B. This is reflected in the constant difference between expected responses when A is applied at levels 1 and 2 over all levels of factor B.

- In Figure 15.7.1(b), there is evidence of interaction, but at all levels of factor B there is a greater response when factor A is at level 1 compared with level 2.

- In Figure 15.7.1(c), there is evidence of interaction characterized by the fact that the higher expected response is not always associated with the same level of factor A.

Where the pattern of Figure 15.7.1(a) applies, it is meaningful to speak of the effect of varying levels of factor A without reference to factor B, and vice versa. The comparison of responses at different levels of factor A is based on the average responses over all levels of factor B. In scientific terms, this same approach is often reasonable when the pattern in Figure 15.7.1(b) applies. The experimenter must make a judgement as to whether the statistical interaction is of scientific importance.

If the pattern displayed in Figure 15.7.1(c) applies, it is misleading to discuss the effect of varying factor A without reference to individual levels of factor B.

15.7.2 Treatment Designs and Model Equations

The description of a design in which a factorial arrangement of treatments has been employed depends simply on establishing the strata in which the different treatment factors have been applied. A range of possibilities is considered in Section 15.2.2.

If there are two factors, A and B, there are three terms to be included in the treatment design equation(s), the *main effects,* A_i and B_j, plus the $A \times B$ *interaction,* $(A.B)_{ij}$. If factors A and B are applied in different strata, the interaction term is always included in the lower stratum.

If there are three factors, A, B and C, there are three *main effects* to be included, A_i, B_j and C_k; three *two-way interactions,* $(A.B)_{ij}$, $(A.C)_{ik}$ and $(B.C)_{jk}$; and one *three-way interaction,* $(A.B.C)_{ijk}$. Each interaction appears in the lowest stratum containing one of the factors included in the interaction.

By way of illustration, the model equations for Examples 15.2, 15.3, 15.7 and 15.9 are stated below. The factors are identified by letters which are descriptive of the factors.

1. Example 15.2 — one stratum design with two factors

$$y_{ijk} = M + S_i + W_j + (S.W)_{ij} + \varepsilon_{ijk}$$

   (S denotes *soil type* and W denotes *moisture level*).

2. Example 15.3 — two strata design (Randomized complete block design) with three factors applied in the second stratum

$$y_{ijkm} = M + B_i + \{S_j + P_k + N_m + (S.P)_{jk} + (S.N)_{jm} + (P.N)_{km} + (S.P.N)_{jkm} + \varepsilon_{i,jkm}\}$$

(S denotes *seeding rate,* P denotes *phosphorus* and N denotes *nitrogen*).

3. Example 15.7 — three strata design with one factor applied in the second stratum and the other in the third stratum

$$y_{ijkm} = M + B_i + \{V_j + \varepsilon_{i,j} + \{D_k + (V.D)_{jk} + \varepsilon'_{i,j,k}\}\}$$

(V denotes *variety* and D denotes *date of cutting*).

Note that, in theory, there could be included interactions between B and V, B and D and between B, V and D. Where B is a non-treatment effect—the 'block' effect—it is customary to assume these interactions do not exist. If these assumptions were not made, there would not be valid tests for examining treatment effects. Experimenters should be aware of these assumptions and use the model checking procedures suggested in Section 15.3 to assess their acceptability. Furthermore, if block differences can be associated with a physical factor (i.e. moisture or fertility variations across the experimental site) the experimenter should question whether prior experience or scientific knowledge supports the assumption that treatment differences are expected to remain constant across the range of conditions defined by inter-block variation. If there is doubt, it may be possible to employ a different design which allows for a block × treatment interaction. However, this possibility can only be explored if advice is sought from a consultant statistician at the design stage.

4. Example 15.9 — One stratum design with a covariate and two factors.

$$y_{ijk} = M + T_i + F_j + (T.F)_{ij} + \beta(x_{ijk} - \overline{x}_{...}) + \varepsilon_{ijk}$$

(T denotes *Time spent on job,* F denotes *Form of production,* x is the covariate, *days of experience*).

Random effects and multi-strata designs. Where subjects or units are randomly selected and factors may be allocated in different strata, there is a vast number of possible designs which might be employed. However, a simple set of rules may be employed to define the correct model equation to describe the design. The rules are stated and illustrated below.

Rule 1: *Where treatments are applied in the second or lower strata, include in the treatment stratum (strata) interactions with all effects in the higher strata.*

Rule 2: *In any stratum in which a treatment (factor) has been randomly allocated among units, include a residual component with subscripts derived from the treatment (factor), replication of the treatment (factor) and all effects appearing in higher strata. Adopt the standard convention of separating subscripts from different strata by commas.*

Rule 3: *Combine random effect terms which have the same subscripts into a single residual term. (A random effect term is any component which includes at least one random effect.)*

Illustrations.

1. *Treatments randomly allocated within each subject*

   (a) no replication of treatments within subjects

   — no treatment structure

$$\begin{aligned} y_{ij} &= M + \mathcal{S}_i + \{T_j + (\mathcal{S}.T)_{ij} + \varepsilon_{i,j}\} \\ &= M + \mathcal{S}_i + \{T_j + \varepsilon'_{i,j}\} \end{aligned}$$

   — factorial arrangement of two factors

$$\begin{aligned} y_{ijk} &= M + \mathcal{S}_i + \{A_j + (\mathcal{S}.A)_{ij} + B_k + (\mathcal{S}.B)_{ik} \\ &\quad + (A.B)_{jk} + (\mathcal{S}.A.B)_{ijk} + \varepsilon_{i,jk}\} \\ &= M + \mathcal{S}_i + \{A_j + (\mathcal{S}.A)_{ij} + B_k + (\mathcal{S}.B)_{ik} \\ &\quad + (A.B)_{jk} + \varepsilon'_{i,jk}\}. \end{aligned}$$

   (b) treatments replicated r times within each subject

   — no treatment structure

$$y_{ijk} = M + \mathcal{S}_i + \{T_j + (\mathcal{S}.T)_{ij} + \varepsilon_{i,jk}\}.$$

   — factorial arrangement of two factors

$$\begin{aligned} y_{ijkm} &= M + \mathcal{S}_i + \{A_j + (\mathcal{S}.A)_{ij} + B_k + (\mathcal{S}.B)_{ik} \\ &\quad + (A.B)_{jk} + (\mathcal{S}.A.B)_{ijk} + \varepsilon_{i,jkm}\}. \end{aligned}$$

2. *Factor A randomly allocated between subjects and factor B randomly allocated within subjects*

$$y_{ijk} = M + A_i + \varepsilon_{ij} + \{B_k + (A.B)_{ik} + \varepsilon'_{ij,k}\}.$$

Non-random treatment allocation. It may happen that a factor cannot be randomly allocated, e.g. *sex*. In this case, Rule 4 applies.

Rule 4: *Factors which are non-randomly allocated between subjects or units are introduced in a new stratum above the units to which they are applied.*

Illustrations.

1. Subjects are grouped on the basis of factor G and within each group there is random allocation of subjects between levels of factor A

$$y_{ijk} = M + G_i + \{A_j + (G.A)_{ij} + \varepsilon_{i,jk}\}.$$

2. Subjects are grouped on the basis of factor G, and within groups there is a repeated measures design with subjects randomly allocated to levels of factor A and times within subjects randomly allocated to levels of factor B.

$$y_{ijkm} = M + G_i + \{A_j + (G.A)_{ij} + \varepsilon_{i,jk} + \{B_m + (G.B)_{im} + (A.B)_{jm} + (G.A.B)_{ijm} + \varepsilon'_{i,jk,m}\}\}.$$

15.7.3 Analysis of Variance

The composition of Analysis of variance tables is determined by model equations. Each term in the model equation generates a row in the Analysis of variance table and that row is positioned within the stratum in which it appears in the model equation. General rules for the construction of degrees of freedom, sums of squares and mean squares are presented below. However, it is presumed that a statistical package is available which can supply the numerical values required.

Degrees of freedom. If A is a factor with a levels, the degrees of freedom for the A row in the Analysis of variance table are $(a-1)$. Furthermore, if B is a factor with b levels, the degrees of freedom for the $A \times B$ (interaction) row are $(a-1)(b-1)$.

Sums of squares. If factor A is represented as the effect A_i in the model equation for which the response variable is y_{ijk}, the sum of squares for the A row in the Analysis of variance table is $\sum_{i=1}^{a} bc(\overline{y}_{i..} - \overline{y}_{...})^2$. Furthermore, if factor B is represented as the effect B_j in the model equation, the sum of squares for the $A \times B$ interaction is $\sum_{i=1}^{a} \sum_{j=1}^{b} c(\overline{y}_{ij.} - \overline{y}_{i..} - \overline{y}_{.j.} + \overline{y}_{...})^2$.

Mean squares. By definition, the mean square for every row in the Analysis of variance table is computed as the ratio of the sum of squares to the degrees of freedom.

The two points which should be checked when using a statistical package are (i) ensuring the rows of the Analysis of variance table correspond to the collection of terms in the model equation, and (ii) the correct F-statistics have been employed. In the latter case, three simple steps are given to determine the correct denominator for the F-statistics.

1. *Note the term in the model equation corresponding to the row in the analysis of variance table. If it is a random component which contains no fixed effect terms, choose as the denominator of the F-statistic, the residual mean square from the next stratum down. Otherwise proceed to Step 2.*

2. *Find all terms in the model equation which include random effects and which include in their sets of subscripts all of the subscripts which appear in the term identified in Step 1.*

3. *If there is only one term identified at Step 2, use the mean square for the corresponding row in the Analysis of variance table as the denominator of the F-statistic. If there is more than one term, select that term which has the smallest number of subscripts.*

The steps are illustrated using the following model equation.

$$y_{ijkm} = M + \mathcal{S}_i + \{A_j + (\mathcal{S}.A)_{ij} + B_k + (\mathcal{S}.B)_{ik} + (A.B)_{jk} + (\mathcal{S}.A.B)_{ijk} + \varepsilon_{i,jkm}\}.$$

Term defining numerator	Possible terms for denominator	Term defining denominator
$\mathcal{S}_i$	$\varepsilon_{i,jkm}$	$\varepsilon_{i,jkm}$
A_j	$(\mathcal{S}.A)_{ij}$ $\varepsilon_{i,jkm}$ $(\mathcal{S}.A.B)_{ijk}$	$(\mathcal{S}.A)_{ij}$
B_k	$(\mathcal{S}.B)_{ik}$ $\varepsilon_{i,jkm}$ $(\mathcal{S}.A.B)_{ijk}$	$(\mathcal{S}.B)_{ik}$
$(\mathcal{S}.A)_{ij}$	$(\mathcal{S}.A.B)_{ijk}$ $\varepsilon_{i,jkm}$	$(\mathcal{S}.A.B)_{ijk}$
$(\mathcal{S}.B)_{ik}$	$(\mathcal{S}.A.B)_{ijk}$ $\varepsilon_{i,jkm}$	$(\mathcal{S}.A.B)_{ijk}$
$(\mathcal{S}.A.B)_{ijk}$	$\varepsilon_{i,jkm}$	$\varepsilon_{i,jkm}$

Analysis of variance tables for Examples 15.2 and 15.7 are shown in Table 15.7.1. The Analysis of variance table for Example 15.9 is shown in Table 15.7.2.

15.7.4 Interpreting the Analysis of Variance Table

In Section 15.7.1, the significance of an interaction between two factors is considered. From that discussion it is apparent that the interpretation of the main effects is dependent on whether there is evidence of an interaction and, if so, the nature of the interaction. The following strategy is recommended for a study in which there are two treatment factors in a factorial arrangement. A consideration of the approach when there are more than two factors appears later in the section.

Table 15.7.1. Analysis of Variance Tables

(a) Analysis for Example 15.2 based on model equation
$$y_{ijk} = M + S_i + W_j + (S.W)_{ij} + \varepsilon_{ijk}$$

Source of Variation	D.F.	S.S.	M.S.	F	p
Moisture levels (W)	3	11.4208	3.8069	22	<0.01
Soil types (S)	3	0.3038	0.1013	0.6	0.6
$S \times W$	9	8.7122	0.9680	5.5	0.002
Residual	16	2.8053	0.1752		

(b) Analysis for Example 15.7 based on model equation
$$y_{ijkm} = M + B_i + \{V_j + \varepsilon_{i,j} + \{D_k + (V.D)_{jk} + \varepsilon'_{i,j,k}\}$$

Source of Variation	D.F.	S.S.	M.S.	F	p
Between block stratum					
Between blocks	5	4.1498	0.8300	6.1	<0.01
Between plots within blocks stratum					
Variety (V)	2	0.1780	0.0890	0.7	>0.1
Residual (plot)	10	1.3623	0.1362	4.9	<0.01
Between sub-plots within plots stratum					
Dates (D)	3	1.9625	0.6542	24	<0.01
$V \times D$	6	0.2106	0.0355	1.2	0.3
Residual (sub-plot)	45	1.2585	0.0279		

Table 15.7.2. Analysis for Example 15.9 Based on Model Equation
$$y_{ijk} = M + T_i + F_j + (T.F)_{ij} + \beta(x_{ijk} - \bar{x}_{...}) + \varepsilon_{ijk}$$

Source of Variation	D.F.	S.S.	M.S.	F	p
Time (T)	1	69.75	69.75	6.9	0.01
Form (F)	1	892.00	892.00	88	<0.01
$T \times F$	1	192.75	192.75	19	<0.01
Covariate	1	1646.49	1646.49	163	<0.01
Residual	55	555.24	10.10		

Procedures for interpreting the Analysis of variance table

1. *When there is a two-factor factorial arrangement of treatments*

Step 1: Examine the p-value for the interaction term.

IF it exceeds 0.05, assume there is no interaction and proceed to a consideration of the p-values for the individual factors. If either p-value suggest evidence for a factor effect, examine the table of means for the factor levels visually and with the aid of a pairwise comparison procedure (Section 15.8.1) to establish where the differences lie.

OTHERWISE, proceed to Step 2.

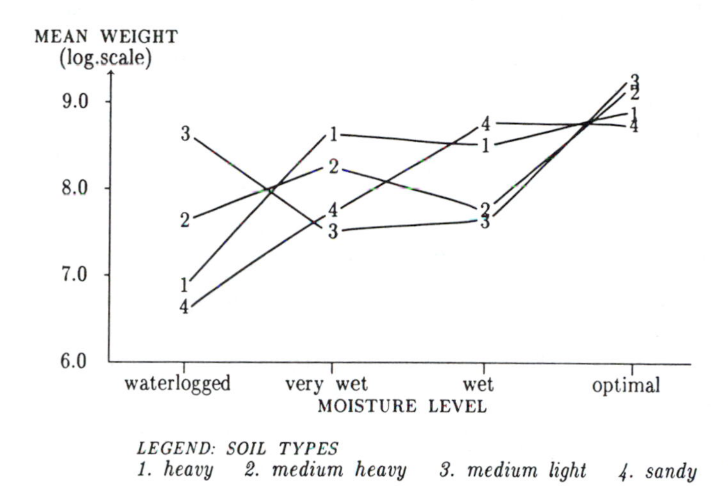

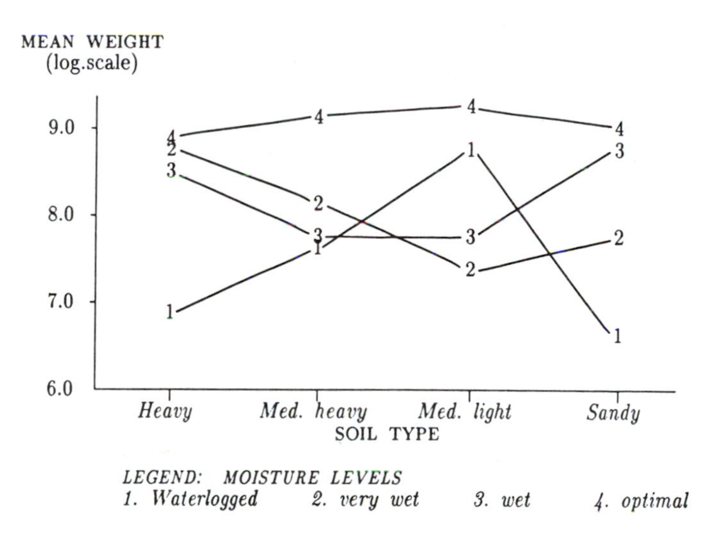

Figure 15.7.2. Two-factor interaction—Example 15.2.

Step 2: If there is evidence of an interaction, use the two-way table of means for the factor combinations to construct a graph which displays the interaction in the manner employed in Figure 15.7.1. (Note that two graphs are possible depending on which factor corresponds to factor A and which corresponds to factor B. Which is the more meaningful view of the interaction depends on the experimental aims.)

Step 3: Interpretation: In general, the interest of the experimenter is in either the comparison of the effect changing levels of factor B within each level of factor A or in the comparison of changing levels of factor A within each level of factor B. For example, when examining the graphs in Figure 15.7.2, the experimenter might consider the effect of varying soil types within each moisture regime **or** the effect of varying moisture levels within each soil regime.

The primary decision to be made relates to the nature of the interaction—is it (i) an interaction in which there is only a change in degree in the pattern (Figure 15.7.1(b)) or (ii) an interaction in which there is a change in form (Figure 15.7.1(c))?

If there is no change in form, i.e. the effect of changing levels of factor A produces the same pattern for all levels of factor B, there is the possibility of simplifying the conclusion by considering the A effect independently of the level of factor B.

If there is a change in form, it is generally meaningless and can be misleading to consider the p-value and tables of means for the individual factors. The correct approach is to consider the effects of varying levels of factor A separately within each level of factor B. Pairwise comparisons of levels of factor A within each level of factor B can be made using the methodology introduced in Section 15.8.1.

Illustrations.

1. *No evidence of interaction* (Example 15.7). The Analysis of variance table in Table 15.7.1(b) provides a p-value of 0.3 for the *variety* × *date* interaction. Hence, there is insufficient evidence to reject the hypothesis that the effect of changing cutting dates is the same for all varieties. In this experiment, it is the effect of changing cutting dates which is of primary interest since varietal comparisons have been made in earlier trials. Therefore, the next step is to determine if changes in cutting dates affect yields by interpreting the p-value for *Dates* in the Analysis of variance table. The p-value is less than 0.01 suggesting that varying the cutting dates has a strong influence on yield. The nature of the effect can be seen in the table of means.

Date	*A*	*B*	*C*	*D*
Mean	1.78	1.34	1.57	1.69

Given that the dates are in a time sequence, a visual inspection suggests that the highest yields are obtained if the cutting takes place either early or late. To attempt to separate treatment effects from sampling variation, pairwise comparisons are used to establish which of the pairs show significant differences. As noted in Section 15.8.1, there are several pairwise comparison procedures in common use. The one which is employed in this book is the l.s.d. procedure. Following the steps presented in Section 15.8.1, the l.s.d. procedure establishes that pairs of means which differ by more than 0.14 are likely to reflect treatment differences. This suggests that yields are lowest if cutting date B is used and that yields at cutting date A are higher than those from cutting date C.

2. *Evidence of interaction* (Example 15.2). In the Analysis of variance table in Table 15.7.1(a), the p-value for the *soil type* × *moisture level* interaction is 0.002 which suggests strong evidence of an interaction. The table of means (logarithms of weights) for the 16 factor combinations is presented in Table 15.7.3, and, based on the figures in this table, two graphs are constructed in Figure 15.7.2 which display the interaction between soil type and moisture level. The experimenter considered the second graph more relevant to his needs since he was primarily interested in the effect of changes in moisture levels. It is apparent from the graph that the effects of changes in moisture level are not consistent from one soil type to another. Hence, the effect of variations in moisture level on root growth should be examined separately within each soil type. (Note that the p-value for Soil types in the Analysis of variance table in Table 15.7.1(a) is 0.6. Given the above information about the interaction, it would be misleading to interpret this p-value as indicating that changes in soil type do not influence root growth.)

Table 15.7.3. Means for Soil Type and Moisture Combinations in Example 15.2 (Based on Data in Table 15.2.2)

Mean Root Growth (log. scale)

Moisture	Soil Type			
Level	Heavy	Med. heavy	Med. light	Sandy
Waterlogged	6.9	7.6	8.7	6.5
Very wet	8.6	8.1	7.4	7.8
Wet	8.4	7.8	7.7	8.8
Optimal	9.0	9.1	9.3	9.0

Using a pairwise comparison procedure (Section 15.8.1) within each soil type, it is possible to establish the manner in which changes in moisture level affect root growth and how the contrasts vary between soil types. The l.s.d. procedure establishes that a difference in means in excess of 0.89 (see Section 15.8.1 for details) provides evidence of a treatment effect. Hence, for example,

- except in medium light soils, root growth is significantly higher when moisture level is optimal than when the soil is waterlogged;
- except in heavy soils, root growth is significantly higher when moisture level is optimal than when it is very wet.

Three-factor or high level interactions. If there is a three-factor interaction or higher level interaction term in the model, the first step is to examine the p-value for the highest level interaction. A significant three-factor interaction between factors A, B and C suggests that the two-factor interaction between A and B has a different form within the levels of factor C. For the same reason as apply in the two-factor case, the next stage is to employ a graphical examination of the changes to establish if the interaction is a consequence of a change in form or merely a change in degree. If a change in form is involved, it is not sensible to discuss the interaction between A and B except within levels of factor C. If the interaction only involves a change in degree, it is for the experimenter to judge if a consideration of the $A \times B$ interaction averaged over levels of factor C is meaningful. Pairwise comparisons may be used (see Section 15.8.1) to establish which mean differences are significant.

Interactions with more than three factors are rare and, where they exist, are usually difficult to interpret.

15.8 More Detailed Comparison of Treatment Differences

15.8.1 Establishing Which Pair of Means Differ

***ad hoc* versus *a priori* comparison.** If, at the time of designing the experiment, there were particular treatment comparisons which were identified as having special significance, these comparisons should be made independently of a general test of the hypothesis that all treatment effects are equal. Such tests come within the general procedure outlined in Section 15.8.2 and can be accommodated within the Analysis of variance framework. Commonly, however, specific comparisons do not provide the motivation for an experiment. A set of treatments are selected with the general aim of establishing if they produce different effects. Where the F-test in an Analysis of variance procedure provides evidence to reject the hypothesis that all treatments have the same effect, there is a need to determine where the differences lie. In this case, a *multiple comparison procedure* may be employed to establish which pairs of treatments appear to be producing differences in mean response.

Multiple comparison procedures. There have been a substantial number of different approaches published which have as their common aim the comparison of all possible pairs of treatments or groups in order to establish which pairs show evidence of difference in mean response. For more than two decades there has been controversy over the best choice and there is not universal agreement today. No attempt is made in this book to survey the proposed methods or to critically compare methods. (The cautionary note is offered that any claim for one method as the best method should be treated with skepticism.) Statistical packages may offer one multiple comparison procedure or a choice of multiple comparison procedures. The more common procedures are the *least significant difference* (*lsd*) procedure, the *Student–Newman–Keuls* procedure, *Duncan's multiple range* procedures and *Tukey's honest significant difference* procedure. Where a choice is offered, use that procedure which is common in your area of research.

One method is described in this section, the *lsd* method. It is offered because it is widely used, has a simple basis and is computationally uncomplicated.

Least significant difference. Under the model proposed in Section 15.2.7, provided a balanced design is employed, the statistic defined by (15.8.1) has a t-distribution with degrees of freedom equal to the residual degrees of freedom from the stratum in the Analysis of variance table which contains the treatment effect.

$$t = [(\overline{y}_{i.} - \overline{y}_{j.}) - (T_i - T_j)]/\sqrt{2s^2/n} \qquad (15.8.1)$$

where $\overline{y}_{i.}$ and $\overline{y}_{j.}$ are means from all units receiving treatments i and j, respectively; T_i and T_j are expected responses under treatments i and j, respectively; n is the number of units receiving the same treatment; and s^2 is the residual mean square from the stratum in which the treatment component appears.

If $T_i = T_j$, (15.8.2) is true.

$$\Pr\{-t_{0.05}(d)\sqrt{2s^2/n} \leq (\overline{y}_{i.} - \overline{y}_{j.}) \leq t_{0.05}(d)\sqrt{2s^2/n}\} = 0.95 \qquad (15.8.2)$$

where d is the residual degrees of freedom from the treatment stratum.

The *least significant difference* or *lsd,* defined by (15.8.3), is used as a yardstick to decide if the observed difference between two treatments, $\overline{y}_{i.} - \overline{y}_{j.}$, has a magnitude which is great enough to provide evidence suggesting the treatment effects T_i and T_j are different. The use of the *lsd* is based on the property displayed in (15.8.2) that, if $T_i = T_j$, the probability that $\overline{y}_{i.} - \overline{y}_{j.}$ will have a magnitude in excess of the *lsd* is 0.05.

$$lsd = t_{0.05}(d)\sqrt{2s^2/n} \qquad (15.8.3)$$

where d, s^2 and n are defined in (15.8.1) and (15.8.2). Variations to the formula for constructing and applying the *lsd* procedure are required if (i) there is a factorial arrangement of treatments and the factors are applied in different strata; or (ii) there is a covariate employed. These cases are considered separately below.

Application of the *lsd* procedure. The *lsd* procedure should only be applied to pairwise comparison of effects where the Analysis of variance test for the effect has a p-value less than 0.05. It can be simply applied by following the steps listed below.

1. *Pairwise comparison of treatment and factor main effects*
   (*The calculations are illustrated using the information in Table* 15.8.1.)

Table 15.8.1. Analysis of Data from Example 14.1

(a) Table of means

Group	G1	G2	G3	G4
Mean	12.80	13.60	13.15	13.70

Note that each mean derives from 20 *observations.*

(b) Analysis of variance table

Source of Variation	D.F.	S.S.	M.S.	F	p
Between groups	3	10.4375	3.4792	2.19	0.10
Residual	76	120.2851	1.5827		

Step 1. Compute a value of the *lsd* using (15.8.3).
From Table 15.8.1

- *the number of observations per group is* $n = 20$;
- *the within group mean square is* $s^2 = 1.5827$;
- *the within group degrees of freedom are* $d = 76$.

From the t-distribution, $t_{0.05}(76) = 1.99$.
Hence, from (15.8.3),

$$lsd = 1.99\sqrt{\frac{2(1.5827)}{20}} = 0.79.$$

Step 2. Rank the t sample means from smallest to largest and, for the first $t-1$ means in the sequence, compute the value of *mean* + *lsd*. Present the information in the manner illustrated below which is based on the table of means in Table 15.8.1 and the *lsd* computed at Step 1.

Group	G1	G3	G2	G4
Mean	12.80	13.15	13.60	13.70
Mean + *lsd*	13.59	13.94	14.39	

Step 3. List the group names or identification in the order of the means (from smallest to largest).

Construct a horizontal line, in the manner displayed below, which begins under the first group and includes all groups with means which have values that are less than or equal to the value of *mean* + *lsd* for the first group.

For the example, starting with G1, the mean + lsd is 13.59. *This exceeds the mean for G3 but not the mean for G2. Hence, the line begun under G1 is extended to include G3.*

$$\underline{G1 \quad G3} \quad G2 \quad G4.$$

Consider the second group in the sequence. Construct a horizontal line which begins with the second group and includes all groups with larger means having values less than or equal to the value of *mean* + *lsd* **provided** the line extends beyond a previously constructed line.

For the example, mean + lsd for G3 is 13.94 *which exceeds the values of means for all groups. Hence, a line is drawn which begins with G3 and extends to G4.*

$$\underline{G1 \quad G3} \quad G2 \quad G4$$
$$ \underline{G3 \quad G2 \quad G4}$$

Consider the third group. Follow the same sequence of steps as are described for the second group.

Continue until *either* the group with the largest mean has been reached *or* the line extends to the group with the largest mean.

For the example, the line drawn at the second stage extends to the group with the largest mean. Hence, the procedure would not continue beyond this stage.

Step 4. Interpretation: any pair of means which are underlined by a unbroken line are said to be *not significantly different.* This may be taken as an indication that the difference in sample means could be explained as sampling variation. Any pairs of means which are not underlined by the same unbroken line are said to be *significantly different* and this suggests evidence of treatment differences.

In the example, G1 and G3 are not significantly different and G3, G2 and G4 are not significantly different. Note that there is not an inconsistency in finding sufficient evidence to conclude that G3 is not significantly different from G1 and G4 even though they are found to be significantly different from each other.

2. *Pairwise comparisons of factor combinations.* Consider an experiment in which there are two factors—factor A with levels $1, 2, \ldots, a$ and B with levels $1, 2, \ldots, b$. For each combination of factors A and B there is a mean response based on n observations. The means are presented in the following table.

		Factor B			
		1	2	$\cdots$	b
	1	$\bar{y}_{11.0}$	$\bar{y}_{12.0}$	$\cdots$	$\bar{y}_{1b.0}$
	2	$\bar{y}_{21.0}$	$\bar{y}_{22.0}$	$\cdots$	$\bar{y}_{2b.0}$
Factor A	$\vdots$	$\vdots$	$\vdots$		$\vdots$
	a	$\bar{y}_{a1.0}$	$\bar{y}_{a2.0}$	$\cdots$	$\bar{y}_{ab.0}$

Comparisons of means may be required across rows, i.e. in an examination of the B effects within a level of A, or down columns, i.e. in an examination of the A effects within levels of B.

If both A and B were applied in the same stratum, the *lsd* is computed using (15.8.3).

Alternatively, consider a situation in which factor A is applied at a higher stratum than factor B. If, in the stratum containing factor A, the residual degrees of freedom and the residual sum of squares are d_a and s_a^2, respectively, while factor B is applied in a stratum in which the residual degrees of freedom and the residual sum of squares are d_b and s_b^2, the *lsd* is defined by (15.8.4).

$$lsd = t_{0.05}(d)\sqrt{2s^2/n} \tag{15.8.4}$$

where

(i) $d = d_b$ and $s^2 = s_b^2$ for the comparison of two means with the same level of factor A; and

(ii) $t_{0.05}(d) = [(b-1)s_b^2 t_{0.05}(d_b) + s_a^2 t_{0.05}(d_a)]/[(b-1)s_b^2 + s_a^2]$ and $s^2 = [(b-1)s_b^2 + s_a^2]/b$ for the comparison of two means with the same level of factor B.

The application of the *lsd* procedure for either a row or column comparison follows the steps outlined in 1.

3. *lsd when there is a covariate in the treatment stratum.* When a covariate is present in the treatment stratum of the model equation, the comparison of treatments is based on adjusted means. If the set of responses is $\{y_{ijo}\}$ with the corresponding covariate values being $\{x_{ijo}\}$ based on the model equation $y_{ij} = M + T_i + \beta(x_{ij} - \overline{x}_{..}) + e_{ij}$, the adjusted mean for the ith treatment is computed from (15.8.5).

$$\overline{y}^a_{i.} = \overline{y}_{i.} - b(\overline{x}_{i.} - \overline{x}_{..}) \qquad \text{for } i = 1, 2, \ldots, t \tag{15.8.5}$$

where b is the sample estimate of β. In other words, the mean is adjusted to the value which is expected if the average value for the covariate is equal to the sample mean $\overline{x}_{..}$.

The estimated difference in adjusted means for treatments i and k is given by (15.8.6)

$$\overline{y}^a_{i.} - \overline{y}^a_{k.} = (\overline{y}_{i.} - \overline{y}_{k.}) - b(\overline{x}_{i.} - \overline{x}_{k.}). \tag{15.8.6}$$

The *lsd* to be applied for comparison of adjusted treatment means must take account of the sampling variability in b and this leads to the formula given in (15.8.7).

$$lsd = t_{0.05}(d)\sqrt{s^2[2/n + (\overline{x}_{i.} - \overline{x}_{k.})^2/RSS_x]} \tag{15.8.7}$$

where

d is the residual degrees of freedom from the stratum in which the treatment term appears;

s^2 is the residual mean square from the stratum in which the treatment term appears;

n is the number of units receiving the same treatment;

$\overline{x}_{i.}$ **and** $\overline{x}_{k.}$ are means for the covariate for treatments i and k, respectively; and

RSS_x is the residual sum of squares from the treatment stratum in the Analysis of variance table for the covariate based on the same design as was employed for the response variable.

Note that a separate *lsd* must be computed for each pair of treatments.

Illustration. The calculation of the *lsd* is illustrated using the data in Example 15.9.

For the purposes of illustrating the calculations, the factorial nature of the design is ignored. Hence, the model equation is $y_{ij} = M + T_i + \beta(x_{ij} - \overline{x}_{..}) + e_{ij}$ where $i = 1, 2, 3, 4$; $j = 1, 2, \ldots, 15$. The information required for the comparison of treatment 1 (full time, serial production) and treatment 2 (full time, parallel production) is as follows:

Time	Form	Treatment	n	$\overline{y}_{i.}$	$\overline{x}_{i.}$
full	serial	1	15	74.867	28.667
full	parallel	2	15	64.733	30.800

$$\overline{x}_{..} = 28.800 \qquad b = 0.5535$$

From Table 15.6.1, $s^2 = 10.10$ and $d = 55$. The fit of the model $x_{ij} = M + T_i + e_{ij}$ provides the values $RSS_x = 5375$.

By (15.8.5),

$$\overline{y}_{1.} = 74.867 - 0.5535(28.667 - 28.800) = 74.94,$$
$$\overline{y}_{2.} = 64.773 - 0.5535(30.800 - 28,800) = 63.63.$$

By (15.8.6),

$$\overline{y}_{1.} - \overline{y}_{2.} = 74.867 - 64.733 - 0.5535(28.667 - 30.800) = 11.3.$$

By (15.8.7), the *lsd* for the comparison of treatments 1 and 2 is

$$lsd = 2.00\sqrt{10.10[2/15 + (28.667 - 30.800)^2/5375]} = 2.3.$$

Interpretation: Since the observed difference between the adjusted means for treatments 1 and 2 has a magnitude in excess of the *lsd*, there is evidence that the treatments are having a different effect.

15.8.2 Analysis When There Is Group Structure

Table 14.1.2 contains a series of questions relating to four groups of rats which are identified as $G1$, $G2$, $G3$ and $G4$. The first of the questions posed is the following.

Does Group 1 appear to differ from the other groups?

In the context of comparing mean responses, the question might be answered by examining the hypothesis

$$H_{01}: T_1 = \frac{(T_2 + T_3 + T_4)}{3}$$

where T_i is the treatment effect for group i $(i = 1, 2, 3, 4)$.

Analysis of variance offers a mechanism for examining this hypothesis by measuring the difference in fit between the model M_{01} which contains the hypothesis H_{01} and the model M_0 which contains the hypothesis that all treatment effects are the same.

There are simple rules given below for applying Analysis of variance in this role. They have application when the treatment can be collected into *sets* and the hypothesis is in terms of the expected responses for the sets.

Thus, the hypothesis H_{01} defined above, equates the expected values of two sets, the first containing treatment 1 and the second containing treatments 2, 3 and 4. An Analysis of variance table is constructed based on the general model which places no restrictions on treatment effects and an additional row is added for each comparison of sets.

The steps in the construction of *Between sets* degrees of freedom and sums of squares are presented below and illustrated in Table 15.8.2 using the questions in Table 14.1.2.

Table 15.8.2. Comparison of Means When There Is Group Structure (Application to Example 14.1 Based on Questions in Table 14.1.2)

(a) Relating hypotheses, degrees of freedom and means (Note that $Y_{i.}$ is the sum of responses for treatment i)

Hypothesis	D.F.	Means			
$T_1 = \frac{T_2+T_3+T_4}{3}$	1	$\bar{y}_{s1} = \bar{y}_{1.}$		$\bar{y}_{s2} = \frac{Y_{2.}+Y_{3.}+Y_{4.}}{n_2+n_3+n_4}$	$\bar{y}_{s..} = \bar{y}_{..}$
$\frac{T_1+T_2}{2} = \frac{T_3+T_4}{2}$	1	$\bar{y}_{s1} = \frac{Y_{1.}+Y_{2.}}{n_1+n_2}$		$\bar{y}_{s2} = \frac{Y_{3.}+Y_{4.}}{n_3+n_4}$	$\bar{y}_{s..} = \bar{y}_{..}$
$T_1 = T_2 = \frac{T_3+T_4}{2}$	2	$\bar{y}_{s1} = \bar{y}_{1.}$	$\bar{y}_{s2} = \bar{y}_{2.}$	$\bar{y}_{s3} = \frac{Y_{3.}+Y_{4.}}{n_3+n_4}$	$\bar{y}_{s..} = \bar{y}_{..}$
$T_2 = T_3 = T_4$	2	$\bar{y}_{s1} = \bar{y}_{2.}$	$\bar{y}_{s2} = \bar{y}_{3.}$	$\bar{y}_{s3} = \bar{y}_{4.}$	$\bar{y}_{s..} = \frac{Y_{2.}+Y_{3.}+Y_{4.}}{n_2+n_3+n_4}$

(b) Analysis of variance table

Source of Variation	D.F.	S.S.	M.S.	F	p
Between treatments	3	10.438	3.479	2.2	0.10
$G1$ vs $G2+G3+G4$	1	7.004	7.004	4.4	0.04
$G1+G2$ vs $G3+G4$	1	9.113	9.113	5.8	0.02
$G1$ vs $G2$ vs $G3+G4$	2	7.413	3.706	2.3	0.11
$G2$ vs $G3$ vs $G4$	2	3.433	1.717	1.1	
Residual	76	120.285	1.582		

Step 1. For each set formed, define the set effect as the average of the group effects. Thus, if groups i and j are combined, the set effect is $\frac{1}{2}(T_i + T_j)$. Express the hypothesis in terms of set effects (as illustrated in Table 15.8.2(a)).

Step 2. If the number of sets which are specified in the hypothesis is S, then compute the corresponding degrees of freedom as $S-1$.

Step 3. Determine (i) the number of observations in set i, n_{si} ($i = 1, 2, \ldots, S$); (ii) the mean for set i, $\overline{y}_{si.}$ ($i = 1, 2, \ldots, S$); and (iii) the mean for all groups which appear in the sets. Denote the overall mean by $\overline{y}_{s..}$.

Step 4. Use the values computed at Step 3 to calculate the *Between set sum of squares* as defined in (15.8.8)

$$BSS = \sum_{i=1}^{S} n_{si}(\overline{y}_{si.} - \overline{y}_{s..})^2. \tag{15.8.8}$$

Step 5. Insert the *Between sets degrees of freedom*, $S-1$, and the *Between sets sum of squares* into the Analysis of variance table in the stratum in which the treatment term appears. Compute a value F_0 for the statistic F = (Between set mean square) / (Residual mean square). Given that the residual degrees of freedom is d_r, compute the p-value $p_0 = \Pr\{F(S-1, d_r) \geq F_0\}$ and use the value of p_0 to establish if there is evidence of differences between the sets.

Illustration. Table 15.8.2(a) contains hypotheses, degrees of freedom and formulae for computing sums of squares for the comparisons presented in Table 14.1.2. The corresponding entries in the Analysis of variance table are contained in Table 15.8.2(b).

Problems

15.1 *Examples* 15.1 *and* 15.8

(a) State the experimental design, treatment design and model design equations which apply in Example 15.1.

(b) Apply relevant model and data checking procedures (see Section 15.3) to determine if the presumed model provides a reasonable fit to the data in Table 15.2.1.

(c) Use the Analysis of variance technique for a one-stratum design (Section 15.4.3, 15.4.4) to establish the acceptability of the hypothesis of no differences in treatment effects based on 'post-test' scores in Table 15.2.1.

(d) Using the additional information in 'pre-test' scores in Table 15.2.8, construct the set of differences 'post-test' – 'pre-test' scores.

Again using the Analysis of variance technique for a one-stratum design, establish that there is evidence of differences in treatment effects.

Apply the l.s.d. procedure (Section 15.8.1) to determine which treatments appear to differ.

Note the importance of the information on 'pre-test' scores in reducing the residual variation (i.e. by the removal of the subject-to-subject variation). This is measured by the reduction in residual mean square from the analysis in (c) to the analysis in (d).

15.2 *Example* 15.2

(a) State the experimental design, treatment design and model design equations which apply in Example 15.2.

(b) Transform the data in Table 15.2.2 to a logarithmic scale.

(c) Apply relevant model and data checking procedures (see Section 15.3) to determine if the presumed model provides a reasonable fit to the data after transformation to a logarithmic scale.

(d) Based on a one-stratum design with factorial arrangement of treatments, reproduce the Analysis of variance table presented in Table 15.7.1(a) and obtain the table of means (on a logarithmic scale) for the 16 factor combinations. Follow the steps in Section 15.7.4 to interpret the information in the Analysis of variance table.

(e) Apply the l.s.d. procedure (Section 15.8.1) to establish the effect of varying moisture level within each soil type.

15.3 *Example* 15.3

(a) State the experimental design, treatment design and model design equations which apply in Example 15.3.

(b) Apply relevant model and data checking procedures (see Section 15.3) to determine if the presumed model provides a reasonable fit to the data in Table 15.2.3.

(c) Assuming a Randomized complete block design with a factorial arrangement of treatments, construct an Analysis of variance table based on the data in Table 15.2.3.

(d) Following the steps outlined in Section 15.7.4, establish evidence of a nitrogen × phosphorus interaction and examine the nature of the interaction.

(e) Describe the effect of differences in seeding rate on yield.

(f) What does the analysis suggest are the optimal conditions for yield?

15.4 *Example* 15.4

(a) State the experimental design, treatment design and model design equations which apply in Example 15.4.

(b) Apply relevant model and data checking procedures (see Section 15.3) to determine if the presumed model provides a reasonable fit to the data in Table 15.2.4.

(c) Reproduce the Analysis of variance table given in Table 15.5.5.

(d) Establish that there is evidence of differences in average responses following different themes.

(e) Use the l.s.d. procedure to determine which themes associate with significantly different attitudes to aggression.

15.5 *Example* 15.5

(a) State the experimental design, treatment design and model design equations which apply in Example 15.5.

(b) Apply relevant model and data checking procedures (see Section 15.3) to determine if the presumed model provides a reasonable fit to the data in Table 15.2.5.

(c) Reproduce the Analysis of variance table given in Table 15.5.7.

(d) Based on the guidelines offered in Section 15.5.3 under the subheading '2. n applications of each treatment per subject,' interpret the three F-values which appear in the Analysis of variance table.

(e) Given there is evidence of differences between exercises, use the l.s.d. procedure (Section 15.8.1) to determine which pairs of exercises show significant differences.

15.6 *Example* 15.6

(a) State the experimental design, treatment design and model design equations which apply in Example 15.6.

(b) Apply relevant model and data checking procedures (see Section 15.3) to determine if the presumed model provides a reasonable fit to the data in Table 15.2.6.

(c) Reproduce the Analysis of variance table given in Table 15.5.2 and note that the hypothesis of no environmental differences is judged to be consistent with the data.

15.7 *Example* 15.7

(a) State the experimental design, treatment design and model design equations which apply in Example 15.7.

(b) Apply relevant model and data checking procedures (see Section 15.3) to determine if the presumed model provides a reasonable fit to the data in Table 15.2.7.

(c) Reproduce the Analysis of variance table given in Table 15.7.1(b).

(d) Follow the steps in Section 15.7.4 to interpret the F-values associated with treatment effects in the Analysis of variance table.

15.8 *Example* 15.9

(a) State the experimental design, treatment design and model design equations which apply in Example 15.9.

(b) Apply relevant model and data checking procedures (see Section 15.3) to determine if the presumed model provides a reasonable fit to the data in Table 15.2.9.

(c) Produce (i) the Analysis of variance table from the fitting of the model defined in (a) to the data in Table 15.2.9 and (ii) a table of means for the four factorial combinations.

(d) Is there evidence of an interaction between *time spent on job* and *form of production?*

(e) Interpret the Analysis of variance table following the steps outlined in Section 15.7.4.

15.9 *Example* 15.10

(a) State the experimental design, treatment design and model design equations which apply in Example 15.10.

(b) Apply relevant model and data checking procedures (see Section 15.3) to the data in Table 15.3.2 to establish that the assumption of Normality and the assumption of equal variances appear to be violated.

(c) Transform the data to a logarithmic scale and establish that the relevant assumptions appear to be met.

(d) Construct an Analysis of variance table based on the model equation in (a) and the logarithmically transformed data.

(e) Is there evidence that the average amount of combustible material per plot varies between treatments?

(f) Are treatment differences, if any, dependent on the nature of the site?

16

Comparing Patterns of Response: Frequency Tables

16.1 The Scope of Applications

1. *Categorical response.* Where the response is categorical, a frequency distribution is defined for each group or treatment. The function of statistical analysis is to compare the frequency distributions for the different groups. Example 16.1 illustrates the typical situation in which data are collected on a categorical response variable with the aim of comparing the patterns of response in that variable between groups.

Example 16.1 (Comparing Spending Patterns Among Different Groups of Tourists). *The data presented in Table* 16.1.1 *derive from a categorical variable labelled 'Amount spent' and come from a survey of the spending patterns of tourists in Tasmania. Based on this data set, the investigator wished to know whether the pattern of expenditure varied according to the place of origin. The places of origin were defined according to the distance travelled to reach Tasmania.*

Contained within the general question of whether spending patterns differ were more specific questions exemplified by

"Is the proportion of high spenders varying with place of origin?"

"Do overseas visitors show a different spending pattern to that shown by Australians?"

Table 16.1.1. Survey Data for Example 16.1

(a) Frequency distributions

	Amount Spent			
Place of Origin*	low	medium	high	total
Vic.	546	1294	375	2215
N.S.W./S.A.	240	827	454	1521
Qld./W.A.	117	320	159	596
Overseas	51	152	48	251
total	954	2593	1036	4583

Table 16.1.1 (*cont.*)

(b) Relative frequencies

Place of Origin*	Amount Spent low	medium	high	total
Vic.	0.25	0.58	0.17	1.00
N.S.W./S.A.	0.16	0.54	0.30	1.00
Qld./W.A.	0.20	0.54	0.26	1.00
Overseas	0.20	0.61	0.19	1.00

(c) Fitted frequencies assuming
(i) the same distribution for all groups
(ii) observed and fitted totals agree

Place of Origin*	Amount Spent low	medium	high
Vic.	$2215 \times \frac{954}{4583} = 461.1$	$2215 \times \frac{2593}{4583} = 1253.2$	$2215 \times \frac{1036}{4583} = 500.7$
N.S.W./S.A.	$1521 \times \frac{954}{4583} = 316.6$	$1521 \times \frac{2593}{4583} = 860.6$	$1521 \times \frac{1036}{4583} = 343.8$
Qld./W.A.	$596 \times \frac{954}{4583} = 124.1$	$596 \times \frac{2593}{4583} = 337.2$	$596 \times \frac{1036}{4583} = 134.7$
Overseas	$251 \times \frac{954}{4583} = 52.2$	$251 \times \frac{2593}{4583} = 142.0$	$= 251 \times \frac{1036}{4583} = 56.8$

*abbreviations used for places of origin are states of Australia (Vic. is Victoria, N.S.W. is New South Wales, S.A. is South Australia, Qld is Queensland and W.A. is Western Australia)

2. *Proportions.* Where the response is categorical but there are only two possible categories for the response variable, the comparison of patterns of response reduces to a comparison of proportions as illustrated in Example 16.2. Relevant statistical methodology is considered in Section 16.2.11.

Example 16.2 (Patronage of a Shopping Complex). *There is a theory that persons living in the suburbs of a large city will be less likely to travel to the nearest shopping complex if that shopping complex is in a direction away from the city center. Rather, they will tend to travel to the nearest shopping complex which is roughly in the direction of the city.*

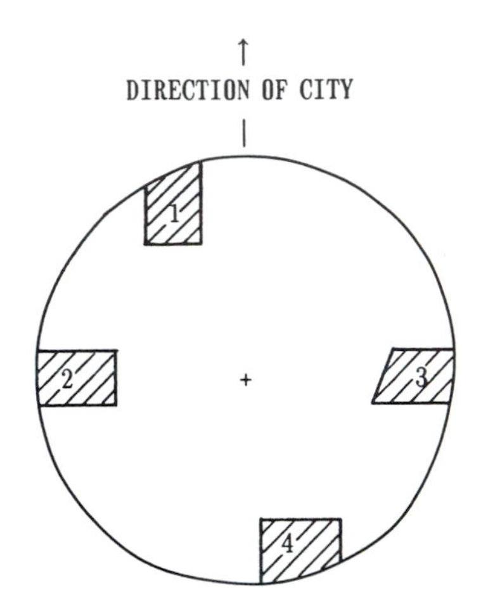

Notes.
1. *Hatched areas are sampled areas.*
2. *The shopping complex is located at the centre of the circle.*
3. *Sampled areas were chosen to be, as near as practical, the same distance from the shopping complex and to involve similar driving conditions.*
4. *The suburb is relatively homogeneous according to socio–economic measures*

Figure 16.1.1. Plan showing location of sampled areas in Example 16.2.

A research team investigating this theory for a development organization, obtained samples of households from different points around a well established shopping complex owned by the company, in the manner illustrated in Figure 16.1.1. *From each area sampled, responses were obtained from* 100 *households and the results are recorded in Table* 16.1.2.

At the simplest level, it might be asked if the expected patronage rates of the different sampling areas are the same. A more pertinent analysis, given the original hypothesis, is one which takes into account the direction of the city in respect of the sampled areas.

Table 16.1.2. Survey Results for Example 16.2

Sample area	Number of households who visited complex in last 7 days	Number of households who did not visit complex in last 7 days
1	28	72
2	43	57
3	34	66
4	56	44

3. *Scaled response.* As a general rule, when the response is scaled it is preferable to compare groups using specific parameters of interest, e.g. the mean, median, variance, etc. A test which is specifically designed to compare group means is normally more sensitive to differences in means than is a test which is designed to detect general distributional differences. The occasions on which general comparisons of distributional patterns might be of value include situations in which

(i) the observed set of responses has been summarized in a frequency table, thereby reducing the information to categorical level;

(ii) the changes from one group to another are expected to involve more than one parameter;

(iii) the changes involve the distributional form.

The methodology of this chapter requires the data for each group to comprise a frequency table. For discrete variables this condition can be met naturally, as Example 16.3 illustrates. When the response is continuous, the variable must be transformed into a categorical variable by grouping responses in the manner described in Section 1.3.3.

Example 16.3 (A Study of Traffic Light Control of Traffic Movement). *Traffic approaching an intersection controlled by traffic lights resulting from workers leaving an industrial complex had virtually the same intensity and pattern from day to day. This gave traffic engineers the opportunity to evaluate computer-constructed settings for the traffic lights under real life conditions.*

The data presented in Table 16.1.3 *were obtained under settings which, in theory, will give different patterns of flow of traffic. The aim of statistical analysis was to establish if there was evidence of differences and, if so, to highlight the manner in which the differences occurred.*

4. *Ordered categorical responses.* The categories of the response variable may be ordered, as is illustrated by the data in Table 16.1.4. The response variable underlying these data is *plant color* and the categories are *green, yellow-green, yellow-brown and brown.* The experimenter who provided the data expected a shift from the green category towards the brown category as pollution levels increase. This shift is clearly evident, for example, in the data from the first three groups, i.e. the data collected from plants exposed to fresh water. Is the shift consistent over all salinity levels? Specific methodology to cater for ordered categorical response exists which is based on the Logistic model (Section 6.6). It is discussed, for example, in Fienberg ([1978], Chap. 6) and Fox ([1984], Chap. 5). A practical alternative is to employ the methods outlined in Sections 16.2.7 and 16.2.8 which ignore the orderings of the categories.

Table 16.1.3. Data Collected in the Traffic Flow Study Reported in Example 16.3

Traffic light setting	Number of cars passing in five minute period									
	≤12	13	14	15	16	17	18	19	20	21
1	4	0	0	3	3	6	2	7	4	10
2	1	0	2	2	0	5	1	4	7	7
3	1	0	1	0	5	0	3	10	4	11
4	3	3	1	1	0	3	7	4	4	6

Traffic light setting	Number of cars passing in five minute period								
	22	23	24	25	26	27	28	29	≥30
1	11	9	15	12	8	11	6	6	3
2	16	13	20	18	11	3	5	1	4
3	8	9	14	21	11	9	4	4	5
4	15	16	9	13	13	6	7	1	7

5. *Factorial arrangement of treatments.* There may be a factorial arrangement of the treatments, as is illustrated in Table 16.1.4. In this case the experimenter's questions are defined in terms of the factors, *Salinity* and *Mercury*, rather than the treatments defined by factor combinations.

The construction of models and the application of methodology to cater for this situation is described in Section 16.2.10.

6. *Comparing observed and expected frequency distributions.* In Example 1.2, Example 3.1 and Table 6.5.1 there are sets of expected frequencies which derive from models in which the hypothesis concerns the form of the frequency distribution. Under the hypothesized model, a set of expected frequencies is defined and the task is to compare this set of frequencies with the table of observed frequencies. The comparison of tables of observed and expected frequencies is discussed in Section 16.2.12.

Example 16.4 (Monitoring Heavy Metal Pollution Using a Plant Indicator). *Example* 14.2 *introduces a research program in which water with different combinations of salinity and different levels of mercury pollution were applied to plants. After treatment, each plant was classified by leaf color into one of four ordered categories—green, yellow-green, yellow-brown and brown. The results obtained in the study are recorded in Table* 16.1.4.

Table 16.1.4. Data from the Study Reported in Example 16.4

Table of Frequencies

Treatment	1	2	3	4	5	6	7	8	9
Salinity level[1]	F	F	F	B	B	B	S	S	S
Mercury level[2]	Z	L	H	Z	L	H	Z	L	H
Colour									
green	40	36	12	49	33	12	43	39	22
yellow-green	9	7	14	1	13	21	6	8	20
yellow-brown	0	5	16	0	1	13	1	3	7
brown	1	2	8	0	3	4	0	0	1

Notes.
[1]Salinity levels are fresh (F), brackish (B) and salty (S).
[2]Mercury levels are zero (Z), low (L) and high (H).

16.2 Statistical Models and Methodology

16.2.1 Data

The data are presumed to be summarized in the form of a frequency table with the structure of Table 16.2.1.

Table 16.2.1. Table of Observed Frequencies

Group	Response Category 1	2	⋯	c	total
1	y_{110}	y_{120}	⋯	y_{1c0}	n_1
2	y_{210}	y_{220}	⋯	y_{2c0}	n_2
⋮					
g	y_{g10}	y_{g20}		y_{gc0}	n_g
total	np_{10}	np_{20}	⋯	np_{c0}	n

where y_{ij0} is the number of sample members in group i which have a response in category j

$p_{j0} = (y_{1j0} + y_{2j0} + \cdots + y_{cj0})/n$ for $j = 1, 2, \ldots, c$

16.2.2 Models

In general, the Multinomial model (Section 6.5) is presumed to apply for the set of responses from each group or treatment. This model is based on the assumption that a pre-chosen number of sample members is obtained by random sampling from each group and that responses from different groups are independent. If the comparison is between two or more treatments, the presumption is that the sample members have been randomly allocated between the treatments.

Less commonly, there is a situation in which a sample is randomly selected from a population which includes all groups, i.e. the experimenter does not control the numbers of sample members from each group. In this case, a single Multinomial distribution is presumed to apply over the complete set of responses. In practice, the basic requirement is that responses from individual sample members are independent.

For the ith group or treatment ($i = 1, 2, \ldots, g$), given a sample of n_i members, there is presumed to be a probability distribution which defines the probability π_{ij} that a response for a member of Group i will be in category j ($j = 1, 2, \ldots, c$). The expected number of sample members in group i which lie in category j is then defined by (16.2.1).

$$E(y_{ij}) = n_i \pi_{ij} \qquad \text{for } i = 1, 2, \ldots, g;\ j = 1, 2, \ldots, c. \qquad (16.2.1)$$

The models which are most commonly employed for group comparisons based on categorical data are *Log-linear models* which are defined in Section 14.2.4.

The assumption that all groups have the same distribution implies that (16.2.2) holds. This is a statement that the expected proportion in category j is constant for all groups.

$$E(y_{ij}) = n_i \pi_j \qquad \text{for } i = 1, 2, \ldots, g;\ j = 1, 2, \ldots, c \qquad (16.2.2)$$

where π_j is the probability that a response will be in category j.

If the model is based on (16.2.2), the Maximum Likelihood estimate of π_j is p_{j0} ($j = 1, 2, \ldots, c$), the proportion of sample members in category j over all groups. By replacing π_j by its sample estimate, the predicted frequencies (i.e. the *fitted values*) under the assumption of no group differences are defined in Table 16.2.2.

Table 16.2.2. Table of Fitted Values if All Groups Have the Same Distribution and Marginal Totals Equal Observed Marginals

Group	Response Category 1	2	$\cdots$	c	total
1	$n_1 p_{10}$	$n_1 p_{20}$	$\cdots$	$n_1 p_{c0}$	n_1
2	$n_2 p_{10}$	$n_2 p_{20}$	$\cdots$	$n_2 p_{c0}$	n_2
$\vdots$					
g	$n_g p_{10}$	$n_g p_{20}$	$\cdots$	$n_g p_{c0}$	n_g
total	$n p_{10}$	$n p_{20}$	$\cdots$	$n p_{c0}$	n

where $p_{j0} = (y_{1j0} + y_{2j0} + \cdots + y_{cj0})/n$ for $j = 1, 2, \ldots, c$

Expected frequencies when there is group or treatment structure. Less restrictive models can be defined by considering the groups or treatments to be collected into sets such that *within* a set the groups are presumed to have the same distribution, but there is allowance for *between* set differences. By way of illustration, consider Example 16.1. There are four groups, three of which are Australian and one of which is not Australian (the *Overseas* group). A model may be proposed which presumes that the distributions are identical for the Australian groups but this distribution may differ from that proposed for the Overseas group.

Suppose that the g groups are divisible into two sets such that Groups $1, 2, \ldots, s$ fall in Set A and Groups $s+1, s+2, \ldots, g$ fall into Set B. If the assumption is that all groups in Set A have the same distribution and all groups in Set B have the same distribution, the probability of a member of the ith group being classified in category j is defined by (16.2.3).

$$\begin{aligned} E(y_{ij}) &= n_i \pi_j^a \qquad \text{for } i = 1, 2, \ldots, s \\ &= n_i \pi_j^b \qquad \text{for } i = s+1, s+2, \ldots, g. \end{aligned} \tag{16.2.3}$$

The fitted values which result under (16.2.3) are presented in Table 16.2.3.

Table 16.2.3. Table of Fitted Values Based on (16.2.3) Assuming Marginal Total Equal Observed Marginals

Set	Group	Response Category 1	2	$\cdots$	c	total
	1	$n_1 p_{10}^a$	$n_1 p_{20}^a$		$n_1 p_{c0}^a$	n_1
	2	$n_2 p_{10}^a$	$n_2 p_{20}^a$		$n_2 p_{c0}^a$	n_2
A	$\vdots$					
	s	$n_s p_{10}^a$	$n_s p_{20}^a$		$n_s p_{c0}^a$	n_s
	$s+1$	$n_{s+1} p_{10}^b$	$n_{s+1} p_{20}^b$		$n_{s+1} p_{c0}^b$	n_{s+1}
	$s+2$	$n_{s+2} p_{10}^b$	$n_{s+2} p_{20}^b$		$n_{s+2} p_{c0}^b$	n_{s+2}
B	$\vdots$					
	g	$n_g p_{10}^b$	$n_g p_{20}^b$		$n_g p_{c0}^b$	n_g

where $p_{j0}^a = (y_{1j} + y_{2j} + \cdots + y_{sj})/(n_1 + n_2 + \cdots + n_s)$
$p_{j0}^b = (y_{s+1,j} + y_{s+2,j} + \cdots + y_{gj})/(n_{s+1} + n_{s+2} + \cdots + n_g)$ for $j = 1, 2, \ldots, k$

A more complex example is provided when there is a factorial arrangement of treatments, as illustrated by Example 16.4 where the combinations of *mercury* and *salinity* define the treatments. In that case, several models might be proposed. One might require that the distribution varies with changes in *mercury* levels but not with changes in *Salinity* levels. Others may allow for variation with changes in either factor. The complete range of possible models is considered in Section 16.2.10.

16.2.3 Choosing a Statistic to Compare Model and Data

Two statistics are defined in Section 14.3.2 for comparing a table of observed frequences $\{y_{ij0}\}$ with a table of fitted frequencies $\{\hat{y}_{ij0}\}$ based on a Log-linear model M. One is the Pearson Chi-squared statistic defined by (16.2.4). The other is the Deviance statistic defined by (16.2.5).

$$X_p^2 = \sum_{i=1}^{g}\sum_{j=1}^{c} \frac{(y_{ij} - \hat{y}_{ij})^2}{\hat{y}_{ij}}, \tag{16.2.4}$$

$$D(y, M) = 2\sum_{i=1}^{g}\sum_{j=1}^{c} y_{ij} \log\left[\frac{y_{ij}}{\hat{y}_{ij}}\right]. \tag{16.2.5}$$

Both statistics have the property that a value of zero implies perfect agreement between model and data and increasing values reflect worsening agreement. In practical terms, the values of the two statistics, when applied to the same sets of observed and fitted values, rarely show a large difference. For the reasons given in Section 14.3.2, the calculations in this book are based on the Deviance statistic.

16.2.4 Experimental Aims

1. *Examining the hypothesis that all groups are the same.* The most basic aim is to establish if there is evidence to reject the hypothesis that all groups have the same distribution. In statistical terms, this question is answered in the affirmative if the variation between the set of observed frequencies defined in Table 16.2.1 and the set of fitted frequencies defined in Table 16.2.2 is too great to be explained as sampling variation. The methodology for this purpose is presented in Section 16.2.7. (In Example 16.1, this approach would be applied to establish if spending patterns appear to vary between the four groups of tourists.)

2. *More detailed examination of group differences.* If there is evidence that groups have different patterns, there is likely to be interest in which pairs of groups show differences and the nature of the differences. Statistical methods which are useful in answering these questions are presented in Section 16.2.8.

Alternatively, there may be structure recognized in the groups. One possibility is that the groups can naturally be formed into two or more sets (as in Example 16.1 where there is a natural division of places of origin into *Australian* and *Overseas*). In this case there is a need to

(i) examine the possibility that the sets have different distributions; and

(ii) establish if there is uniformity within the sets.

The methodology appropriate for answering these questions is described in Section 16.2.9.

3. *Factorial arrangement of treatments.* In a study in which the treatments are identified by combinations of factors A and B, there are three questions which may be of interest. They are

(i) Does the distribution vary with changing levels of factor A?

(ii) Does the distribution vary with changing levels of factor B?

(iii) Does the effect of changing levels of factor A depend on the level of factor B?

(These questions may be related to Example 16.4 by equating factor A with mercury level and factor B with salinity level.)

The methodology appropriate for answering these questions is contained in Section 16.2.10.

4. *Comparing an observed distribution with an expected distribution.* The above comparisons have related to comparisons between different groups. There is another type of problem in which only one group is identified and a theoretical form is postulated for the expected frequency distribution. A range of examples are introduced in Section 16.2.12 together with the method of analysis.

5. *The comparison of proportions, success rates and odds.* If there are only two categories for the response variable, the comparison may be based on proportions, success rates or odds. The choice of methodology is discussed in Section 16.2.11.

16.2.5 Analysis of Deviance

As explained in Section 14.3.3, the statistical analysis based on the Deviance statistic is termed *Analysis of deviance* and the form of presentation of results is termed the *Analysis of deviance table.* Both are formally introduced in Section 14.3.3.

The primary application of Analysis of deviance in this chapter is to test the hypothesis that all groups have the same distribution. The test is based on the comparison of the table of observed frequencies in Table 16.2.1 with the set of fitted frequencies in Table 16.2.2. The application of the test is described in Section 16.2.7.

More complex comparisons can be performed by invoking the additivity property of the Deviance statistic defined in (14.3.10). For example, if M_0 is the model which is based on the assumption that all groups have the same distribution and M_1 is the model based on the assumption that the groups can be formed into sets such that within sets the groups have the same distributions, but there may be differences between sets, the Deviance $D(y, M_0)$ may be partitioned according to (16.2.6).

$$\begin{array}{ccccc} D(y, M_0) & = & D(M_1, M_0) & + & D(y, M_1) \\ \text{TOTAL DEVIANCE} & = & \text{BETWEEN SET DEVIANCE} & + & \text{WITHIN SET DEVIANCE} \end{array} \quad (16.2.6)$$

The Analysis of deviance table corresponding to this partitioning is presented in Table 16.2.4.

Table 16.2.4. Analysis of Deviance Table Based on (16.2.6)

Source of Variation	D.F.	Deviance	p
Residual (M_0)	d_0	$D(y, M_0)$	p_{00}
Residual (M_1)	d_1	$D(y, M_1)$	p_{01}
Gain from M_1 over M_0	$d_0 - d_1$	$D(y, M_0) - D(y, M_1)$	p_{02}

Degrees of freedom. The definition of *Degrees of freedom* (d) is given in Section 14.3.1 and the evaluation of d for a given model and data set is based on (14.3.5). For the tables and models considered in this chapter there are some simple rules for establishing the degrees of freedom associated with a specified model.

The expression of (14.3.5) in the current context is given by (16.2.7).

$$d = gc - p \quad (16.2.7)$$

where g is the number of groups; c is the number of categories; and p is the number of independent parameters to be estimated. The minimal set of parameters to be assigned values are those which appear in (16.2.2). They constitute the g sample sizes, $n_1, n_2, \ldots, n_g$ and the expected proportions in the c categories, $\pi_1, \pi_2, \ldots, \pi_c$. Note, however, that the expected proportions are not independent since they must sum to one. Hence, there are only $c - 1$ independent parameters for which values must be assigned. Under M_0, the degrees of freedom are given by (16.2.8).

$$d = gc - [g + (c - 1)] = (g - 1)(c - 1). \quad (16.2.8)$$

If (16.2.3) applies, i.e. the groups are formed into two sets and the same distribution is assumed for all groups within a set, there are two sets of expected proportions to be estimated, one for Set A and one for Set B. Thus,

$$d = gc - [g + 2(c - 1)] = (g - 2)(c - 1).$$

By following the same reasoning, if there were t rather than two sets, the degrees of freedom would be $d = (g - t)(c - 1)$.

With a little common sense and if care is taken to identify constraints on the parameters (such as the requirement that the category proportions must sum to one), the determination of degrees of freedom associated with a specified model is not difficult.

16.2.6 Studying Residuals

If there is evidence of group differences or, more generally, if a model does not provide an adequate description of the data, the next step in analysis would typically be to establish the manner in which the model fails to fit the data. This secondary analysis can be based on an examination of *standardized residuals.* Under the Log-linear models proposed in Section 16.2.2, the standardized residuals are values of the statistics defined in (16.2.9).

$$r_{s_{ij}} = \frac{y_{ij} - \hat{y}_{ij}}{\sqrt{\{\hat{y}_{ij}\}}} \qquad \text{for } i = 1, 2, \ldots, g; \; j = 1, 2, \ldots, c \tag{16.2.9}$$

where $\{\hat{y}_{ij}\}$ is the set of estimators of the fitted values under the proposed model.

If the correct model has been employed, the sampling distribution of $r_{s_{ij}}$ in large samples is well approximated by a $N(0, 1)$ distribution. Thus, a value of $r_{s_{ij}}$ has approximately a 95% chance of lying within the range -2 to $+2$ if the correct model is fitted. When the model is judged inadequate by a test based on the Deviance statistic, the manner in which the model fails can be determined by noting which standardized residuals are of surprisingly large magnitude. Commonly, magnitudes in excess of 2.0 are taken as indicators of cells in which there is a poor fit of the model to the data.

The examination may be extended by use of the statistics S_i ($i = 1, 2, \ldots, g$) which are defined in (16.2.10).

$$S_i = \sum_{j=1}^{c} r_{s_{ij}}^2 \qquad \text{for } i = 1, 2, \ldots, g. \tag{16.2.10}$$

Increasing values of S_i reflect worsening agreement between model and data. Thus, if a model has been judged inadequate by Analysis of deviance, it is the groups which have the largest values of S that can be identified as making the greatest contributions to the failure of the model. By examining the pattern of signs of the residuals it is possible to establish the manner in which groups differ. The application of residual analysis is illustrated in Section 16.2.8.

16.2.7 Seeking Evidence of Group or Treatment Differences

If there is no group structure identified prior to analysis, the first stage in analysis is to examine the general hypothesis that all groups have the same distribution. This requires the comparison of the observed frequency table represented by Table 16.2.1 and the table of fitted values represented by Table 16.2.2. The steps in the analysis are described below.

Step 1 — Compute the table of fitted values using the formulae in Table 16.2.2.

Step 2 — Check that no fitted value is less than 1.0. Where this condition is not met, pool observed frequencies from different categories to ensure the condition is met. (This is a necessary condition for the acceptability of the Chi-squared approximation in the computation of a p-value from the value of the Deviance statistic.) There are no formal rules to establish which categories should be grouped. If the categories have a natural ordering, it is sensible to pool frequencies from neighboring categories. If there are certain categories which are of interest in their own right, they should not be pooled (so long as they have no frequencies below 1.0). If it is inconvenient to combine categories, the alternative is to combine groups. Obviously, this would mean the loss of information about differences between the pooled groups.

Step 3 — Compute a value of $D(y, M_0)$ from (16.2.5) based on the fitted values obtained at Steps 1 and 2. Call this value X_0^2.

Step 4 — Use the Chi-squared tables with $d = (g-1)(c-1)$ degrees of freedom to convert X_0^2 to a p-value by obtaining a value for $p_0 = \Pr\{\chi^2(d) \geq X_0^2\}$.

The value of p which results is an approximate p-value for assessing the acceptability of the hypothesis that all groups have the same distribution. If there is evidence against the hypothesis, more detailed examination to establish the nature of the differences in the patterns should follow. Section 16.2.8 provides steps for identifying and presenting group differences.

Illustration. The calculations are applid to the data in Table 16.1.1(a). The fitted values are shown in Table 16.1.1(c). The value of X_0^2 computed using (16.2.5) is 111.9. From Chi-squared tables with $(4-1)(3-1) = 6$ degrees of freedom, $p_0 = \Pr\{\chi^2(6) \geq 111.9\}$ is found to be less than 0.01. Hence, there is strong evidence against the hypothesis that the distributions are the same for all groups.

16.2.8 Secondary Analysis: Identifying Group Differences

When a model fails to provide an adequate fit to the data, there is generally a need to make a detailed examination to identify the manner in which the model is inadequate. This is particularly so when there is no identifiable group structure and the model which is rejected has assumed all groups have the same distribution.

There is no formal test which is useful to establish which pairs of groups are different. The steps outlined below provide a simple, yet practical, means of identifying where differences lie.

Step 1. Prepare a table of standardized residuals using (16.2.9) and include the sums of squares defined by (16.2.10). Table 16.2.5(a) provides an illustration.

Table 16.2.5. Analysis of the Data in Example 16.1 Assuming All Groups Have the Same Distribution

(a) Table of standardized residuals

Place of Origin	Amount Spent			
	low	medium	high	S_i
Vic.	3.95	1.15	−5.62	48.5
N.S.W./S.A.	−4.31	−1.14	5.94	55.1
Qld./W.A.	−0.63	−0.94	2.09	5.6
Overseas	−0.17	0.84	−1.16	2.1

(b) Information extracted from (a)

Place of Origin	Amount Spent			
	low	medium	high	S_i
Vic.	⊕	+	⊖	48.5
N.S.W./S.A.	⊖	−	⊕	55.1
Qld./W.A.	−	−	+	5.6
Overseas	−	+	−	2.1

⊕ denotes a value of $r_{s_{ij}}$ greater than 2.0.
⊖ denotes a value of $r_{s_{ij}}$ less than −2.0.

Step 2. From the information computed at Step 1 construct a table of signs of standardized residuals in which (i) signs associated with standardized residuals with magnitudes in excess of 2.0 are clearly distinguished; and (ii) values of S_i $(i = 1, 2, \ldots, g)$ are listed for groups. Table 16.2.5(b) provides an illustration.

The information obtained at Step 2 summarizes in a simple form the relevant differences between groups. In effect, the pattern in each group is being compared with the average pattern for all the groups. Where one group deviates from the average in a certain direction there will be another group or groups which deviates in the opposite direction, where 'direction' is identified by the pattern of the signs. Table 16.2.5(b) provides a clear example. The group labelled *Vic.*, that is the Victorians, differ substantially from the average, as judged by the high S value. They are characterized by having a surplus of low spenders and a deficiency of high spenders. (The 'Scrooges' among the visitors?) Visitors labelled *N.S.W./S.A.*, that is the visitors from New South Wales and South Australia, also deviate in a substantial way from the average, but, judging from the pattern of signs, in the opposite direction to the Victorians. The group labelled *Qld./W.A.*, that is the visitors from Queensland and Western Australia and the group labelled *Overseas*, have relatively low S values which suggests they are the *average* groups in respect of the variable *Amount spent.*

16.2.9 Analysis When There Is Group Structure

Structure may appear in two forms. They may be the collection of groups or treatments into sets for the purpose of examining between sets and within sets variation. Alternatively, the groups or treatments may have an ordering and the aim is to establish if the frequency distribution varies in a systematic fashion which is related to the ordering. These two forms appear in the following applications:

- in Example 16.1, the four groups may be divided into two sets—the *Australian* set and the *Overseas* set for the purpose of (i) comparing spending patterns of Australian and overseas tourists; and (ii) comparing spending patterns between groups within the Australian set;

- in Example 16.4, the nine treatments are arranged in a factorial arrangement and there is interest in the comparisons of sets based on factor levels. An added degree of complexity is introduced by the need to consider the effect of the two factors simultaneously; and

- in Example 16.2, there is an ordering of the sites from *Site* 1 which is nearest to the city to *Site* 4 which is furthest from the city. There is interest in establishing if the proportion of households using the local shopping center increases with increasing distance from the city, i.e. from site 1 to site 4.

The steps in analysis when interest lies in *between sets* and *within sets* comparisons are presented below. The analysis when there is a factorial arrangement of treatments is described in Section 16.2.10. Where the groups are ordered, the analysis of residuals described in Section 16.2.8 provides a simple, yet effective, way of seeking evidence of a systematic change in the distribution.

1. *Between and within set comparisons.* The two basic questions to be answered are:

1. *Do the patterns differ between the sets?*

2. *Do the patterns differ within the sets?*

To answer these questions, two models must be fitted, the model M_0 which includes the hypothesis that all groups have the same distribution, and the model M_1 which is less restrictive and includes the hypothesis that all groups in the same set have identical distributions. A comparison of M_0 with M_1 provides a test of the first hypothesis and a comparison of M_1 with the data provides a test of the second hypothesis. The steps in analysis are presented below.

Step 1 — *Fitting the model M_0.* Follow the steps in Section 16.2.7 to obtain a value for the Deviance, $D(y, M_0)$. Enter this value as D_0 and the residual degrees of freedom (d_0) in an Analysis of deviance table with the format of Table 16.2.6(a).

Table 16.2.6. Analysis of Deviance Based on (16.2.6)

(a) Between and within set comparisons

Source of Variation	D.F.	Deviance	p
Residual (Groups)	d_0	D_0	p_{00}
Residual (Sets)	d_1	D_1	p_{01}
Between sets	$d_g = d_0 - d_1$	$D(M_1, M_0) = D_0 - D_1$	p_{02}

(b) Comparing Australian and Overseas spending patterns (Example 16.1)

Source of Variation	D.F.	Deviance	p
Residual (Groups)	6	111.9	<0.01
Residual (Sets)	4	109.4	<0.01
Between sets	2	2.5	>0.05

Step 2 — *Fitting the model M_1.* **If** a statistical package is being used for the computations, and

(i) data are held case-by-case, define a new factor (variable) with values identifying the *set* to which each sample belongs; or

(ii) data are held in a frequency table classified by *group*, condense the table to provide a frequency table which is classified by *set*.

Hence, follow the steps in Section 16.2.7 using the factor *set* rather than *group* as the classifying factor to obtain a value of the Deviance, $D(y, M_1)$. Enter this value, D_1, plus the corresponding residual degrees of freedom, d_1, into the Analysis of deviance table begun in Step 1.

Otherwise, form a frequency table classified by *set* and follow the steps in Section 16.2.7 to compute a value of the Deviance, $D(y, M_1)$, and residual degrees of freedom (d_1). Enter this value, D_1, plus the corresponding residual degrees of freedom d_1, into the Analysis of deviance table begun in Step 2.

Step 3 — ***Completing the Analysis of deviance table.*** Compute the values of $D(M_1, M_0)$ and d_g using the formulae given in Table 16.2.6(a). Enter these values in the Analysis of deviance table.

Step 4 — ***Computing and interpreting p-values.*** Where p-values have not been computed automatically as part of the analysis by a statistical package, use one of the methods described in Section 7.5.4 to obtain values for p_{00}, p_{01} and p_{02}. The p-values have the following applications and can be interpreted by following the usual guidelines.

> p_{00} provides a test of the hypothesis that there are no group differences.
>
> p_{01} provides a test of the hypothesis that the distributions are identical within sets.
>
> p_{02} provides a test of the hypothesis that the distributions are identical between sets.

Illustration. In Example 16.1, the four groups were collected into two sets—the *Australian* set comprising groups 1 to 3 and the *Overseas* set composed only of group 4. The set membership is documented in Table 16.2.7. Table 16.2.6(b) contains the Analysis of deviance table produced in Steps 1 to 3. Note that the fitted frequencies under the model M_1 are provided in Table 16.2.7 and are computed from formulae given in Table 16.2.3 with $s = 3$ and $g = 4$. The p-values lead to the following conclusion.

> p_{00} is less than 0.01 and hence, there is strong evidence that the four groups have different distributions.
>
> p_{01} is less than 0.01 indicating that there is variation between the groups within the Australian set.
>
> p_{02} is greater than 0.05 indicating that the spending pattern for the average Australian visitor is not significantly different from that of Overseas visitors.

Table 16.2.7. Fitted Frequencies Based on Data in Example 16.1 and Model M_1

Place of origin	Set	Amount Spent: low	medium	high
Vic.	A	$2215 \times \frac{903}{4332}$ $= 461.7$	$2215 \times \frac{2441}{4332}$ $= 1248.1$	$2215 \times \frac{988}{4332}$ $= 505.2$
N.S.W./S.A.	A	$1521 \times \frac{903}{4332}$ $= 317.1$	$1521 \times \frac{2441}{4332}$ $= 857.1$	$1521 \times \frac{988}{4332}$ $= 346.9$
Qld./W.A.	A	$596 \times \frac{903}{4332}$ $= 124.2$	$596 \times \frac{2441}{4332}$ $= 335.8$	$596 \times \frac{988}{4332}$ $= 135.9$
Overseas	B	$251 \times \frac{51}{251}$ $= 51.0$	$251 \times \frac{152}{251}$ $= 152.0$	$251 \times \frac{48}{251}$ $= 48.0$

Notes

1. Sets A and B are the Australian and Overseas sets respectively.
2. 4332 is the total count in Set A.
3. 903, 2441 and 988 are the total counts in Set A for the low, medium and high categories.
4. Since there is only one group in Set B, the fitted frequencies must match the observed frequencies.

The manner in which spending patterns vary between Australian visitors from different places of origin (i.e. different groups) has already been considered in Section 16.2.8.

16.2.10 Factorial Arrangement of Treatments

If there are two factors, A and B, the general questions of interest are

1. *Does a change in the level of factor A affect the frequency distribution?*
2. *Does a change in the level of factor B affect the frequency distribution?*
3. *Does the effect of varying factor A depend on the level of factor B?*

Methodology based on Log-linear models is available to provide statistical examination of these questions. It is based on the general rules applying to Analysis of deviance presented in Sections 14.3.3 and 16.2.5. The steps in analyzing data are provided below. For more detail on the Log-linear format for model equations, the interpretation of parameters and the construction of the Analysis of deviance table, refer to Bishop et al. [1975].

Model identification. The identification of models is based on the following rules:

1. If the expected distribution may vary with changes in levels of factor A, the letter A appears in the model description, otherwise it is absent.

2. If factors A and B are both required in the equation, there is the possibility that

   (i) their contributions are independent, in which case their joint contribution is represented as $A + B$; or

   (ii) the contribution from A depends on the level of B (and vice versa) in which case there is an $A \times B$ *interaction* and their joint contribution is represented as $A + B + A.B$.

Notation

M_0 All treatments (i.e. factor combinations) have the same effect.

M_A The expected distribution may not be the same over all levels of factors A but is presumed to be identical for all levels of factor B.

M_{A+B} The expected distribution may vary with changes in levels of either factor A or factor B. However, the effect of changing factor A is independent of the level of factor B.

$M_{A+B+A.B}$ The expected distribution may vary with changes in levels of either factor A or factor B. However, the effect of changing factor A may not be the same at all levels of factor B.

Where the number of factors exceeds two, the notation is expanded in an obvious manner with the three factor interaction between factors A, B and C requiring a component $A.B.C$ in the specification.

Analysis of deviance table. The format of the Analysis of deviance table is shown in Table 16.2.8 for the situation in which there are two factors, A with a levels and B with b levels. The response variable is presumed to have c categories.

Table 16.2.8. Analysis of Deviance Table for a Two-Factor Factorial Arrangement of Treatments

Source of Variation	D.F.	Deviance	p
Residual (M_0)	$r-1$	$D(y, M_0)$	p_{00}
Residual (M_A)	$r-(a-1)$	$D(y, M_A)$	p_{A0}
Residual (M_B)	$r-(b-1)$	$D(y, M_B)$	p_{B0}
Residual ($M_A + M_B$)	$r-a-b+1$	$D(y, M_{A+B})$	p_{AB}
Gain adding A ($M_0 \to M_A$)	$a-1$	$D(y, M_0) - D(y, M_A)$	p_{a1}
Gain adding B ($M_0 \to M_B$)	$b-1$	$D(y, M_0) - D(y, M_B)$	$p_{b,1}$
Gain adding A ($M_B \to M_{A+B}$)	$a-1$	$D(y, M_B) - D(y, M_{A+B})$	p'_{a1}
Gain adding B ($M_A \to M_{A+B}$)	$b-1$	$D(y, M_A) - D(y, M_{A+B})$	p'_{b1}

where $r = (ab-1)(c-1)$

Stages in Analysis. The following strategy is recommended for analysis when there is a factorial arrangement of treatments with two factors. The notation employed is taken from Table 16.2.8.

In each of the following steps in which a model is fitted to the data, the set of fitted frequencies under the model should be examined to ensure there are no frequencies below 1.0. *In this event, it is necessary to combine categories and refit the model until the condition is met. Failure to meet the condition brings into suspicion the p-value from the comparison of model and data since this is based on a Chi-squared approximation which loses validity when one or more fitted frequencies is small. If the condition cannot be met it is likely that the sample size is too small for application of a model of the desired complexity. If condensation of categories destroys an important component of the experiment, seek advice from a consultant statistician. There may be an alternative approach to analysis.*

Step 1 — *Is the variation in frequency distributions between factor combinations merely sampling variation?*

Fit M_0 to the data.

If this model fits the data, i.e. p_{00} is at least 0.05, it is reasonable to conclude that neither factor A nor factor B influences response. Unless there is an independent basis for examining more specific hypotheses, no further analysis is recommended.

Otherwise, proceed to Step 2.

Step 2 — *Is there an $A \times B$ interaction?*

Fit the model M_{A+B}.

If p_{AB0} is less than 0.05, the need is established for the interaction term $A + B$ in the model, i.e. there is evidence that both factors are having an effect, and the effect of factor A is dependent on the level of factor B. In this case, the method outlined in Section 16.2.8 should be employed to examine the set of standardized residuals from the fit of the model M_{A+B}. This examination can provide information on the manner in which factors A and B affect the pattern of response and the effect of their interaction. The application is illustrated below.

Otherwise, proceed to Step 3.

Step 3 — *Does factor A affect response?*

Assuming there is no evidence of an $A \times B$ interaction, fit the model M_A to the data. There are two possibilities.

1. The model M_A fits the data.

   If the model M_A fits the data, i.e. if p_{A0} exceeds 0.05, compare the fit of M_A with the fit of M_0 using p_{a1}. If it is found that M_A provides a significantly better fit than does M_0, this may be taken as evidence that the response pattern varies with changes in the level of factor A. The pattern in the standardized residuals (Section 16.2.8) based on fitted frequencies computed under the model M_0 allow the nature of the change to be determined. Since there is no $A \times B$ interaction, it is valid (Fienberg [1978], Theorem 3–1) to collapse the tables of observed and fitted values over the B categories, i.e. to sum frequencies in a response category which have the same level of factor A and to compute standardized residuals from the reduced table.

2. The model M_A does not fit the data.

   In this case, it is the comparison of the model M_{A+B} with the model M_B, i.e. the value of p'_{a1}, which establishes if there is evidence of an effect of factor A. If p'_{a1} is less than 0.05, there is evidence of an A effect. The nature of the effect can be determined from the pattern in the residuals which are based on fitted values determined under the model M_B. As noted in 1., the table of observed and fitted frequencies may be collapsed over the categories of B before the computation of standardizes residuals.

— *Does factor B affect response?*

The approach described above applies equally to a consideration of the effect of factor B with the symbols for A and B interchanged in the notation for models and p-values.

Illustration. The analysis of data in Table 16.1.4 is presented below. The data arise from a study in which there are two factors—*mercury* (*pollution*) and *salinity*, each with three levels—and the response variable has four categories. The Analysis of deviance table based on the format of Table 16.2.8 is given in Table 16.2.9. (Note, however, that Table 16.2.9 is based on data for which the *brown* and *yellow-brown* categories are combined—see reasons given below for the combination of categories.)

1. *Analysis of deviance*

   (a) *Fitting the model* M_0

   The Deviance $D(y, M_0)$ based on a four category response has a value of 167.0 and $p_{00} = \Pr\{\chi^2(24) \geq 167\}$ is less than 0.01. Hence, there is evidence that at least one of the factors is having an effect.

   (b) *Fitting the model* M_{M+S}

   Problems with small fitted values

   The set of fitted values under this model includes several values which are less than 1.0. Hence, there is a need to condense at least two categories to meet the requirements of the Chi-squared approximation which is used to give a p-value. Since the small fitted values all fall in the *brown* and *yellow-brown* categories, and since these are neighboring categories on the scale from green to brown, they are the obvious categories to combine.

After combination of these categories, there were no further problems with small fitted values. The Deviance from the fit of M_{M+S} to the data based on three response categories is given in Table 16.2.9.

Table 16.2.9. Analysis of Deviance Table for Example 16.4

Based on three categories—categories brown and yellow-brown are combined

Source of Variation	D.F.	Deviance	p
Residual (M_0)	16	156.5	<0.01
Residual (M_S)	12	145.2	<0.01
Residual (M_M)	12	28.6	<0.01
Residual (M_{M+S})	8	15.0	0.05
Gain adding S ($M_0 \to M_S$)	4	11.3	0.02
Gain adding M ($M_0 \to M_M$)	4	127.9	<0.01
Gain adding S ($M_M \to M_{M+S}$)	4	13.6	<0.01
Gain adding M ($M_S \to M_{M+S}$)	4	130.2	<0.01

(c) *Fitting the model M_{M+S}*

The p-value is 0.05 which is on the borderline of the need to consider the Mercury × Salinity interaction. The two options are considered below. Firstly, examination is based on the assumption of an interaction, and secondly, examination is based on the assumption of no interaction.

(d) *Examining the nature of the Mercury × Salinity interaction.*

A table of standardized residuals was constructed by following Steps 1 and 2 in Section 16.2.8 and revealed only two standardized residuals with magnitudes in excess of 2.0 and no clear cut pattern of interest. The scientist who conducted the experiment was primarily interested in the effect of changes in mercury levels. Hence, it was decided to examine the pattern of standardized residuals derived under the model which presumes no variation with changes in mercury but does allow for variation with changes in salinity, i.e. the pattern under the model M_S. By this means, both the effects of changing mercury levels, and the manner in which such changes are influenced by different levels of salinity are highlighted. This set of standardized residuals is presented in Table 16.2.10.

Interest in the examination of patterns of residuals in Table 16.2.10 centers on the effect of changing mercury levels from zero to low to high within each salinity level. It is apparent that within all salinity levels the same pattern emerges, namely, the proportion of green plants is higher than expected when the mercury level is low and is lower than expected when the mercury is high. The converse is true at the other end of the color spectrum (i.e. in the combined group). The basic evidence for the interaction effect is seen in the relative magnitudes of the sums of squares of the residuals. The (relatively) lower set of S values for the highest level of salinity (6.8 1.8 15.2) suggests less differentiation between mercury levels in salt water than is found in fresh and brackish water.

Table 16.2.10. Signs of Standardized Residuals[1] and Related Information from the Fit of M_S to the Data in Table 16.1.4[2]

Treatment	Mercury	Salinity	Response: green	yellow-green	combined	S_i
1	zero	fresh	+	−	⊖	12.7
2	low	fresh	+	−	−	3.7
3	high	fresh	⊖[3]	+	⊕[4]	28.4
4	zero	brackish	⊕	⊖	⊖	26.7
5	low	brackish	+	+	−	1.5
6	high	brackish	⊖	⊕	⊕	33.6
7	zero	salty	+	−	−	6.8
8	low	salty	+	−	−	1.8
9	high	salty	⊖	⊕	+	15.2

Notes.

[1] based on a log-linear model fitted to the data

[2] the yellow-brown and brown categories have been combined

[3] ⊖ denotes a standardized residual less than −2.0

[4] ⊕ denotes a standardized residual in excess of 2.0

(e) *Examining the effect of mercury assuming no $M \times S$ interaction.*

It can be seen in Table 16.2.9 that the p-value which measures the fit of the model M_M to the data is less than 0.01, i.e. the model M_M does not fit the data. Hence, to examine the mercury effect it is necessary to compare the models M_{M+S} and M_S. The p-value for this comparison is seen from Table 16.2.9 to be less than 0.01, thereby suggesting a significant mercury effect.

Table 16.2.11(a) is derived from Table 16.1.4 by summing over the categories of the factor *salinity.* Table 16.2.11(b) is obtained from the fitted values under the model M_S by summing over categories of *salinity*. Following the steps in Section 16.2.8, Table 16.2.11(c) was constructed. It clearly shows the substantial effect of changes in mercury level on the pattern of response. There is evidence of a significant decrease in the proportion of plants at the green end of the spectrum with increasing mercury levels and a corresponding increase in plants at the brown end of the spectrum. Based on the values of S_1, S_2 and S_3 and the signs of thc residuals, it is apparent that the *low* level of mercury is seen to be intermediate between the *zero* and *high* levels.

In this example the results are clear cut and Table 16.2.11(d), which provides the proportions of each response category for the three levels of mercury, displays the effect in an easily interpreted form.

Table 16.2.11. Information on the Effect of Mercury on Plant Colour

(a) observed frequencies (obtained from Table 16.1.4[1])

Mercury Level	Response green	yellow-green	combined
zero	132	16	2
low	108	28	14
high	46	55	49

(b) Fitted values under M_S[2]

Mercury Level	Response green	yellow-green	combined
zero	95.3	33.0	21.7
low	95.3	33.0	21.7
high	95.3	33.0	21.7

(c) Pattern of standardized residuals

Mercury Level	Response green	yellow-green	combined	S_i
zero	$\oplus$[3]	$\ominus$[4]	$\ominus$	41
low	$-$	$-$	$+$	5
high	$\ominus$	$\oplus$	$\oplus$	75

Table 16.2.11 (*cont.*)
(d) relative frequencies

Mercury Level	Response		
	green	yellow-green	combined
zero	0.88	0.11	0.01
low	0.72	0.19	0.09
high	0.31	0.37	0.33

Notes.
[1]the yellow-brown and brown categories have been combined
[2]based on a log-linear model fitted to the data
[3]$\oplus$ denotes a standardized residual in excess of 2.0
[4]$\ominus$ denotes a standardized residual less than −2.0

16.2.11 Comparing Proportions, Success Rates and Odds

When the response variable has only two categories, the specification of a frequency distribution for a group or treatment can be reduced to a single parameter. This may be the expected proportion in category 1 (or category 2), the expected number of occurrences in category 1 per, say, 1,000 trials or the odds of a response in category 1 (relative to category 2). These possibilities are summarized in the following table.

Group	1	2	···	g
Proportion	π_1	π_2		π_g
Rate/1,000	$1{,}000\pi_1$	$1{,}000\pi_2$		$1{,}000\pi_g$
Odds	$\omega_1 = \pi_1/(1-\pi_1)$	$\omega_2 = \pi_2/(1-\pi_2)$		$\omega_g = \pi_g/(1-\pi_g)$

Provided individual responses are independent, the Binomial model applies to the data collected from each group.

Seeking evidence of treatment or group differences. If the hypothesis $\pi_1 = \pi_2 = \cdots = \pi_g$ is correct, then expected success rates are identical and odds are identical for all groups or treatments. A test that all treatments have the same effect, or all groups are identical can be based on the general method described in Section 16.2.7. Where there is evidence of treatment or group differences, a simple assessment of the nature of the differences can be based on the set of standardized residuals computed from (16.2.11).

$$r_{s_i} = [p_i - p_.]/\sqrt{p_.(1-p_.)/n_i} \qquad \text{for } i = 1, 2, \ldots, g \qquad (16.2.11)$$

where p_i is the proportion of sample members in group i who recorded successes; n_i is the number of sample members in group i; and $p_.$ is the proportion of sample members who recorded successes.

Groups which have negative standardized residuals have below average proportions or rates of success. The magnitudes of the standardized residuals indicate the relative extents of departure from the average.

Illustration. The following information comes from the survey results in Table 16.1.2.

$$p_. = (28 + 43 + 34 + 56)/400 = 0.4025$$

Site (i)	1	2	3	4
p_i	0.28	0.43	0.34	0.56
n_i	100	100	100	100
r_{s_i}	−2.5	0.6	−1.3	3.2

It is apparent that sites 1 and 4 are the extreme sites with site 1 being below average and site 4 above average.

Factorial arrangement of treatments. The situation changes when there is group or treatment structure. The experimenter must then consider the nature of the comparison between groups. The possibilities are discussed in Chapter 12 and comprise

- the study of differences in expected proportions (or success rates);
- the study of ratios of expected proportions (or success rates); and
- the study of odds ratios.

As McCullagh and Nelder [1983] explain, the form of comparison can be made independently of the nature of the distributional form. With the GLIM and GENSTAT statistical packages this independence is maintained and the appropriate link function can be employed in conjunction with the Binomial distribution included in the model (identity link for differences, log. link for ratios and logit link for odds ratios). However, most packages do not offer this flexibility. In general, the available options are

(i) the use of a Log-linear model for the study of ratios; and

(ii) the use of the Logistic model for the study of odds ratios.

16.2.12 Examining a Distributional Hypothesis

An experimenter may summarize data from a single group in a frequency table with the aim of determining if the observed pattern in the frequencies is consistent with a set of hypothesized frequencies. For example,

- in Example 1.2, the distribution of bomb hits on London was obtained to establish if the pattern of hits fits the pattern expected under the assumption of a Poisson model;

- in Example 3.1, the distribution of the number of cases of *thyrotoxicosis* in different seasons of the year was examined to establish if there was evidence of seasonal variation;
- in Table 6.5.1, observed frequencies are listed together with expected frequencies under the assumption of a Normal distribution for the purpose of establishing if the assumption of Normality is reasonable.

Given a sample size n, the general problem is one of comparing a set of observed frequencies $\{y_{i0}\}$ with a set of fitted frequencies $\{\hat{y}_{i0}\}$ which are estimates of expected frequencies $\{n\pi_i\}$ where the structure of the $\{\pi_i\}$ is defined by a statistical model. The values of the expected proportions may be

(i) completely defined by the statistical model, as in Example 3.1 where the hypothesis states that $\pi_1 = \pi_2 = \pi_3 = \pi_4 = \frac{1}{4}$; or

(ii) defined by distributional forms in which the value of one or more parameters is not specified, as in Example 1.2 where the Poisson distribution is assumed and $\pi_i = \exp(-M)M^i/i!$ for $i = 0, 1, 2, \ldots$.

In the latter case it is necessary to provide estimates of values of parameters from the data.

The Chi-squared statistic or the Deviance defined in (16.2.4) and (16.2.5), respectively, can be used for model and data comparisons. The appropriate degrees of freedom are defined by (16.2.12).

$$d = c - p \tag{16.2.12}$$

where c is the number of categories for which frequencies are provided; p is the number of independent parameters to be given values. The minimal number of parameters to be given a value from the data is one, this being the sample size n. In the case of the Poisson distribution there is one additional parameter which must usually be estimated, namely the mean, M. For the Normal distribution both the mean (M) and the variance (σ^2) may require estimation.

The steps in applying the methodology are those described in Section 16.2.7. Where the hypothesized model is rejected, the manner in which the observed frequencies and expected frequencies differ can be assessed by constructing and examining the signs and magnitudes of the standardized residuals in an analogous manner to that described in Section 16.2.8.

Problems

16.1 *Example* 16.2

Use the methods in Section 16.2.7 and 16.2.11 applied to the data in Table 16.1.2 to provide support for the theory that people in a large city tend to shop at a shopping center which is in the direction of the city center.

16.2 *Survey on farmers' attitudes to a tree planting scheme*

The following data were obtained in a survey of a sample of Tasmanian farmers which sought information on their attitude to a suggested scheme for increasing the amount of land in farming areas devoted to native vegetation.

(a) Establish if there is evidence of differences in attitudes based on property size, and if so, determine the manner in which the patterns of response differ.

(b) Establish if attitude varies with number of years farming, and if so, how the attitudes differ.

	Observed frequencies		
Property size	benefits outweight disadvantages	benefits equal disadvantages	disadvantages outweigh benefits
less than 200 hectares	75	32	12
200 to 500 hectares	36	16	9
more than 500 hectares	47	14	8

	Observed frequencies		
No. of years farming	benefits outweigh disadvantages	benefits equal disadvantages	disadvantages outweigh benefits
10 or fewer	43	8	1
more than 10	113	54	25

16.3 *Comparing energy intakes*

(a) Using the data in Table 1.3.5, construct separate frequency tables for energy intakes of boys and girls by grouping intakes into the following categories:

$$0-499 \quad 500-999 \quad 1000-1499 \quad 1500-1999 \quad 2000-2499 \quad \geq 2500.$$

(b) Use the test described in Section 16.2.7 to establish evidence of different patterns in energy intakes between boys and girls. Use the methodology of Section 16.2.8 based on the signs and magnitudes of standardized residuals to identify the manner in which the patterns differ.

16.4 *Comparison of responses in national park survey* (*Problem* 9.3)

(a) Establish that the three groups defined in Problem 9.3 show different patterns of response to the two questions for which survey results are given. (Compare the sets of four responses presented for the three groups using the method described in Section 16.2.7.)

(b) Identify the source(s) of the differences by interpreting the set of standardized residuals which are computed under the assumption that the patterns are the same in all groups (Section 16.2.8).

16.5 *Looking for evidence of a spatial pattern* (*Example* 1.2)

Table 1.2.2 contains a set of observed frequencies and a set of fitted frequencies under the assumption of a Poisson distribution. If the data has a pattern expected under a Poisson distribution, it can reasonably be concluded that the bombs landed randomly over the area from which data were collected. If not, there is evidence of greater concentrations of bombs in some regions than in others. Apply the relevant steps in Section 16.2.7 to test the hypothesis that the distribution is Poisson.

16.6 *Seasonal variation in thyrotoxicosis*

Do the data in Table 3.2.2 provide support for the theory that there is seasonal variation in the incidence of thyrotoxicosis in northern Tasmania?

17

Studying Association and Correlation

17.1 Relations Between Two Categorical Variables

17.1.1 Areas of Application

The methods and discussion in this section have application when there is interest in

(i) establishing if there is evidence of an association between two variables;

(ii) examining the nature of the relationship between two variables when there is evidence of an association;

(iii) measuring the degree of association between two variables.

The methodology which is presented in this section has application when the data can be presented in the form of a frequency table such as is illustrated in Table 17.1.1. Most commonly, however, when data are presented in a frequency table classified by two variables, one variable has the role of a factor which identifies groups and the experimental aim is seek evidence of group differences. The methods of Chapter 12 (studying proportions or odds) or Chapter 16 (comparing frequency distributions) should be applied in these situations.

Table 17.1.1. Examples of Data Presented in a Frequency Table

(a) Genetic association — Linkage
(Data from Example 17.1)

	Plant Colour	
Aleurone Colour	Green	Yellow
Purple	127	19
White	67	44

Table 17.1.1. (*cont.*)

(b) Length of stay versus Age
(Data from the study reported in Example 16.1)

	Length of Stay		
Age	Short	Medium	Long
Young	364	612	109
Middle	603	420	151
Old	976	705	447

Cases where there is a genuine interest in the study of association between two variables arise (i) when there is a parameter which has a physical interpretation as a measure of association, as illustrated by the *linkage* parameter in Example 17.1; (ii) where there is interest in comparing the level of association between two variables in different groups; and (iii) when there is a battery of tests or a series of questions asked when have equal status and there is interest in determining the relative levels of associations between pairs of responses.

Example 17.1 (Genetic Association — Linkage). *The basic idea of the Mendelian Theory of genetic inheritance assumes that genes segregate independently. However, it has been known for many years that genes on the same choromosome are more likely to segregate as a group than as independent entities. Yet the connection does not always take place. As genes are more separated on a chromosome, the chance that two genes on the same chromosome will remain together decreases.*

The extent to which two genes tend to stay together is termed the 'linkage' between the genes. Where the genes each have an unambiguous outward expression, as in the experiment situation which provided the data of Table 17.1.1(a), *it is possible both to define a measure of association between the variables based on the linkage and to give a physical meaning to the number which is obtained.*

17.1.2 Statistical Models

For the purposes of constructing inferential methods, the Multinomial model (Section 6.4) is presumed. Under the Multinomial model the probability that a randomly selected individual will have outcomes i ($i = 1, 2, \ldots, a$) for variable A and j ($j = 1, 2, \ldots, b$) for variable B is π_{ij}. Hence, in a sample of size n, the expected number of sample members with outcomes i and j for variables A and B, respectively, is given by (17.1.1.). In this representation, y_{ij} is a statistic which yields the value y_{ij0} in the sample.

$$E(y_{ij}) = n\pi_{ij} \qquad \text{for } i = 1, 2, \ldots, a;\ j = 1, 2, \ldots, b \tag{17.1.1}$$

where n is the sample size; and π_{ij} is the probability that a randomly selected sample member will have outcomes i and j, respectively, for variables A and B.

The *Independence* Model. The most restricted model which is considered is generally the *Independence model* which arises when knowledge of an individual's response for one variable gives no information as to the likely response for the other variable. This is represented by (17.1.2).

$$\pi_{ij} = \pi_{i.}\pi_{.j} \qquad \text{for } i = 1,2,\ldots,a;\ j = 1,2,\ldots,b \tag{17.1.2}$$

where $\pi_{i.}$ is the probability of obtaining an outcome i for variable A; and $\pi_{.j}$ is the probability of obtaining an outcome j for variable B.

Equation (17.1.2) is a statement of the probabilistic definition of *independence.* Based on the Independence model, a set of predicted frequencies, termed *fitted values,* can be constructed from the data. Where the data are summarized in the frequency table represented by Table 17.1.2, the fitted values are defined in Table 17.1.3.

Table 17.1.2. Summary of Data — Frequency Table

Notation

Variable A	Variable B 1	2	$\cdots$	b	Total
1	y_{110}	y_{120}		y_{1b0}	$Y_{1.0}$
2	Y_{210}	Y_{220}		Y_{2b0}	$Y_{2.0}$
$\vdots$					
a	y_{a10}	y_{a20}		y_{ab0}	$Y_{a.0}$
Total	$Y_{.10}$	$Y_{.20}$		$Y_{.b0}$	n

Standardized Residuals. The set of values $\{r_{s_{ijo}}\}$, where $r_{s_{ijo}}$ is defined by (17.1.3), is called the set of *standardized residuals.* If the Independence model is correct then $r_{s_{ijo}}$ is the value of a statistic $r_{s_{ij}}$ which has a sampling distribution which is approximately $N(0,1)$. If the variables A and B are related, the expected value of at least some of the $r_{s_{ij}}$'s are different from zero. When the Independence model has been rejected, the patterns of the signs and magnitudes of the standardized residuals may be used to establish the nature of the relationship between the variables.

$$r_{s_{ijo}} = \frac{y_{ij0} - np_{i.0}p_{.j0}}{\sqrt{np_{i.0}p_{.j0}}} \qquad \text{for } i = 1,2,\ldots,a;\ j = 1,2,\ldots,b. \tag{17.1.3}$$

Table 17.1.3. Fitted Values Under the Independence Model

Variable A	Variable B 1	2	$\cdots$	b	total
1	$np_{1.0}p_{.10}$	$np_{1.0}p_{.20}$		$np_{1.0}p_{.b0}$	$Y_{1.0}$
2	$np_{2.0}p_{.10}$	$np_{2.0}p_{.20}$		$np_{2.0}p_{.b0}$	$Y_{2.0}$
$\vdots$					
a	$np_{a.0}p_{.10}$	$np_{a.0}p_{.20}$		$np_{a.0}p_{.b0}$	$Y_{a.0}$
total	$Y_{.10}$	$Y_{.20}$		$Y_{.b0}$	n

where

$p_{i.0} = Y_{i.0}/n$ for $i = 1, 2, \ldots, a$
$p_{.j0} = Y_{.j0}/n$ for $j = 1, 2, \ldots, b$.

17.1.3 Seeking Evidence of Association

There is general agreement that the absence of association corresponds with the probabilistic notion of independence as defined by (17.1.2). If the experimental aim is simply to test for evidence of association, the usual way to do so is by use of the Pearson Chi-squared statistic defined in (17.1.4).

$$X_p^2 = \sum_{i=1}^{a} \sum_{j=1}^{b} \frac{(y_{ij} - np_{i.}p_{.j})^2}{np_{i.}p_{.j}} \tag{17.1.4}$$

where y_{ij} is the statistic which provides the value y_{ij0} in Table 17.1.2; and $p_{i.}$ and $p_{.j}$ are the Maximum Likelihood estimators of $\pi_{i.}$ and $\pi_{.j}$ in (17.1.2) which provide the estimates $p_{i.0}$ and $p_{.j0}$ in Table 17.1.3.

Where there is evidence of an association, the nature of the association can be established by examining the pattern in the signs and magnitudes of the set of standardized residuals $\{r_{s_{ij0}}\}$ which are defined in (17.1.3). The steps are presented below.

Testing the hypothesis of independence. The statistic X_p^2 has the property that a zero value shows greatest agreement between the Independence model and the data. Increasing values of X_p^2 reflect lessening evidence for independence and hence, increasing evidence of association. Given an observed value of X_0^2 for the statistic X_p^2, the measure of agreement between model and data is approximated by p_0 in (17.1.5).

$$p_0 = \Pr\{\chi^2(d) \geq X_0^2\} \tag{17.1.5}$$

where $d = (a-1)(b-1)$.

Application. The steps in testing for evidence of association and establishing the nature of the association are presented below.

Step 1 — *check on the adequacy of the Chi-squared approximation*
Compute fitted values using the formulae in Table 17.1.3. If none of the fitted values are less than 1.0 proceed to Step 2. Otherwise follow the appropriate procedure described below.

(i) If both variables have two categories it is necessary to employ *Fisher's exact test.* Details are provided, for example, in Siegel [1956].

(ii) If at least one variable has more than two categories, it is necessary to combine categories in order to meet the requirement that no fitted values are less than 1.0. There are no formal rules for the method of combining categories. Common sense dictates that where categories are ordered, it is adjacent categories in the ordering which should be combined. More generally, the combination should be based on meaningful groupings of categories where this is possible.

Step 2. Compute a value of X_p^2 from (17.1.4) and convert it to a p-value using (17.1.5). Interpret the p-value in the usual way, noting that smaller p-values equate with lesser evidence of independence and hence, greater evidence of association between variables A and B.

Step 3. If there is evidence of an association between the variables, the set of standardized residuals $\{r_{s_{ij0}}\}$ computed from (17.1.3) may be examined

(i) for any pattern in the signs of the standardized residuals which has meaning; and

(ii) for any standardized residuals with magnitudes in excess of 2.0 since these are suggestive of observed frequencies which are substantially different from the values expected under the Independence model.

To aid in the interpretation of patterns of residuals, it is useful to construct the table of residuals in a way which places the categories for each variable in a natural ordering or natural groupings. Then it is possible to see if the signs cluster in a significant way. For example, if the categories are ordered, a pattern in which positive residuals appear down a diagonal and negative signs appear off the diagonal may be meaningful.

Illustration. The steps are illustrated using the data in Table 17.1.1(b).

Step 1 — Fitted values computed from formulae in Table 17.1.3 using data in Table 17.1.1(b) are given below.

Step 2 — The value of X_p^2 is computed from (17.1.4) to be 208 and $p_0 = \Pr\{\chi^2(4) \geq 208\}$ is found to be less than 0.001. Hence, there is strong evidence of association between the variables *Length of stay* and *Age.*

Table of fitted values

Age	Length of Stay		
	Short	Medium	Long
Young	480.6	429.6	174.9
Middle	520.0	464.8	189.2
Old	942.5	842.6	342.9

Note that all fitted values exceed 1.0.

Step 3 — The table of standardized residuals computed using (17.1.3) is given below.

Table of standardized residuals

Age	Length of Stay		
	Short	Medium	Long
Young	−5.3	8.8	−5.0
Middle	3.6	−2.1	−2.8
Old	1.1	−4.7	5.6

Since positive values indicate combinations for which higher than expected frequencies were observed, it is apparent (for example) that young people are more inclined to have a medium length stay than are the other groups; that the middle age group is more likely to have a short stay and the older people are more likely to have a long stay.

17.1.4 Choosing, Applying and Interpreting a Measure of Association

There have been many proposals for measures of association. Goodman and Kruskal [1979] provides a classification. If (i) a model can be formulated which includes a parameter that can be given a scientific meaning as a measure of association, or (ii) there is a measure of association which is traditionally used in a discipline, these are sound reasons for employing that measure.

Few measures have been defined for which numerical values have scientific meaning. The only point on most scales which has meaning is the value of zero. This is absence of association and, almost universally, represents independence (in a probabilistic sense). While there are a number of parameters which have upper limits on the scale of association, there is not generally a meaningful interpretation of *complete association*. Thus, a maximum reading on one scale will not necessarily correspond to a maximum reading on another scale.

Except for the special cases where (i) a measure of association has a clear cut scientific meaning, or (ii) the measure of association is being used as a summary statistic, the recommended approach is to follow the steps outlined in Section 17.1.3 and test for evidence of association. If there is evidence of association, the nature of the association may then be examined by attempting to match the pattern in the table of standardized residuals with a plausible scientific explanation.

17.2 Relations Between Two Scaled Variables

17.2.1 Areas of Application

The study of the relationship between two scaled variables has one of two basic purposes.

1. *Prediction or estimation.* The most common purpose is represented by the situation in which one variable is considered to be a predictor or estimator of the other. Example 18.1 provides an illustration. In that example, interest lies in the value of using the performance in mathematics in the final high school examination (*high school mathematics score*) as an indicator of likely performance in mathematics in first year university (*university mathematics score*). In this context, there is only one *response* variable, namely *university mathematics score. High school mathematics score* has the role of an *explanatory* variable (see Section 2.2). The statistical approaches designed for problems of this type are discussed in Chapter 18.

2. *Correlation.* There are circumstances in which both variables are response variables and the function of statistical analysis is to establish if

(i) there is evidence of a relationship between the variables; and, if so,

(ii) to measure the strength of the relationship or to compare the strength of the relationship between different groups or under different treatments.

Example 17.2 provides an illustration of the type of problem in which the relationship between two response variables is of interest.

Example 17.2 (Tests of Mathematical Aptitude). *Table* 17.2.1 *contains the scores obtained on two mathematics tests by three groups of school children. The tests, labelled S and U, have been designed to measure different aspects of mathematical ability. There is interest in the extent to which scores on the two tests are related and whether the extent of the relationship varies between the three groups.*

Table 17.2.1. Scores on Mathematical Aptitude Tests (Example 17.1)

12 Year olds		12 Year old special group		14 Year olds	
U	S	U	S	U	S
58	23	55	36	85	36
61	22	57	34	89	37
63	23	61	36	89	38
63	25	64	42	90	37
64	26	65	38	91	36
67	28	66	38	92	37
68	26	66	42	93	37
69	27	68	37	93	42
69	28	69	39	94	36
70	32	70	37	94	40
71	27	71	39	96	37
71	28	71	41	96	38
71	25	71	43	96	38
73	28	72	40	97	39
73	31	73	37	97	40
74	29	75	41	98	41
74	31	76	42	99	37
74	32	77	41	99	38
76	28	78	43	100	38
76	31	81	44	100	42
77	30	85	46	101	38
77	32			101	40
77	26			101	42
79	30			103	42
80	32			105	42
80	34			106	41
81	31			107	44
82	34			108	41
82	33			108	43
84	33			110	44
86	34			115	44
88	37				
89	34				
93	36				

The statistical measurement of the strength of a relationship between two variables is termed the *correlation* between the variables and its measurement is considered in Section 17.2.3.

17.2.2 DATA

The data are presumed to be in the form of paired responses to variables x and y obtained from n sample members and are typically recorded in a table with the following form.

Sample member	1	2	...	n
Variable x	x_{10}	x_{20}	...	x_{n0}
Variable y	y_{10}	y_{20}	...	y_{n0}

17.2.3 TREND LINES AND CORRELATION

Consider the scatter plot presented in Figure 17.2.1(a). It is apparent from visual observation that higher scores on Test S tend to be associated with higher scores on Test U. More specifically, it might be said that there is a *linear trend* in the relationship since a straight line could be superimposed on the points to reflect the general manner in which increases in values of one variable are related to changes in the other.

Figure 17.2.1(b) is also suggestive of a relationship between two variables. However, there are some points of distinction between plots (a) and (b). Whereas the trend in (a) could be represented by a straight line, clearly a curved line is required in (b). The relationship in (a) is *positive* since increasing values of one variable are associated with increasing values of the other variable, while in (b) it is *negative* since increasing values in one variable are associated with decreasing values of the other variable. The relationship in (a) is *weaker* than in (b) since the points cluster less closely about the trend line.

Correlation. If, for two variables x and y, a trend line exists which is either continually increasing or continually decreasing, there is said to be a *correlation* between x and y. In qualitative terms, the correlation is said to become stronger as the points cluster more closely about the trend line.

Absence of correlation is portrayed visually by either the absence of a trend line [Figure 17.2.2(a)] or a trend line which runs parallel to one or other axis [Figure 17.2.2(b)]. The implication is that an increase in the value of the x variable gives no information as to whether the y variable is likely to have shown a corresponding increase or decrease, or vice versa.

In a descriptive sense, perfect correlation might be portrayed by (i) all points lying on the trend line, or (ii) by an increase in value of the x variable always being accompanied by an increase in the value of the y variable (or always being accompanied by a decrease in the value of the y variable). The latter interpretation is more general than the former.

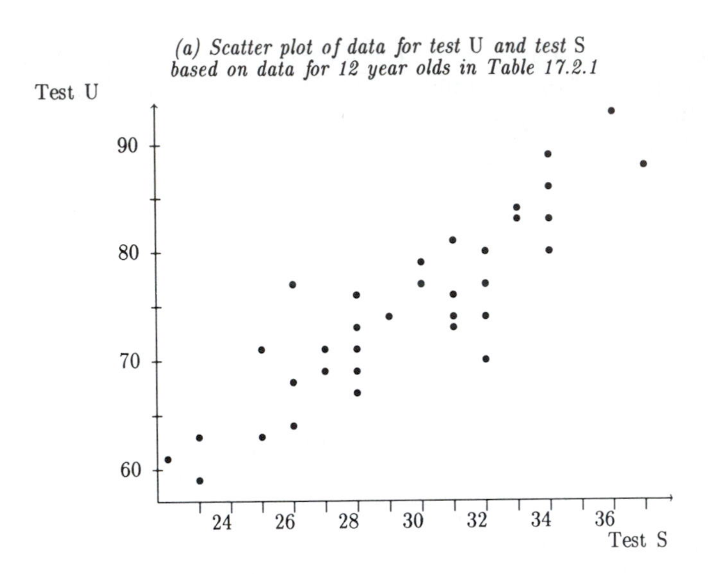

(a) Scatter plot of data for test U and test S

(b) Scatter plot illustrating negative correlation
and a non–linear trend line

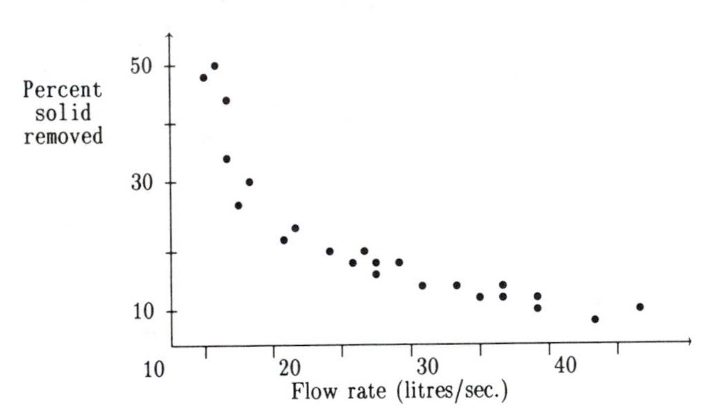

Figure 17.2.1. Scatter plots displaying relationships between variables.

17.2.4 Correlation Coefficients

Statistics or parameters which quantify the strength of correlation between two variables are called *correlation coefficients* and it is conventional that they meet the following conditions.

1. A value of zero indicates the absence of correlation.
2. Increasing magnitude implies increasing strength of correlation.
3. The maximum magnitude which a correlation may take is one.

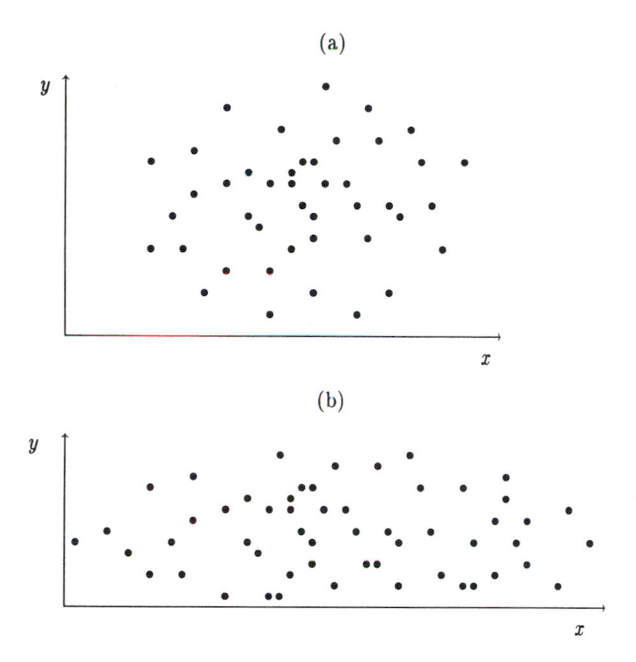

Figure 17.2.2. Plots displaying low correlations between variables.

4. A negative correlation implies increases in one variable tend to be associated with decreases in the other variable, whereas a positive correlation implies increases in one variable tend to be associated with increases in the other variable.

The two correlation coefficients introduced below have wide application in science.

Product-moment correlation coefficient. The statistic defined by (17.2.1) is known as *(Pearson's) Product-moment correlation coefficient.*

$$r = \frac{s_{xy}}{s_x s_y} \tag{17.2.1}$$

where s_x and s_y are the sample standard deviations for variables x and y; and s_{xy} is the sample *covariance* and is defined by (17.2.2).

$$s_{xy} = \sum_{i=1}^{n}[(x_i - \overline{x})(y_i - \overline{y})]/(n-1). \tag{17.2.2}$$

If all points $(x_1, y_1), (x_2, y_2), \ldots, (x_n, y_n)$ lie on a straight line when plotted, then r has a value of $+1$ if the line has a positive slope and -1 if the line has a negative slope. (It is undefined if the line is either horizontal or vertical. These cases correspond to situations in which either x or y shows no variation. If either of the variables is constant, nothing can be learned from the data about the manner in which the two variables covary.)

In general, r will have a large magnitude when the products $(x_i - \overline{x})(y_i - \overline{y})$ are of constant sign. This situation is illustrated in Figure 17.2.3(a). Low magnitude for r will result if the plots have the form illustrated in Figure 17.2.3(b).

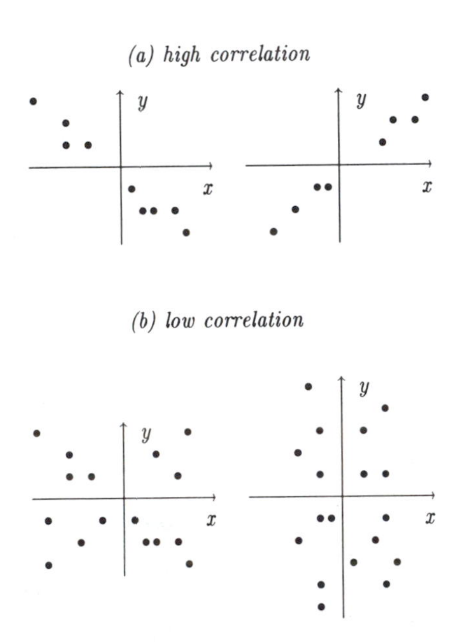

Figure 17.2.3. Scatter plots illustrating different levels of correlation.

Kendell's Tau (τ) correlation coefficient. Application of the Product-moment correlation coefficient is limited by the requirement that the trend line must be linear. A less restrictive measure of correlation is based on the probabilistic notion that correlation between variables x and y is strong if, on average, there is a high probability that an increase in x will be accompanied by an increase in y (or a decrease in y). Given this basis, the only limitation imposed on the trend line is that it should be either continually increasing or continually decreasing.

One statistic which is based on this notion of correlation is *Kendall's Tau correlation coefficient* (t_τ). Given a data set comprising n paired values, $(x_1, y_1), (x_2, y_2), \ldots, (x_n, y_n)$, a value for Kendall's Tau coefficient can be constructed by the following steps.

Step 1 — Order the values of variable x, $x_1, x_2, \ldots, x_n$, from smallest to largest. Denote the ordered values by $x_{(1)}, x_{(2)}, \ldots, x_{(n)}$ and the value of y corresponding to $x_{(i)}$ by y_i' $(i = 1, 2, \ldots, n)$.

Step 2 — Compare y'_1 with $y'_2, y'_3, \ldots, y'_n$ and note the number of values which are greater than y'_1 and the number which are less than y'_1. Record these numbers as c_1 and d_1, respectively.

Compare y'_2 with $y'_3, y'_4, \ldots, y'_n$ and note the number of values which are greater than y'_2 and the number which are less than y'_2. Record these numbers as c_2 and d_2, respectively.

Repeat this procedure for each y'_i up to y'_{n-1} and hence, obtain sets of numbers $\{c_i\}$ and $\{d_i\}$ $(i = 1, 2, \ldots, n-1)$.

Step 3 — Compute a value for t_τ from (17.2.3).

$$t_\tau = \left[\sum_{i=1}^{n-1}(c_i - d_i)\right] / \left[\frac{1}{2}n(n-1)\right]. \qquad (17.2.3)$$

*If there are tied values for either variable a modification is required—see, for example, Siegel [1956]. In practical terms, the modification only assumes importance if there is more than about ten percent of ties in either set of values.

Computations. The calculations are illustrated in Table 17.2.3 using data from Example 17.3.

Properties. If the sequence $y'_1, y'_2, \ldots, y'_n$ is the ordering of the y-values from smallest to largest, then $d_i = 0$ for all i and, in the absence of tied values, $c_1 = n-1$, $c_2 = n-2, \ldots, c_{n-1} = 1$. The sum of $c_1, c_2, \ldots, c_{n-1}$ is $\frac{1}{2}n(n-1)$. Hence, t_τ takes the value $+1.0$ and this can be interpreted as perfect positive correlation. Perfect negative correlation arises when the sequence $y'_1, y'_2, \ldots, y'_n$ is the ordering of the y-values from largest to smallest.

Absence of correlation, i.e. $t_\tau = 0$, arises when, on average, a y-value in the sequence has as many observations above it as below it.

An alternative formulation with the same basis as Kendall's Tau is *Spearman's rank correlation coefficient.* Its definition and method of computation are widely available (see, for example, Conover [1980]). It is less general in its application since it cannot be extended to include the influence of a third variable on the relation between x and y, its value is arguably less easy to interpret, and it has the same power as Kendall's Tau in testing the hypothesis that there is no correlation between two variables.

Example 17.3 (Air Quality in the Tamar Valley). *The Tamar valley in northern Tasmania contains an aluminum smelter, a ferro-alloy plant and an oil fired power station. There has been concern over the levels of pollution in the valley as a consequence of these sources of pollution. Using dust fallout gauges, data were obtained from dust samples collected at* 16 *sites in the valley over monthly periods. The information obtained from* *those gauges for one of the monthly periods is presented in Table* 17.2.2. *Of interest are the correlations between pairs of readings.*

Table 17.2.2. Data Collected in the Month of October 1982 (Example 17.3) (Data courtesy of Dr. Pak Sum, School of Environ. Studies, Univ. of Tasmania)

Site	Total dust[1]	Al[2]	Fe[2]	Zn[2] Sol.	Zn[2] Insol.	Mn[2] Sol.	Mn[2] Insol.	F[3]	pH
1	17	221	174	7	5	10	24	0.11	5.5
2	27	500	563	12	8	102	100	0.38	5.8
3	16	402	666	118	14	131	160	0.36	5.9
4	25	574	830	131	12	207	201	0.83	5.7
5	118	1688	2812	6	36	16	300	1.60	8.0
6	8	563	298	15	5	110	77	0.43	5.7
7	153	1235	1369	399	444	409	1744	0.25	6.1
8	163	1816	1787	–[4]	97	55	2387	0.70	8.2
9	–[5]	–[5]	–[5]	–[5]	–[5]	–[5]	–[5]	–[5]	–[5]
10	68	1007	921	21	251	125	513	0.11	6.1
11	62	796	1691	391	49	38	426	0.09	6.6
12	129	238	338	12	23	11	16	0.04	8.7
13	86	778	1791	326	976	25	231	0.10	6.1
14	28	404	631	–[4]	15	16	27	0.06	6.2
15	19	610	445	–[4]	11	232	361	0.18	6.7
16	18	340	324	21	12	49	61	0.25	5.6

Notes.
[1]units are mg/m^2/day
[2]units are $\mu g/m^2$/day
[3]units are ppm
[4]concentration below detection limits of analysis
[5]not available because sample was damaged.

17.2.5 Statistical Models

The sample product-moment correlation coefficient (r) is an estimator of the population product-moment correlation coefficient, ρ, in the *Bivariate Normal distribution model* which is defined below. In this context, there have been developed tests of the hypothesis $\rho = 0$, a confidence interval for ρ which implies the capacity to test an hypothesis $\rho = \rho_0$ for any value of ρ in the range -1 to $+1$, and a test that ρ is identical in different groups or under different treatments.

Kendall's Tau correlation coefficient finds greatest use as a basis for testing the hypothesis that the variables are uncorrelated. For this purpose a simple *distribution-free model* is required. It is defined in Section 17.2.6.

Table 17.2.3. Steps in Calculating a Value of Kendall's Tau Illustrated Using Dust and Aluminium Scores from Table 17.2.2

Site	$x_{(i)}$	y'_i	c_i	d_i	$c_i - d_i$
6	8	563	8	6	2
3	16	402	10	3	7
1	17	221	12	0	12
16	18	340	10	1	9
15	19	610	6	4	2
4	25	574	6	3	3
2	27	500	6	2	4
14	28	404	6	1	5
11	62	796	4	2	2
10	68	1007	3	2	1
13	86	778	3	1	2
5	118	1688	1	2	−1
12	129	238	2	0	2
7	153	1235	1	0	1
8	163	1816			
Total					51

$$t_\tau = 51 / \left[\frac{1}{2} \times 15 \times 14\right] = 0.49.$$

Independence and absence of correlation. Variables may be uncorrelated but not independent. For example, if the relation between *Time of sunrise* and *Day of year* were considered over a period of 365 consecutive days, there is clearly a relationship but the calculation of the value of either r or t_τ would yield a value of zero.

Bivariate Normal distribution model. Paralleling the Normal distribution model (Section 6.5) which is defined for a single response variable y, is the *Bivariate Normal distribution* model for a pair of response variables (x, y). The assumption is made that a sample is randomly selected and responses are recorded for both variables x and y on each sample member. The probability distribution of (x, y) is defined by (17.2.4) and includes the parameter ρ which is the population equivalent of the (sample) product-moment correlation coefficient r.

$$p(x,y) = \frac{1}{2\pi\sigma_x\sigma_y\sqrt{\{1-\rho^2\}}} \exp\left\{\frac{-1}{2(1-\rho^2)}\left[\left[\frac{x-M_x}{\sigma_x}\right]^2 - 2\rho\left[\frac{x-M_x}{\sigma_x}\right]\left[\frac{y-M_y}{\sigma_y}\right] + \left[\frac{y-M_y}{\sigma_y}\right]^2\right]\right\} \quad \text{for all real } x, y \qquad (17.2.4)$$

where M_x and M_y are the expected values of x and y, respectively; σ_x and σ_y are the standard deviations of x and y, respectively; and ρ is the (population) product-moment correlation coefficient.

The existence of a linear trend line is a characteristic of all Bivariate Normal distributions for correlated variables.

Absence of correlation ($\rho = 0$) reduces (17.2.4) to the product of two Normal distributions, $p_x(x)$ and $p_y(y)$ which are $N(M_x, \sigma_x^2)$ and $N(M_y, \sigma_y^2)$, respectively. The fact that the joint distribution can be represented as the product of the individual distributions for x and y implies that x and y are independent.

17.2.6 Seeking Evidence of Correlation

Two situations are considered. They are (i) situations in which there is no assumed distributional form and no presumed form for the trend line, and (ii) situations in which the Bivariate Normal distribution is presumed with the implied assumption that the trend line is linear.

Distribution-free approach

Model. A population or experimental process is defined from which a sample is presumed to be drawn by random selection. Two scaled response variables, x and y are defined which have a probability distribution $p(x, y)$.

The hypothesis states that the variables are independent and hence, uncorrelated. This equates with a value of zero for the parameter τ which is Kendall's Tau correlation coefficient computed from the set of population responses.

Data. The data are presumed to be in the form of n paired observations (x_i, y_i) for $i = 1, 2, \ldots, n$ or n paired rankings (r_i^x, r_i^y) for $i = 1, 2, \ldots, n$, where r_i^x is the rank of x_i when the X-values are ranked from smallest to largest and r_i^y is similarly defined for the Y-values. (The latter situation arises when there is a latent variable which is not directly measurable or when values for the scaled variable have been grouped by dividing the scale into intervals and noting into which interval a response falls.)

Measure of agreement. The comparison of the data with the model which includes the assumption of independence is based on the statistic t_τ defined by (17.2.3). A value of zero for t_τ reflects the greatest agreement between model and data and increasing magnitude reflects worsening agreement. Thus, if the observed value for t_τ is t_0, the p-value is computed from (17.2.5). In large samples, an approximate value for p is provided by (17.2.6).

$$p_0 = \Pr(|t_\tau| \geq |t_0|) \tag{17.2.5}$$

$$p_0 \cong \Pr\left\{|z| \geq |t_0| \Big/ \sqrt{\frac{2(2n+5)}{9n(n-1)}}\right\}. \tag{17.2.6}$$

Application. The steps below are based on the use of Kendall's Tau (Section 17.2.4). They could equally be based on Spearman's rank correlation coefficient.

Step 1 — *checking the trend line*

If both x and y are scaled variables (i.e. not ranks), plot a scatter diagram using the n pairs of readings, $\{(x_i, y_i)\}$. Check that, if there is evidence of a trend line, it appears to be either continually increasing or continually decreasing. If this requirement is not met, the concept of correlation as defined in this section is not applicable.

Step 2 — *constructing the value of* t_τ

By using a statistical package or by following the steps describing the construction of Kendall's Tau in Section 17.2.4, obtain a value of t_τ which is denoted by t_0.

Step 3 — *constructing and interpreting a p-value*

Convert the value t_0 to a p-value for comparing the data with the Independence model by

(i) obtaining a value for (17.2.5) from the use of a statistical computing package; or

(ii) if n is greater than 10 by obtaining an approximate value of p from (17.2.6); or

(iii) if n is 10 or less, using tables which are available, for example, in Conover [1980].

Interpret the p-value in the usual way, noting that small values of p provide support for acceptance of a correlation between the variables.

Illustration. The calculations of t_0 are illustrated in Table 17.2.3 using data from Table 17.2.2 for the variables *Total dust* and *Aluminum* (Al). A value of 0.50 is obtained based on 15 pairs of readings. Application of (17.2.6) yields

$$p \cong \Pr\left\{|z| \geq |0.50| / \sqrt{\frac{2(30+5)}{9(15)(15-1)}}\right\} = \Pr\{|z| \geq 2.60\} = 0.01.$$

Hence, there is evidence against the hypothesis that the variables are independent. The positive sign of t_τ indicates that the variables are positively correlated, i.e. that an increase in the value of one variable tends to be accompanied by an increase in the value of the other variable.

A test of the hypothesis $\rho = 0$

Model. The Bivariate Normal distribution model is assumed with the value of ρ set equal to zero.

Data. The data are presumed to comprise a set of n pairs of scaled responses (x_i, y_i) for $i = 1, 2, \ldots, n$.

Statistics. Two statistics, t and z which are defined below, may be employed to test the hypothesis $\rho = 0$.

The t-statistic. Given a Bivariate Normal distribution model with $\rho = 0$, the statistic t as defined by (17.2.7) is a t-statistic with $n - 2$ degrees of freedom.

$$t = r\sqrt{(n-2)}/\sqrt{(1-r^2)} \tag{17.2.7}$$

where r is defined by (17.2.1).

The z-statistic. Given a Bivariate Normal distribution model with correlation coefficient ρ, the statistic z_r defined by (17.2.8) is approximately Normally distributed with mean, M_z, and variance, σ_z^2, defined by (17.2.9). The Normal approximation is adequate if the sample size exceeds 20.

$$z_r = \frac{1}{2}[\log(1+r) - \log(1-r)] \tag{17.2.8}$$

$$\begin{aligned} M_z &= \tfrac{1}{2}[\log(1+\rho) - \log(1-\rho)] \\ \sigma_z^2 &= 1/(n-3). \end{aligned} \tag{17.2.9}$$

The limitation of the t statistic is that its sampling distribution is known only when $\rho = 0$. Thus, t may be used to test the hypothesis $\rho = 0$ but not to construct a confidence interval for ρ. By applying (8.3.3), approximate confidence limits for ρ may be constructed using the statistic z_r—see the Estimation section below.

Model and data checking.

1. Construct a scatter plot based on the two variables and check that the trend line appears linear. Also look for evidence of an error in the data in the form of a point which is far removed from the other points in the plot.

2. Since both hypothesis testing and estimation rely on the assumption of Normality, it is advisable to construct a Normal probability plot for each variable (as described in Section 10.3.4). If there is strong evidence of non-Normality from the Normal probability plots, the inferential methods described below should not be employed and a test of the hypothesis of no correlation should be based on the distribution-free method which uses Kendall's Tau.

Estimation. By application of (8.3.3), approximate $100(1-\alpha)\%$ confidence limits for M_z are given by (17.2.10).

$$L_z = z_r - z_{\frac{1}{2}\alpha}/\sqrt{\{n-3\}}$$
$$U_z = z_r + z_{\frac{1}{2}\alpha}/\sqrt{\{n-3\}} \qquad (17.2.10)$$

where z_r is determined from (17.2.8). The limits L_z and U_z can be transformed into approximate upper and lower $100(1-\alpha)\%$ confidence limits for ρ using (17.2.11) which is the inverse function to (17.2.8).

$$L_\rho = [\exp(2L_z) - 1]/[\exp(2L_z) + 1]$$
$$U_\rho = [\exp(2U_z) - 1]/[\exp(2U_z) + 1]. \qquad (17.2.11)$$

Hypothesis testing. Provided the sample size is at least 20, the construction of a confidence interval for ρ is recommended. Evidence against the hypothesis $\rho = 0$ is provided if the confidence interval does not include the value zero. Where the sample size is less than twenty, a test may be based on a value t_0 computed from (17.2.7) and a p-value computed from the formula $p = \Pr\{|t(n-2)| \geq |t_0|\}$.

Application. If the value of r is computed as r_0 from a sample of size n, convert the value r_0 to z_{r0} using (17.2.8) and proceed to the computation of upper and lower confidence limits for ρ using (17.2.11) or by use of a statistical computing package. (Unless there are reasons for choosing a different level of confidence, the convention is to use a level of 95%.) If the value $\rho = 0$ is contained between the limits L_ρ and U_ρ, the assumption that the variables are uncorrelated is reasonable.

Illustration. The calculations are illustrated for the variables U and S in Example 17.2 using the data for the 12 year olds in Table 17.2.1.

The scatter plot is produced as Figure 17.2.1(a) and visual inspection suggests a linear trend line is reasonable. There is no evidence from the scatter plot or the Normal probability plots of an error in the data and the assumption of Normality is judged to be reasonable for both variables (from the Normality probability plots).

The product-moment correlation coefficient has a value $r_0 = 0.887$ and the sample size $n = 34$. From (17.2.8), the value of z_r is computed to be $z_0 = 1.41$.

> Applying (17.2.10) yields values $L_z = 1.06$ and $U_z = 1.76$.
>
> Conversion to confidence limits for ρ using (17.2.11) yields values $L_\rho = 0.78$ and $U_\rho = 0.94$.

Since the confidence interval does not include zero, there is evidence that the test scores are correlated. The fact that the limits are positive suggests that the variables are positively correlated.

17.2.7 Comparing Correlations Between Groups

A test is described below of the hypothesis that the correlations between variables x and y are identical in groups 1 and 2. The test is based on the Product-moment correlation coefficient.

Model. For each group the joint probability distribution of variables x and y is presumed to be a Bivariate Normal distribution which implies the trend line is linear in each case. Samples are presumed to be randomly selected from each group and there is independence of selection of samples between groups.

The model includes the hypothesis that the correlations between x and y in groups 1 and 2 are identical.

Data. The data comprise a set of n_1 paired responses from group 1 and a set of n_2 paired responses from group 2.

Statistic. If variables x and y have correlations r_1 and r_2 in groups 1 and 2, respectively, the statistics z_1 and z_2 can be defined as functions of r_1 and r_2 by (17.2.8). Under the proposed model, z_1 and z_2 are independently and approximately Normally distributed with identical means and with variances $1/(n_1 - 3)$ and $1/(n_2 - 3)$, respectively. It follows from property 2(b) in Section 7.3.3 that the statistic $z_d = z_1 - z_2$ is also approximately Normally distributed with mean zero and variance $1/(n_1 - 3) + 1/(n_2 - 3)$.

Measure of agreement. If the hypothesis $\rho_1 = \rho_2$ is correct, increasing magnitude of z_d implies worsening agreement between model and data. Given the value of z_d computed from the data is z_{d0}, the p-value is computed as

$$p_0 = \Pr(|z_d| \geq |z_{d0}|).$$

Application. The steps required to apply the method are provided below.

Step 1 — Construct a scatter plot for each group and check that the trend lines appear linear and there is no evidence of error in the data. Construct separate Normal probability plots for each variable from the data in each group.

If there is evidence of an error in the data, the offending point(s) should be deleted before proceeding with the analysis.

If there is clear evidence of non-Normality and/or a non-linear trend line, it is unwise to proceed with the test.

Step 2 — Compute values of z_{r1} and z_{r2} using (17.2.8) and hence, a value of $z_d = z_{r1} - z_{r2}$. Denote this difference in z-values by z_{d0}.

Step 3 — Based on the assumption that z_d has a $N(0, [1/(n_1-3)+1/(n_2-3)])$ distribution, determine $p_0 = \Pr(|z_d| \geq |z_{d0}|)$.

(Computational methods are given in Section 7.3.4.)

Interpretation. If p_0 is less than 0.01 there is strong evidence the correlations are different; if p_0 lies between 0.01 and 0.05 there is weak evidence the correlations are different; if p_0 exceeds 0.05 the data are consistent with the assumption that the correlations are the same for the two groups.

Illustration. Table 17.2.1 contains data on test scores U and S for 12 *year olds* and 14 *year olds.* Model and data checking suggests the conditions for application of the method are met.

Product-moment correlations of $r_1 = 0.8868$ and $r_2 = 0.7955$ are computed from the data for the 12 year olds and the 14 year olds based on samples of 34 and 31, respectively.

From the value of r_1 a value of 1.407 is computed for z_{r1} and from the value of r_2 a value of 1.086 is computed for z_{r2}. Hence, z_{d0} is 0.321.

For a Normal distribution with mean zero and variance $\sigma_{zd}^2 = [1/31 + 1/28] = 0.0680$, the probability of a value outside the range -0.321 to 0.321 is 0.22. Since p_0 is greater than 0.05, the data are consistent with the assumption that the correlations between the test scores are the same for both groups.

17.2.8 Interpretation—The Sensible Use of Correlation Analysis

Scatter plot. The scatter plot is an invaluable aid in the study of correlation. As is evident in the plots in Figure 17.2.1, the scatter diagram gives a clear picture of the trend line and the scatter of the points about the trend line. Furthermore, the comparative nature and strength of association in different groups may be easily assessed by comparing the plots for data from the same pair of variables taken from different groups.

For the Bivariate Normal distribution to have application, the scatter plot may be used to judge if the trend line appears linear and the points form a single grouping. Evidence of two or more groupings may be an indication that the sample is drawn from more than one population.

Influence. Figure 17.2.4 illustrates a situation in which

(i) an error has been made in a plotted point; or

(ii) an individual with an unusually high readings for the variables has been included in the sample (perhaps indicating the presence of two distinct groups within the sampled population).

Whatever the explanation, the single isolated point derives from a pair of readings which have a large *influence* on the value of the Product-moment correlation coefficient. If the point is included, there is a substantial positive correlation between the variables; if the point is excluded, the data would suggest little or no correlation. The scatter diagram clearly displays the role of the point of high influence.

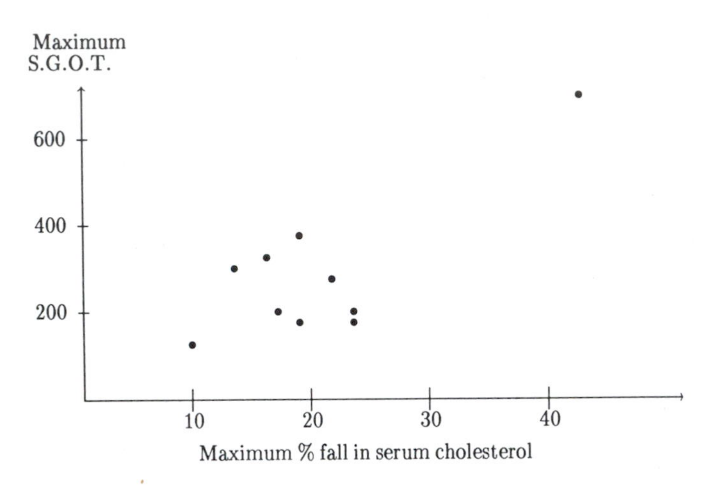

Figure 17.2.4. Scatter plot displaying a point of high influence.

Measuring correlation. The Product-moment correlation coefficient is the most widely used of the statistics which are employed to measure the level of correlation between two scaled variables. Why is this? One argument which might be advanced is that a realistic statistical model can be proposed in the form of the Bivariate Normal distribution which includes a parameter ρ for which r is an estimator. This argument supposes that values of ρ have a scientific interpretation. There is some sense in which this is so if one of the variables is seen as a predictor of the other. That situation arises in the context discussed in Chapter 18 and is considered there. It is not relevant in the current context. The best which can be said is that within a field of application there may develop a consensus as to what constitutes *low* and *high* levels of correlation. The ranges may vary widely between disciplines. For example, a value of ρ of 0.3 may be regarded as *high* in some areas of the social sciences but as *low* in areas of biological or physical sciences.

More valuable may be the use of numerical values in a comparative sense. The fact that a correlation may be higher in one group than in another or higher in one environment than another may have relevance, although there ultimately arises the question "*How much higher*?" and the problem of giving a meaning to a numerical quantity again arises.

Kendall's Tau has relevance in the following context. Suppose two randomly selected members of a population yield values for the pair of variables (x, y) of (x_1, y_1) and (x_2, y_2). If x_1 is greater than x_2 what is the probability that y_1 will exceed y_2? If the variables are uncorrelated, the probability is one half. If the correlation is positive, then increasing correlation implies the probability that y_1 will exceed y_2 tends towards one. If the correlation is negative, then increasing correlation implies that the probability will tend towards zero. Kendall's Tau reflects this trend in probability. The limitation is that a value of say 0.5 for τ does not transform into a specific value for the probability.

There is merit in quoting numerical values for correlation coefficients (subject to the comments below on sampling variation) for possible use in qualitative comparisons. Beyond this application, the contribution of numerical values to scientific knowledge is doubtful.

Inferential procedures. Using hypothesis testing rather than estimation to test the hypothesis $\rho = 0$ raises the usual note of caution. In a small sample, the power of the tests is low and with samples of less than ten it is quite likely that a test will fail to reject the hypothesis of no correlation when the variables are correlated. For example, if the correct value of ρ were 0.8, it would not be surprising in a sample of size 5 to obtain a value of r anywhere in the range -0.3 to $+0.95$. In large samples, a *statistically significant* correlation may be found but its physical or biological significance may be doubtful. For example, in a sample of 500, a value of 0.1 for r would lead to a p-value which is less than 0.05. There exist very limited scientific applications in which a correlation of 0.1 would have any scientific significance.

When the product-moment correlation coefficient is used, the construction of a 95% confidence interval for ρ is strongly recommended to identify the likely range of values within which ρ might lie. By this means, the investigator can assess the emphasis to be attached to the results of the correlation analysis. In partnership with a scatter plot, this would seem to be the best way to avoid misconceptions about the relevance of the findings.

17.3 Relations Between Three or More Categorical Variables

17.3.1 The Range of Applications

Exploratory studies. In an exploratory study, many variables may be included with the aim of establishing broad groupings of related variables. Ecologists, educational researchers and social scientists when studying an environment may seek to gain a broad view of the interrelations between a collection of variables which characterize the environment. Rarely is there sufficient knowledge to consider the construction of a complex model to describe the interrelations between the variables. Methods of Data analysis are generally employed which are based on the set of associations between each pair of variables included in the study. The method of *Cluster Analysis* is perhaps the most widely used of all methods. The basic elements of clustering methods are (i) the measure of association employed; and (ii) the strategy adopted in the formation of groups of variables from the table of associations. (See, for example, Gordon [1981].)

Most comprehensive statistical packages include routines for applying methods of cluster analysis.

An alternative approach is to view the set of associations between pairs of variables as providing measures of distances between the variables and to attempt to portray the set of distances geometrically in a two-dimensional or perhaps three-dimensional picture without introducing too great a distortion in the relative distances. The techniques which fall within this approach are termed methods of *ordination. Principal coordinate analysis* (see Chatfield and Collins [1980]) and *Multi-dimensional scaling* (see Schiffman et al. [1981]) are methods which come within this province.

Model building. When the number of variables is small, there is scope for defining and examining statistical models which portray the nature of interrelations between the complete set of variables. A systematic development in the statistical approach has become feasible with the ready availability of high speed computing facilities. In particular, the introduction of methodology based on Log-linear models has provided scientists with an important tool in the study of interrelations between categorical variables. This application is considered further in Sections 17.3.2 and 17.3.3.

Examining pre-defined questions. There is methodology for answering pre-defined questions, although the application of this methodology is necessarily linked to the process of model building, as is explained in Section 17.3.3.

17.3.2 Independence and Assocation

1. *Two variables*

(a) *Defining absence of association.* Given two variables x and y, the condition under which there is no association between x and y is defined by the probabilistic condition of *independence.* If x and y are categorical variables which can take values $1, 2, \ldots, a$ and $1, 2, \ldots, b$, respectively, and π_{ij} is the probability that x and y have classifications i and j, respectively, then x and y are independent if (17.3.1) is true.

$$\pi_{ij} = \pi_i^x \pi_j^y \qquad \text{for } x = 1, 2, \ldots, a;\ y = 1, 2, \ldots, b \tag{17.3.1}$$

where π_i^x and π_j^y are the probabilities that x and y are classified as i and j, respectively.

The definition of absence of association provided by (17.3.1) does not readily generalize to three variables. Statisticians most commonly adopt an alternative, but equivalent, definition of absence of association which is based on odds ratios. It is developed in the following way. The odds that x is classified into category i rather than category k $(i, k = 1, 2, \ldots, a)$ when y is classified as j is π_{ij}/π_{kj}. Independence (or absence of association) is equivalent to the requirement that, for any fixed value of j (i.e. $j = 1, 2, \ldots, b$) this odds is constant for all i (i.e. $i = 1, 2, \ldots, a$). In other words, the odds that the x-classification will be i rather than k is independent of knowledge of the y-classification.

(b) *Notation.* Two types of models are identified. They are:

M_{X+Y} which indicates that the variables are independent; and

$M_{X+Y+X.Y}$ which represents the more general model and places no restrictions on the relation between x and y.

The term $X.Y$ reflects the association between x and y and has an analogy with an interaction between two factors in the model equations for designed experiments.

To simplify this notation, it is common practice to introduce the contraction X^*Y for $X+Y+X.Y$. Thus, the model $M_{X+Y+X.Y}$ is represented as M_{X*Y}.

2. *Three variables*

(a) *Defining absence of association.* With the introduction of a third categorical variable, z, with categories $1, 2, \ldots, c$, the definition of association between the variables x and y is not uniquely linked to the independence condition. Two approaches have been proposed:

- the *additive* approach in which lack of association between x and y arises when information about z is ignored, i.e. the table of probabilities is pooled over the c categories for z and (17.3.1) is applied to define lack of association; and
- the *multiplicative* approach in which lack of association between x and y is tied to independence of x and y at every level of z.

Darroch [1974] discusses the merits of these two approaches. It is the multiplicative approach which has gained almost universal acceptance and which can be defined in terms of parameters in Log-linear models.

(b) *Three-factor association.* Under the multiplicative approach, the definition of no three-way association between variables x, y and z is based on the sets of odds ratios which can be constructed for the variables x and y at each level of z. Absence of a three-way association between x, y and z arises if the set of odds ratios which can be constructed for x and y at one level of z is identical for all levels of z. This is analogous to the definition of no three-way interaction in linear models as introduced in Chapter 15, with odds ratios taking the place of differences.

(c) *Two-factor association in the presence of a third variable.* In the presence of the third factor, z, lack of association between x and y cannot generally be determined by collapsing the frequency table over all levels of z and applying the independence condition to the two-way table for x and y. Absence of association between x and y arises when (i) there is no three-way association; **and** (ii) there is independence between x and y at all levels of z.

The fact that the definition of zero association between x and y relies on properties of the **three**-way table for x, y and z indicates that, when information on factor z is being considered, zero association between x and y is not equivalent to independence of x and y as defined in (17.3.1). Furthermore, this definition of absence of two-way association is not analogous to the definition of no two-way interaction employed with linear models. Plackett [1981] formalizes the above concepts.

Notation

M_{X+Y+Z} denotes absence of association between any pair of variables.

$M_{X \cdot Y+Z}$ denotes possible association between x and y but no association between x and z or between y and z.

$M_{X \cdot Y+X \cdot Z}$ denotes possible association between x and y and between x and z but no association between y and z.

$M_{X \cdot Y \cdot Z}$ denotes a possible three-way association between x, y and z and possible two-way associations between each pair of variables.

17.3.3 Experimental Aims and Statistical Analysis

Examining the association between two factors. In a study in which the only information available is on the pair of variables, x and y, there is a unique pair of models which may be compared to establish evidence of an association, namely $M_{X \cdot Y}$ versus M_{X+Y}. In the presence of a third variable, z, more than one comparison is available. For example, $M_{X \cdot Y+Z}$ versus M_{X+Y+Z} and $M_{X \cdot Y+X \cdot Z}$ versus $M_{X \cdot Z+Y}$ are both comparisons distinguished only by the presence or absence of the term $X.Y$. Figure 17.3.1 displays all possible models which may be fitted when there are three factors. A perusal of the figure establishes the existence of four possible model comparisons in which the only difference relates to the term $X.Y$. To establish which comparison(s) is (are) appropriate to seek evidence of an association between x and y, it is necessary to firstly fit the possible models containing the term $X{*}Y$ to the data and determine the simplest model(s) which provides an adequate fit to the data. The sequence of models to be fitted, in order or complexity, would be

1. $M_{X \cdot Y+Z}$ 2. $M_{X \cdot Y+X \cdot Z}$ and 3. $M_{X \cdot Y+X \cdot Z+Y \cdot Z}$.

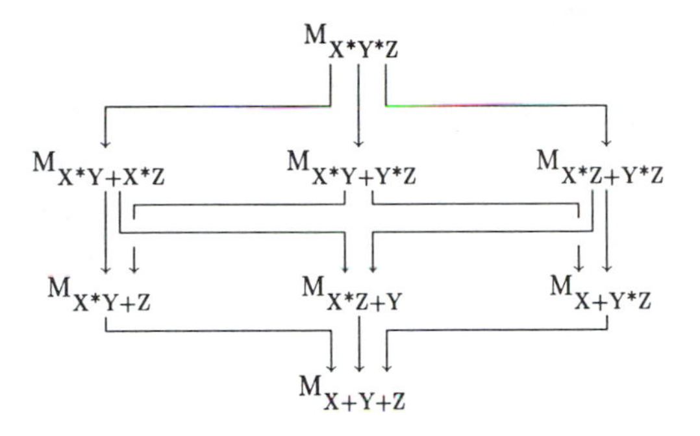

Figure 17.3.1. Possible models in a three variable study.

In other words, the statistical analysis is a two-stage process, the first being a selection stage in which an acceptable model is determined, and the second being a model comparison stage in which the question of the association between x and y is explored. If there are more than three variables, the number of possible models to be considered increases substantially and there is need for a consideration of model fitting strategies. The approach to the selection of models and the subsequent model comparisons is further complicated by the nature of the definition of association in multiplicative models (see Section 17.3.2). The absence of a two-factor association is not only dependent on the rejection of the term $X.Y$ in the model, but also on the absence of higher order terms.

The practical application of log-linear methodology to study relations between three or more variables requires lengthy discussion of the steps in model fitting and model selection and consideration of a wide variety of examples to give insight into the interpretation of findings. Fienberg [1978] offers a sound introduction and Bishop et al. [1975] an in depth discussion.

17.4 Relations Between Three or More Scaled Variables

17.4.1 The Range of Applications

The most widely employed applications are based on interpretation of the set of product-moment correlations formed from the pairwise comparison of every pair of variables. They fall into the three broad categories which are outlined below.

1. *Cause-effect structures.* Determining that x and y have a high correlation may not constitute the end-point of a study. The more important question may be "*Why do x and y have a high correlation*?" A possible

explanation is that both x and y are highly correlated with z, and it is this common connection which explains the high correlation between x and y. This explanation could be examined by the statistical technique of *partial correlation analysis* which is described in Section 17.4.2.

More complex cause-effect structures are considered in Chapter 20.

2. *Seeking collections of related variables.* Where all variables have equal status, the scientific aim is commonly to seek groupings of related variables. A data-analytic technique known as *Principal component analysis* is widely used for this purpose. A description of the technique and the manner in which it is applied can be found in many books on Multivariate Statistics (see, for example, Jolliffe [1986]).

3. *Latent variables.* An area of Statistics which assumes some importance in the social sciences is *factor analysis.* It is based on a model which supposes that correlations between pairs of measured variables can be explained by the connections of the measured variables to a small number of non-measurable, but meaningful variables, which are termed *factors.* For example, the concept of intelligence is based on this premise. While intelligence cannot be explicitly measured, there is a belief that a numerical value for intelligence can be constructed from sources on a battery of tests.

The aim of factor analysis are to

- identify the number of underlying factors required to describe the interrelationships between the measured variables;
- define the underlying variables as functions of the measured variables; and
- study the factors which have been defined.

There are numerous books written on factor analysis. See, for example, Harman [1976].

17.4.2 Partial Correlation Analysis

Two variables are correlated because either (i) one variable influences the values taken by the other variable, or (ii) the values taken by both variables are influenced or controlled by one or more supplementary variables. If an investigator hypothesizes that the latter situation applies, there is a statistical technique, *Partial correlation analysis,* which may be used to examine the proposition. Consider the situation described in Example 17.4.

Example 17.4 (Biochemical Studies Made During Fetal Development). *Table* 17.4.1 *contains sets of measurements made on liquid extracted from mothers, fetuses and amniotic fluid* (*the fluid which surrounds the sac holding the fetus*). *In each sample collected, the information obtained comprises the levels of sodium* (Na), *potassium* (K), *chloride* (Cl) *and osmotic pressure* (OP). *Calculations based on the data for 'Mothers' yielded the following product-moment correlations:*

$$r_{Na,Cl} = 0.7618 \qquad r_{Na,OP} = 0.8580 \qquad r_{Cl,OP} = 0.7587.$$

These results suggest a high correlation between sodium and chloride. The researcher who collected the data believed that both sodium and chloride levels would rise as a consequence of an increase in osmotic pressure. Hence, the correlation between sodium and chloride levels is postulated to be at least partly attributable to variations in osmotic pressure.

One way of determining if this is the case would be to obtain measurements for sodium and chloride levels on women who all have the same level of osmotic pressure. A correlation between sodium and chloride computed from these data which is substantially lower than the figure of 0.7518 *given above would support the researcher's claim. Such an approach is impractical since it is not feasible to seek a sample of women who give the same reading for osmotic pressure. Partial correlation analysis offers an alternative which meets the same objective without placing constraints on the values of osmotic pressure taken by the women in the sample.*

Basis. To obtain the partial correlation between x and y adjusted for z, both x and y are presumed to be related to z by linear equations

$$x = A + Bz + e_x \qquad \text{and} \qquad y = C + Dz + e_y$$

where e_x and e_y are the components of x and y, respectively, which are independent of z. The partial correlation between x and y adjusted for z is then the correlation between e_x and e_y.

Data. For each of n sample members, measurements are presumed to be available on p scaled variables. For each pair of variables, the Product-moment correlation is presumed to have been calculated using (17.2.1). This yields a set of Product-moment correlation coefficients $\{r_{ij}\}$ where i and j identify the variables $(i, j = 1, 2, \ldots, p)$.

Model. Since the method is based on the Product-moment correlation coefficient, it is presumed that a linear trend line exists in the relation between each pair of variables. Apart from this restriction, the only other condition imposed is that the sample be randomly selected.

Application. The partial correlation between variables i and j after eliminating the effects of variable k can be computed from (17.4.1).

$$r_{ij.k} = \frac{r_{ij} - r_{ik}r_{jk}}{\sqrt{(1 - r_{ik}^2)(1 - r_{jk}^2)}}. \tag{17.4.1}$$

Table 17.4.1. Chemical Measurements Made in Fetal Studies (Example 17.4)

Mother				Baby				Amniotic fluid			
Na	K	Cl	OP	Na	K	Cl	OP	Na	K	Cl	OP
133	4.0	105	276	141	6.6	97	282	124	4.5	100	261
135	4.4	106	279	138	6.9	104	281	118	4.7	98	244
135	3.9	105	280	134	5.2	107	261	130	4.6	110	273
139	3.9	108	288	140	6.0	106	285	122	4.0	100	251
140	4.2	108	286	137	7.3	106	283	134	4.6	113	272
132	4.9	99	263	135	6.3	102	278	112	5.0	93	232
134	4.8	108	281	135	4.2	106	281	120	5.6	105	250
136	4.7	107	286	139	6.3	102	280	128	4.7	110	265
135	3.9	104	287	138	4.9	105	278	126	4.3	106	265
137	4.3	101	284	133	5.0	101	273	115	3.5	99	235
134	4.6	106	282	137	4.8	107	282	118	4.1	102	250
135	4.0	108	280	139	6.4	104	280	124	5.5	107	158
135	4.6	107	284	136	6.6	107	280	123	5.1	105	260
135	4.2	104	281	132	5.8	107	276	119	4.2	101	245
136	4.2	108	284	134	7.3	102	285	118	4.8	101	246
134	4.5	99	275	141	6.6	97	282	123	4.6	100	251
140	6.4	100	286	138	6.9	104	281	122	4.2	103	240
138	4.0	107	288	136	4.9	101	280	124	3.8	102	250
136	3.7	106	285	140	4.6	104	280	126	4.6	104	255
134	3.8	102	283	140	6.1	96	284	126	4.6	105	259
136	4.4	106	279	138	5.6	105	286	127	4.5	100	258
134	4.7	99	274	135	6.3	102	278	123	4.2	100	251
135	4.5	99	291	136	4.8	104	275	122	3.8	100	251
137	4.2	108	282	135	4.7	103	271	128	5.2	110	263
137	3.8	104	279	137	4.8	107	282	128	5.2	108	259
135	3.8	104	279	*	*	*	*	122	4.3	105	254
136	5.7	103	280	136	6.6	107	280	118	4.8	100	242
136	4.3	109	278	138	4.7	109	287	117	4.2	97	236
136	4.1	107	276	*	*	*	*	128	4.0	108	255
137	4.3	104	285	*	*	*	*	115	3.5	102	256
137	4.4	102	279	*	*	*	*	125	4.4	104	252
135	4.5	105	287	*	*	*	*	123	4.4	105	254
135	3.7	115	283	*	*	*	*	131	4.4	105	271
139	3.6	105	276	*	*	*	*	130	4.4	104	264
137	4.6	104	286	*	*	*	*	131	4.1	112	265
138	5.0	103	273	*	*	*	*	123	4.6	103	253
137	4.3	106	276	*	*	*	*	129	4.6	110	261
139	3.7	104	278	*	*	*	*	129	4.0	105	257
139	4.5	104	275	*	*	*	*	130	5.0	107	259
138	4.6	105	280	*	*	*	*	119	4.2	101	244

Table 17.4.1 (*cont.*)

Mother				Baby				Amniotic fluid			
Na	K	Cl	OP	Na	K	Cl	OP	Na	K	Cl	OP
136	6.4	106	284	*	*	*	*	124	3.9	103	250
136	4.2	97	277	*	*	*	*	129	5.0	104	262
137	4.3	101	283	*	*	*	*	133	4.0	109	266
138	4.0	106	280	*	*	*	*	126	4.5	106	264
141	4.7	100	291	*	*	*	*	116	5.6	94	240
136	3.8	104	275	*	*	*	*	124	3.9	101	255
135	4.6	101	276	*	*	*	*	118	4.7	97	247
137	4.4	104	278	*	*	*	*	124	4.3	107	256
137	3.4	100	272	*	*	*	*	120	3.6	93	243
139	4.7	104	275	*	*	*	*	129	5.1	109	260
136	4.8	106	285	*	*	*	*	126	4.1	105	263
129	5.9	93	265	*	*	*	*	131	4.8	105	266
140	3.7	103	283	*	*	*	*	122	4.6	103	253
136	4.5	101	279	*	*	*	*	128	4.4	102	258
134	4.8	108	275	*	*	*	*	128	3.9	108	257
136	4.5	101	276	*	*	*	*	125	4.1	103	254
135	3.5	100	276	*	*	*	*	123	4.0	106	245
135	4.5	101	280	*	*	*	*	126	4.6	106	258
140	3.6	108	273	*	*	*	*	125	4.3	104	262
136	5.3	103	282	*	*	*	*	126	5.2	105	252
138	4.9	103	282	*	*	*	*	126	4.1	114	263

The partial correlation between i and j after eliminating the effects of variables k and m can be computed from (17.4.2.).

$$r_{ij.km} = \frac{r_{ij.m} - r_{ik.m}r_{jk.m}}{\sqrt{(1 - r_{ik.m}^2)(1 - r_{jk.m}^2)}}. \tag{17.4.2}$$

Illustration. To compute the partial correlation between *Sodium* and *Chloride* after elimination of the effects of *Osmotic pressure* based on the data for *Mothers* which is contained in Table 17.4.1, the product-moment correlations required for input into (17.4.1) are

$$r_{Na,Cl} = 0.7618 \qquad r_{Na,OP} = 0.8580 \qquad r_{Cl.OP} = 0.7587.$$

By (17.4.1), the partial correlation between *Sodium* and *Chloride* after eliminating the effects of *Osmotic pressure* is

$$r_{Na,Cl.OP} = \frac{0.7618 - (0.8580)(0.7587)}{\sqrt{[1 - (0.8580)^2][1 - (0.7587)^2]}} = 0.33.$$

Interpretation. As with other correlation coefficients, partial correlation coefficients may vary between −1 and +1. A value of zero implies no correlation and increasing magnitude implies an increasing level of correlation.

However, it tends not only to be the size of the partial correlation coefficient which is of interest. The extent to which, and the manner in which, the partial correlation coefficient differs from the correlation coefficient is equally important. Of primary interest is whether the partial correlation coefficient is substantially different from the correlation coefficient. If not, then the relation between the variables of interest is little affected by the variables for which adjustment has been made. If the correlation coefficient is high, two situations of particular interest may be distinguished.

1. The partial correlation is much closer to zero than the correlation, thereby indicating that the observed strong relationship between the variables may be associated with the relation each has with the variable(s) for which adjustment has been made. (This was the case in the partial correlation coefficient computed above from the data in Example 17.4. The correlation between *Sodium* and *Chloride* is 0.76 before adjustment for variations in *Osmotic pressure,* but drops to 0.33 after adjustment for variations in *Osmotic pressure.*)

2. The partial correlation has a large magnitude but is of opposite sign. The use of partial correlation analysis in this situation often provides the key to understanding an apparently anomalous correlation between two variables.

Practical constraints on interpretation

1. Statistical analysis can never establish the existence of a cause-effect relation. If, however, there is strong scientific evidence to suggest a causal relationship, the statistical finding can be used to support the scientific evidence.

2. There is no formal test available to establish when a partial correlation is significantly different from the correlation. Particularly when sample sizes are small (less than 20 is a reasonable rule-of-thumb) findings should be treated with caution.

Problems

17.1 *Table* 17.1.1(b)

(a) Following the steps in Section 17.1.3, reproduce the results from testing the hypothesis of no association between *length of stay* and *age* based on the data in Table 17.1.1(b). Obtain tables of fitted values based on the assumption of no association.

(b) Hence, obtain and interpret the table of standardized residuals to establish the nature of the relationship between the variables.

17.2 *Survey of national park users* (*Problem* 9.3)

In Problem 9.3, there are tables of frequencies for three groups of users of national parks. For each group, apply the steps in Section 17.1.3 to establish if there is evidence of association between the variables *facilities* and *information*.

17.3 *Relations between performance in science and social science for students with different mathematics ability*

Data were collected on 1602 high school students for grades in science, mathematics and social science. The three frequency tables below were constructed from the data.

For each table, test for evidence of association between science and social science using the steps given in Section 17.1.3. Where there is evidence of association, construct and examine the set of standardized residuals based on the fit of the Independence model to establish the nature of the association. Does the analysis suggest that the findings are consistent between the three groups? If not, describe how the relationships differ.

Mathematics

	C			B			A		
	Social Science			Social Science			Social Science		
Science	C	B	A	C	B	A	C	B	A
C	21	52	10	27	75	12	6	6	1
B	50	108	21	69	344	102	12	37	30
A	25	70	12	20	92	136	3	83	178

17.4 *Example* 17.2 — *single group studies*

For each group, 12 *years olds*, 12 *year old special group and* 14 *year olds*,

(a) construct a scatter plot for test U versus test S (as illustrated in Figure 17.2.1(a) for the twelve year olds) and check for evidence that the trend line is linear and there are no apparent errors in the data; and

(b) construct a Normal probability plot for both variables (see Section 10.3.4) to check on the assumption of Normality.

(c) If the examination of the scatter plot suggests the trend line is linear and there is no evidence of errors in the data, compute a value of the sample product moment correlation coefficient (r) from (17.2.1).

(d) If model checking in (a) and (b) suggests the assumption of a Bivariate Normal distribution is acceptable, use (17.2.8), (17.2.10) and (17.2.11) to construct confidence intervals for the population product-moment correlation coefficient (ρ).

17.5 *Example* 17.2 — *comparing groups*

Follow the steps in 17.2.7 to

(a) reproduce the test of the hypothesis that the twelve year olds and the fourteen year olds have the same correlation for variables U and S.

(b) Test the hypothesis that the twelve year olds and the twelve year old special group have the same correlation for variables U and S.

17.6 *Example* 17.3

For each variable in Table 17.2.2 from *aluminum* to pH

(a) construct a scatter plot versus *dust*.

(b) If the scatter plot suggests the trend line is either continually increasing or continually decreasing, follow the steps in Section 17.2.4 to compute a value for Kendall's τ statistic. Hence, test the hypothesis of no association between the two variables.

(c) On the basis of the values of t_τ, rank the variables in order of the level of association with *dust*.

17.7 *Example* 17.4

In Section 17.4.2, there is an examination of the correlations between sodium and chloride measurements based on data for *mothers* in Table 17.4.1. The correlation is shown to be substantially reduced after adjustment for variations in *osmotic pressure*. Determine if the same effect applies for *babies* and *amniotic fluid*.

18

Prediction and Estimation: The Role of Explanatory Variables

18.1 Regression Analysis

18.1.1 Mathematical Relations and Trend Lines

In mathematics and mathematically related disciplines, scientists speak of *functional relations* between variables. The statement $y = 3 + x$ *for all positive values of* x is an example of a functional relation between variables x and y. Through this relation it is possible to establish the exact value of y for any value of x. For example, if x takes the value 3 then y takes the value 6. In the physical sciences, functional relations may be referred to as *Laws.* Students of physics would recognize the relation between pressure (P) and volume (V) of a gas, $P = k/V$ or $PV = k$ as *Boyle's Law.* This relation establishes that the product of the pressure and volume of an ideal gas is constant.

The mathematical statement of the relation is exact. *Boyle's Law* is based on theoretical premises which are believed to be true, and which, up to a point, are verifiable by experimentation. However, it cannot be expected that experimental results will produce a set of pressure and volume readings which exactly satisfy Boyle's Law. There are three reasons why theory and experimental results may not agree. They are

- the existence of *measurement error* which arises because of the finite limit to accuracy in any measuring device;

- change in the conditions or environment under which measurements are made; and

- the fact that Boyle's Law applies only to *ideal* gases.

From an experimental viewpoint, the representation of the pressure-volume relationship is correctly represented by the set of equations defined in (18.1.1).

$$\begin{aligned} p_i &= P_i + e_i \\ v_i &= V_i + f_i \end{aligned} \qquad (i = 1, 2, \ldots, n) \qquad (18.1.1)$$

where

p_i, v_i is the measured pair of pressure and volume values made at the ith reading;

P_i, V_i are the pressure and volume readings which would have been obtained under standard conditions in the absence of measurement errors if the gas were ideal (and which satisfy Boyle's Law, i.e. $P_iV_i = k$); and

e_i, f_i are the effects of changes from the ideal or expected readings in pressure and volume values, respectively, at the ith reading which are attributable to the causes listed above.

If the aim of the study is to make a judgment on the acceptability of Boyle's Law, it is necessary to include in the model some structure to allow for the variations arising from measurement error and other factors since the experimenter has only the observed set of values, $\{(p_i, v_i)\}$, rather than the 'true' values for pressure and volume. In statistical models this structure may take the form of a probability distribution which describes the pattern of variation expected in the pairs (e_i, f_i) under a specified sampling scheme.

Furthermore, it is unusual for the theoretical understanding of the relation to have reached a stage where the functional form of the relationship is known. Thus, in Example 18.1, which provides the data in Figure 18.1.1, it is expected that students who gained higher marks in the high school mathematics examination will perform better in the university mathematics examination. However, there is no educational theory available to define the form of the relationship between *university mathematics mark* and *high school mathematics mark*. Commonly the choice is based on empirical evidence—in Figure 18.1.1 there is evidence to support the choice of a linear trend line.

Equation (18.1.2) contains equations which might apply to the relation portrayed in Figure 18.1.1. The variables are the *university mathematics mark* (Y) and the *high school mathematics mark* (X).

$$\begin{aligned} Y &= A + BX \\ y_i &= Y_i + e_i \qquad \text{for } i = 1, 2, \ldots, 50 \\ x_i &= X_i + f_i \end{aligned} \tag{18.1.2}$$

where (y_i, x_i) is the pair of readings for *university mathematics mark* and *high school mathematics mark;* and (e_i, f_i) is the pair of deviations between the observed values and the 'true' values for *university mathematics mark* and *high school mathematics mark* for the ith student.

There are conceptual problems associated with the variables Y and X in (18.1.2). In the pressure, volume example, the 'true' conditions could be envisaged as values under stated environmental conditions and for an ideal gas. In the educational example, there is no obvious meaning for 'standard' conditions. Nevertheless, the assumption of a linear trend line

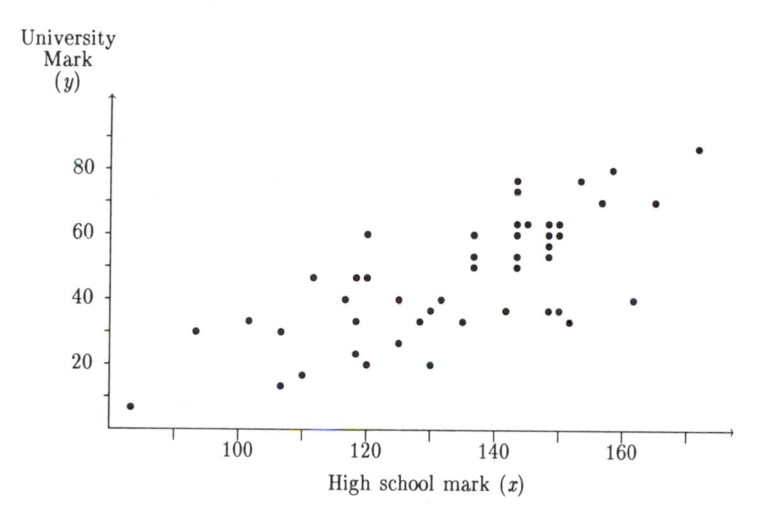

Figure 18.1.1. Plot of university maths mark vs. high school maths mark based on data from Table 18.1.1.

for the relation between the two examination marks has benefit both in providing a simple description of information in the figure and to answer meaningful scientific questions.

Statistical models can be developed from the basic equations of (18.1.1) and (18.1.2). Such models are termed *Functional relation models.* They have application when there is a law-like relation between variables Y and X or when there is a cause-effect structure which involves a complex of variables.

A simpler statistical methodology is available for problems in which Y is presented as the response variable of interest and the aim is to examine the manner in which variation in values of Y are related to changes in the explanatory variable X. The technique is known as *Regression analysis* and the scope of its application is illustrated in Section 18.1.2. The basic structure of statistical models for such applications is considered in Section 18.2 and the methods based on those models are introduced and illustrated in Sections 18.3 to 18.6.

18.1.2 Experimental Objectives — Illustrations

An indication of the scope of regression analysis is provided by the following examples.

Example 18.1 (A Study of University Entrance Requirements). *Data presented in Table* 18.1.1 *are a portion of the data collected in a study of the usefulness of high school mathematics examinations as the basis for* *determining conditions for admission to the first year university course in The University of Tasmania.*

The aim of the study was to examine the usefulness of the high school mathematics (x) as a predictor of the first year university mathematics mark (y).

Table 18.1.1. Mathematics Performance in High School and University (Example 18.1)

Student	High school maths. mark	Univ. maths. mark	Student	High school maths. mark	Univ. maths. mark
1	143	55	26	130	38
2	148	66	27	116	46
3	119	45	28	137	57
4	118	41	29	158	70
5	119	24	30	147	38
6	125	28	31	84	6
7	93	31	32	142	35
8	158	80	33	115	16
9	137	50	34	131	19
10	106	14	35	151	39
11	128	35	36	143	49
12	119	61	37	143	64
13	132	41	38	135	33
14	165	72	39	118	33
15	143	65	40	119	47
16	145	65	41	149	62
17	119	20	42	136	57
18	145	73	43	149	60
19	149	54	44	162	41
20	151	35	45	109	30
21	153	74	46	143	59
22	143	76	47	171	87
23	145	62	48	111	32
24	123	40	49	147	55
25	124	41	50	124	32

Example 18.2 (Studies of Seaweed as Home to Animals). *Data in Table* 18.1.2 *were collected from samples of two species of seaweed which have different morphological characteristics. The figures obtained from each plant comprise biomass (in the form of dry weight) and abundance of selected animal species who use the plant as a host. The experimental aim is to establish if the use of seaweed plants as a host by the selected animal*

species varies between the two species of seaweed. As part of this examination, the relation between abundance and biomass was compared between the two species.

Table 18.1.2. Biomass (Dry Weight) and Abundance of Selected Animal Species on Two Species of Seaweed Plants (Example 18.2) (Data by courtesy of Dr. G. Edgar, Dept. of Zoology, Univ. of Tasmania)

Caulocystis cephalornithos		*Sargassum verruculosum*	
Abundance	Dry weight	Abundance	Dry weight
681	28.8	67	2.5
320	15.1	222	12.5
244	7.5	131	8.6
219	6.5	225	10.8
422	9.9	71	3.7
441	12.7	391	12.0
1160	11.8	142	4.5
347	7.6	123	4.9
176	8.5	186	8.0
229	7.3	239	9.7
363	14.1	294	4.1
310	11.0	256	6.7
161	9.6	435	11.2
337	12.0	460	10.5
331	9.2	222	4.8
337	21.5	572	10.1
825	30.4	224	6.4
707	24.4	344	4.6
514	11.1	661	14.3
901	28.5	439	14.2
957	26.0	277	8.7
		129	3.6
		430	8.6
		338	4.5
		530	11.5

Example 18.3 (Characterizing the Growth of Pea Pods from the Garden Pea (*Genus Pisum*). *The average length of a pea pod* (M_L) *when plotted against time* (t) *can be well approximated by the sigmoidal curve represented by* (18.1.3) *and displayed in Figure* 18.1.2.

$$M_L = \frac{B_1}{1 + \exp(B_2 - B_3 t)} \tag{18.1.3}$$

where B_1, B_2, B_3 *are parameters whose values are unknown.*

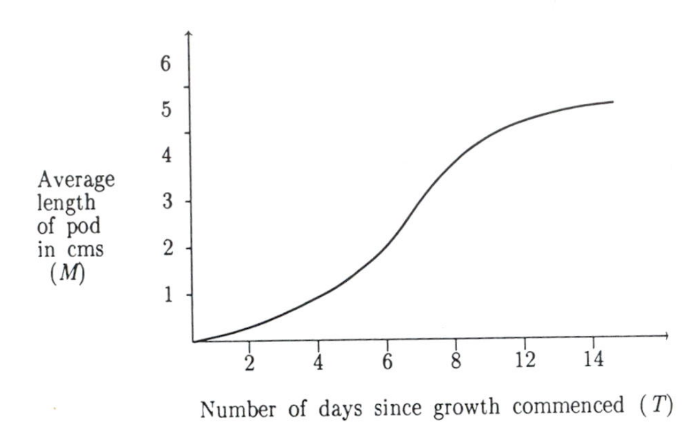

Figure 18.1.2. An illustration of a sigmoidal relationship.

The parameters of the function defined in (18.1.3) *can be used to define the following properties:*

- *the maximum length of the pod* (B_1);
- *the maximum growth rate achieved* ($B_1B_3/4$); *and*
- *the time at which maximum growth rate is achieved* (B_2/B_3).

Regression analysis is employed to provide estimates of the parameters and hence estimates of the characteristics of experimental interest. Further analysis might be used to provide comparisons of these characteristics between species or under different growing conditions.

There may be more than one explanatory variable which can account for the observed variation in the response variable. Example 18.4 provides an illustration.

Example 18.4 (Profitability of a Supermarket Chain). *A national supermarket chain was examining the structure of its operations and, as part of the investigation, sought to establish which factors might be important in determining profitability in its stores. To assist in this aspect of the study, statistical analysis was employed to assess the extent to which changes in profitability could be related to variations in a range of explanatory variables. Data from this part of the study are presented in Table* 18.1.3.

Table 18.1.3. Data Collected in the Study Reported in Example 18.4

Store	Profit[1]	Floor area	No. of employees[2]	Age of manager	No. of years	Grading of suburb[3]	No. of competitors[4]
1	22	278	5	29	1	6	2
2	10	1814	10	30	1	3	1
3	25	207	7	36	2	5	1
4	47	2074	11	33	3	5	1
5	66	181	8	34	5	1	4
6	71	181	8	34	5	2	3
7	46	1063	11	41	4	7	1
8	37	2696	13	38	5	8	5
9	89	441	8	39	4	6	0
10	88	1607	12	43	5	2	2
11	123	1322	10	36	5	3	0
12	106	1089	10	33	5	5	1
13	107	2540	13	40	6	8	2
14	82	3162	15	47	7	2	3
15	105	2773	15	44	6	3	4
16	111	3903	13	38	5	3	5
17	99	2281	12	42	7	5	3
18	103	1244	9	43	6	5	1
19	130	2618	15	47	6	6	2
20	123	1892	12	40	7	3	1
21	143	1218	9	38	7	2	1
22	149	3240	17	46	9	1	4
23	132	2851	15	44	7	7	5
24	106	4873	16	45	8	5	5
25	124	3862	18	50	9	3	3
26	97	2488	12	45	5	4	2
27	171	3292	16	49	10	6	3
28	115	2722	15	47	9	4	3
29	142	4406	19	54	9	6	5
30	194	4199	15	50	10	10	3

Notes.

[1]'Profit' is a measure of return on investment

[2]Effective full time employees

[3]A socio-economic grading based on a standard scale

[4]Major competitors in the same shopping area

18.2 Statistical Models

18.2.1 The Concept of a Regression Equation

The intuitive notion of a trend line to describe the relation between the response variable y and the explanatory variable x must be given a precise meaning if statistical models and methodology are to be developed. Statisticians have developed a theory and associated methodology in which the trend line is defined as the relation between the *average* values of y and the values of x. Formally, there is defined a probability distribution for y for each value of x, the conditional probability distribution of y given x, which has an expected value $M_{y.x}$. *In regression analysis, the trend line in the relation between y and x is defined as the functional relation between $M_{y.x}$ and x for all values of x.* This relationship defines the *regression equation of y on x.* Figure 18.2.1 provides a pictorial representation of a regression equation.

Note that the relation is asymmetric. If the regression equation of x on y is also constructed, then, in general, the graph of $M_{y.x}$ versus x and the graph of $M_{x.y}$ versus y will not produce the same trend line. Thus, the regression relation is quite different from functional and law like relations (e.g. Boyle's Law). The two types of relations have different applications and are based on different statistical models.

The definition of a regression equation may be extended to cover the situation in which there is more than one explanatory variable. If the explanatory variables are denoted by $x_1, x_2, \ldots, x_k$, the mean or *expected value* of y when $x_1, x_2, \ldots, x_k$ take the values $x_{10}, x_{20}, \ldots, x_{k0}$, respectively, is denoted by $M_{y.x_{10},x_{20},\ldots,x_{k0}}$.

18.2.2 Component Equations

The statistical development of regression analysis is simplified by assuming that the response (y) is composed of two distinct parts, a component determined from the relation with the explanatory variable x, represented by the value of the regression function $M_{y.x}$, and a second component representing the collective contribution of all other sources. When random selection is employed this is commonly referred to as the *random error.* It is denoted by e. The manner in which these components are combined to produce the response is defined by a *component equation.* The additive equation presented in (18.2.1) is commonly assumed.

$$y = M_{y.x} + e. \tag{18.2.1}$$

Variations in the component equation are defined by the different forms of the regression equation (Section 18.2.3) and the imposition of a design structure (Section 18.7.1).

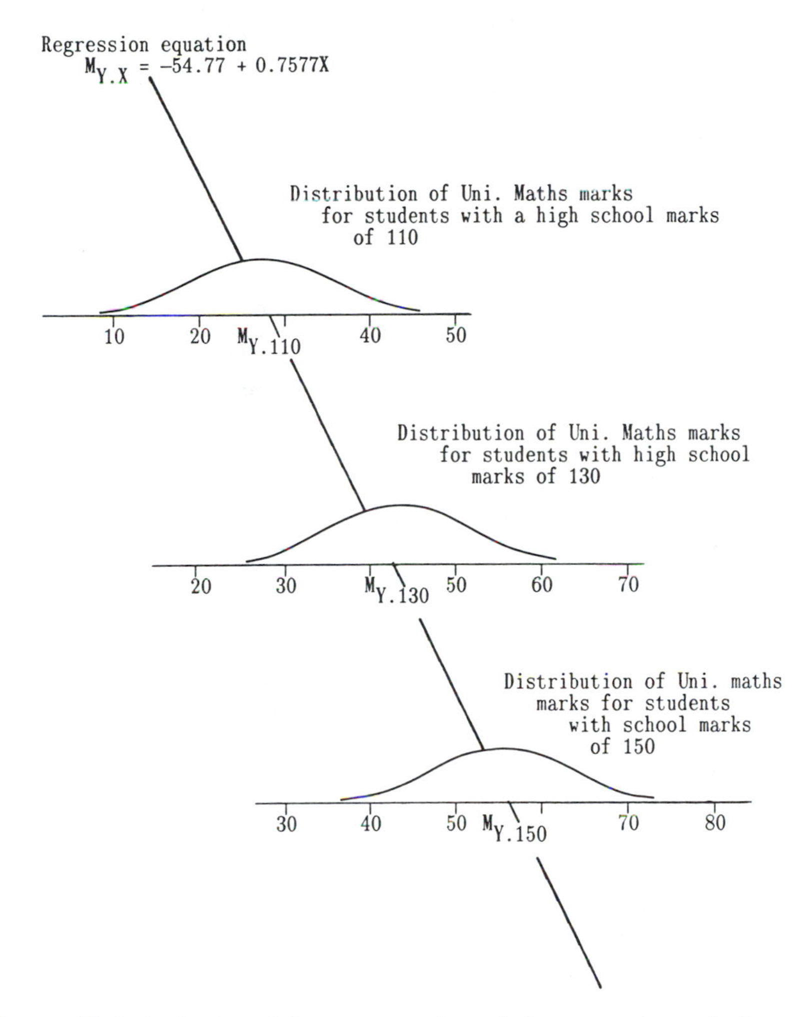

Figure 18.2.1. A pictorial representation of the regression relation as it applies in Example 18.1.

18.2.3 Forms of Regression Equations

For statistical purposes, regression equations are classified into the following types, with each type identifying a family of relations, the members of which are distinguished by parameters which are known as *regression coefficients.*

1. Regression equations including only one explanatory variable

   (a) *Linear equations* which include

   (i) *Simple linear* $M_{y.x} = B_0 + B_1 x$

   (ii) *Curvilinear* $M_{y.x} = B_0 + B_1 x + B_2 x^2 + \cdots + B_p x^p$

(b) *Non-linear* e.g. $M_{y.x} = \dfrac{B_0}{1 + \exp\{B_1 - B_2 x\}}$

2. Regression equations including more than one explanatory variable

(a) *Multilinear*

$$\text{e.g. } M_{y.x_1,x_2} = B_0 + B_1 x_1 + B_2 x_2 + B_3 x_1 x_2$$
$$M_{y.x_1,x_2} = B_0 + B_1 x_1 + B_2 x_2 + B_3 x_1 x_2 + B_4 x_1^2 + B_5 x_2^2$$

(b) *Multiple non-linear*

$$\text{e.g. } M_{y.x_1,x_2} = \frac{B_0}{1 + \exp\{B_1 - B_2 x_1 - B_3 x_2\}}$$

A broader class of regression equations which encompass linear equations are *Generalized linear* regression equations which satisfy the equation

$$\varphi[M_{y.x_1,x_2,\ldots,x_p}] = B_0 + B_1 x_1 + B_2 x_2 + \cdots + B_p x_p$$

where φ is a function referred to as the *link* function (since it links the original scale of measurement to a scale on which the regression equation has a linear form).

From a theoretical point of view, a *linear* regression equation is one in which the parameters appear linearly. Thus, the regression equation $M_{y.x} = B_0 + B_1 \log(x)$ would be termed a *linear* equation even though the relation between y and x is clearly non-linear. *Curvilinear* regression equations are a subgroup of *Linear* regression equations.

18.2.4 Regression Models

There are four basic elements to consider in a regression model. They are identified and discussed below. For simplicity and brevity of expression, the conditions are expressed in terms of a single explanatory variable x. They apply equally when more than one explanatory variable is included in the regression equation.

1. *Component equations.* If $y_1, y_2, \ldots, y_n$ are random variables which represent the response variable y when the explanatory variable x takes values $x_1, x_2, \ldots, x_n$, the relationship between y and x is expressed through the component equations in (18.2.2).

$$y_i = M_{y.x_i} + e_i \qquad \text{for } i = 1, 2, \ldots, n \tag{18.2.2}$$

where $M_{y.x_i}$ defines the regression equation for the ith response; and e_i is the unexplained component for the response y_i.

2. *Hypotheses.* Each regression equation in (18.2.2) is defined in terms of a set of regression coefficients and statistical hypotheses are defined in terms of these parameters.

3. *Sampling assumptions.* Sample selection may take one of two forms:

- either random selection of sample members with a pair of measurements being made on each sample member to provide values for y and x;
- or pre-determined values for x with the presumption that the ith sample member ($i = 1, 2, \ldots, n$) is randomly selected from all sample members which have a value x_i for x.

In practical terms, the requirements are that (i) the sample members are not chosen on the basis of the values taken for the response variable; and (ii) the responses of the sample members are independent. There are two important conditions in which the independence condition is likely to be violated. The first occurs when the same unit or subject is repeatedly sampled to obtain the sequence of responses and the second occurs when the units being sampled are spatially connected. In either case, neighboring units are expected to produce correlated responses. Specialized methodology is required in such cases. Where responses are being regressed on time, the statistical analysis is called *time series analysis.* This is the topic of Chapter 21.

4. *Distributional assumptions.* There has been limited development of regression methods based on distribution-free models and they are not considered in this book. The majority of applications are based on the assumption that $y_1, y_2, \ldots, y_n$ are Normally distributed. The simplest models assume the response variables are independent and all distributions have the same variance. Lack of dependence arises when there are spatial or temporal connections between the sample members (as indicated above).

18.3 Statistical Methods

18.3.1 Data

The data are presumed to be obtained from n sample members, with each sample member contributing a value of the scaled response variable y and values for each of the scaled explanatory variables $x_1, x_2, \ldots, x_p$. If there is only one explanatory variable it is referred to as x, i.e. no subscript is used.

The data are presumed to have one of the following forms:

- if there is only one explanatory variable (as illustrated in Table 18.1.1);

Sample Member	1	2	$\cdots$	n
x	x_{10}	x_{20}		x_{n0}
y	y_{10}	y_{20}		y_{n0}

– if there is more than one explanatory variable (as illustrated in Table 18.1.3)

Sample Member	1	2	⋯	n
x_1	x_{110}	x_{120}		x_{1n0}
x_2	x_{210}	x_{220}		x_{2n0}
⋮				
x_p	x_{p10}	x_{p20}		x_{pn0}
y	y_{10}	y_{20}		y_{n0}.

18.3.2 Statistical Models

The models for which methodology is provided in this chapter differ only in the regression equations employed. The assumptions they have in common are:

1. Sampling assumptions: the responses are presumed to be independent.

2. Form of distribution: the assumption of Normality is made with the additional requirement of equal variance for the response variable at all values of the explanatory variable(s).

Three basic models are considered. They each include the two assumptions quoted above and are distinguished by the form of the regression equation.

The **Simple linear regression model** which applies when there is a single explanatory variable and the regression equation is defined by (18.3.1).

$$M_{y.x_i} = A + Bx_i \qquad \text{for } i = 1, 2, \ldots, n. \tag{18.3.1}$$

The **Simple quadratic regression model** which applies when there is a single explanatory variable and the regression equation is defined by (18.3.2).

$$M_{y.x_i} = A + Bx_i + Cx_i^2 \qquad \text{for } i = 1, 2, \ldots, n. \tag{18.3.2}$$

The **Multiple linear regression model** which applies when there is more than one explanatory variable and the regression equation is defined by (18.3.3).

$$M_{y.x_{1i},x_{2i},\ldots,x_{pi}} = B_0 + B_1x_{1i} + B_2x_{2i} + \cdots + B_px_{pi} \text{ for } i = 1, 2, \ldots, n. \tag{18.3.3}$$

Note that

(i) the Simple linear regression model is a special case of the Multiple linear regression model in which p equals one;

(ii) the Simple quadratic regression model is a special case of the Multiple linear regression model in which x_{1i} is equated with x_i and x_{2i} is constructed as x_i^2; and

(iii) the Simple linear and Simple quadratic regression models are special cases of the *Simple polynomial regression model,* which has the regression equation (18.3.4) for the pth order polynomial ($p = 1, 2, \ldots, n$).

$$M_{y.x_i} = B_0 + B_1 x_i + B_2 x_i^2 + \cdots + B_p x_i^p \qquad \text{for } i = 1, 2, \ldots, n. \quad (18.3.4)$$

18.3.3 Fitted Values, Residuals and Parameter Estimation

The regression equations presented in Section 18.3.2 identify families of regression models rather than individual models. The members of the family are distinguished by the set of values of the regression coefficients, e.g. A and B in the Simple linear regression model. The first step in statistical analysis is to estimate values for those parameters in the equation which have not been previously assigned values either as a result of information supplied by the experimenter or by hypotheses in the statistical model.

Parameter estimation. Two methods of parameter estimation are widely used in regression analysis. They are based respectively on the distribution-free *principle of least squares estimation* and the distribution-based *principle of maximum likelihood estimation.* Under the sampling and distributional assumptions proposed in Section 18.3.2, the two methods provide the same estimators for the parameters. These are the set of statistics which minimize the sum of squares in (18.3.5).

$$\begin{aligned}
\sum_{i=1}^{n} e_i^2 &= \sum_{i=1}^{n} [y_i - (A + Bx_{i0})]^2 && \text{Simple linear} \\
&= \sum_{i=1}^{n} [y_i - (A + Bx_{i0} + Cx_{i0}^2)]^2 && \text{Simple quadratic} \qquad (18.3.5) \\
&= \sum_{i=1}^{n} [y_i - (B_0 + B_1 x_{1i0} + \cdots + B_p x_{pi0})]^2 && \text{Multiple linear}
\end{aligned}$$

For all regression equations which are represented as *linear* functions of the parameters, i.e. which have the general form expressed in (18.3.3), the parameter estimators are obtained by solving a system of linear equations which yield solutions that express each parameter estimator as a linear function of the set of response variables, $\{y_i\}$. (The theoretical development is straightforward and widely available in texts on mathematical statistics.) The solution for the parameters A and B which minimizes (18.3.5) for the Simple Linear regression equation are given by (18.3.6). For the other linear models, analytic expressions are available but are cumbersome to present without resorting to matrix notation. In practice, computer-based calculations are recommended for the computation of parameter estimates.

$$\begin{aligned} \hat{A} &= \overline{y} - \hat{B}\overline{x}_0 \\ \hat{B} &= \left[\sum_{i=1}^{n}(y_i - \overline{y})(x_{i0} - \overline{x}_0)\right] / \left[\sum_{i=1}^{n}(x_{i0} - \overline{x}_0)^2\right] \end{aligned} \tag{18.3.6}$$

where

$$\overline{y} = \sum_{i=1}^{n} y_i/n \quad \text{and} \quad \overline{x}_0 = \sum_{i=1}^{n} x_{i0}/n.$$

Fitted values. Corresponding to each response variable, y_i $(i = 1, 2, \ldots, n)$, there is a statistic $\hat{y}_i$ which is defined by the regression equation in which the parameters are replaced by their Maximum Likelihood estimators. Values for these statistics computed from the data are termed the *fitted values,* $\{\hat{y}_{i0}\}$, under the proposed model. If a, b, c and $b_0, b_1, \ldots, b_p$ are the Maximum Likelihood estimates for the parameters A, B, C and $B_0, B_1, \ldots, B_p$, respectively, then the fitted values for Linear regression models may be computed from the relevant equation in (18.3.7).

$$\begin{aligned} \hat{y}_{i0} &= a + bx_{i0} && \text{Simple linear} \\ \hat{y}_{i0} &= a + bx_{i0} + cx_{i0}^2 && \text{Simple quadratic} \\ \hat{y}_{i0} &= b_0 + b_1 x_{1i0} + \cdots + b_p x_{pi0} && \text{Multiple linear} \end{aligned} \tag{18.3.7}$$

for $i = 1, 2, \ldots, n$.

Residuals are formed as the difference between observed responses and fitted values as defined in (18.3.8).

$$\hat{e}_{i0} = y_{i0} - \hat{y}_{i0} \qquad \text{for } i = 1, 2, \ldots, n. \tag{18.3.8}$$

As in other areas of statistical application, fitted values and residuals play an important role in model and data checking.

18.3.4 Model Comparisons and Experimental Aims

Many of the basic experimental questions which are answered using regression analysis require model comparisons. The models to be compared differ in that one includes a more restricted form of the regression equation than does the other. Some examples will illustrate.

1. To establish if there is evidence of a relation between y and x, a comparison might be made between the model with the regression equations $M_{y.x} = A$ and the model with the regression equation $M_{y.x} = A + Bx$. If the latter model provides a significantly better fit, evidence is provided of a relation between y and x.

2. Where a linear regression equation is assumed, is it reasonable to assume the regression line passes through the origin? To answer this question, models with equations $M_{y.x} = Bx$ and $M_{y.x} = A + Bx$ are compared. If there is not a significant difference in the fits of the two models to the data, it is reasonable to accept the hypothesis that the regression line passes through the origin.

3. In the analysis of the data in Example 18.2, it is of interest to the experimenter to determine whether the same regression relation fits the two species. On the assumption of linear regression equations, the model comparison is based on regression equations

$$M_{y.x} = A + Bx \quad \text{versus} \quad M_{y.x} = A_m + B_m x \qquad \text{for } m = 1, 2$$

where y and x represent *abundance* and *dry weight* respectively; and $m = 1$ and 2 denote species *C. Cephalornithos* and *S. verruculosum* respectively.

The first of these equations assumes a common line fits the data whereas the second allows the intercepts and slopes to vary between the species. If the model containing the second equation provides a significantly better fit, evidence is provided that the relation between abundance and dry weight is not the same for both species.

18.3.5 Analysis of Variance

To cater for the many scientifically important model comparisons which may be undertaken using regression analysis, it is necessary to have a generally applicable methodology. The technique of *Analysis of variance*, which is defined in Section 14.3.4, serves that need. The specific application of Analysis of variance in regression analysis is described in this section and its application illustrated in Sections 18.5, 18.6 and 18.7.

Notation. The notation to distinguish models which is introduced in Section 14.3.3 is employed. Thus,

M_0 is the model which presumes no relation exists between the response variable and the explanatory variable(s);

M_* is the most general model which is under consideration and is presumed to include the correct regression equation; and

$M_1, M_2, M_3, \ldots$ are intermediate models of increasing generality.

To illustrate,

– the most common sequence of models fitted are the polynomials in which

$$\begin{array}{llcl} M_0 \text{ includes the equation} & M_{y.x} & = & A \\ M_1 \text{ includes the equation} & M_{y.x} & = & A + Bx \\ M_2 \text{ includes the equation} & M_{y.x} & = & A + Bx + Cx^2 \end{array}$$

– when there is more than one explanatory variable, a possible sequence is

$$\begin{array}{llcl} M_0 \text{ includes the equation} & M_{y.x_1,x_2} & = & B_0 \\ M_1 \text{ includes the equation} & M_{y.x_1,x_2} & = & B_0 + B_1x_1 \\ M_2 \text{ includes the equation} & M_{y.x_1,x_2} & = & B_0 + B_1x_1 + B_2x_2. \end{array}$$

Sums of squares. For each model, M_a, which is fitted to the data, there can be constructed a set of fitted values, $\{\hat{y}_i^a\}$, and a measure of the fit of the model to the data is provided by the sum of squares statistic defined in (18.3.9). This is the *residual sum of squares* under the model M_a.

$$SS(y, M_a) = \sum_{i=1}^{n} (y_i - \hat{y}_i^a)^2. \quad (18.3.9)$$

When the model M_0 is fitted to the data, all fitted values are equal to the (sample) mean response and the sum of squares defined by (18.3.10) results. This is given the special name *Total sum of squares* in regression analysis.

$$SS(y, M_0) = \sum_{i=1}^{n} (y_i - \bar{y})^2. \quad (18.3.10)$$

The sum of squares statistic is also defined for the comparison of the fits of two models to the data. If the models are M_a and M_{a+1}, the sum of squares is $SS(M_{a+1}, M_a)$ which is defined by (18.3.11).

$$SS(M_{a+1}, M_a) = \sum_{i=1}^{n} (\hat{y}_i^{a+1} - \hat{y}_i^a)^2 = SS(y, M_a) - SS(y, M_{a+1}). \quad (8.3.11)$$

Equations (18.3.9) and (18.3.10) are connected by a fundamental identity of regression analysis which arises from the additivity property of sums of squares and is expressed in (18.3.12). The identity establishes that the lack of fit of the model M_0 to the data can be decomposed into

— the gain from fitting M_a rather than M_0; and

— the lack of fit of M_a.

$$\begin{array}{ccccc} \text{TOTAL} & & \text{REGRESSION} & & \text{RESIDUAL} \\ \text{SUM OF SQUARES} & = & \text{SUM OF SQUARES} & + & \text{SUM OF SQUARES} \\ \sum_{i=1}^{n}(y_i - \bar{y})^2 & = & \sum_{i=1}^{n}(\hat{y}_i^a - \bar{y})^2 & + & \sum_{i=1}^{n}(y_i - \hat{y}_i^a)^2 \end{array} \quad (18.3.12)$$

Analysis of variance table. Equation (18.3.9) corresponds to equation (14.3.12) in the formal development of Analysis of variance which appears in Section 14.3.4. By continuing the correspondence, a comparison of the fits of models M_a and a more general model, M_{a+1}, may be based on the Analysis of variance table presented in Table 18.3.1(a) which derives from Table 14.3.2. The Analysis of variance table based on the identity (18.3.12) has the standard format displayed in Table 18.3.1(b).

Table 18.3.1. Analysis of Variance Table for the Comparison of Two Regression Models

(a) General formation

Source of Variation	D.F.	S.S.	M.S.	F	p
Gain from M_{a+1} over M_a	$d = d_a - d_{a+1}$	$SS = SS_a - SS_{a+1}$	$MS = \frac{SS}{d}$	$F = \frac{MS}{MS_*}$	p_0
Residual under M_*	d_*	SS_*	$MS_* = \frac{SS_*}{d_*}$		

(b) Standard representation when M_a is M_0 and M_{a+1} is M_*

Source of Variation	D.F.	S.S.	M.S.	F	p
Regression	$d = d_0 - d_*$	$SS = SS_0 - SS_*$	$MS = \frac{SS}{d}$	$F = \frac{MS}{MS_*}$	p_0
Residual	d_*	SS_*	$MS_* = \frac{SS_*}{d_*}$		
Total	d_0	SS_0			

Legend.

d.f. — Degrees of freedom
S.S. — Sum of squares
M.S. — Mean square
F — F-statistic
p — p-value
SS_a and SS_{a+1} are values of $SS(y, M_a)$ and $SS(y, M_{a+1})$
SS and SS_* are values of $SS(M_{a+1}, M_a)$ and $SS(y, M_*)$
MS and MS_* are values of $MS(M_{a+1}, M_a)$ and $MS(y, M_*)$

Degrees of freedom, mean squares and F-tests. If, in the fit of the model M_a to the data, there are p regression coefficients to be estimated from a data set containing n responses, the *degrees of freedom*, d_a, associated with the sum of squares $SS(y, M_a)$, is equal to $n-p$ and the *mean square* $MS(y, M_a)$ is equal to $SS(y, M_a)/d_a$. When comparing the fits of the models M_a and M_{a+1}, the sum of squares $SS(M_{a+1}, M_a)$ has associated degrees of freedom $d_a - d_{a+1}$ and mean square $MS(M_{a+1}, M_a)$ is equal to $SS(M_{a+1}, M_a)/(d_a - d_{a+1})$.

As explained in Section 14.3.4, the primary function of the Analysis of variance technique is to compare the fits of models M_a and M_{a+1} to the data. The difference in the two models rests with an hypothesis H_a included in M_a. The model comparison is undertaken to examine this hypothesis. There are two properties of the mean square statistics which are relevant to this examination.

1. Since the model M_* is presumed to include the correct regression equation, $MS(y, M_*)$ is presumed to be an unbiased estimator of the residual variance σ^2 irrespective of the correctness or otherwise of H_a.
2. $MS(M_{a+1}, M_a)$ is an unbiased estimator of σ^2 if H_a is correct, otherwise it has an expectation which is greater than σ^2.

These properties are exploited through the use of the F-statistic defined in (18.3.13) by noting that increasing values of F represent lessening support for H_a.

$$F = MS(M_{a+1}, M_a)/MS(y, M_*). \tag{18.3.13}$$

A value of F computed from the data, F_0, may be converted to a p-value using (18.3.14).

$$p_0 = \Pr\{F(d_a - d_{a+1}, d_*) \geq F_0\} \tag{18.3.14}$$

where d_a, d_{a+1} and d_* are degrees of freedom for models M_a, M_{a+1} and M_*, respectively.

Interpretation of the p-value follows the usual guidelines with low values of p_0 offering evidence against the hypothesis H_a.

Note that the model M_{a+1} is frequently also the model M_* in regression problems. This is the assumption made in Table 18.3.1(b).

A wide range of applications of the Analysis of variance technique based on Table 18.3.1(a) are presented in Section 18.6.

18.3.6 Computational Considerations

The estimation of values for regression coefficients based on (18.3.5) reduces to the solution of a system of linear equations. For Simple linear regression equations, there is a unique solution which is expressed by (18.3.6). When there is more than one explanatory variable, there may arise situations in which there is not a unique solution. This occurs when the set of values for one explanatory variable can be expressed as a linear combination of the values for other explanatory variables. When this situation arises, there is said to be *collinearity* between the variables.

If there is collinearity or near collinearity, standard algorithms used to compute values for the regression coefficients will fail. It is likely that a statistical package when encountering this problem will present a message to the effect that collinearity has been detected and action is required to eliminate the cause of the collinearity before the estimation of the regression coefficients can proceed.

The practical implications of collinearity and near collinearity are considered in Section 18.6.1. For the present, attention is focused on the methods available for continuing the regression analysis when collinearity has been detected. The three common causes of collinearity are

(i) a mathematical relationship between two or more explanatory variables—for example, when a set of variables are percentages which must necessarily add to one hundred;

(ii) the use of an unnecessarily complex equation;

(iii) the inclusion of many explanatory variables in the regression equation.

If there is a known mathematical relation between the variables, the obvious solution is to remove one of the variables and repeat the parameter estimation. Recognize, however, that (i) while the solution obtained is unique in respect of the parameters employed in the estimation, it is always possible to replace any one of these parameters by the parameter which was excluded and obtain a different regression equation; and (ii) whichever set of parameters is employed, the same set of fitted values are obtained.

Frequently the form of the regression relation is unknown and an attempt is made to approximate the relation using a polynomial equation. It is unwise to initially postulate a relation with higher order terms, e.g. quadratic, cubic or quartic terms and with cross-product terms, e.g. $\beta x_i x_j$, $\beta x_i^2 x_j$, etc., unless there is clear evidence they are needed. By starting with a simple equation and adding components when there is evidence of lack of fit of the equation to the data, the likelihood of introducing collinearity or near collinearity is minimized.

If there are p variables in a Multiple linear regression equation, and the number of sample members, n, is less than p, a unique solution for the regression coefficient is not possible. For most practical applications of regression analysis it is necessary for p to be substantially less than n if the technique is to provide reliable estimation. Hence, there may be a need to reduce the number of explanatory variables to be included in the regression equation. The investigator may undertake this reduction without statistical assistance by (i) discarding variables or sets of variables which are of peripheral interest or which, on independent grounds, are unlikely to have a significant explanatory role; (ii) combining variables, e.g. by summing scores from different examinations to give a single combined result. Alternatively, statistical analysis may be employed to undertake either of these tasks. The table of pairwise product-moment correlations is an aid in determining relationships between the variables. Where the number of variables to be deleted is large, the application of *Principal component analysis* (see, for example Jolliffe [1986]), to the set of explanatory variables has been suggested as a basis for either selection of correlated sets of variables or for transforming the problem into one in which a new smaller set of explanatory variables is formed.

The remarks and suggestions of the previous paragraph apply equally when p is less than n but there is evidence of collinearity. If there are facilities for employing *ridge regression* (see Fox [1984]), it is a useful tool for combining the roles of detecting the sources of collinearity and providing parameter estimates.

18.4 Practical Considerations

18.4.1 Choosing a Regression Equation

The choice of a mathematical form for the regression equation in the model may be made in one of the following ways:

- by an hypothesis of the experimenter;
- from scientific theory;
- from examination of trend lines in past studies;
- from examination of the trend line in the current study; or
- by assuming a simple form, e.g. a linear or simple curvilinear relation.

The selection of a correct form for the regression equation is of some importance when the equation is to be used for prediction or for the estimation of values of regression coefficients. Inferential methods based on Normal distribution models rely on the correct form being fitted, since it is necessary to have an unbiased estimator of the residual variance σ^2.

A set of rules cannot be provided to identify the best choice for the form of the regression equation in given experimental situations. The only general guideline is to employ a form which is suggested independently of the current data set, where feasible. If not, careful attention should be paid in the model checking operation to diagnostics which may indicate an incorrect model.

As a general rule, if the current data set is used as the basis for suggesting a mathematical form for the regression equation or specific values for the parameters, it is unwise to use the same data set to test whether the suggested form or suggested parameter values are acceptable. Sometimes it is feasible to randomly split the data set into two groups and use one group to suggest a form for the regression equation and the other group to examine the acceptability of that form.

18.4.2 Experimental Aims

The wide ranging scientific uses of regression analysis are summarized below and references given to the sections in which relevant statistical methodology can be found. In respect to seeking a method of statistical analysis, the experimental aims may be considered in the following groupings:

- questions about the form of the regression equation;
- questions about parameters in a regression equation;
- the comparison of regression equations between groups; and
- prediction or estimation using the regression equation.

These are considered in turn.

1. *The form of the regression equation*

(a) *Is there a relation between the response variable and the explanatory variable(s)?*

If there is no relation, the regression equation is a constant, i.e. $M_{y.x} = A$ if there is a single explanatory variable or $M_{y.x_1,x_2,\ldots,x_p} = B_0$ is there are p explanatory variables. Given there is a single explanatory variable, x, examination of the scatter plot of y versus x provides a quick and simple means of checking for evidence of a relationship. A formal statistical test involves the comparison of the fit of a model which includes one of the above equations with the fit of a model which includes an equation that assumes a relationship exists—typically a linear equation is assumed. The appropriate test is provided by the F-test in the Analysis of variance table in Table 18.3.1(b) and is described in Section 18.5.2.

(b) *Does an hypothesized form for the regression equation fit the data?*

If the experimenter approaches the study with an hypothesized form for the regression equation, the statistical task is to compare the fit of this equation to the data with the fit of a more general equation. The F-test in Table 18.3.1(a) may be employed for this purpose. The method is described and illustrated in Section 18.6.1. The most common application arises when a Simple linear regression equation is assumed, i.e. $M_{y.x} = A + Bx$, in which case the alternative equation is usually presumed to be the Simple quadratic equation $M_{y.x} = A+Bx+Cx^2$. If the quadratic equation provides a significantly better fit than the linear, the hypothesis of a linear regression relation can be rejected.

(c) *What is the simplest form of regression equation which fits the data?*

In many experimental situations, the basic aim is to establish a regression equation which accurately predicts responses from a knowledge of values of the explanatory variable(s). There is a considerable body of statistical information which relates to this need under the heading of *Response surface designs and methodology* (see, for example, Box and Draper [1987]).

Alternatively, or additionally, there may be interest in exploring the effects on response of varying one or more specified explanatory variables. There are methods available which systematically examine the relation between the response variable and subsets of the explanatory variables to identify individual variables and subsets of variables which are judged to be important in explaining variation in the response variable. The methods are discussed in Sections 18.6.1 and 18.6.5.

2. *Regression coefficients.* There may be interest in estimating or testing hypotheses about the regression coefficients. This is particularly the case in Simple linear regression where the parameters can be interpreted as the intercept and slope of the regression line. The slope can also be regarded as the expected rate of change in response for a unit change in the explanatory variable.

Estimates of these parameters and the standard errors of the estimates are typically provided in the basic output from statistical computing packages (see Section 18.5.2). From this information, it is possible to construct confidence intervals and test hypotheses about the regression coefficients (see Section 18.6.2).

3. *Comparing regression equations between groups.* Suppose there are g groups and in group m ($m = 1, 2, \ldots, g$) the regression equation is $M_{y.x} = A_m + B_m x$. There may be interest in

(i) examining the hypothesis that the regression lines are identical for all groups by comparing the fit of the model which allows for the possibility of different equations (i.e. includes $M_{y.x} = A_m + B_m x$) with the model which assumes identical equations (i.e. includes $M_{y.x} = A + Bx$);

(ii) examining the hypothesis that the regression lines have the same slope by comparing the fit of a model which allows for different slopes (i.e. includes $M_{y.x} = A_m + B_m x$) with the model which requires the slopes to be identical (i.e. includes $M_{y.x} = A_m + Bx$); and

(iii) examining the hypothesis that the regression lines have the same intercept by comparing the fit of a model which allows for different intercepts (i.e. includes $M_{y.x} = A_m + B_m x$) with the model which requires the intercepts to be identical (i.e. includes $M_{y.x} = A + B_m x$).

Such comparisons may be made using the Analysis of variance procedure which is described and illustrated in Section 18.6.4. The same approach may be used for Curvilinear and Multiple linear regression equations.

4. *Prediction and estimation.* An important application of regression analysis is the provision of an equation which can be used to predict values of the response variable from a known value(s) of the explanatory variable(s). The methodology can be found in Section 18.6.3.

Less commonly, when there is a single explanatory variable, the aim is to predict a value of an explanatory variable (x) from a known value of the response variable (y). Special methodology is required which is widely available when the assumed regression equation is linear (see, for example, Draper and Smith [1981]).

18.4.3 Selecting a Method

The methodology presented in this chapter is restricted to the following conditions:

- the regression equation is linear (includes curvilinear and multiple linear);
- the assumed distribution is the Normal distribution; and
- the random components are presumed to be independently and identically distributed.

Non-linear regression equations have important roles in many specialized areas of application. There are both computational variations and important questions of parameterization of equations which must be considered in Non-linear regression analysis. Ratkowsky [1983] offers advice on the choice of models while Bates and Watts [1988] provides a good coverage of the practical application of the methodology.

Generalized linear models allow for extension of regression methods to important areas of application in which non-Normal distributions are required. These include:

(i) *logistic regression* which applies when the response is binary, i.e. success and failure are the possible responses, and the aim is to explore the manner in which the odds of success are related to one or more explanatory variables (See Section 6.6 for a description of a model which underlies logistic regression.); and

(ii) studies in which the response is either the number of independent events occurring in a fixed period of time or the time between occurrences of independent events (see Section 6.3).

McCullagh and Nelder [1983] provides an extensive description of Generalized linear regression analysis.

Not to be confused with Generalized linear regression is *Generalized least squares* analysis, often abbreviated to GLS, which has application in situations in which the random components are not independent and/or are not identically distributed. (Note that the method of least squares referred to in Section 18.3.3 is termed *Ordinary least squares* (OLS) to distinguish it from Generalized least squares. There is yet another variant, *Weighted least squares* (WLS), which assumes random components are independently distributed but are not necessarily identically distributed.) See, for example, Theil [1971] for a development of GLS regression analysis.

18.5 Statistical Analysis

In the discussion of statistical analysis in this section, it is presumed that a statistical package is available for performing computations.

18.5.1 Model and Data Checking

Simple tools which may be used for model and data checking are described below.

Scatter plot. If there is only one explanatory variable, a scatter plot should always be constructed. The following checks should be made.

1. If a Linear regression equation is proposed, check whether the trend evident in the points appears to be linear.

2. Look for evidence of one or more points which do not appear to be consistent with the trend of the remaining points, i.e. evidence of a possible error in the data.

Plot of fitted values versus residuals. If the correct regression equation is employed and variability in response does not change with changing levels of the explanatory variable, there should be no evidence of a pattern in the plot of residuals versus fitted values. Look for evidence of

1. a trend in the plot as an indication that the regression equation is incorrect (as illustrated in Figure 18.5.1);

2. non-constant variability of the residuals (the most common pattern is illustrated in Figure 18.5.2).

If the wrong regression equation has been fitted, an alternative form must be selected. If the equation is linear, a quadratic may be considered. In general, if the equation is a polynomial, a higher order polynomial might be considered. When a Multiple linear regression equation fails to fit the data, plot the set of residuals against values for each explanatory variable and look for evidence of a relationship.

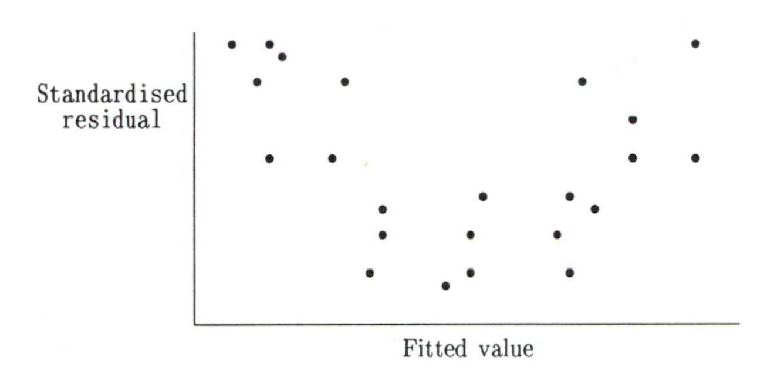
Note the presence of a curved trend line.
If the model is correct there should be no relation evident
Standardised residual
Fitted value

Figure 18.5.1. Plot of residuals versus fitted values displaying evidence of an incorrect trend line.

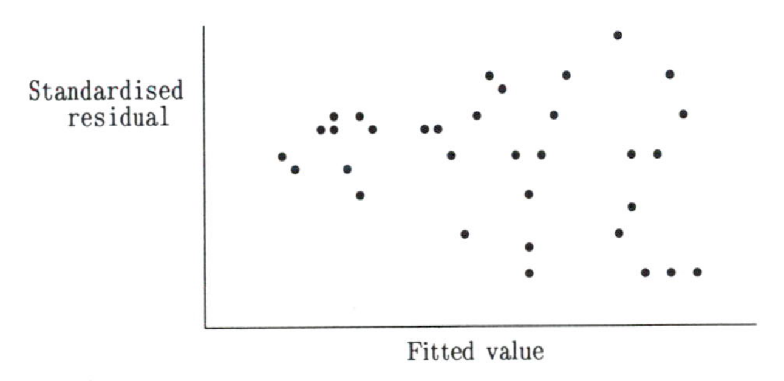
Note the increasing spread of residuals with increasing fitted values
Standardised residual
Fitted value

Figure 18.5.2. Plot of residuals versus fitted values displaying evidence of unequal variance.

Non-constant variance indicates a distributional problem. If it occurs in conjunction with evidence of non-Normality, a transformation of the scale of measurement for the response variable should be considered. Where variability increases with increasing response, a logarithmic transformation should be tried. If the Normality assumption appears acceptable, the model must be changed to accommodate unequal variances. Weighted least squares is then required. Williams [1959] discusses models of this type.

Normal probability plot. Construct a Normal probability plot based on the set of standardized residuals following the steps outlined in Section 10.3.4. Use the plot to check the assumption of Normality following guidelines given in Section 10.3.4 and to check for possible errors in the data (Section 10.4.1).

Standardized residuals. The set of standardized residuals is useful to complement the above procedures. Where there is an error in the data detected in one or more of the plots, the corresponding standarized residual will typically have a large magnitude. It may prove a useful pointer to the offending value in the data set.

If there are errors in the data, it is important to either correct the values or delete them before proceeding with model fitting and examination. If the Normality assumption is not met, there is the possibility of transforming to another scale of measurement on which it is met or seeking an alternative model for which methodology is available. This may be a distribution-free model or a Generalized linear model. Seek guidance from a consultant statistician to determine the best course of action.

18.5.2 Basic Statistical Information Provided in Regression Analysis

There is a traditional form of presentation of the basic information which is output in regression analysis. It is presented below and illustrated in Table 18.5.1.

1. *Predictive equation:*

$$\hat{y} = a + bx \qquad \text{or}$$
$$\hat{y} = b_0 + b_1x_1 + b_2x_2 + \cdots + b_px_p$$

where $\hat{y}$ is the estimated or predicted response; a, b or $b_0, b_1, \ldots, b_p$ are estimated regression coefficients; and x or $x_1, x_2, \ldots, x_p$ are explanatory variables.

2. *Estimates of regression coefficents and their standard errors.*

Parameter	A	B	or	B_0	B_1	$\cdots$	B_p
Parameter estimate	a	b	or	b_0	b_1	$\cdots$	b_p
Standard error of parameter estimate	s_a	s_b	or	s_{b0}	s_{b1}	$\cdots$	s_{bp}

If the correct model has been employed, the Maximum Likelihood estimator of a parameter P, denoted by $\hat{P}$, (where P may represent one of the parameters A, B, $B_0, B_1, \ldots, B_p$) is Normally distributed with mean P and variance σ_P^2. If s_P^2 is the sample estimator of σ_P^2 it follows from (7.4.1) that the statistic $t = [\hat{P} - P]/s_P$ has a $t(n-p)$ distribution where n is the sample size and p is the number of parameters estimated from the data.

Commonly, in the output from statistical packages, the value of t is provided for each parameter on the assumption that the parameter takes the value zero. By converting the value (t_0) to a p-value using the formula $p_0 = \Pr\{t(n-p) \geq |t_0|\}$ the acceptability of the assumption $P = 0$ can be assessed. If the value of $n-p$ exceeds 30, a simple rule-of-thumb is to reject the hypothesis $P = 0$ when the magnitude of t_0 exceeds 2.0 Rejection of the hypothesis implies that the variable for which P is the coefficient does

Table 18.5.1. Basic Information from the Analysis of Data in Example 18.1 Assuming a Simple Linear Regression Equation

(a) Predictive equation

$\hat{y} = -54.77 + 0.7577x$ where

y is the University mathematics mark;
x is the High school mathematics mark.

(b) Estimates of regression coefficients and their standard errors

Parameter	A	B
Parameter estimate	−54.77	0.7577
Standard error of parameter estimate	13.55	0.09990

(c) Analysis of variance

Source of Variation	D.F.	S.S.	M.S.	F	p
Regression	1	9328.0	9328.0	57.5	<0.01
Residual	48	7784.0	162.2		
Total	49	17112.8			

(d) Percentage of variation explained by explanatory variable

$$R^2 = 100(9328.0/17112.8) = 55\%$$
$$R^2_{\text{adj}} = 100[1 - (162.2/349.2)] = 54\%$$

(e) Data of special interest

Subject (i)	High school maths mark (x_i)	University maths mark (y_i)	Fitted value ($\hat{y}_i$)	Standardized residual (r_{s_i})
7	93	31	15.7	1.3 *
12	119	61	35.4	2.1 †
31	84	6	8.9	−0.3 *
44	162	41	68.0	−2.2 †

Notes.
*point of high influence
†point with standardized residual having magnitude in excess of 2.0

provide significant additional information about variation in the response variable over and above that provided by other explanatory variables in the regression equation.

3. *Analysis of variance table.* A common form of summary of the information deriving from regression analysis is the *Analysis of variance* table presented in Table 18.3.1(b).

On the assumption that the presumed regression equation defines the correct form of the regression equation, p_0 may be used

(i) to test the hypothesis $B = 0$ for the Simple linear regression equation $M_{y.x} = A + Bx$; or

(ii) to test the hypothesis $B_1 = B_2 = \cdots = B_p = 0$ for the Multiple linear regression equation $M_{y.x_1,x_2,\ldots,x_p} = B_0 + B_1x_1 + \cdots + B_px_p$.

If the hypothesis is correct, F_0 is the value of an $F(p, n-p)$ statistic and the p-value can be computed as $p_0 = \Pr\{F(p, n-p) \geq F_0\}$. Following the usual interpretation of p-values, a value of p_0 below 0.05 (0.01) provides weak (strong) evidence against the hypothesis.

4. *Percentage of variation explained by explanatory variables.* The ratio of the regression sum of squares to the total sum of squares is known as the *coefficient of multiple determination.* It is commonly expressed as a percentage when it is referred to as a value of the *R-squared statistic* defined by (18.5.1). In the present context it is widely interpreted as the percentage (or proportion) of variation in the set of responses which can be explained by variation in the explanatory variable(s).

$$R^2 = 100(TSS - ESS)/TSS \qquad (18.5.1)$$

where TSS and ESS are the total sum of squares and the residual sum of squares respectively as defined in (18.3.12).

A more logical measure is provided by R^2_{adj} which is defined in (18.5.2) since it measures the percentage reduction in the variance of the response variable which results from the use of the information in the explanatory variable(s). Whereas R^2 cannot decrease with the addition of explanatory variables which provide no additional information about y, R^2_{adj} has the capacity to reflect the lack of gain from adding an extra explanatory variable by decreasing in value.

$$R^2_{\text{adj}} = 100(TMS - EMS)/(TMS) \qquad (18.5.2)$$

where TMS and EMS are the total mean square and the residual mean square, respectively.

5. *Data of special interest.* In the course of computing the data for the Analysis of variance table, the fitted values and residuals can be computed and it is useful to

(i) present observed values, fitted values and standardized residuals for any sample members which have standardized residuals with magnitudes in excess of 2.0; and

(ii) identify any points of large influence if the option to do so is available (*Influence* is discussed in Section 10.4.4.)

18.6 Applications of Regression Analysis

18.6.1 Questions About the Form of the Regression Equation

Two possible situations arise in respect to the choice of the form of regression equation. In one situation, a form is proposed for the regression equation which is decided independently from the current experimental information and the task is to establish if this form is acceptable. The alternative situation arises when a suitable form is decided by examining the current data set. In the latter case, the one data set is employed both in the suggestion of a possible form and the examination of that suggestion. The two situations are considered separately below.

1. *Examining a proposed equation.*

Notation. For brevity, the hypothesized equation is denoted by H_a, e.g. H_a represents $M_{y.x} = A + Bx$ if a Simple linear regression equation is proposed, and the model containing the hypothesized form is denoted by M_a.

A simple, visual check on the acceptability of H_a can be made by fitting M_a to the data and examining the plot of fitted values versus residuals. There should be no trend in the plot if the correct regression equation has been fitted. Figure 18.5.1 illustrates a situation in which there is evidence of an incorrect model.

The plot of fitted values versus residuals provides a simple and sensitive check on the acceptability of the hypothesized form of the regression equation. If a formal test is required, it involves the comparison of the fit of M_a to the data with the fit of a model (M_{a+1}) which includes a more general form for the regression equation.

The test is based on the Analysis of variance table in Table 18.3.1(a) and can be applied using the following steps.

Step 1 — Fit the model M_a and determine the Analysis of variance table. From the Analysis of variance table obtain the residual degrees of freedom (d_a) and sum of squares (SS_a). Insert these values into Table 18.6.1.

Step 2 — Seek a model M_{a+1} which is identical with M_a except for the form of the regression equation. Include a regression equation which is a generalization of the hypothesized form and is likely to provide a correct representation of the regression equation should H_a be in error.

Table 18.6.1. Analysis of Variance Tables for the Comparison of M_a and M_{a+1}

Source of Variation	D.F.	S.S.	M.S.	F	p
Residual under M_a	d_a	SS_a			
Gain from M_{a+1} over M_a	$d =$ $d_a - d_{a+1}$	$SS =$ $SS_a - SS_{a+1}$	$MS = \frac{SS}{d}$	$F = \frac{MS}{MS_{a+1}}$	p_0
Residual under M_{a+1}	d_{a+1}	SS_{a+1}	$MS_{a+1} = \frac{SS_{a+1}}{d_{a+1}}$		

Step 3 — Fit the model M_{a+1} to the data and determine the Analysis of variance table. Obtain values for the residual degree of freedom (d_{a+1}) and sum of squares (SS_{a+1}). Insert these values into Table 18.6.1.

(*The presumption is made that M_{a+1} includes a correct regression equation. This is a necessary part of the analysis since it is the residual mean square under M_{a+1} which provides the estimate of σ^2 that serves as a yardstick in the F-test employed at Step 4. If M_{a+1} does not provide a good fit to the data, the residual mean square will provide an inflated estimate of σ^2 and hence, the p-value at Step 4 will also be inflated. In practice, it is common to take a rather liberal view when performing model and data checking on the grounds that a significant improvement in fit for M_{a+1} over M_a is likely to be detected provided M_{a+1} is not grossly incorrect.*)

Step 4 — Complete Table 18.6.1. Based on the value of p_0 in Table 18.6.1, the decision may be taken to accept H_a if p_0 is at least 0.05.

Illustration. The steps are illustrated using the data in Table 3.2.5 which arise from a study reported in Example 3.3. The purpose of the study is to examine the hypothesis that drop in pressure of blood flowing along an artery (y) is linearly related to flow rate (x). The assumed regression equation is $M_{y.x} = A + Bx$. While there is clear visual evidence of a nonlinear relation from the plot of residuals versus fitted values, a formal test was requested. The analysis involved the fitting of the linear equation $M_{y.x} = A + Bx$ and the quadratic equation $M_{y.x} = A' + B'x + C'x^2$. These yield the Analysis of variance tables in Tables 18.6.2(a) and 18.6.2(b), respectively. From the information in these tables, Table 18.6.2(c) is constructed. Since the value of p_0 in Table 18.6.2(c) is less than 0.01, there is strong evidence to reject the hypothesis of a linear relation between pressure drop (y) and flow rate (x).

Table 18.6.2. Analysis of Variance Tables (Example 3.3)

(a) Analysis based on a Simple linear equation

Source of Variation	D.F.	S.S.	M.S.	F	p
Regression	1	1843.03	1843.0	3400	
Residual	12	6.50	0.54		
Total	13	1849.53			

(b) Analysis based on a Simple quadratic equation

Source of Variation	D.F.	S.S.	M.S.	F	p
Regression	2	1847.75	923.88	5800	
Residual	11	1.78	0.16		
Total	13	1849.53			

(c) Analysis of variance table based on formulae in Table 18.6.1 and data from Tables 18.6.2(a) and (b)

Source of Variation	D.F.	S.S.	M.S.	F	p
Residual (linear)	12	6.50			
Gain from quadratic	1	4.72	4.72	29	<0.01
Residual (quadratic)	11	1.78	0.16		

There may be grounds for querying the scientific, as distinct from the statistical, significance of this finding. Based on the R^2_{adj} statistic, the linear equation accounts for 99.9% of total variance. From a practical point of view, it may be reasonable to question whether any improvement in fit could be of practical significance in these circumstances. This is a question which lies outside the province of Statistics, but it is a question which should be addressed when interpreting the findings of the statistical analysis.

2. *The selection of a regression equation.* In many studies which use the methods of regression analysis, there is no prior expectation for the form of the regression equation. There is an enormous literature on the selection of regression equations in this situation, but the end results of the deliberations is a scarcity of simple, clear-cut guidelines. This is not a consequence of poor research. Rather, it is arises primarily because of the fact that the one data set is being called upon to both provide suggestions for regression equations which might fit the data and examine those suggestions to judge which of them are reasonable.

Where possible, it is highly recommended that the data be randomly split into two equal-sized groups, with one group being used to suggest regression equations and the second group to judge the worth of the selection(s). This proposal attracts the criticism that much, perhaps scarce, data are being lost at both the selection and testing stages. Such criticism ignores the fact that to use the one data set for both phases is highly likely to lead to an erroneous or inefficient selection.

Suggestions are offered below for possible paths to follow in seeking the *best,* or a collection of *good* regression equations for a given data set. The suggestions are in the form of a series of steps which represent roughly the sequence in which the process of selection should proceed.

Step 1 — *Experimental objectives.* Establish whether the major purpose of the study is (i) to provide an equation for use in predicting responses; or (ii) to understand the manner in which the explanatory variables influence response.

Step 2 — *Choosing the explanatory variables.* With the advent of high speed computers has come an ease of storage and manipulation of large data sets. It is tempting to obtain information on many explanatory variables and leave the computer to sort out which are important as explanatory variables in a regression study. This way of thinking is false on two counts. Firstly, the statistical analysis can, at best, only be as good as the information which is provided. The researcher has, or should have, much knowledge over and beyond the set of numbers which form the data. Effective and efficient scientific experimentation demands full use of scientific knowledge. Secondly, as the number of explanatory variables approaches the number of cases from which data are collected, the end result of regression analysis becomes highly sample dependent—take a second sample and possibly obtain a completely different result.

Where many explanatory variables are being considered, exclude those variables which are judged to be of peripheral interest or which include information contained in other variables. Seek to reduce the number of variables by combination of variables. If, for example, there is a sequence of ten variables which are test scores during a course, can these be reduced to one or two variables by combining results from the different tests?

At the other end of the scale, make sure that no important variables have been excluded. There is much to be gained by determining the likely set of important explanatory variables before planning the data collection.

Step 3 — *Data checking.* If there is an error in the data, it may have a serious effect on the selection of a regression equation. As a first step in data checking, examine the maximum and minimum values for all variables for evidence of an unusually high or unusually low reading.

More detailed examination of data hinges on a knowledge of the correct model, which points to an obvious dilemma. A practical solution is to make a guess at the likely set of important explanatory variables, fit a Multiple linear regression equation based on these variables and employ the model and data checking techniques defined in Section 18.5.1 to seek evidence of erroneous data. Additionally, when the final choice of one or more regression equations is made, the process of model and data checking should again be employed.

Step 4 — *Preliminary examination.* Construct a scatter plot for the response variable versus each explanatory variable. If there is evidence of a non-linear trend line, consider the possibility of transforming the values of the explanatory variable to a new scale of measurement, e.g. by taking logs or inverses. (Perhaps there is evidence from previous studies to suggest a transformation.) Less desirable is the addition of higher order polynomial terms (quadratic, etc.) since this introduces additional parameters.

Step 5 — *Computational considerations.* There are many computational procedures which have been proposed for selection of subsets of explanatory variables on the assumption of a Multiple linear regression equation. Some of the more common procedures are identified and briefly discussed below. In a practical situation, which of these is available is determined by the statistical package to which a scientist has access.

There are three criteria which must be established when seeking the *best* regression equation. They are (i) a strategy for systematically examining the different possible equations; (ii) the criterion for model selection; and (iii) a stopping rule in sequential selection procedures. These are considered in turn.

1. *Strategies for systematically examining regression equations.* There are three strategies which are widely available in statistical packages for systematically examining regression equations based on different subsets of explanatory variables. These strategies are described below.

1. The *forward selection* procedure which commences with an examination of the contributions of individual explanatory variables, and selects the variable which produces the greatest reduction in residual sum of squares. Next the addition of a second variable is considered using the same criterion, and so on. At each stage, a stopping rule is applied which is used to judge if the addition of a further variable makes a significant improvement in the fit of the model to the data. The process stops when it is not possible to make a significant improvement in fit by the addition of any of the remaining explanatory variables. A variation is offered by the *forward stepwise selection* procedure in which there is the possibility of deleting one or more variables at each stage.

2. The *backward elimination* procedure reverses the preceding strategy. The starting point is a regression equation based on the complete set of explanatory variables. Each variable is considered in turn to establish the increase in residual sum of squares from the removal of the variable. The variable whose removal causes the smallest increase in residual sum of squares is removed provided the loss of information is not judged significant by the stopping criterion. The process continues with the consideration of

dropping a second variable, and so on. A variation is offered by ***backward stepwise elimination*** in which there is the possibility of reintroducing a variable eliminated at an earlier point.

3. All possible regression equations might be fitted and the best equation selected.

The stepwise procedures have the advantage that they only require examination of a small subset of the total number of possible regression equations. Indeed, looking at all possible regression equations becomes impractical when the number of explanatory variables is large. (If there are p explanatory variables, there are 2^p possible regression equations, e.g. if $p = 10$, there are 1,024 possible regression equations.) Unfortunately, there is no guarantee that any stepwise procedure will find the optimal regression equation or that the different procedures will reach the same choice. However, they are often the only practical methods.

Recommendations. In the absence of a more specialized procedure which has a sound reason for its use, the following choices are recommended.

1. Fit all possible regression equations when practical. (If there is a mechanism for automatically computing and storing a value of the statistic used as a measure of goodness of fit of the model to the data for each equation, this is a feasible option for p less than or equal to about ten.)

2. Use the backward stepwise elimination procedure for p less than about 30. (There comes a point at which the process is too time consuming. Where this point lies is dependent on the power and availability of the computer which is being used.)

3. For larger values of p, use (i) forward stepwise selection; or (ii) Principal component analysis applied to the set of explanatory variables prior to regression analysis to reduce the number of explanatory variables—see Jolliffe [1986] for details.

Selection criteria. The most commonly used selection criteria are based on a residual mean square statistic. In forward selection, the best possible additional variable is that which causes the greatest reduction in residual mean square; in backward elimination, the most desirable variable to discard is the variable whose loss causes the least increase in residual mean square.

The disadvantage of this criterion is that it takes no account of the consequences of selecting an incorrect model. Where the aim of selection is to choose a regression equation which provides the best predictor of the response variable, there is scope for using a criterion which is developed for this purpose. *Tukey's rule* is a simple and well respected choice. It states that the best regression equation is that which has the smallest value of

s^2/d where s^2 is the residual mean square based on the regression equation and d is the corresponding residual degrees of freedom.

Recommendation. Use a criterion based on the residual mean square except where the aim is prediction, when Tukey's rule is recommended if it can be easily implemented.

Stopping rules. When sequential strategies are employed, it is necessary to have a rule by which the decision can be made that the search need proceed no further. To decide if the inclusion of a variable significantly improves the fit of the regression equation, it is customary to rely on an F-statistic which measures the gain from including the variable. Three factors combine to make a theoretical basis for selecting a critical value for the F-statistic difficult, if not impossible. The first is the general unavailability of an unbiased estimator of σ^2 for the denominator of the F-statistic. The second is the fact that the same data are being used for both the selection and testing of equations. The third is the multiple comparison nature of the problem—the various tests are not independent. The consequence is the use of *ad hoc* grounds for making a decision. Commonly, statistical packages offer a choice in the value which may be used. Such research as has been done suggests it is wise to err on the side of using cut-off values which encourage more rather than fewer equations to be checked.

Recommendation. Typically, if there is an option in a statistical computing program for sequential examination of regression equations, the stopping rule is based on tabulated values of the F-distribution which are decided by the choice of significance levels. Where the user is given control of the probability level to employ, use a large value, say 0.25, in order that a greater number of equations will be examined.

General comments. There are disturbing findings from research into the selection of regression equations which suggest that not only is routine application of the stepwise methods unlikely to find the optimal solution, but also that spurious relations between unrelated variables are likely to appear when many explanatory variables are included in the initial set (see Problem 18.6). There are very persuasive reasons for

(i) seeking to exclude on independent grounds, variables which are unlikely to play a significant role; and

(ii) seeking to reproduce the findings on an independent data set.

Hocking [1976] provides a comprehensive coverage of the important aspects of variable selection.

18.6.2 Questions About Parameters

The general questions raised about a parameter in a linear regression equation based on the model described in Section 18.3.2, may be answered by constructing a confidence interval for the parameter using formulae given below. Commonly, the aim is to test the hypothesis $B_i = 0$ where B_i is the coefficient of x_i in the regression equation. Acceptance of this hypothesis implies that the inclusion of the term $B_i x_i$ in the regression equation is not improving the explanatory powers of the regression equation. A test of this hypothesis is most easily based on the t-test described below. It can equally be based on a comparison of models using the F-test in the Analysis of variance.

Statistic. Given that the regression equation $M_{y.x_1,\ldots,x_p} = B_0 + B_1 x_1 + \cdots + B_p x_p$ provides an acceptable fit to the data $y_1, y_2, \ldots, y_n$, and $\hat{B}_i$ is the Maximum likelihood estimator of B_i ($i = 1, 2, \ldots, p$), the statistic t defined in (18.6.1) has a t-distribution with $n - p$ degrees of freedom.

$$t = (\hat{B}_i - B_i)/s_{b_i} \tag{18.6.1}$$

where s_{b_i} is the standard error of $\hat{B}_i$.

Hypothesis testing. To test the hypothesis $B_i = 0$, compute a value of the statistic $t = \hat{B}_i/s_{b_i}$. Call this value t_0 and obtain a p-value from (18.6.2).

$$p_0 = \Pr\{|t(n-p)| \geq |t_0|\}. \tag{18.6.2}$$

Interpret the value of p_0 in the usual manner, i.e. if p_0 is less than 0.05 (0.01) there is weak (strong) evidence supporting the inclusion of the term $B_i x_i$ in the equation; if p_0 is 0.05 or greater, there is support for the proposition that the term is not contributing to the explanation of variation in values of y. In Simple linear regression, the test establishes if there is evidence for a relation between y and x. In Multiple linear regression, the test establishes whether there is evidence that the variable x_i is making a contribution to the explanation of variation in y over and above that made by the other variables which are included in the equation. This point is expanded upon in Section 18.6.5.

Test of the intercept. The above test may be applied without modification to test the hypothesis $B_0 = 0$ (Multiple linear regression) or $A = 0$ (Simple linear regression).

An illustration is provided below based on a Simple linear regression equation. For the Multiple linear regression case, see also the discussion in Section 18.6.5. Note that

(i) a test of the hypothesis $B_i = B_0$ may be based on the statistic t in (18.6.1) by computing a value t_0 and converting this to a p-value using (18.6.2);

(ii) an equivalent test of the hypothesis $B_i = 0$ may be based on the F-test described in Section 18.3.5 by employing two models which differ only in the presence or absence of the term $B_i x_i$.

Estimation. Based on the statistic t defined in (18.6.1) and on (8.3.4), confidence intervals for B_i may be constructed using (18.6.3).

$$
\begin{array}{l}
95\% \text{ confidence interval } \hat{B}_i - t_{0.05}(n-p)s_{b_i} \text{ to } \hat{B}_i + t_{0.05}(n-p)s_{b_i} \\
100(1-\alpha)\% \text{ confidence interval } \hat{B}_i - t_{\alpha}(n-p)s_{b_i} \text{ to } \hat{B}_i + t_{\alpha}(n-p)s_{b_i}
\end{array}
\qquad (18.6.3)
$$

Illustration. Table 18.5.1 provides the Analysis of variance table and basic summary statistics necessary to

(i) compute confidence intervals for A and B;

(ii) test the hypothesis $A = 0$ and $B = 0$ using the t-statistic; or

(iii) test the hypothesis $B = 0$ using the F-test in the Analysis of variance.

Estimation. The values for $\hat{B}$ and s_b of 0.7577 and 0.09990 are obtained from Table 18.5.1. The degrees of freedom required for the t-distribution may be computed from the formula $d = n - p$ or, more conveniently, are equal to the residual degrees of freedom in the Analysis of variance table, i.e. $d = 48$. The only additional information required for the construction of confidence intervals is the value for $t_{0.05}(48)$ for a 95% confidence interval, or $t_{\alpha}(48)$ for a $100(1-\alpha)\%$ confidence interval. The methods for obtaining one of these values can be found in Section 7.4.4. A value of 2.01 is obtained for $t_{0.05}(48)$. Application of (18.6.3) yields the following 95% confidence interval for B.

$$
\begin{array}{c}
0.7577 - 2.01 \times 0.0990 \text{ to } 0.7577 + 2.01 \times 0.09990 \\
0.56 \text{ to } 0.96
\end{array}
$$

The hypothesis $B = 0$ can be rejected and, hence, it can be concluded that variations in marks in the university examination can be related to variation in marks in the high school exam.

Hypothesis testing — t-test. To test the hypothesis $B = 0$, the values for $\hat{B}$ and s_b of 0.7577 and 0.09990 are obtained from Table 18.5.1. Additionally, the degrees of freedom for the t-distribution are required. These may be computed from the formula $d = n - p$ or, more conveniently, are equal to the residual degrees of freedom in the Analysis of variance table, i.e. $d = 48$. Hence, $t_0 = 0.7577/0.09990 = 7.6$. Since $p_0 = \Pr\{|t(48)| \geq |7.6|\}$ is less than 0.01, there is strong evidence to reject the hypothesis $B = 0$.

Hypothesis testing — F-test. Since the value of p_0 in the Analysis of variance table in Table 18.5.1 is less than 0.01, there is strong evidence to reject the hypothesis $B = 0$.

Comment. In this example, the numerical value of the coefficient is of little intrinsic interest. Hence, the confidence interval provides little, if any, additional information over and above that provided by the hypothesis testing procedures. Since the aim is to test the hypothesis $B = 0$, and where the p-value for the F-test is automatically provided in the Analysis of variance table, it is the obvious choice. For Multiple linear regression, the t-test is usually the most convenient form of establishing if $B_i = 0$ or, more generally, $B_i = B_0$ is an acceptable hypothesis.

18.6.3 Estimation and Prediction

An important area of application of regression analysis is in the prediction of values of the response variable y from known values of the explanatory variable(s). In fact, there are two types of problems which are encountered. They are distinguished in this book by the names *estimation* and *prediction.* The distinction is as follows.

Estimation. As explained when the concept of a regression equation was introduced in Section 18.2.1, the equation $M_{y.x} = A + Bx$ is a relationship between the explanatory variable, x, and the average value of the response y. An *estimate* of $M_{y.x}$ when $x = x_0$ is, therefore, an estimate of the average response from that the subset of population members who record a value x_0 for the variable x. Since $M_{y.x_0}$ is a parameter of the population, there is scope for employing standard methods for constructing confidence intervals and testing hypotheses about $M_{y.x_0}$.

Prediction. A different problem is posed when there is a value x_0 recorded for an individual member of the population and the aim is to estimate the response (y_0) for that individual from a knowledge of the regression equation and the value of x. The term *prediction* is employed in this case. The assumption in the statistical model is that y_0 is the sum of two components—the regression component $M_{y.x_0}$ and the unexplained component e_0. Since the latter is a random variable, y_0 is not a parameter to be estimated. The approach adopted in estimation of y_0 is to replace e_0 by its expected or average value, which, by definition, is zero. Hence, the best estimate of y_0 is identical with the best estimate of $M_{y.x}$. However, the precision with which y_0 can be estimated is less than the precision with which $M_{y.x}$ can be estimated since allowance must be made for the variability in e_0.

While the above introduction has been couched in terms of Simple linear regression the distinction applies equally when there is more than one explanatory variable.

Assessing the usefulness of an estimator or a predictor. Generally, when a regression equation is being considered for estimation or prediction, there are good grounds for believing that the equation has some capacity for estimation and there is interest in assessing the likely worth of the equation. A crude, but easily computed measure is provided by the R^2_{adj} statistic defined in (18.5.2) which can be interpreted as the percentage of variance of the response variable which can be explained by variation in the explanatory variable(s). The value provided must be interpreted within the context of the scientific environment and the purpose of the application. In the social sciences, a value of R^2_{adj} in excess of 20% may have scientific significance whereas in the biological sciences a value of 50% may be required before there is judged to be scientific significance. If the aim is to predict individual responses, a value of R^2_{adj} which is less than 90% suggests the predictive power of the regression equation will almost certainly be too poor to have practical value.

Illustration. From the analysis of the data in Example 18.1, which is provided in Table 18.5.1, the value of R^2_{adj} is 54% which indicates that a little more than half of the variation observed in university examination makrs can be explained by variation in high school examination marks.

Statistical methods. The statistical methods required for estimation and prediction when a Simple linear regression equation is employed are presented below and illustrated by data collected in Example 18.1. The theory underlying the methods and the extension to Multiple linear regression equations may be found, for example, in Draper and Smith [1981].

Theory. Given a value of x_0 for the explanatory variable x, the best estimator of $M_{y.x_0} = A + Bx_0$ is defined by the statistic $\hat{y}_e(x_0)$ in (18.6.4).

$$\hat{y}_e(x_0) = \hat{A} + \hat{B}x_0. \tag{18.6.4}$$

If the sample regression equation is based on values of the response variables $y_1, y_2, \ldots, y_n$ where $y_i = M_{y.x_i} + e_i$ $(i = 1, 2, \ldots, n)$ and $e_1, e_2, \ldots, e_n$ are independent $N(0, \sigma^2)$ variables (i.e. the conditions described in Section 18.3.2 apply), $\hat{y}_e(x_0)$ is $N(M_{y.x_0}, \sigma_e^2(x_0))$ where $\sigma_e^2(x_0)$ is defined by (18.6.5).

$$\sigma_e^2(x_0) = \sigma^2 \left[1/n + (x_0 - \bar{x})^2 / \sum_{i=1}^{n} (x_i - \bar{x})^2 \right]. \tag{18.6.5}$$

In practice, the value of σ^2 is usually unknown and must be replaced with an estimate. The sample variance of the estimator is defined by (18.6.6).

$$s_e^2(x_0) = s^2 \left[1/n + (x_0 - \bar{x})^2 / \sum_{i=1}^{n} (x_i - \bar{x})^2 \right] \tag{18.6.6}$$

where s^2 is the residual mean square from the Analysis of variance table based on the fit of the model containing $M_{y.x}$ to the data.

To predict a value of $y_0 = M_{y.x_0} + e_0$ requires the estimation of both $M_{y.x_0}$ and e_0. Furthermore, the variance of the predictor must take into account not only the uncertainty of the estimation of $M_{y.x_0}$, but also the variability in e_0. The estimator of $M_{y.x_0}$ is provided by (18.6.4). Since e_0 is random variable, the theory of estimation of parameters which has elsewhere been employed is not applicable. We rely on the reasonable approach of replacing e_0 in the estimation by its expected value, namely zero. Thus, the predictor of y_0 denoted by $\hat{y}_p(x_0)$, can be expressed as $\hat{y}_p(x_0) = \hat{A}+\hat{B}x_0$, i.e. by an expression which is identical with that in (18.6.4). The variance of the predictor, $\sigma_p^2(x_0)$, is the sum of the variances of $\hat{y}_e(x_0)$ and e_0, and can be estimated by $s_p^2(x_0)$ as defined in (18.6.7)

$$s_p^2(x_0) = s_e^2(x_0) + s^2 \qquad (18.6.7)$$

where $s_e^2(x_0)$ and s^2 are defined in (18.6.6).

1. *Estimation.* The statistic $t_e = [\hat{y}_e(x_0) - M_{y.x_0}]/s_e(x_0)$ has a t-distribution with $n-2$ degrees of freedom. Hence, by invoking (8.3.4), confidence intervals for $M_{y.x_0}$ are provided by (18.6.8).

$$\begin{array}{lll} 95\% \text{ C.I.} & \hat{y}_e(x_0) - t_{0.05}(n-2)s_e(x_0) & \text{to} \quad \hat{y}_e(x_0) + t_{0.05}(n-2)s_e(x_0) \\ 100(1-\alpha)\% \text{ C.I.} & \hat{y}_e(x_0) - t_{\alpha}(n-2)(s_e(x_0) & \text{to} \quad \hat{y}_e(x_0) + t_{\alpha}(n-2)s_e(x_0). \end{array} \qquad (18.6.8)$$

Illustration. The calculations are illustrated in the computation of the 95% confidence interval for the expected university mathematics mark for students who gained a mark of 120 in the Matriculation examination based on a Simple linear regression equation computed from the data in Table 18.1.1.

The sample size, n, is 50 and $\overline{x} = 134.40$. From Table 18.5.1, the estimated regression equation is $\hat{y}(x_0) = -54.77 + 0.7577x_0$ and the value of s^2 is determined from the Analysis of variance table to be 162.2. Based on data in Table 18.1.1, the sum of squares for the high school mathematics marks, SSx, is computed to be 16249.9. Applying (18.6.7), the value of $s_e^2(120)$ is

$$s_e^2(120) = 162.2\left[\frac{1}{50} + \frac{(120-134.40)^2}{16249.9}\right] = 5.314.$$

Therefore, $s_e(120)$ has the value 2.31. The value of $t_{0.05}(48)$ is 2.01. Applying (18.6.8) to obtain the 95% confidence interval for $M_{y.120}$ yields

$$-54.77 + 0.7577 \times 120 - 2.01 \times 2.31 \quad \text{to} \quad -54.77 + 0.7577 \times 120 + 2.01 \times 2.31$$
$$31.5 \quad \text{to} \quad 40.8.$$

2. *Prediction.* Since y_0 is not a parameter it is not meaningful to speak of a confidence interval for y_0. It is, however, possible to construct an interval on the y scale which has a given probability of including the true value. Such an interval is known as a *tolerance interval* and can be interpreted in the same manner as a confidence interval for a parameter. Based on the fact that the statistic $t_p = [\hat{y}_p(x_0) - M_{y.x_0}]/s_p(x_0)$ has a t-distribution with $n-2$ degrees of freedom, (18.6.9) defines intervals which have specified probabilities of including the true value of y_0.

$$\begin{array}{lll} 95\% \text{ T.I.} & \hat{y}_p(x_0) - t_{0.05}(n-2)s_p(x_0) & \text{to} \quad \hat{y}_p(x_0) + t_{0.05}(n-2)s_p(x_0) \\ 100(1-\alpha)\% \text{ T.I.} & \hat{y}_p(x_0) - t_{\alpha}(n-2)s_p(x_0) & \text{to} \quad \hat{y}_p(x_0) + t_{\alpha}(n-2)s_p(x_0). \end{array} \qquad (18.6.9)$$

Illustration. The application of (18.6.9) is illustrated using the data from Example 18.1 to obtain 95% tolerance limits on the university mathematics mark for a student who obtained a mark of 120 in the high school examination. In the illustration given above, it has been established that

$$t_{0.05}(48) = 2.01 \quad s^2 = 162.2 \quad s_e^2(120) = 5.314.$$

Since $\hat{y}_p(120) = -54.77 + 0.7577 \times 120 = 36.15$, 95% tolerance limits are

$$36.15 - 2.01\sqrt{\{162.2 + 5.314\}} \text{ to } 36.15 + 2.01\sqrt{\{162.2 + 5.314\}}$$
$$10 \text{ to } 62.$$

The width of these limits suggests that the information contained in the high school mark alone is insufficient to provide an accurate prediction of the university mathematics mark which an individual student is likely to obtain. By contrast, the 95% confidence limits for the average value of the university mark for students who obtain a mark of 120 in the high school examination is 31.5 to 40.8. The greater width in the prediction is a consequence of the large component in variability of university scores which cannot be explained by knowledge of high school marks.

18.6.4 Comparing Regression Equations Between Groups

The statistical task of comparing regression equations between groups can be simply accomplished using the Analysis of variance procedure. The process is explained below on the assumption of a Simple linear regression equation and is illustrated by the comparison of the regression equations for the two groups in Example 18.2.

Notation. If there are g groups, the regression equation for the mth group is denoted by $M_{y.x} = A_m + B_m x$ for $m = 1, 2, \ldots, g$.

Comparisons and their experimental significance. The diagram below shows a set of possible models which might be fitted and some questions which might be answered by comparing the models.

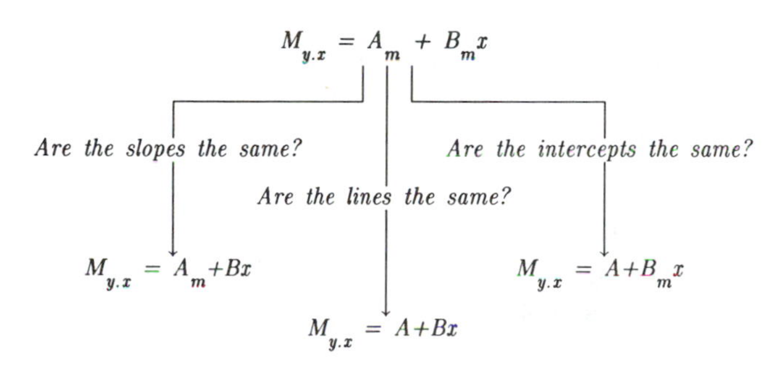

Statistical approach. The technique of Analysis of variance is applicable under assumptions in the model described in Section 18.3.2. The steps in applying the technique are listed below.

Step 1 — Fit the model containing the most general regression equation (M_*) and check that it provides an acceptable fit to the data by applying the methods of model and data checking described in Section 18.5.1.

Step 2 — Construct an Analysis of variance table with the structure of Table 18.3.1(a), and enter the residual degrees of freedom d_* and residual sum of squares SS_* obtained by fitting the general model to the data. Hence, compute the residual mean square MS_* for use as the denominator in F-tests.

Step 3 — Fit the two models to be compared (one of which may be the most general model) and obtain the two Analysis of variance tables. In the notation of Table 18.3.1(a) these are referred to as models M_a and M_{a+1}. Compute the difference in residual degrees, d, and the difference in residual sum of squares, SS, and enter them into the Analysis of variance table being constructed.

Step 4 — Complete the Analysis of variance table. Interpret the p-value in the usual way, remembering that small p-values offer evidence that the equations fitted to the groups differ in respect of the parameter(s) being compared.

> *Note. If your statistical package does not offer a means of fitting models which allow for group differences, it is possible to obtain information to answer the general question 'Do the groups have the same equation' by following the steps given below.*
>
> 1. *Fit separate regression equations to each group and* (i) *sum the residual degrees of freedom to obtain* d_{a+1}*; and* (ii) *sum the residual sums of squares to obtain* SS_{a+1}.
>
> 2. *Ignore the group structure and fit a single equation to the complete data set. This corresponds to fitting* $M_{y.x} = A + Bx$. *Denote the residual degrees of freedom by* d_a *and the residual sum of squares by* SS_a. *Compute* $d = d_a - d_{a+1}$ *and* $SS = SS_a - SS_{a+1}$ *and enter them into the Analysis of variance table which is being constructed. Complete the Analysis of variance table and interpret the p-value in the manner suggested in Step 4 above.*

Illustration. The Analysis of variance table presented below provides the information necessary to answer both of the following questions as they apply to the comparison of the two species in Example 18.2. Simple linear regression equations are assumed and the analysis is based on data modified by the removal of an odd value which was detected in model and data checking—see Problem 18.1.

Question 1: Do the lines have the same slope?
Question 2: Are the lines identical?

Source of Variation	D.F.	S.S.	M.S.	F	p
Residual ($A_m + Bx$)	42	593218			
Common slope[1]	1	5589	5589	0.4	>0.1
Residual ($A + Bx$)	43	609353			
Same line[2]	2	21724	10862	0.8	>0.1
Residual ($A_m + B_m x$)	41	587629	14332		

Notes.
[1]comparison of equations $M_{y.x} = A_m + B_m x$ and $M_{y.x} = A_m + Bx$
[2]comparison of equations $M_{y.x} = A_m + B_m x$ and $M_{y.x} = A + Bx$

Both p-values exceed 0.05 thereby supporting the view that (i) the lines have common slopes; and (ii) the same linear equation applies for both species.

Multiple linear regression equations. While the method described above has been related to Simple linear regression equations, the principle applies equally to comparisons of groups when Polynomial or Multiple linear regression equations are involved. Table 18.3.1(a) provides a universal format for applying the Analysis of variance technique. The selection of models for a given experimental question is basically a matter of common sense.

18.6.5 INTERRELATIONS BETWEEN EXPLANATORY VARIABLES

Where two explanatory variables x_1 and x_2 are highly correlated, it can be anticipated that they duplicate information about the response variable. Hence, if x_1 is a good predictor of the response variable when a Simple linear regression equation is fitted, it can be expected that x_2 will also be judged a good predictor in a Simple linear regression equation. Yet, if both variables are included in the one equation, it is likely that one, or both, of the variables will be judged to be making little contribution to the prediction of the response variable in the presence of the other variable. There is no inconsistency in these findings. In the multiple linear equation $M_{y.x_1,x_2} = B_0 + B_1x_1 + B_2x_2$, what is being judged by examining the coefficient B_1 is the *additional* information being supplied by x_1 over and above that which is supplied by x_2. (Formally, the coefficient B_1 can be written $B_{y.x_1/x_2}$ to indicate this fact. This establishes the distinction between the coefficient in the Multiple linear regression equation based on x_1 and x_2 from the coefficient in the Simple linear regression equation which would be written $B_{y.x_1}$. The fact that the two coefficients may be substantially different can be illustrated by reference to Example 18.4 in which the regressions of *profitability* (y) on *age of manager* (x_1) and *years of experience* (x_2) and *profitability* on *age of manager* yield the equations

$$y = 19.99 - 0.5545x_1 + 17.12x_2 \quad \text{and} \quad y = -104.02 + 4.926x_1,$$

respectively. The value of -0.5545 has an associated standard error of 1.2 while the value 4.926 has an associated standard error of 0.926. Based on the t-test described in Section 18.5.1, the conclusion would be reached that *age* is not making a significant contribution when employed in conjunction with *years of experience* but it is make a significant contribution when fitted alone.)

Examination based on an Analysis of variance table. A general method of studying the contributions from either a single explanatory variable or a collection of variables is based on the Analysis of variance technique. By examining (i) the gain from fitting a variable (or collection of variables) separately from remaining variables, and (ii) the gain from adding the variable (or collection of variables) after inclusion of the remaining variables, it is possible to establish the extent to which information about the response variable is duplicated in the two variables or collections of variables. Table 18.6.3 presents the basic Analysis of variance table which is required to make the necessary model comparisons and Table 18.6.4 provides numerical values based on the variables *age of manager* and *number of years experience* in Example 18.4.

The useful information which may be extracted from the Analysis of variance table in Table 18.6.3 comprises

- p-values which indicate whether explanatory variables are providing any information about variation in the response variable; and
- R^2_{adj} values which provide quantitative measures of the gain made by including a specific variable or variables.

Of particular interest is the comparison of the contribution made by including x_1 in the absence of x_2 with the contribution made by including x_1 after x_2 has been included. This can be based on comparisons of (i) p^r_{01} and p^a_{01}; and (ii) $R^2_1(\text{adj}) = 100(MS_t - MS_1)/MS_t$ and $R^2_{1.2}(\text{adj}) = 100(MS_2 - MS_*)/MS_t$ which measure the gain made by including x_1 in the absence of x_2 and then in the presence of x_2. These sets of statistics may be interpreted in the following way:

Table 18.6.3. Analysis of Variance Table Used in the Examination of the Interrelations Between Explanatory Variables

Source of Variation	D.F.	S.S.	M.S.	F	p
Residual under x_1[1]	d_1	SS_1	$MS_1 = \frac{SS_1}{d_1}$		
Residual under x_2[2]	d_2	SS_2	$MS_2 = \frac{SS_2}{d_2}$		
Regression x_1[1]	d_1^r	SS_1^r	$MS_1^r = \frac{SS_1^r}{d_1^r}$	$F_1 = \frac{MS_1^r}{MS_*}$	p_{01}^r
Regression x_2[2]	d_2^r	SS_2^r	$MS_2^r = \frac{SS_2^r}{d_2^r}$	$F_2 = \frac{MS_2^r}{MS_*}$	p_{02}^r
Adding x_1 after x_2	$d_1^a = d_2 - d_*$	$SS_1^a = SS_2 - SS_*$	$MS_1^a = \frac{SS_1^a}{d_1^a}$	$F_1 = \frac{MS_1^a}{MS_*}$	p_{01}^a
Adding x_2 after x_1	$d_2^a = d_1 - d_*$	$SS_2^a = SS_1 - SS_*$	$MS_2^a = \frac{SS_2^a}{d_2^a}$	$F_2 = \frac{MS_2^a}{MS_*}$	p_{02}^a
Residual under x_1, x_2[3]	d_*	SS_*	$MS_* = \frac{SS_*}{d_*}$		
Total[4]	d_t	SS_t	$MS_t = \frac{SS_t}{d_t}$		

Notes.

[1] d_1^r, d_1, SS_1^r, SS_1 from the regression of y on x_1 alone

[2] d_2^r, d_2, SS_2^r, SS_2 from the regression of y on x_2 alone

[3] d_*, SS_* from the regression of y on x_1, x_2

[4] d_t, SS_t are common to all above regressions

Table 18.6.4. Analysis of Variance Table Used in the Examination of the Interrelations Between the Variables *Age of Manager* (x_1) and *Years of Experience* (x_2)

(Based on data in Example 18.4)

Source of Variation	D.F.	S.S.	M.S.	F	p
Residual under x_1	28	27849	995		
Residual under x_2	28	13583	485		
Regression x_1	1	28154	28154	56.4	<0.01
Regression x_2	1	42420	42420	85.0	<0.01
Adding x_1 after x_2	1	105	105	0.2	>0.1
Adding x_2 after x_1	1	14371	14371	28.7	<0.01
Residual under x_1, x_2	27	13478	499.2		
Total	29	56003	1931		

- if both p-values are below 0.05, there is evidence that x_1 makes a significant contribution to the explanation of variation in the responses;

- if only p_{01}^r is less than 0.05, there is evidence that the information contained in values of x_1 about variation in responses is also provided by values of x_2;

- the difference between $R_{1.2}^2(\text{adj})$ and $R_1^2(\text{adj})$ is often portrayed as measuring the extent to which information in x_1 about variation in responses is duplicated in x_2.

Illustration. Based on the contents of Table 18.6.4, it is possible to examine the extent to which information in the variable *age of manager* is also found in the variable *years of experience* and vice versa. Based on an examination of p-values ($p_{01} < 0.01$ and $p_{01}^a > 0.1$), it is apparent that, while *age* makes a significant contribution in the absence of *experience*, it does not add information about variation in the response, *profitability*, after taking into account information on *experience*. This conclusion is reinforced by a comparison of R_{adj}^2 values. In the absence of information on *experience*, approximately one half of the variance of *proftability* is explained by variation in *age of manager* ($R_1^2 = 100 \times (1931 - 995)/1931 = 49\%$). If the information on managerial experience is first introduced, there is an **increase** in variance of profitability from the addition of information about *age of manager*, i.e. $R_{1.2}^2 = 100 \times (485 - 499)/1931 = -0.7\%$. Note that the situation is quite different if the two variables are reversed. Based on the p-values, there is a significant contribution from *experience* both in the absence and the presence of *age*. After utilizing the information in *age*, there remains approximately 26% ($R_{2.1}^2(\text{adj}) = 100 \times (995 - 499)/1931$) of the variation in *profitability* which can be explained by *experience*.

It is tempting from the results of this analysis to conclude that *experience* is the more important attribute when looking to the selection of managers to lift *profitability*. The wisdom of drawing such a conclusion from the above analysis is considered in Chapter 20.

18.7 Experimental Design and Regression Analysis

Note: in the discussion of experimental designs in this section, the representation of designs and the format of Analysis of variance tables introduced in Chapter 15 *are assumed.*

18.7.1 Design Components in Regression Models

In the regression models considered in Sections 18.2 and 18.3, the basic component equation is represented by $y_i = M_{y_i.x_i} + e_i$ $(i = 1, 2, \ldots, n)$. From a design point of view, this equation describes a situation in which the only identified source of variation is the regression component. There is no reason why an experimental design structure should not be imposed in addition to the regression structure. Indeed, one example of such a combination is provided by experimental designs which include a covariate (Section 15.2.4). Example 18.5 provides an illustration in which a simple design structure is employed in conjunction with a regression component.

Example 18.5 (Relating Oestriol Levels in Pregnant Women to Period of Gestation). *There is evidence that, during normal pregnancies, the level of oestriol remains within limits which are a function of the length of pregnancy. Where there is a problem with fetal development, this can sometimes be detected by abnormally low levels of oestriol relative to what is predicted for the given point in the pregnancy. In a study to construct a relation between oestriol level and length of gestation for normal women, there was a classification of women into groups which were defined by the gestation period as estimated to the nearest week. Women were randomly selected from each group to form the sample of* 761 *subjects for the experiment. For our purposes, interest focuses on the suitability of a linear regression relation between oestriol level* (y) *and length of gestation* (x). *The data collected are too extensive to tabulate, but the summary provided in Table* 18.7.1 *gives sufficient information for the analysis which is proposed.*

Table 18.7.1. Summary Information for Example 18.5
(Means and variances based on logarithms of oestriol readings)

Group (i)	x_i	n_i	$\bar{y}_{i.}$	s_i^2
1	30	18	1.003	0.0204
2	31	24	1.020	0.0280
3	32	50	1.036	0.0337
4	33	57	1.074	0.0193
5	34	83	1.096	0.0237
6	35	62	1.133	0.0273
7	36	60	1.125	0.0290
8	37	100	1.200	0.0230
9	38	81	1.237	0.0299
10	39	79	1.248	0.0261
11	40	69	1.236	0.0280
12	41	44	1.268	0.0266
13	42	34	1.254	0.0299

The division of the population into groups based on gestation period is described by the one stratum experimental design equation defined in (18.7.1).

$$y_{ij} = G_i + \varepsilon_{ij} \qquad \text{for } i = 1, 2, \ldots, g; \; j = 1, 2, \ldots, n_i \qquad (18.7.1)$$

where

y_{ij} is the oestriol level for the jth subject in group i;

G_i is the contribution arising from the fact that the subject is in the ith gestation group;

ε_{ij} is the contribution from unidentified sources to the oestriol level for the jth subject in group i;

g_i is the number of groups—13 in this example; and

n_i is the number of subjects in group i—see Table 18.7.1.

In the design formulation, no connection is proposed between the group means. A regression component is included when the set of means are presumed to show variation which can be related to differences in values of a scaled variable, x. In the case of Example 18.5 this is the length of gestation (in weeks). If a simple linear regression relation is assumed, the appropriate equation is $G_i = M_{y_i.x_i} + D_i$ (for $i = 1, 2, \ldots, g$) where D_i measures the deviation of the mean for the ith group from the value determined by the regression equation. Irrespective of the form of the regression equation, the model equation can be expressed by (18.7.2).

$$y_{ij} = M_{y_i.x_i} + D_i + \varepsilon_{ij} \qquad \text{for } i = 1, 2, \ldots, g; \; j = 1, 2, \ldots, n_i. \quad (18.7.2)$$

If the regression equation defines exactly the relation between group means $\{G_i\}$ and values of the explanatory variable $\{x_i\}$, the values of the parameters $D_1, D_2, \ldots, D_g$ are all zero. Hence, by comparing the fit of the model which assumes $D_1 = D_2 = \ldots = D_g = 0$ with the fit of the model which places no restriction on the values of the D_i's (i.e. which includes the equation (18.7.2) or, equivalently (18.7.1)), a test that $M_{y_i.x_i}$ is the correct regression equation is possible. The format of the required Analysis of variance table is presented in Table 18.7.2. Note that the design structure ensures there is an unbiased estimator of sampling variation, σ^2, provided by the within group mean square irrespective of the correctness or otherwise of the chosen regression equation.

Two tests are possible. The first, which is based on p_{01}, tests the hypothesis that there is no relation between y and x. The second, based on p_{02}, tests the hypothesis that the proposed regression equation provides an adequate fit to the data.

Table 18.7.2. Analysis of Variance Table Based on (18.7.2)

Source of Variation	D.F.	S.S.	M.S.	F	p
Between groups[1]	$g-1$	GSS			
Regression[2]	p	RSS	$RMS = \frac{RSS}{p}$	$F_r = \frac{RMS}{WMS}$	p_{01}
Deviations	$d = g-1-p$	$DSS = GSS - RSS$	$DMS = \frac{DSS}{d}$	$F_d = \frac{DMS}{WMS}$	p_{02}
Within groups[1]	$n-g$	WSS	$WMS = \frac{WSS}{n-g}$		
Total[1]	$n-1$	TSS			

Notes.

[1]Degrees of freedom and sums of squares can be obtained from the Analysis of variance table in Table 15.4.2(a).

[2]Degrees of freedom and sum of squares can be obtained from the Analysis of variance table in Table 18.3.1(b) based on the fit of $M_{y.x}$ to the complete data set, i.e. ignoring group structure.

Fitting polynomials. Commonly, there is no scientific basis for the selection of the form of regression equation which connects the group means. It is traditional in such cases to fit a sequence of equations based on polynomial regression relations. Firstly, a linear equation is fitted. If that provides an inadequate fit, a quadratic is fitted, and so on. This leads to an Analysis of variance structure with the following partitionings of the between treatment components.

Source of Variation	D.F.
Between treatments	$g-1$
Linear component	1
Deviations from linear	$g-2$
Gain from adding quadratic term	1
Deviations from quadratic	$g-3$
etc.	

Statistical computing packages which provide analyses for designed experiments, generally offer a simple option for performing such an analysis.

In the case of Example 18.5, the Analysis of variance table in Table 18.7.3 provides an F-test which indicates (i) the linear relation is required ($p_{01} < 0.01$); and (ii) the linear equation provides an adequate fit ($p_{02} > 0.1$).

More complex designs. While the above introduction is based on a simple design, the principle applies to all designs. The task is to partition the treatment effect into regression and deviation components. This is conveyed in the Analysis of variance table by a partitioning of the between treatment sum of squares into corresponding components.

Table 18.7.3. Analysis of Variance Table Based on a Linear Regression Equation for Oestriol Regressed on Gestation Period (Example 18.5)

Source of Variation	D.F.	S.S.	M.S.	F	p
Between groups	12	5.02			
Regression (linear)	1	4.66	4.66	179	<0.01
Deviations	11	0.36	0.033	1.3	>0.1
Within groups	748	19.52	0.026		
Total	760	24.57			

Comparing response curves. The graph of a regression equation constructed in the above circumstances is often referred to as a *response curve.* A common form of designed experiment is one in which there is a factorial arrangement of treatments with one factor (A) having quantitative levels to which a response curve is fitted, and with the experimental aims (i) to determine the form of the response curve; and (ii) to establish if the response curve varies over levels of a second factor (B). Such an experiment arises, for example

- in agricultural studies in which factor A represents levels of application of a fertilizer, while B represents different types of fertilizer;
- in a psychology experiment in which A represents levels of a stimulus, while B represents different sources for the stimulus.

Table 18.7.4 derives from a two-strata design with a factorial arrangement of treatments at the first stratum. The model equation is

$$y_{ijkm} = M + A_i + B_j + (A.B)_{ij} + \varepsilon_{ijk} + \{e'_{ijk,m}\}$$

where $i = 1, 2, \ldots, 5$; $j = 1, 2, 3$; $k = 1, 2, 3, 4$; $m = 1, 2$. The levels of factor A relate to values of a stimulus (x) and regression equations fitted are,

$$\begin{aligned} A_i &= Bx_i + D_i && \text{on the assumption of a linear fit} \\ &= B'x_i + C'x_i^2 + D'_i && \text{on the assumption of a quadratic fit.} \end{aligned}$$

The parameters B, B' and C' may be interpreted as average values over all levels of factor B. There is need to establish if there is evidence that different regression equations are required for different levels of factor B.

To answer this question, the interaction term $(A.B)_{ij}$ can be expressed in a regression formulation which includes parameters representing variations in one or both parameters across levels of B, i.e.

$$\begin{aligned} (A.B)_{ij} &= \beta_j x_i + \delta_{ij} && \text{on the assumption of a linear equation} \\ &= \beta'_j x_i + \gamma'_i x_i^2 + \delta'_{ij} && \text{on the assumption of a quadratic equation.} \end{aligned}$$

Table 18.7.4. Analysis of Variance Table Based on a Two-Strata Design with a Two-Factor Treatment Design and Regression Equations Fitted to Levels of Factor A

Source of Variation	D.F.	S.S.	M.S.	F	p
Between units stratum					
Factor A	4	388.25	77.65	177	<0.01
– linear	1	382.54	382.54	840	<0.01
– deviations (linear)	3	5.71	1.90	4.2	0.01
– add quadratic	1	4.07	4.07	8.9	<0.01
– deviations (quadratic)	2	1.64	0.82	1.8	0.2
Factor B	2	10.62	5.31	49	<0.01
$A \times B$ interaction	8	3.80	0.54	1.2	0.3
– common linear	2	0.30	0.15	0.3	0.7
– deviations (linear)	6	3.50	0.58	1.3	0.3
– add common quadratic	2	2.60	1.30	2.8	0.07
– deviations (quadratic)	4	0.90	0.23	0.5	0.7
Residual	45	20.50	0.4556	0.5	0.6
Between sub-units within units stratum					
Between sub-units	60	54.00	0.9000		

The Analysis of variance procedure provides a means of testing (i) the hypothesis $\delta_{ij} = 0$ ($i = 1, 2, \ldots, a$; $j = 1, 2, \ldots, b$) to determine if the linear fit is adequate; and (ii) the hypothesis $\beta_j = 0$ ($i = 1, 2, \ldots, b$) to establish if the linear component is constant over all levels of factor B. Where the linear fit is inadequate, the analysis can proceed with a test of the hypothesis $\delta'_{ij} = 0$ ($i = 1, 2, \ldots, a$; $j = 1, 2, \ldots, b$) to judge the adequacy of a quadratic fit, and so on.

Interpretation. When there is a two-factor factorial arrangement of treatments, the first task in examination of the Analysis of variance table is to consider the interaction. Where a regression structure is applied to the levels of factor A, the recommended sequence of steps is as follows:

Step 1 — Determine the p-value for the *deviations* (*linear*) term for the interaction. A p-value which is less than 0.05 suggests a higher order polynomial might be required and hence, there is need to fit a quadratic equation. Consider the deviations term for the quadratic, and so on.

Based on Table 18.7.4, *the p-value for the deviations (linear) is* 0.3 *and the conclusion would be reached that it is unnecessary to consider higher order polynomials.*

Step 2 — Examine the p-value for the common linear component (and common quadratic, etc. if higher order terms were required from the examination at Step 1) for the interaction term to establish if there is evidence of an interaction. This is suggested by a p-value which is below 0.05. If there is no evidence of interaction, proceed to Step 3.

If there is evidence of interaction, the conclusion to be reached is that different response curves are required for different levels of factor B. It is customary for an analysis provided by a statistical package to provide the values of the regression coefficients. Based on these values, the simplest means of determining how the response curves vary between levels of factor B is to plot the curves and visually compare the curves. There is also scope for pairwise comparisons of coefficients based on a t-test.

From Table 18.7.4, *the conclusion would be reached that there is no evidence of interaction.*

Step 3 — In the absence of interaction, it does not necessarily follow that the response curves are identical for all levels of factor B. That conclusion is reasonable only if the p-value for Factor B exceeds 0.05. If not, the conclusion to be reached is that the curves are parallel, i.e. the differences in response between the curves are constant over all levels of factor A. The magnitudes of the differences could be determined from differences in means for factor B. Application of a multiple comparison procedure (Section 15.8.1) would establish which levels have significantly different response curves.

There is evidence from Table 18.7.4 *of variations between levels of factor B since the p-value is less than* 0.01. *Hence, there is evidence that the response curves differ.*

Step 4 — To establish the nature of the response curve, the sequence of steps begins with an examination of the *deviations* (*linear*) terms in the partition of the factor A variation. If the p-value is less than 0.05, add the quadratic term to the regression equation and examine the *deviations* (*quadratic*) term, and so on. Select the simplest possible equation. (Some judgment is required to meet this last condition. In Table 18.7.4, the deviations (linear) term is significant and there is evidence from the statistical test of a gain from adding the quadratic. However, note the relative sums of squares for adding the linear component (382.54) and for adding the quadratic component (4.07). The fact that the addition of the linear component produces a gain which is approximately one hundred times that of adding the quadratic suggests the linear fit might be adequate for practical purposes. This decision can only be made by the experimenter after examining the deviations from the linear fit and after considering the importance of the possible bias from accepting the simpler equation.)

The design on which Table 18.7.4 is based is obviously only one possible design. There may be more than two factors or there may be more than one factor to which a regression equation can be fitted. These possibilities are discussed in books on *Response surface design*—see, for example, Box and Draper [1987].

18.7.2 Regression Formulation of Experimental Design Models

It is always possible to represent the model equation for an experimental design in a regression format. For example, a one-stratum design structure which is represented by the component equation

$$y_{ij} = M + T_i + e_{ij} \qquad \text{for } i = 1, 2; \; j = 1, 2, 3$$

could be expressed in the regression format

$$y_k = B_0 + B_1 x_{1k} + B_2 x_{2k} + e_k \qquad \text{for } k = 1, 2, \ldots, 6$$

by defining variables x_1 and x_2 which take the values given below and by adopting the correspondence between k and (i, j) which is shown below.

i	1	1	1	2	2	2
j	1	2	3	1	2	3
k	1	2	3	4	5	6
x_1	1	1	1	0	0	0
x_2	0	0	0	1	1	1

By setting $M = B_0$, $T_1 = B_1$ and $T_2 = B_2$, the regression formulation is equivalent to the design format.

Parameterizations. From a regression viewpoint, the above representation cannot be employed for model fitting because it violates computational restrictions. Note that $x_1 + x_2$ is constant and hence, for the reasons given in Section 18.3.6, it is not possible to obtain unique estimators of B_1 and B_2. In the design formulation it is implicitly assumed that $T_1 + T_2 = 0$ since the parameters T_1 and T_2 are deviations from the mean, M. With this added restriction, unique estimators of these parameters are possible. There is generally no option in a regression formulation to input this restriction explicitly into the instructions for a statistical computing package. Hence, where the regression formulation must be used, it is necessary to modify the formulation of the regression equation to include an additional restriction. There is not a unique parameterization to include this restriction. One possibility is to simply exclude variable x_2 and fit the equation using the stated values for x_1. In that case there is the following correspondence between parameter sets:

$$B_0 + B_1 = M + T_1 \quad \text{and} \quad B_0 = M + T_2.$$

With the additional restriction $T_1+T_2 = 0$, it follows that $M = \frac{1}{2}(2B_0+B_1)$ and $T_1-T_2 = B_1$. Another possibility is to give x_1 the values, $1, 1, 1, -1, -1, -1$. This leads to the correspondence

$$B_0 + B_1 = M + T_1 \quad \text{and} \quad B_0 - B_1 = M + T_2.$$

With the additional restriction $T_1 + T_2 = 0$, it follows that $M = B_0$ and $T_1 - T_2 = 2B_1$. A third possibility is to exclude the parameter B_0 from the regression equation. Given the initial sets of values for x_1 and x_2, this approach produces the correspondence:

$$B_1 = M + T_1 \quad \text{and} \quad B_2 = M + T_2.$$

In this case, $T_1 - T_2$ is estimated by $B_1 - B_2$.

Given that the aim is the comparison of treatment effects, the different parameterizations do not affect treatment comparisons. Provided the analyst knowns which parameterization is being used, it is possible to employ output from regression analysis for the purpose of treatment comparison.

Application of the regression formulation in designed experiments. Simple formulae exist for the analysis of designed experiments when the design is balanced. Otherwise, the regression formulation must be used. In the more comprehensive packages, the user is unaware of the need to employ a more complex structure because it is not visible. However, there are packages, and there may be some designs, for which the only approach to the analysis of the design is through the use of the regression approach.

Problems

18.1 *Example* 18.2 — *Model and data checking followed by basic analysis*

Based on the data in Table 18.1.2, for each species,

(a) Construct a plot of *abundance* versus *dry weight*.

(Note the odd point in the plot for *C. cephalornithos*. This was identified by the experimenter as an error and the reading of 1160 for *abundance* should be treated as a missing value in regression analysis.)

(b) Fit a Simple linear regression equation for *abundance* on *dry weight*.

(c) Obtain fitted values and residuals and produce a plot of fitted values versus residuals.

(d) Construct a Normal probability plot using the standardized residuals.

Following the guidelines given in Section 18.5.1, establish that the assumption of a linear trend line and the Normality assumption are reasonable based on the information obtained in (a) to (c) after the removal of the odd value.

(e) Obtain the estimated linear regression equation for the regression of *abundance* on *dry weight*, estimates of regression coefficients and produce an Analysis of variance table. Determine the value of R^2_{adj} as a measure of the proportion of the variance of *abundance* which can be associated with variation in *dry weight*.

18.2 *Example* 18.2 — *comparison of regression relations*

(a) By treating the data in Table 18.1.2 as a single sample (and after elimination of the odd value detected in Problem 18.1) regress *abundance* on *dry weight*.

(b) Using the information in the Analysis of variance table obtained in (a) plus the information in the Analysis of variance tables obtained in Problem 18.1, follow the steps in Section 18.6.4 to test the hypothesis that the same regression equation fits both species.

(c) If you have the computing facilities, test (i) the hypothesis of common slopes; and (ii) the hypothesis of common intercepts by following the steps given in Section 18.6.4.

18.3 *Gas mixtures in the pouches of marsupials*

Marsupials (e.g. kangaroos) are characterized by the fact that the young live in pouches on the underside of their mothers. Data are provided below on the levels of carbon dioxide (CO_2) and oxygen (O_2) in the gas mixture in the pouches of 24 *potoroos* which were carrying young. It is known that carbon dioxide levels rise and oxygen levels fall with increasing age of the young. By considering the regression of CO_2 and O_2, determine if the data are consistent with the theory that the rate of increase in the percentage of carbon dioxide equals the rate of decline in the percentage of oxygen in the pouch. (In other words, fit a linear regression equation for the regression of CO_2 on O_2 and establish if a slope of $B = -1$ is consistent with the data. This may be done by computing a confidence interval for B using (18.6.3) or testing the hypothesis $B = -1$ using (18.6.1) provided a linear regression equation and the assumption of Normality are reasonable.)

Table of O_2 and CO_2 Readings in the Pouches of 24 Potoroos
(Data by courtesy of Dr. D. Megirian, Dept. of Physiology,
Univ. of Tasmania)

Animal	1	2	3	4	5	6	7	8
% CO_2	1.0	1.2	1.1	1.4	2.3	1.7	1.7	2.4
% O_2	20.0	19.6	19.6	19.4	18.4	19.0	19.0	18.3

Animal	9	10	11	12	13	14	15	16
% CO_2	2.1	2.1	1.2	2.3	1.9	2.4	2.6	2.9
% O_2	18.2	18.6	19.2	18.2	18.7	18.5	18.0	17.4

Animal	17	18	19	20	21	22	23	24
% CO_2	4.0	4.2	3.3	3.0	3.4	2.9	1.9	3.9
% O_2	16.5	*	17.2	17.3	17.8	17.3	18.4	16.9

18.4 *Example 3.3 — further data and a comparison of regression relations*

In Example 3.3, data are presented from a study of the relation between *pressure drop* and *flow rate* for liquid flowing along a tube which simulates blood flowing along a vein. The data provided in Table 3.2.5 are the results for one viscosity level. The complete data set are provided below.

(a) Based on these data, fit separate quadratic regression equations for *pressure drop* versus *flow rate* at each level of viscosity.

(b) By employing the model and data checking procedures suggested in Section 18.5.1, establish that there is evidence of errors in the data. Remove the offending points by treating them as missing values. Remember to repeat the checking procedure after deleting the odd points in case further possible errors are detected. (In the original analysis, the experimenter was able to provide sound reasons for excluding the odd values.)

(c) By following the steps in Section 18.6.4, establish that a common quadratic equation provides a significantly worse fit to the data than does a separate quadratic for each viscosity level.

Is there evidence that the curvature of the regression lines increases with decreasing viscosity?

Data Obtained in the Study Reported in Example 3.3
(f denotes flow rate and p denotes pressure drop)

Viscosity

39.3		33.9		31.1		28.7		22.7	
f	p	f	p	f	p	f	p	f	p
0.11	1.1	0.31	3.0	0.84	11.5	0.36	2.8	0.11	0.7
0.19	2.1	0.10	1.4	0.07	0.7	0.26	2.0	0.25	1.5
0.51	5.3	0.24	2.1	0.20	1.5	0.95	6.0	0.89	4.6
0.75	6.8	0.79	5.9	0.40	2.6	1.13	6.9	0.24	7.1
0.95	8.4	0.49	3.9	0.55	3.8	1.46	8.9	1.36	7.4
1.13	9.8	1.30	9.7	0.79	5.1	1.76	11.0	1.59	9.0
1.45	12.9	1.76	12.3	1.27	8.2	2.24	13.8	1.89	11.0
1.65	14.7	2.60	18.8	1.61	10.6	2.94	19.8	2.25	12.8
1.99	16.7	3.64	28.0	2.25	14.9	4.03	28.5	2.67	16.2
2.37	20.6	4.28	36.3	2.86	19.6	5.05	38.2	3.14	19.5
2.69	23.3			3.67	27.2			3.98	28.3
3.31	29.5			4.81	37.0			4.90	37.5
3.66	33.3			2.99	21.0			5.30	44.5
4.25	39.9								

18.5 *Lifetime of manufacturing components*

The table below records lifetimes, in number of 'pours,' of containers which hold molten metal together with information on variables thought to influence lifetime. The containers come from two different suppliers, labelled A and B in the table below.

(a) Based on the information collected on containers from Supplier A, use (i) the *forward selection* and *backward elimination methods* (Section 18.6.1) to select the subset of variables which appear to give the simplest regression equation which utilizes the information in the data.

(b) Repeat the steps in (a) based on the information collected on containers from Supplier B.

Note that the same subset of variables is found in both cases.

(c) Establish that there is evidence the regression relations differ for containers from the two different suppliers.

The company who supplied the data wished to know which manufacturer they should choose to supply containers. Assuming costs, supply times and related factors are identical, what, if any, guidance is offered by the results of the above analysis?

Supplier A					Supplier B				
Container	y	x_1	x_2	x_3	Container	y	x_1	x_2	x_3
1	150	47	2900	28	1	153	46	2800	28
2	151	45	2800	26	2	156	48	3200	28
3	153	47	2600	26	3	157	49	3000	26
4	154	50	3500	26	4	160	49	3300	27
5	154	46	3100	29	5	163	49	3200	29
6	155	48	3600	30	6	165	50	3200	28
7	157	46	2800	28	7	165	50	3400	30
8	158	50	3000	26	8	166	53	3500	29
9	159	54	3400	27	9	167	53	4000	29
10	160	48	3300	31	10	170	52	3400	28
11	161	49	2800	29	11	171	50	2500	28
12	161	54	3500	28	12	171	51	3600	30
13	161	51	3400	27	13	172	53	3900	32
14	163	58	4200	31	14	172	55	4000	30
15	164	52	3200	31	15	174	55	3700	29
16	164	54	4000	29	16	175	57	4200	32
17	164	55	3700	28	17	175	52	3400	30
18	166	55	3700	28	18	176	54	3700	31
19	167	54	4200	31	19	176	55	4100	31
20	167	51	3500	29	20	178	56	4300	31
21	167	53	3600	28	21	178	56	3800	30
22	168	57	4300	33	22	178	56	4300	33
23	168	54	3700	31	23	179	56	4100	32
24	168	55	3600	33	24	180	55	4100	31
25	170	57	4600	33	25	180	53	3600	30
26	170	58	4200	31	26	181	56	4200	33
27	170	58	4400	32	27	182	58	4600	32
28	170	55	4400	32	28	182	56	4100	31
29	173	58	4600	31	29	184	55	4000	32
30	174	56	4300	31	30	187	57	4400	33
31	178	60	4600	32	31	187	58	4600	33
					32	190	60	4600	32

18.6 *Simulation study*

Freedman [1983] carried out a simulation study in which 5,100 pseudo-random numbers were generated from a $N(0,1)$ distribution and presumed to form 100 observations from 51 variables. The last variable was labelled the response variable and was regressed on the previous fifty variables. From among the fifty, those which appeared to contribute significantly to explanation of variation in the 'response' variable were found and included in a regression equation. The results, which Freedman repeated a number of times with different sets of pseudo-random numbers, suggested the existence of a regression equation which could explain a significant proportion of the variation in the values of the 'response' variable.

Repeat portion of Freedman's experiment on a smaller scale—produce 630 pseudo-random numbers from a $N(0,1)$ distribution; allocate them to 20 'explanatory variables' and 1 'response' variable; apply the forward selection and backward elimination methods to determine if there is a subset of explanatory variables which can explain a statistically significant proportion of the variation in the response variable.

19

Questions About Variability

19.1 Variability—Its Measurement and Application

19.1.1 Measuring Variability

The word *variability* is vague and requires precise definition in a statistical context. In a sample, the simplest measurement of variability is the *range,* i.e. the difference between the largest and smallest values. This statistic has limited application because it is strongly dependent on sample size, ignores much information in the data and is highly sensitive to odd values in the data. More appealing are contrasts among percentiles, the most widely used being the *interquartile range* (Section 1.3.8) which has the useful property that it is the spread of the middle fifty percent of sample members. Other percentiles may also have a place in the definition of variability, a point touched upon in Section 1.3.10. *Sample variance* and *standard deviation* are widely used statistics to describe variability in a sample. Their interpretive value is limited to circumstances in which the Normal distribution can be regarded as a reasonable approximation to the underlying frequency or probability distribution, a point considered further in Section 19.1.2.

It is for the experimenter to decide which measure of variability to employ. The decision should, however, pay due regard to interpretation of any measure which is chosen, and, in particular, to the general superiority of percentile-based measures in this regard.

Parameters of probability distributions which define the spread or variability of the distribution are commonly referred to as *scale* parameters. The parameter σ^2 of the Normal distribution is a well known example of a scale parameter. The Normal distribution is unique in having a scale parameter which is independent of the parameter which defines the mean of the distribution. In the Poisson distribution, by contrast, the parameter M, which is the mean of the distribution, is also a scale parameter since the variance of the distribution is M.

The estimation of any parameter of a probability distribution relies on a statistic. Every statistic has a sampling distribution and the variability of the sampling distribution provides a basis for measuring the precision of estimation of parameters. Section 19.1.4 provides some insight into the manner in which measures of variability are employed to define precision of estimation.

Where two or more factors are presumed to contribute to the variability observed in values of a response variable, there may be interest in seeking to establish the relative sizes of contributions from different sources. There are some areas of special importance. These include:

- the separation of genetic and environmental sources by biologists; and
- the separation of contributions from inherent skills or ability and acquired skills or ability.

The methods used for partitioning variation and the application of these methods are briefly discussed in Section 19.1.5 and are developed more extensively in Section 19.2.

19.1.2 Scale Parameters and Their Estimation

1. *Variance* (σ^2). While the variance of a distribution, σ^2, can be described as the averaged squared deviation of values from the mean or expected value, a numerical value of σ^2 has no intuitive meaning, at least in its own right. It is only when coupled with the Normal distribution that numerical values of σ^2 may have value. In that case, it is possible to relate useful probability statements to the value of the standard deviation, σ. In particular,

approximately 95% *of responses lie within* 2σ *of the mean*

approximately 99.9% *of responses lie within* 3σ *of the mean.*

Statistic. When the Normal distribution is assumed, the unbiased Maximum Likelihood estimator of σ^2 is the sample variance (Section 1.3.9) in the case where there is no design structure, or the *residual mean square* from the Analysis of variance table (Chapter 15) if there is design structure. The sample estimator of σ^2 is generally denoted by s^2 and has an associated parameter known as the *degrees of freedom, d*, which is equal to $n-1$ in the case of an unstructured data set of n responses or is equal to the *residual degrees of freedom* from the Analysis of variance table when there is design structure. Under a model which assumes responses are values of Normally and independently distributed variables, the sampling distribution of ds^2/σ^2 is a Chi-squared distribution with degrees of freedom d.

Estimation. If ds^2/σ^2 has a $\chi^2(d)$ distribution, (19.1.1) is true. Rearrangement of (19.1.1) provides the confidence intervals for σ^2 in (19.1.2).

$$\Pr(X^2_{1-\frac{1}{2}\alpha} \le ds^2/\sigma^2 \le X^2_{\frac{1}{2}\alpha}) = 1-\alpha \qquad (19.1.1)$$

$$\begin{array}{ll} ds^2/X^2_{0.025}(d) \text{ to } ds^2/X^2_{0.975}(d) & 95\% \text{ confidence interval} \\ ds^2/X^2_{\frac{1}{2}\alpha}(d) \text{ to } ds^2/X^2_{1-\frac{1}{2}\alpha}(d) & 100(1-\alpha)\% \text{ confidence interval} \end{array} \qquad (19.1.2)$$

Applications. The sample variance, s^2, provides a point estimate of σ^2. A confidence interval is provided by (19.1.2) and this may be used to test an hypothesis $\sigma^2 = \sigma_0^2$.

Computations. The formulae given are sensitive to the assumption of Normality and the first step in analysis should be to examine the adequacy of this assumption by constructing a Normal probability plot using the method provided in Section 10.3.4. If the Normality assumption is unacceptable, the variance is unlikely to be a useful measure of variability.

Values for $X^2_{\frac{1}{2}\alpha}$ and $X^2_{1-\frac{1}{2}\alpha}$ can be obtained using one of the approaches in Section 7.5.4. Hence, a confidence interval for σ^2 can be computed using (19.1.2).

2. *The Interquartile Range* (ω). If y is a continuous random variable with a distribution function Π, the interquartile range, ω, is defined by (19.1.3).

$$\omega = P_{75} - P_{25} \tag{19.1.3}$$

where P_{25} and P_{75} are, respectively, the 25th and 75th percentiles of the probability distribution of y and hence, satisfy $\Pi(P_{25}) = 0.25$ and $\Pi(P_{75}) = 0.75$. For any distributional form, the interquartile range is the distance over which the middle fifth percent of the distribution is spread.

A point estimate is provided by the sample interquartile range (Section 1.3.8).

3. *Coefficient of Variation* (γ). A variation on the standard deviation which is useful in some instances is the *coefficient of variation* which is defined by (19.1.4).

$$\gamma = 100\sigma/M \tag{19.1.4}$$

where M is the mean of the distribution; and σ is the standard deviation of the distribution. The coefficient of variation has a role when the variability is expected to increase with increasing mean and comparisons of variability between groups are required where the means might be expected to differ. Empirical evidence suggests that in many laboratory studies of the same material or in many field trials involving similar crops, the coefficient of variation is stable if the same level of control of extraneous sources of variation is maintained. Thus, the coefficient of variation has a role in comparing quality of experimentation in different institutions or at different field stations. The experimenter who is repeatedly studying the same material may also use the coefficient of variation to check that a constant level of quality is being maintained. A sudden jump in the coefficient of variation might indicate a problem with laboratory staff, experimental material or experimental practice.

Use of the coefficient of variation is best restricted to variables which do not take negative values.

The sample coefficient of variation (c) as defined by (19.1.5) provides a point estimate of γ.

$$c = 100s/\bar{y} \tag{19.1.5}$$

where $\bar{y}$ and s are the sample mean and variance, respectively.

19.1.3 Comparing Variances Between Groups

A method is presented below for pairwise comparison of variances under the assumption of a Normal distribution.

Model. The probability distributions for the two groups are presumed to be Normal distributions with variances σ_1^2 and σ_2^2, respectively. Samples are presumed to be obtained by random selection from both groups and sample selection is presumed to be independent between groups.

Statistic. Samples of size n_1 and n_2 are presumed to have been drawn from Groups 1 and 2, respectively, and from these samples, variances s_1^2 and s_2^2 computed. The statistic $r = s_1^2/s_2^2$ is employed to test hypotheses about the ratio $R = \sigma_1^2/\sigma_2^2$ and to construct a confidence interval for R. The sampling distribution of $[\sigma_2^2/\sigma_1^2]r$ is an $F(n_1 - 1, n_2 - 1)$ distribution.

Estimation. Based on the fact that the statistic $[\sigma_2^2/\sigma_1^2]r$ has an $F(n_1 - 1, n_2 - 1)$ distribution and provided α is less than 0.5,

$$\Pr\{F_{1-\frac{1}{2}\alpha}(n_1 - 1, n_2 - 1) \leq [\sigma_2^2/\sigma_1^2]r \leq F_{\frac{1}{2}\alpha}(n_1 - 1, n_2 - 1)\} = 1 - \alpha.$$

On rearrangement,

$$\Pr\{r/F_{\frac{1}{2}\alpha}(n_1 - 1, n_2 - 1) \leq [\sigma_1^2/\sigma_2^2] \leq r/F_{1-\frac{1}{2}\alpha}(n_1 - 1, n_2 - 1)\} = 1 - \alpha.$$

This leads to the confidence intervals for σ_1^2/σ_2^2 provided in (19.1.6).

$$r/F_{0.025}(n_1 - 1, n_2 - 1) \text{ to } r/F_{0.975}(n_1 - 1, n_2 - 1) \text{ 95\% C.I.}$$

$$r/F_{\frac{1}{2}\alpha}(n_1 - 1, n_2 - 1) \text{ to } r/F_{1-\frac{1}{2}\alpha}(n_1 - 1, n_2 - 1) \text{ } 100(1 - \alpha)\% \text{ C.I.} \tag{19.1.6}$$

Application. Most commonly, the aim is to examine whether the hypothesis $\sigma_1^2 = \sigma_2^2$ is acceptable.

Computation. A confidence interval for the ratio $R = \sigma_1^2/\sigma_2^2$ can be constructed by following the steps below.

Step 1 — Assuming sample of size n_1 and n_2 are obtained from Groups 1 and 2, respectively, compute sample variances, s_1^2 and s_2^2 and *degrees of freedom* $d_1 = n_1 - 1$ and $d_2 = n_2 - 1$.

Step 2 — Construct Normal probability plots for both samples using the method described in Section 10.3.4. Proceed to Step 3 if the assumption of Normality appears reasonable and there is no evidence of an odd value in the data. It is not wise to proceed to the construction of a confidence interval for the ratio of variances if there is evidence of non-Normality since the procedure is highly sensitive to an incorrect distributional assumption.

Step 3 — Compute the ratio $r = s_1^2/s_2^2$, determine the values of $F_{0.025}(d_1, d_2)$ and $F_{0.975}(d_1, d_2)$ (or $F_{\frac{1}{2}\alpha}(d_1, d_2)$ and $F_{1-\frac{1}{2}\alpha}(d_1, d_2)$ for a specified α) using one of the methods described in Section 7.6.4. Hence, compute a confidence interval for $R = \sigma_1^2/\sigma_2^2$ from (19.1.6).

If the confidence interval includes a value of $R = 1$, the hypothesis $\sigma_1^2 = \sigma_2^2$ is reasonable. The calculations are illustrated using the data in Example 19.1.

Example 19.1 (Comparing Methods of Variability in Chemical Determinations). *Two sets of numbers are presented below which represent measurements on the level of salt in a solution. As conducted, the laboratory study involves the taking of 22 samples from the salt solution and random allocation of the samples between the two methods such that each method was to be applied to 11 samples. One sample to be used with method 2 was lost before a measurement was made.*

The aim is to establish if there is evidence that one method is more precise than the other. In this situation it is only variability in measurement which is important since an adjustment can be made for bias in a method.

Method 1	16.20	16.19	16.22	16.18	16.28	16.21
Method 2	16.26	16.33	16.47	16.44	16.32	16.44

Method 1	16.11	16.20	16.20	16.19	16.16
Method 2	16.44	16.43	16.54	16.30	

Illustration.

Step 1 — Calculation of sample variances yields

$$s_1^2 = 0.0016873 \qquad s_2^2 = 0.0078899.$$

The sample sizes for Methods 1 and 2, respectively, are 11 and 10 and hence, the degrees of freedom are $d_1 = 10$ and $d_2 = 9$.

Step 2 — Normal probability plots for both methods are consistent with the assumption of Normality.

Step 3 — The value of r is $0.0016873/0.0078899 = 0.214$, the value of $F_{0.025}(10, 9)$ is 3.964 and the value of $F_{0.975}(10, 9)$ is 0.2647. Hence, by (19.1.6), the 95% confidence interval for $R = \sigma_1^2/\sigma_2^2$ is

$$0.214/3.964 \text{ to } 0.214/0.2647$$
$$0.054 \text{ to } 0.81.$$

Since the interval does not include the value one, there is evidence the variances are not equal. Furthermore, the fact that s_1^2 is smaller than s_2^2 indicates that Method 1 is likely to be more precise (i.e. less variable) than Method 2.

19.1.4 Precision of Estimation

If the statistic x is an unbiased estimator of a parameter M, the standard deviation of x (denoted by σ_x) is often used as a measure of the precision of the estimator. If the value of σ_x is unknown, the value of its sample estimator, s_x, is generally employed. In the context that x is an *estimator* of M, σ_x is commonly termed the *standard error of the estimator* and the computed value of s_x is often called the *standard error of the estimate.* If two competing estimators are available and both are unbiased estimators of M, the estimator with the smaller standard error would be termed the more precise estimator. In other words, precision is inversely related to the standard error.

In general, the numerical value of σ_x or s_x has no obvious application except for the situation in which the sample distribution of x is a Normal distribution or is well approximated by a Normal distribution. Where the Normality assumption is applied, (8.3.2) or (8.3.3) may be invoked to provide confidence intervals for M. In this circumstance, it is the value of σ_x or s_x which determines the width of the confidence interval.

A quite different application of precision arises in designed experiments or observational studies in which there is variation at more than one stratum level. For example,

1. there may be variation in response due to differences between students (σ_b^2) and variation between measurements made repeatedly on the one student (σ_w^2) due to changing environmental conditions;

2. there may be variation in response due to differences between plots of ground (σ_b^2) and variation in response between samples drawn from the one plot due to non-uniformity of the plot (σ_w^2);

3. there may be variation between chemical readings made on samples from different sources (σ_b^2) and variation between duplicates from the one sample (σ_w^2).

If there are n randomly selected samples and r randomly selected subsamples taken from each sample, and a response is recorded for each subsample, the variance of the mean response from the nr readings is

$$\text{var}(\overline{y}) = \frac{\sigma_b^2}{n} + \frac{\sigma_w^2}{nr}.$$

When designing experiments in which the precision of estimation of the mean is of importance, it is apparent that the relative numbers of samples

(n) and sub-samples (r) collected will influence the variance and hence, the precision of estimation. This point is considered in some detail in Section 19.2.7.

19.1.5 Reliability

Precision, as discussed in Section 19.1.4, is an example of a broader class of measures of *reliability.* There are situations in which reliability reflects accuracy of measurement. In the physical sciences, for example, it may be possible to place absolute bounds on the accuracy of measurement—the statement 16.2 ± 0.1 implies that the true value lies within the limits 16.1 and 16.3.

There are situations in the biological and social sciences when reliability can only be expressed in relative terms. A classic example is found in plant breeding where reliability is measured by the proportion of variability which is genetic in origin, i.e. can be passed from one generation to the next. The measurement of reliability is based on the division of variance in (19.1.7).

$$\begin{array}{ccccc} \text{Total variation} & = & \text{Genetic variation} & + & \text{Environmental variation} \\ \sigma_t^2 & = & \sigma_g^2 & + & \sigma_e^2 \end{array} \qquad (19.1.7)$$

The quantity $h = \sigma_g^2/\sigma_t^2$ provides a measure of reliability which is termed *heritability.*

The same concept of reliability is required by social scientists who wish to measure the reliability of a method for measuring an attribute. For example, a test may be proposed for measuring mathematical ability. When the test is applied to a group of students, there is likely to be a variation in scores. Portion of that variability is due to differences in mathematical ability between the students. The remaining variation is due to differences in factors which are not related to mathematical ability. For instance, one student might be suffering from an illness on the day of the test and perform below his or her true ability. Reliability is a measure of the proportion of variability in the measurements which is attributable to variation in mathematical ability.

The measurement of heritability is relatively straightforward if it is possible to obtain measurements on the subjects which are genetically identical, since variability between genetically identical subjects must be solely environmental and hence, can provide an estimate of σ_e^2 in (19.1.7). In the social science context, the measurement of reliability is generally much more difficult since it is usually not possible to obtain subjects who are known to have the same value for the attribute of interest, e.g. mathematical ability. Ideally, the 'environmental' component of variability can be determined from multiple measurements on the same subject. The danger of this approach is obvious—if the measurement of the response is through a test, then repeatedly testing the one subject is likely to produce responses which are affected by learning which resulted from the previous tests. Furthermore, the conditions under which the successive tests are done may have factors in common which causes the responses to be correlated and hence, bias the estimate of non-attribute variability. There have been lengthy discussions in the psychology literature for several decades as to the 'best' way to define and measure reliability. Jensen [1980] provides an in-depth consideration of the topic.

The end result is a proliferation of procedures and measures. Those who are concerned with the measurement of reliability in the social science context should seek reviews on these procedures and methods and seek to understand their bases and the conditions under which they may have valid and sensible application. Most statistical texts in the area of psychology (e.g. Winer [1971]) include sections on the computational aspects of measuring reliability.

19.2 Variance Components

19.2.1 Partitioning Variation

Where there is a design structure which can be represented by a linear experimental design equation, there is scope for partitioning the variance of the response variable into *variance components.* Example 19.2 provides an illustration.

Example 19.2 (Factors Influencing the Variability in Frost Tolerance of Eucalypts). *Trees of a Eucalypt species which is found over a diverse range of climatic conditions show a wide range of frost tolerances. Interest lies in establishing the sources of the variability. To this end, four possible sources of variability were identified. They were*

- *changes in aspect, which may range from north-facing to south-facing;*
- *differences in altitude;*
- *tree-to-tree variation; and*
- *variation between seedlings from the same tree.*

A study was instituted in which six aspects were selected; at each aspect two altitudes were selected; from each altitude at each aspect three trees were selected; and from each tree growing in each altitude at each aspect, a sample of seven seedlings were selected from among the seeds which germinated. This experimental design is represented by the experimental design equation in (19.2.1) *on the assumption that random selection was employed at each stage. (See Chapter* 15 *if the form of specifying the equation is unfamiliar.) Data from the study appear in Table* 19.2.1.

Table 19.2.1. Data for Example 19.2
(Data by courtesy of Dr. B. Potts, Dept. of Botany, Univ. of Tas.)

Frost Tolerances

			Seedling						
Region	Altitude	Tree	1	2	3	4	5	6	7
1	1	1	−55	−57	−51	−61	−46	−58	−56
1	1	2	−65	−50	−46	−57	−55	−53	−63
1	1	3	−56	−65	−64	−57	−61	−50	−48
1	2	1	−60	−61	−60	−58	−73	−54	−66
1	2	2	−56	−56	−59	−46	−58	−62	−51
1	2	3	−57	−58	−56	−55	−54	−58	−60
2	1	1	−50	−47	−45	−45	−51	−53	−57
2	1	2	−55	−45	−46	−61	−51	−61	−47
2	1	3	−51	−59	−54	−62	−55	−49	−50
2	2	1	−60	−65	−60	−64	−55	−52	−53
2	2	2	−55	−55	−60	−51	−64	−53	−42
2	2	3	−48	−50	−48	−51	−52	−45	−54
3	1	1	−52	−59	−52	−57	−65	−64	−68
3	1	2	−62	−55	−61	−60	−56	−52	−51
3	1	3	−52	−56	−52	−56	−54	−50	−49
3	2	1	−59	−54	−58	−60	−53	−51	−65
3	2	2	−56	−55	−50	−58	−56	−54	−62
3	2	3	−65	−56	−45	−56	−56	−54	−57
4	1	1	−41	−53	−43	−43	−50	−48	−44
4	1	2	−56	−46	−53	−55	−64	−59	−46
4	1	3	−46	−50	−42	−53	−44	−48	−41
4	2	1	−52	−52	−52	−56	−58	−53	−48
4	2	2	−63	−53	−50	−60	−60	−60	−60
4	2	3	−50	−53	−42	−50	−53	−50	−44
5	1	1	−44	−41	−46	−56	−45	−50	−48
5	1	2	−42	−52	−42	−48	−52	−48	−42
5	1	3	−45	−44	−46	−54	−53	−54	−41
5	2	1	−63	−53	−48	−57	−53	−55	−54
5	2	2	−48	−50	−49	−59	−50	−59	−51
5	2	3	−55	−45	−51	−56	−52	−52	−50
6	1	1	−60	−55	−56	−63	−54	−58	−66
6	1	2	−57	−52	−53	−60	−57	−56	−54
6	1	3	−56	−54	−51	−54	−53	−53	−58
6	2	1	−58	−53	−64	−60	−63	−56	−56
6	2	2	−56	−53	−50	−57	−56	−59	−54
6	2	3	−52	−60	−52	−57	−56	−54	−54

$$y_{ijkm} = M + \mathcal{R}_i + \{\mathcal{U}_{i,j} + \{\mathcal{T}_{i,j,k} + \{\mathcal{S}_{i,j,k,m}\}\}\} \qquad (19.2.1)$$

for $i = 1, 2, \ldots, 6$; $j = 1, 2$; $k = 1, 2, 3$; $m = 1, 2, \ldots, 7$ where

y is the response variable which has mean M and variance σ_y^2;

$\mathcal{R}$ is the adjustment reflecting variations in aspects which has mean 0 and variance σ_r^2;

$\mathcal{U}$ is the adjustment which reflects variations in altitudes within aspects and has mean of 0 and variance σ_a^2;

$\mathcal{T}$ is the adjustment which reflects variations between trees within altitudes within aspects and has mean 0 and variance σ_t^2; and

$\mathcal{S}$ is the adjustment reflecting variations between seedlings from the same tree within altitudes within aspects and has mean 0 and variance σ_s^2.

Provided the contributions from the different sources are independent, the expected variance in response can be partitioned according to (19.2.2)

$$\sigma_y^2 = \sigma_r^2 + \sigma_a^2 + \sigma_t^2 + \sigma_s^2. \qquad (19.2.2)$$

It is in this context that the variances on the right-hand side of (19.2.2) are meaningfully called *variance components.* From the experimental viewpoint, it is usually not the absolute values of the components which are important, but the relative contributions they make to the total variance, σ_y^2.

Heritability. One of the classic applications of variance component analysis lies in the area of plant and animal breeding where the aim is to separate genetic and environmental effects. The aim of plant and animal breeders is to produce high yielding varieties or breeds which have the capacity to pass the high yielding ability on to their progeny. Example 19.3 provides an illustration.

Example 19.3 (Heritability in Time from Bud Formation to Overmaturation in a Plant Species). *Plant species are distinguished by having different genetic make-ups, i.e. different genotypes. In this experiment, six genotypes were randomly selected from a large pool of genotypes and three plants cloned within each genotype (i.e. produced with the same genotype). The* 18 *plants selected were randomly allocated between* 18 *plots. From each plant, five flowers were selected for study. The time taken for each flower to proceed from bud formation to overmaturation was determined and is recorded in Table* 19.2.2.

Table 19.2.2. Data for Example 19.3

Days from bud formation to overmaturation

Genotype	Plant	Flower 1	2	3	4	5
	1	12	13	13	12	12
A	2	11	13	12	12	13
	3	12	11	11	9	10
	1	14	14	16	16	14
B	2	15	16	16	16	15
	3	14	16	14	16	16
	1	10	13	10	11	13
C	2	10	14	15	10	14
	3	13	13	12	10	11
	1	13	13	15	13	15
D	2	15	14	13	12	12
	3	15	13	14	13	14
	1	15	15	15	15	15
E	2	13	13	13	13	12
	3	13	13	13	13	13
	1	15	15	14	15	15
F	2	15	15	14	14	14
	3	14	15	15	14	14

The experimental design equation employed in Example 19.3 is defined by (19.2.3).

$$y_{ijk} = M + \mathcal{G}_i + \{\mathcal{B}_{i,j} + \{\mathcal{F}_{i,j,k}\}\} \text{ for } i = 1,2,\ldots,6; j = 1,2,3; k = 1,2,\ldots,5 \quad (19.2.3)$$

where

y_{ijk} is the number of days from bud formation to overmaturation for flower k on plant j from genotype i;

$\mathcal{G}_i$ is the deviation of the ith genotype from the mean M;

$\mathcal{B}_{i,j}$ is the deviation of the jth plant from the average for the ith genotype; and

$\mathcal{F}_{i,j,k}$ is the deviation of the kth flower from the average for the jth plant from the ith genotype.

The genotypes used in this study are presumed to be obtained by random selection from a large pool of possible genotypes. The variability between average values for the different genotypes is a measure of the genetic component of variability in the time to overmaturation which is measured by the variance of $\mathcal{G}_i$ (σ_g^2). Non-genetic variability is found in plant-to-plant variability within a genotype and flower-to-flower variability within a plant. These are measured by the variances of $\mathcal{B}_{i,j}(\sigma_p^2)$ and $\mathcal{F}_{i,j,k}(\sigma_f^2)$, respectively. Of particular interest is the ratio $\sigma_g^2/(\sigma_g^2+\sigma_p^2+\sigma_f^2)$ which is the proportion of variability which is genetic in origin and is known as the *heritability*.

Sub-sampling and variance components. Sampling is a commonly used method for improving precision of estimation of a treatment or group mean. It has application in circumstances where there is an opportunity to repeatedly sample each experimental unit. Two sources of variation are defined—the unit-to-unit variation and the variation between samples taken from the same unit. The precision with which a treatment mean is estimated is dependent on the relative contributions made by each source according to (19.2.4).

$$\text{var}(\overline{y}) = \frac{\sigma_u^2}{n} + \frac{\sigma_s^2}{nr} \qquad (19.2.4)$$

where

σ_u^2 is the between unit variance;

σ_s^2 is the between sample variance within units;

n is the number of units per treatment; and

r is the number of subsamples from each unit.

If the values of the variance components are known or can be estimated, it is possible to determine the effects of varying either n or r on the precision of estimation. By including costs associated with unit and sample collection, it is possible to identify an optimal combination of n and r. The details are presented in Section 19.2.7. Example 19.4 is a study in which the method found application.

Example 19.4 (Activity of Children at Play in Different Environments). *Example* 15.6 *introduces a study in child behaviour in which nine treatments were being compared. To each treatment a number of children were assigned. Sampling took the form of repeated recordings for each child. In the initial design, two children were assigned per treatment and two recordings obtained from each child. The methodology for examining the possible gain from varying the number of children per treatment and/or the number of recordings per child is considered in Section* 19.2.7.

19.2.2 Statistical Models

Variance component models have application where there is a multi-strata design and a linear component equation is employed. The contributions to the response, y, from strata $1, 2, \ldots$ are denoted by $\mathcal{U}_1, \mathcal{U}_2, \ldots$. Thus, for a two-strata design, the experimental design equation is defined by (19.2.5) and, more generally, the experimental design equation is defined by (19.2.6).

$$y_{ij} = M + \mathcal{U}_{1_i} + \{\mathcal{U}_{2_{i,j}}\} \tag{19.2.5}$$

$$y_{ijk\ldots} = M + \mathcal{U}_{1_i} + \{\mathcal{U}_{2_{i,j}} + \{\mathcal{U}_{3_{i,j,k}} + \ldots\}\}. \tag{19.2.6}$$

In *Variance component models* the units are presumed to be randomly selected in those strata which contain no contributions from treatment effects. Consequently, each component in (19.2.5) and (19.2.6) is presumed to be a random variable. Within each stratum the contributory variables are presumed to be identically distributed with the variances at strata $1, 2, \ldots$ being denoted by $\sigma_1^2, \sigma_2^2, \ldots$, respectively.

At one or more strata levels, units may be allocated between treatments for the purpose of establishing if the treatments produce different mean effects. Example 19.4 illustrates this situation. There remains, however, at each stratum level in which treatment effects appear, a component which reflects the contribution from unidentified factors. Provided that treatments have been randomly allocated between units, this component is a random variable and the variance of its probability distribution is the variance component of interest in the stratum.

Hypotheses included in the model are stated in terms of $\sigma_1^2, \sigma_2^2, \ldots$, either for individual parameters or functions of two or more parameters, e.g. $\sigma_1^2/(\sigma_1^2 + \sigma_2^2)$.

The inferential methods considered in this chapter are all based on the assumption that the random components are Normally and independently distributed.

19.2.3 Experimental Aims

The statistical methodology in this section has application to the following studies.

1. The estimation of values of variance components;
2. the estimation of values of functions of variance components, e.g. heritability and reliability;
3. testing hypotheses about variance components; and
4. the application of variance component analysis in the design of experiments.

19.2.4 Analysis of Variance and Expected Mean Squares

The methodology discussed in this section is based on the partitioning of variation in the set of responses into components using the technique of

Analysis of variance and assumes a balanced hierarchical design. (An hierarchical design arises when the process of selection proceeds in a sequential fashion—firstly select units, then select sub-units within units, then sub-sub-units within sub-units, and so on. The hierarchical design is *balanced* if the number of sub-units per unit is the same for all units, the number of sub-sub-units is the same for all sub-units, etc.)

Example 19.2 provides an illustration of a balanced, hierarchical design since the selection process involved (i) the selection of genotypes, (ii) the selection of plants within genotypes, and (iii) the selection of flowers within plants within genotypes. For each genotype there were three plants selected and from each plant, five flowers were selected. Searle [1971] provides a coverage of methods for unbalanced designs.

Consider the design defined by (19.2.7).

$$y_{ij} = M + \mathcal{U}_{1_i} + \{\mathcal{U}_{2_{i,j}}\} \qquad \text{for } i = 1, 2, \ldots, a; \; j = 1, 2, \ldots, b. \tag{19.2.7}$$

Application of Analysis of variance partitions the variation in the set $\{y_{ij}\}$ into the two sources of variation shown in the Analysis of variance table in Table 19.2.3. If (i) the components in the set $\{\mathcal{U}_{1_i}\}$ are independently and identically distributed with variance σ_1^2; (ii) the components in the set $\{\mathcal{U}_{2_{i,j}}\}$ are independently and identically distributed with variance σ_2^2; and (iii) the components in one set are independent of components in the other set, then statistical theory establishes that expected values for the mean squares for the *Between units* term and *Between sub-units within units* term can be expressed by the formulae in Table 19.2.3. Based on these formulae, the variance components σ_1^2 and σ_2^2 are defined in (19.2.8).

Table 19.2.3. Analysis of Variance Table and Expected Mean Squares for a Two-Strata Balanced Hierarchical Design

Source of Variation	D.F.	S.S.	Expected M.S.
Between units	$a-1$	$\sum_{i=1}^{a} b(\overline{y}_{i.} - \overline{y}_{..})^2$	$\sigma_2^2 + b\sigma_1^2$
Between sub-units in units	$b(a-1)$	$\sum_{i=1}^{a} \sum_{j=1}^{b} (y_{ij} - \overline{y}_{i.})^2$	σ_2^2

$$\begin{aligned} \sigma_1^2 &= (E[MS_1] - E[MS_2])/b \\ \sigma_2^2 &= E[MS_2] \end{aligned} \tag{19.2.8}$$

where $E[MS_1]$ is the expected mean square for the units stratum; and $E[MS_2]$ is the expected mean square for the sub-units stratum. For balanced, hierarchical designs, the extension to three or more strata is straightforward. Table 19.2.4 provides the expected mean squares for the three strata design employed in Example 19.3.

Where there is one or more treatment effects included in the component equation in a stratum, it is the residual mean square for that stratum which is employed in variance component analysis. Table 19.2.5 provides an illustration of an Analysis of variance table in which there is a treatment effect included in the first stratum.

Table 19.2.4. Analysis of Variance Table for Example 19.3 Displaying the Expected Values for Mean Squares

Source of Variation	Expected M.S.
Between genotypes	$\sigma_f^2 + f\sigma_p^2 + pf\sigma_g^2$
Between plants within genotypes	$\sigma_f^2 + f\sigma_p^2$
Between flowers within plants within genotypes	σ_f^2

Legend.
p is the number of plants per genotype ($p = 3$)
f is the number of flowers per plant ($f = 5$)

Table 19.2.5. Analysis of Variance Table and Expected Mean Squares for a Two-Strata Balanced Hierarchical Design with a Treatment Component in the First Stratum

Source of Variation	D.F.	S.S.	Exp. M.S.
Units stratum			
Between treatments	$t-1$	$\sum_{i=1}^{t} rs(\bar{y}_{i..} - \bar{y}_{...})^2$	
Residual	$t(r-1)$	$\sum_{i=1}^{t}\sum_{j=1}^{r} s(\bar{y}_{ij.} - \bar{y}_{i..})^2$	$\sigma_2^2 + s\sigma_1^2$
Sub-units stratum			
Between sub-units in units	$tr(s-1)$	$\sum_{i=1}^{t}\sum_{j=1}^{r}\sum_{k=1}^{s}(y_{ijk} - \bar{y}_{ij.})^2$	σ_2^2

Legend.
t is the number of treatments;
r is the number of replications of each treatment;
s is the number of sub-samples per unit.

For the purpose of analyzing variance components, the presence of treatments in the design for Example 19.4 has no effect on the relation between the variance components and the expected values of mean squares. A comparison of Tables 19.2.3 and 19.2.5 reveal the same structure for expected mean squares of the random components.

19.2.5 Estimation of Variance Components

Point estimation. Given a balanced, hierarchical design, the simple connection between variance components and expected mean squares is clearly established in Table 19.2.3 (two strata design) and Table 19.2.4 (three strata design). Equation (19.2.8) defines the variance components for a two-strata design in terms of expected values of mean squares. By replacing the expected mean squares by the computed mean squares in these equations, point estimates of the variance components are obtained and can be computed from (19.2.9).

$$\begin{aligned}\hat{\sigma}_2^2 &= MS_2 \\ \hat{\sigma}_1^2 &= [MS_1 - MS_2]/b\end{aligned} \tag{19.2.9}$$

where

$$MS_1 = \sum_{i=1}^{a} b(\overline{y}_{i.} - \overline{y}_{..})^2/(a-1)$$

$$MS_2 = \sum_{i=1}^{a}\sum_{j=1}^{b}(y_{ij} - \overline{y}_{i.})^2/a(b-1).$$

If a third stratum is added to the design, with the number of repetitions at the third stratum level being c, the point estimators of the variance components are given by (19.2.10).

$$\begin{aligned}\hat{\sigma}_3^2 &= MS_3 \\ \hat{\sigma}_2^2 &= [MS_2 - MS_3]/c \\ \hat{\sigma}_1^2 &= [MS_1 - MS_2]/bc\end{aligned} \tag{19.2.10}$$

where

$$MS_1 = \sum_{i=1}^{a} bc(\overline{y}_{i..} - \overline{y}_{...})^2/(a-1)$$

$$MS_2 = \sum_{i=1}^{a}\sum_{j=1}^{b} c(\overline{y}_{ij.} - \overline{y}_{i..})^2/a(b-1)$$

$$MS_3 = \sum_{i=1}^{a}\sum_{j=1}^{b}\sum_{k=1}^{c}(y_{ijk} - \overline{y}_{ij.})^2/ab(c-1).$$

Interval estimation. There are formulae for computing confidence limits for variance components and ratios for variance components. Searle [1971] provides formulae and references.

19.2.6 Testing Hypotheses About Variance Components

When a balanced, hierarchical design is employed in which there is a single random effect in each stratum, it is apparent from Tables 19.2.3 and 19.2.4

that expected mean squares in successive strata differ only in respect of a single variance component. In general, if the variance component for the effect defining the ith stratum is σ_i^2 and the mean square for the ith stratum in the Analysis of variance table is MS_i ($i = 1, 2, \ldots, p$), the statistic $F = MS_i/MS_{i+1}$ may be used to test the hypothesis $\sigma_i^2 = 0$. If the hypothesis is true, F has an $F(d_i, d_{i+1})$ distribution where d_i and d_{i+1} are the degrees of freedom corresponding to MS_i and MS_{i+1}, respectively. Given an observed value F_0 for the statistic F, $p_0 = \Pr\{F(d_i, d_{i+1}) \geq F_0\}$ can be used to judge the acceptability of the hypothesis $\sigma_i^2 = 0$.

Illustration. In Table 19.2.6, $F_0 = 28.58/2.63 = 10.9$ is the ratio of the Between genotypes mean square to the Between plants within genotypes mean square. Given the expected mean squares in Table 19.2.4, a test of the hypothesis $\sigma_g^2 = 0$ may be based on F_0. If the hypothesis is true, F_0 is a value from an $F(5, 12)$ distribution. Since $p_0 = \Pr\{F(5, 12) \geq 10.9\} < 0.01$, there is strong evidence to reject the hypothesis $\sigma_g^2 = 0$.

Table 19.2.6. Analysis of Variance Table for Example 19.3 Based on (19.2.3)

Source of Variation	D.F.	S.S.	M.S.
Between genotypes	5	142.90	28.58
Between plants within genotypes	12	31.60	2.63
Between flowers within plants	72	77.60	1.08

19.2.7 Variance Components and the Design of Experiments

Consider the case in which there are tn units (or subjects) and each unit is sampled r times. If the units are randomly allocated among t treatments such that each treatment is allocated n units, the variance of a treatment mean is defined by (19.2.11).

$$\text{var}(\overline{y}) = \frac{\sigma_u^2}{n} + \frac{\sigma_s^2}{nr} \qquad (19.2.11)$$

where σ_u^2 and σ_s^2 are variance components from the units and sampling strata, respectively. The percentage of variance which can be attributed to sampling is determined by (19.2.12). Evaluating P(r) for different values of r permits a judgment of the effect of varying the number of samples per unit on the precision of estimation of the sample mean.

$$\text{P}(r) = 100/(r\sigma_u^2/\sigma_s^2 + 1). \qquad (19.2.12)$$

If the total cost per treatment of conducting the experiment is C, then a function relating total cost to costs of providing units and costs of providing samples per unit can be obtained. A simple form which is widely applied is given by (19.2.13).

$$C = nC_u + nrC_s \qquad (19.2.13)$$

where C_u is the cost of providing a single unit; and C_s is the cost of providing a single sample per unit. Based on (19.2.13), the optimum number of units (n_0) and samples per unit (r_0) for a fixed total cost are found from (19.2.14).

$$r_0 = \sqrt{C_u\sigma_s^2/C_s\sigma_u^2}$$
$$n_0 = C/(C_u + r_0C_s). \qquad (19.2.14)$$

Application and illustration. The manner in which the above formulae may be used in the planning of experiments is considered by references to an experiment in which the units correspond to pots which contain plants and the samples represent portions of plant roots from each pot. The values of the variance components are, of course, unknown. However, estimates of the two components are available from an earlier experiment. (Clearly, it is not possible to apply this procedure in the absence of estimates of the variance components.)

Experimental information. In a previous study, the number of pots used was three (i.e. $n = 3$) and the number of samples taken per pot was nine (i.e. $r = 9$).

From application of Analysis of variance,

- the unit mean square is 4528.6
- the sample mean square is 235.1.

Hence, by (19.2.9),

- the estimated sample variance component is 235.1.
- the estimated unit variance component is $(4528.6 - 235.1)/9 = 477.1$.

Based on (19.2.11) in which sample estimates replace variance components, the estimated variance of a treatment mean is $\frac{477.1}{3} + \frac{235.1}{3\times 9} = 167.7$.

The contribution of sampling to overall precision. Based on (19.2.12), the percentage of variation in the estimated variance of the treatment mean which can be attributed to sample variability is given by the formula $\mathrm{P}(r) = 100/(2.029r + 1)$. This formula provides the contents of the following table.

r	1	2	3	4	5	6
P(r)	33	20	14	11	9	8

Optimal design. For the purposes of cost comparisons it is convenient to set C_s equal to one, i.e. to regard the cost of providing one sample per pot as the unit cost. Then C_u is the relative cost of providing one pot per treatment. Based on (19.2.14), the optimal number of samples per pot is estimated to be $r_0 = \sqrt{\{0.4928 C_u\}}$ where $0.4928 = 235.1/477.1$ is the ratio of the variance components.

The following table gives values of r_0 for selected values of C_u.

C_u	2	5	10	20	50
r_0	1.0	1.6	2.2	3.1	5.0

Thus, for example, if the cost of providing a pot is ten times the cost of providing one sample per pot, 2 samples per pot is optimal.

For purposes of illustration, suppose the cost of including a pot is ten times the cost of providing one sample per pot, i.e. $C_u = 10$. Then, the optimal number of samples per pot is computed to be 2.2.

For comparative purposes, suppose the total cost is equal to the cost of operating the previous experiment in which 3 pots were used and 9 samples taken from each pot. Then, with C_u set equal to 10 and C_s equal to one,

$$C = 3 \times 10 + 3 \times 9 \times 1 = 57.$$

Based on (19.2.14) and with $r = 2$, the optimal value of n is

$$n_0 = 57/(10 + 2 \times 1) \cong 5.$$

Precision of estimation of a treatment mean. In the previous experiment, 3 pots and 9 samples per pot were employed for each treatment. Based on (19.2.11), the estimated variance of a treatment mean is 167.7. With 5 pots per treatment and 2 samples per pot, the estimated variance of a treatment mean is 118.9. A significant reduction would be made by changing to this combination. If 4 pots and 3 samples per pot are employed, the estimated variance of a treatment mean is 138.9.

While the best choice, based strictly on cost considerations, is found to be 5 pots per treatment and 2 samples per pot, there are practical limitations to be considered. For example, in this experiment, it is necessary that the number of pots per treatment be a multiple of three because of constraints imposed by the experimental design. In experiments where the optimal choice is one sample per pot, this combination might be regarded as undesirable since the loss of a result from one sample means all information from the pot is lost.

The formulae given in this section allow other combinations to be contrasted and the best choice obtained within the totality of constraints which are imposed.

Problems

19.1 *Coefficient of variation*

The data presented below was obtained from twelve successive experiments in a laboratory. The ten readings in each column were taken from samples chosen from a single unit. As is indicated, the readings were made over a two week period.

Week	1	1	1	1	1	1
Day	M	T	W	T	F	S
	131	327	102	35	90	240
	120	341	66	41	108	322
	127	321	66	59	149	227
	136	277	84	33	108	212
	94	315	84	36	153	222
	102	248	78	58	106	178
	141	320	54	32	137	184
	122	329	96	42	131	250
	154	340	73	51	140	177
	131	263	78	46	107	211

Week	2	2	2	2	2	2
Day	M	T	W	T	F	S
	135	360	128	106	202	37
	143	402	127	185	247	31
	56	19	161	4	212	27
	80	187	118	112	281	25
	65	270	173	49	266	24
	124	198	141	34	146	12
	140	178	65	205	251	33
	168	216	70	141	225	31
	70	247	151	89	63	33
	161	174	144	26	50	36

(a) For each day, compute the mean and standard deviation and hence, the coefficient of variation as defined by (19.1.5).

(b) Plot the coefficients of variation versus the numbers $1, 2, \ldots, 12$ to reflect the sequence of days on which the sets of measurements were taken. What does the graph suggest?

19.2 *Example* 12.2 — *comparing variances*

Apply the steps in Section 19.1.3 to the data in Example 12.2 to determine if the assumption of equality of variances is reasonable for the supplement and control groups.

19.3 *Example* 19.2 — *estimation of variance components*

(a) By noting the patterns in (19.2.9) and (19.2.10), produce formulae for estimating variance components in a balanced hierarchical design in which there are four strata.

(b) Based on the model equation given in (19.2.1), produce an Analysis of variance table using the data in Table 19.2.1 which were collected in the study reported in Example 19.2.

(c) Use the formulae obtained in (a) to estimate variance components. Express as percentages, the contributions to the total variance made by the components.

19.4 *Example* 19.4 — *design of experiments*

(a) Based on information in Table 15.5.2 and using (19.2.9), obtain estimates of the between children and between times within child variance components.

(b) Using (19.2.11) in which sample estimates replace variance components, compute the estimated variance of a treatment mean assuming n and r take the values used in the experiment reported in Example 15.6.

(c) For $r = 1, 2, \ldots, 5$, compute values of $\mathrm{P}(r)$ using (19.2.12).

(d) For $C_u = 2, 3, 4, 5$ and 10, compute the optimal values of r by employing (19.2.14).

(e) Assuming (i) the cost of including an extra child in the experiment is ten times the cost of making an extra observation per child and (ii) the total cost of a subsequent experiment cannot exceed the cost of the experiment reported in Example 15.6, determine the optimal number of children to be included in the experiment from (19.2.14).

(f) Compute the estimated variances of a treatment mean based on possible combinations for values of n and r. What is your recommended combination for n and r if the experiment is to be repeated?

20

Cause and Effect: Statistical Perspectives

20.1 The Allocation of Causality: Scientific Aims and Statistical Approaches

20.1.1 Scientific Aims

Evidence of relationships between variables commonly leads to an interest in the reasons for and the nature of the relationships. In some cases, scientists can provide a graphical picture which displays a cause-effect structure for the variables and the aim is to establish the relative importance of the different connections. Figure 20.1.1 provides an illustration. Of primary interest to the researchers is the degree of disability suffered by a person afflicted with rheumatoid arthritis and the importance of the identified factors which are believed or known to influence the degree of disability.

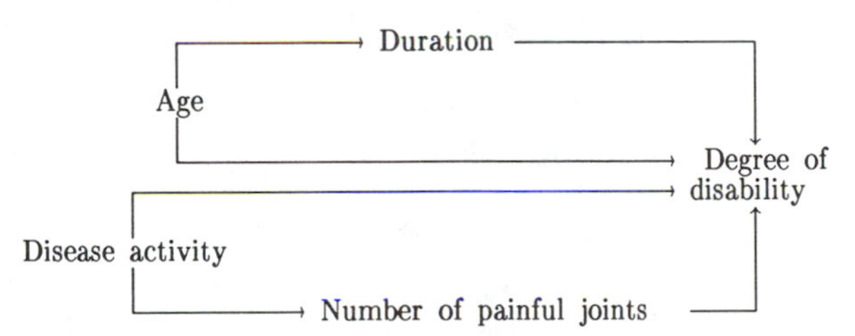

Figure 20.1.1. A postulated cause-effect structure for degree of disability suffered by rheumatoid arthritis sufferers.

From a scientific viewpoint, given the cause-effect structure portrayed in Figure 20.1.1, it is reasonable to ask questions such as

> *"Is age an important contributory factor?"*
>
> *"Is age an important factor in its own right, or does it merely reflect the fact that older people are more likely to have had the disease for a longer period?"*

Answers to these questions require a statistical approach which can partition causality among the different identified factors.

20.1.2 Statistical Approaches

Regression analysis. In Chapter 18, component equations are introduced to partition a response, y, into two components—$M_{y.x}$, which represents the contribution from an explanatory variable x, and e, which is the contribution from all other factors or variables. The response is represented as the sum of these components, i.e. $y = M_{y.x} + e$. Consequently, the variation in y can be partitioned according to (20.1.1).

$$\begin{array}{c} \text{Total variation} \\ \text{in} \\ \text{y} \end{array} = \begin{array}{c} \text{Variation in } y \\ \text{attributable} \\ \text{to variation in } x \end{array} + \begin{array}{c} \text{Unexplained} \\ \text{variation} \end{array} \qquad (20.1.1)$$

If x and e are presumed to represent the factors which cause variation in y, the situation can be expressed graphically in the following manner

$$x \longrightarrow y \longleftarrow e.$$

The arrows symbolize the cause-effect relation.

If there are two explanatory variables, x_1 and x_2, and the component equation is $y = B_0 + B_1x_1 + B_2x_2 + e$, it may appear reasonable to partition the variation in y into three components which reflect contributions from x_1, x_2 and e and to graphically represent the relation in the following manner.

x_1, x_2 ⟶ y ⟵ e

On the left hand side of the graph are the identified variables with contributions from x_1 and x_2 distinguished, while on the right hand side is the contribution from factors which are distinct from the effects of x_1 and x_2.

It has been established in Section 18.6.5 that, in general, the contributions from x_1 and x_2 are not separable. There is generally some duplication of information about y in the x-variables. This suggests a need to modify the model which describes the relationship. In graphical terms, this relationship is most commonly represented by an arrow between x_1 and x_2. The arrow is double-headed if there is no identified cause-effect relation between the two explanatory variables. Figure 20.1.2 is the common representation.

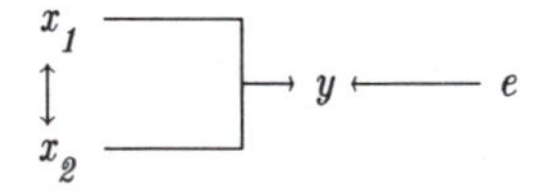

Figure 20.1.2. A cause-effect structure based on correlated explanatory variables.

Clearly, the equation $y = B_0 + B_1x_1 + B_2x_2 + e$ fails to provide a complete picture of the interrelations. Additional information can be provided by the following table of correlations.

$$\begin{array}{c} y \\ x_1 \\ x_2 \\ e \end{array} \begin{bmatrix} 1 & & & \\ \rho_{y1} & 1 & & \\ \rho_{y2} & \rho_{12} & 1 & \\ \rho_{ye} & 0 & 0 & 1 \end{bmatrix}$$
$$\quad\quad y \quad\quad x_1 \quad\quad x_2 \quad\quad e$$

In this table, ρ_{y1} is the correlation between y and x_1, etc. Note that the two zeros in the bottom row reflect the absence of correlation between x_1 and e and between x_2 and e.

Even with this additional information, it is not apparent how a separation of the effects of x_1 and x_2 might be achieved, or if such a separation is possible. Two methods have been proposed which some proponents have claimed can meet these objectives—*Path analysis* and *Analysis of commonality.* The models and methodological bases of these techniques are considered in Section 20.2. For the present, the possibility of defining cause-effect models which might underlie these techniques is explored.

Analysis of commonality. To allow for duplication of information in x_1 and x_2, there is a need to recognize three distinct sources which explain the joint effect of x_1 and x_2. These are

- a unique contribution from x_1 (the part which is not duplicated in x_2);
- a unique contribution from x_2 (the part which is not duplicated in x_1);
- a common contribution which arises from sources which are common to x_1 and x_2.

Graphically, this construction has the representation portrayed in Figure 20.1.3.

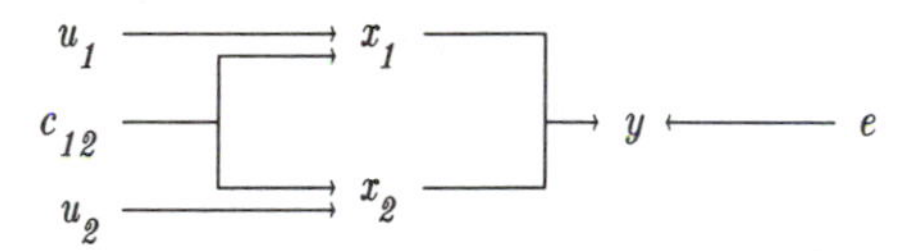

Figure 20.1.3. Analysis of commonality—the presumed cause-effect structure when there are two explanatory variables.

The variables u_1 and u_2 represent contributions from factors which operate solely through x_1 and x_2, respectively, while c_{12} is the contribution from factors which operate through both x_1 and x_2. Hence, the duplication of information about y which is contained in x_1 and x_2 arises because the contribution from c_{12} is included in both. From a scientific viewpoint, the aim is to partition the variation in y into independent contributions from u_1, u_2, c_{12} and e. This is the purported function of the technique of Analysis of commonality.

Path analysis. There is a widely used technique for operating on cause-effect structures known as *Path analysis.* It has application beyond a single response variable. However, to maintain a comparability with Analysis of commonality, it is discussed in this section as it would apply to the problem considered above, namely the separation of the causality between two explanatory variables.

In Path analysis, the contribution from x_1 and x_2 is split into four components or *paths* which are

- *direct paths* from x_1 to y and x_2 to y (in the spirit of the unique contributions in Analysis of commonality); and
- *indirect paths* from x_1 through x_2 to y and x_2 through x_1 to y.

The four paths which are defined are

$$x_1 \rightarrow y \quad x_2 \rightarrow y \quad x_1 \rightarrow x_2 \rightarrow y \quad x_2 \rightarrow x_1 \rightarrow y.$$

As with Analysis of commonality, the scientific aim is to partition the combined effects of x_1 and x_2 into mutually exclusive and meaningful components.

20.2 Statistical Methods in Use

The informal introductions to the methods of Analysis of commonality and Path analysis presented in Section 20.1 suggest the use of statistical analysis in helping scientists establish the relative importance of different sources of causality. The fact is that there is no reasonable basis for partitioning causality between correlated explanatory variables which has a meaningful scientific interpretation. In both Analysis of commonality and Path analysis, mathematical identities (which are correct) have been assigned interpretations which are not soundly based. Plausible interpretations have been given a status that cannot be substantiated. The widespread use of Path analysis and Analysis of commonality in studies which attempt to apportion causality among different sources or to different paths is a consequence of a natural scientific desire to discover why there is a relationship among variables. Unfortunately, this desire has overriden sound model-based statistical development.

20.2.1 Regression Analysis and Analysis of Commonality

Where there is a single response variable, there is scope for employing a regression model which includes an equation expressing the response in terms of one or more explanatory variables and an unexplained component. Unless there is reason for believing otherwise, a linear regression equation is usually assumed. If there are two explanatory variables, x_1 and x_2, the equation can be expressed as $y = B_0 + B_1x_1 + B_2x_2 + e$. As explained in Section 20.1.2, this is an inadequate representation if the aim is to partition causality between the two explanatory variables. The Analysis of commonality approach, introduced in Section 20.1.2, seeks to overcome this deficiency by replacing the combined contribution from x_1 to x_2 by independent contributions which are termed the *unique* contributions from the variables plus a *common* contribution. The connection can be expressed in (20.2.1).

$$B_1x_1 + B_2x_2 = \beta_1u_1 + \beta_2u_2 + \beta_3c \tag{20.2.1}$$

where u_1 and u_2 are the unique contributions from x_1 and x_2; and c is the common contribution.

The regression sum of squares from fitting x_1 and x_2 can be partitioned into three components which correspond to the three contributions on the right hand side of (20.2.1), i.e. that (20.2.2) holds. By application of the R^2 statistic defined by (18.5.1), the claim is then made that the relative importance of each contributory factor can be measured.

$$RegSS(x_1, x_2) = RegSS(u_1) + RegSS(u_2) + RegSS(c). \tag{20.2.2}$$

Since u_1, u_2 and c are latent variables, i.e. are not measurable, it is not possible to directly regress these variables on y to determine the regression sums of squares on the right hand side of (20.2.2). However, the presumption made in Analysis of commonality is that the regression sums of squares for the unique and common contributory variables can be computed using the following formulae:

$$Reg(u_1) = Reg(x_1, x_2) - Reg(x_2)$$

i.e. represents the gain from fitting x_1 after x_2 has been fitted;

$$Reg(u_2) = Reg(x_1, x_2) - Reg(x_1)$$

i.e. represents the gain from fitting x_2 after x_1 has been fitted;

$$Reg(c) = Reg(x_1) + Reg(x_2) - Reg(x_1, x_2).$$

This *ad hoc* definition of contributions from unique and common components was suggested by Mood [1971]. It has not, and cannot have, a basis in theory since there is no unique partitioning of the regression sum of squares $Reg(x_1, x_2)$ without additional restrictions on the model. Furthermore, there have been no additional restrictions on the model which have been shown to lead to the above definitions of contributions from unique and common effects. The fact is that Analysis of commonality is not a method of partitioning causality which offers meaningful scientific interpretation. Its use is not recommended.

20.2.2 Path Analysis

The model and methodology of Path Analysis have been extensively studied and there is a sound mathematical development of the technique offered by Kang and Seneta [1980]. The model underlying Path analysis is based on the assumption that all variables are random variables and that a path diagram of the form illustrated in Figure 20.1.1 is provided which defines the cause-effect structure of the variables. The variables are two types—the *exogenous* variables which have no identified causes (i.e. have no single-headed arrows directed at them) and the *endogenous* variables which are presumed to be controlled by other variables. Each endogenous variable is presumed to be representable as a linear function of causal variables plus an unexplained component. Consider the cause-effect structure represented by Figure 20.1.2. The variable y has arrows directed at it from x_1, x_2 and e. Hence,

$$y = B_0 + B_1x_1 + B_2x_2 + B_3e.$$

If the variance of y is denoted by σ_y^2, the x-variables are scaled to have unit variance and the correlation between y and the identified causal component $B_0 + B_1x_1 + B_2x_2$ is represented as $\rho_{y,12}$, there is a fundamental identity, (20.2.3), which partitions variability in y between the identified and unidentified causal factors.

$$\sigma_y^2 = \sigma_y^2\rho_{y.12}^2 + \sigma_y^2(1 - \rho_{y.12}^2). \tag{20.2.3}$$

On the (reasonable) assumption that $B_0 + B_1x_1 + B_2x_2$ and e are independently distributed, the quantity $\rho_{y.12}^2$ is the proportion of variation in y which can be associated with the identified sources. At this stage there is not a decomposition of the contribution from x_1 and x_2. That is introduced by (20.2.4).

$$\rho_{y.12}^2 = \rho_{y1.2}^2 + \rho_{y2.1}^2 + 2\rho_{y1.2}\rho_{y2.1}\rho_{12} \tag{20.2.4}$$

where $\rho_{y1.2}$ and $\rho_{y2.1}$ are partial correlation coefficients (which are termed *path coefficients*); and ρ_{12} is the (simple) correlation between x_1 and x_2. Wright [1923, p. 245]—the originator of Path analysis—makes the unsubstantiated statement

> *Because of this property* (equations (20.2.3) and (20.2.4)) *the squares of the path coefficients give a useful measure of the degree of determination. Each one measures the portion of the squared standard deviation for which the factor in question is responsible.*

Despite reasoned arguments about this interpretation by Tukey [1954] and Duncan [1975] and a general understanding among statisticians that it is not justified, there remains a widespread belief in the correctness of Wright's interpretation.

Path analysis allows the partitioning of causality to be carried further. Note that x_1 has both a direct causal path to y and an indirect path via the double-headed arrow to x_2. Wright suggested a means of partitioning the correlation between y and x_1 into components which reflected these two paths. The proposed relation is represented by (20.2.5).

$$\rho_{y1} = P_{1y} + \rho_{12}P_{2y} \tag{20.2.5}$$

where ρ_{y1} and ρ_{12} are correlation coefficients; and P_{1y} and P_{2y} are titled *path coefficients* by Wright. The relative magnitudes of the components on the right-hand side are claimed to reflect the relative levels of importance of the *direct* and *indirect* paths by which x_1 is said to affect y.

Equation (20.2.5) is a special case of a general identity which allows for any number of indirect paths between the two variables, x_1 and y. This claim of interpretability is extended to the components in this general identity.

In the application of Path analysis, the task is to express all possible correlations in terms of direct and indirect paths, thereby producing a system of linear equations in which the correlation coefficients are replaced by sample estimates and the path coefficients become the unknowns for which values are assigned by solving the system of linear equations. Kang and Seneta [1980] provide a method for ensuring unique solutions exist for the path coefficients. However, neither they nor any other author has established a justification for Wright's interpretation of the numerical values which result.

21

Studying Changes in Response Over Time

21.1 Applications

The study of changes in a response variable as a function of time or a time-related variable has many facets. In some instances, the models and methodology of previous chapters may be applied with modification. In a significant number of applications, special models and methodology are required to cater for the lack of independence of responses which arises because of repeated sampling of the one unit or subject. The purpose of this section is to provide guidance in the identification of the correct statistical model and methodology. The remaining sections introduce a special area of Statistics which is known as *Time series analysis.*

Time as a factor in designed experiments and observational studies. Organisms may be classified into time-related categories, e.g. *young, middle-aged* and *old;* tests or attempts may be classified as *first, second, third,* etc.; stages in development may be graded as *one, two, three* etc. What is common in these examples is the use of an ordered categorical variable for which the categories follow a time sequence. Provided the same units or subjects are not repeatedly sampled to obtain the responses in the different categories, the methodology of Chapters 15 and 16 may be applied to compare changes in average response or response patterns between categories. If there is an underlying scaled variable, the Logistic model (Section 6.6) may be used in conjunction with Generalized linear regression models.

If the responses are repeatedly obtained from the same units or subjects, e.g. if the response at first attempt, second attempt, etc. are recorded for the same subject, it is likely that the successive responses will not be independent and hence, the statistical models underlying the methods in Chapters 15 and 16 are not directly applicable. Two approaches are available in this case. They are

1. model modification in which a correlation structure is imposed on the set of responses; and

2. data modification in which the set of responses are transformed into a set of values of independent statistics, thereby satisfying the independence conditions in the models of earlier chapters.

An easily applied form of data modification for scaled response variables is the conversion of pairs of scaled responses to differences, e.g. form the difference between responses at the first and second attempts. The method may be extended to more general weighted contrasts, e.g. does the average for the first and second attempts differ from the average response at the third attempt? Any approach which reduces the set of correlated responses to a single number eliminates the problem of correlated responses and makes accessible the methodology of earlier chapters.

If the response is categorical and there are repeated measurements on the same unit or subject, the summarization of the data in a frequency table and the associated analysis must reflect the fact. For example, if the response is success (S) or failure (F) and 60 observations are made at each of two times, the frequency tables have the following forms dependent on whether the responses are collected from the same or different subjects.

Independent Responses

	Time		
Response	1	2	Total
S	30	40	70
F	30	20	50
	60	60	120

Related Responses

	Time 1		
Time 2	S	F	Total
S	25	15	40
F	5	15	20
	30	30	60

If the responses are independent and there are 60 observations at each time, data are collected from 120 subjects, whereas if each subject is sampled at two times there is only a total of 60 subjects. When the same subjects are sampled at both times, it is necessary to identify the pair of responses for each subject. If measurements were made at a third time, in the independent case, the frequency table would remain a two-way table with the addition of a further column. In the case of repeated sampling, a third *variable* would be added, thereby giving a three-way table. The methodology appropriate for independent responses can be found in Chapters 12 and 16. For repeated sampling, Fleiss ([1981], Chap. 8) provides methodology for two-category response variables. For the multi-category response, general methodology based on Log-linear models is required.

Note that in the above situation, the *experimental* aim may be the same in both cases, namely to establish if the success rate changes with successive attempts. Yet the statistical approach is not determined solely by the experimental aim. This is a valuable illustration of the need for scientists to understand the basis of methodology which they may have to employ and to appreciate that it is the complete statistical model which determines the appropriate methodology, not merely the statistical hypothesis which is being examined.

Time as an explanatory variable in designed experiments. There arise many experimental situations in which the levels of a factor which is time-related are scaled values. For example, temperature may be recorded at thirty minute intervals after an operation, yield of a fruit tree might be recorded on a yearly basis, measurements may be made on children at different ages. Commonly, the aim is to fit a regression relation with time as the explanatory variable and to compare the relation under different treatments or between different groups. If the responses at different times are made on different units or subjects, the approach described in Section 18.7.1 is applicable. Where each unit or subject is repeatedly sampled, the lack of independence of responses necessitates a different statistical model and alternative methodology. One approach which is simply applied when a linear regression equation is employed is to fit the regression equation separately to each unit or subject and base the analysis on the set of slopes of the regression equations. In this way, the data has been reduced to a single value for each unit (see Snedecor and Cochran [1980] for an illustration). Other approaches are possible. The choice depends on the nature of the study, the amount of information available and the form of the regression relation. Seek the advice of a consultant statistician in this case.

Time as an explanatory variable. Over many scientific disciplines, there has been and remains an interest in exploring the way in which responses change with time, or in using established relations to make future predictions. This is particularly so in the study of economic indicators, in the study of physical phenomena such as temperature, sun spot activity and in production processes where quality control is constantly applied. Interest has grown rapidly in biological areas where pollution measurements, animal population numbers, etc. are increasingly being monitored, and in medical areas both at an individual and a community level. Town planners and traffic engineers monitor development using series of measurements collected over time.

A common characteristic of many of these applications is the collection of responses over time from the same set of subjects or units. The consequence is a need for statistical models which include the possibility of correlations between responses. Furthermore, the deterministic component of the response is commonly more complex than the usual equations applied in regression analysis (Chapter 18). In addition to a trend line, the graph of the data may reflect the presence of one or more cyclical components, e.g. seasonal effects, which must be separated from the trend line. A consequence of the specialized nature of the deterministic component and the correlations between random (stochastic) components in the model has resulted in a specialized class of models and associated methodology which are loosely collected under the title of *Time series analysis,* the elements of which are introduced in Section 21.2.1. The applications of these models and methods are in two broad areas:

— the *time-related* applications in which the basic aim is to model the manner in which the response changes with time, commonly with the intention of predicting future response; and

— the *frequency-related* applications in which the basic aim is to explore the mechanism which produces the changes in response over time.

The distinction is of some importance to scientists since the methodology for the two applications is substantially different. Nelson [1973] provides a good introduction to the time-related approach while Jenkins and Watts [1969] provides an introduction to the frequency-related approach for scientists.

21.2 Time Series

21.2.1 DEFINITIONS

A *time series* is a data set which records values of a response variable and corresponding values of time or a time-related variable over an interval of time. If the data comprise values of a response variable y recorded at discrete points in time, e.g. (y_{10}, t_1), $(y_{20}, t_2), \ldots, (y_{n0}, t_n)$, the series is said to be a *discrete time series.* Most commonly, the times at which recordings are made are equally spaced. Figure 21.2.1 is a graphical representation of a discrete time series with equally spaced points in time.

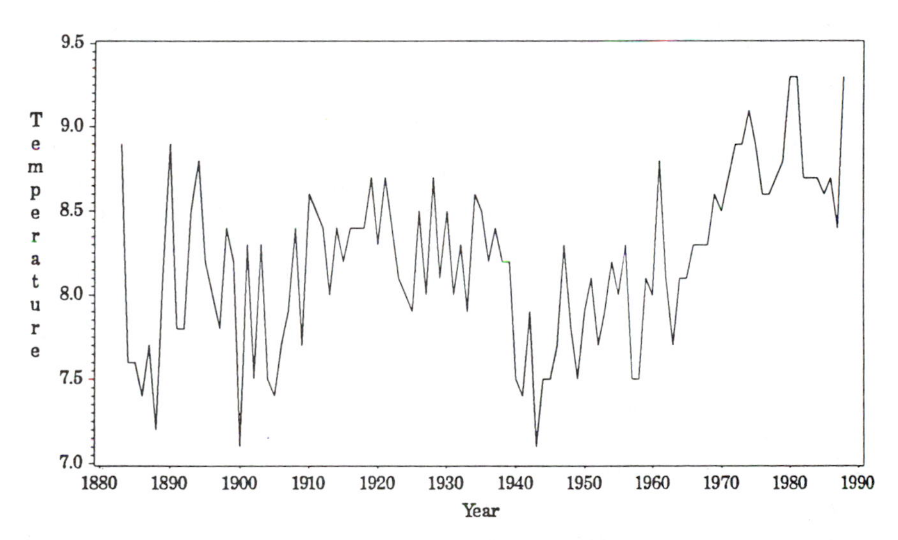

Figure 21.2.1. A graph of a discrete time series based on data in Table 21.2.1.

Alternatively, a *continuous time series* may be defined in which a value of y is defined for every value of t in a specified interval of time. In practice, continuous time series arise when (i) a recording device monitors values of variable over the time interval and records the results as a continuous graph; or (ii) the response variable is discrete and the times at which changes take place are noted. This produces the graphical representation shown in Figure 21.2.2. Note that the words *continuous* and *discrete* as applied to times series carry no implication for the type of response variable involved. Both discrete and continuous time series may arise from either discrete or continuous response variables. Continuous time series have a limited and specialized application and are not considered further in this book.

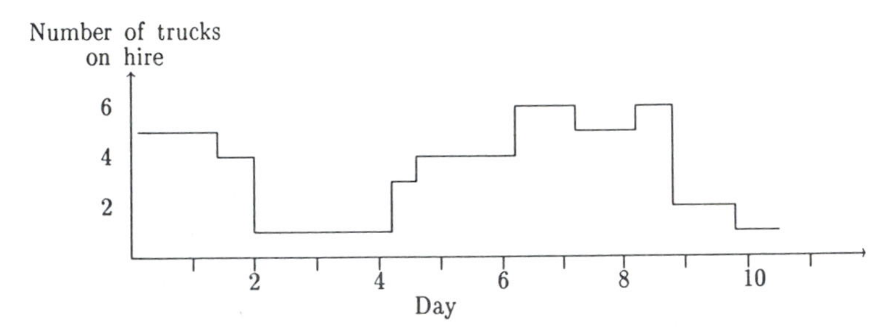

Figure 21.2.2. A graph of a continuous time series displaying the number of trucks on hire from a vehicle hire co. over a 10 day period.

21.2.2 Examples

Three examples of discrete time series are provided below.

Example 21.1 (Temperature and the Greenhouse Effect). *There is considerable interest in the theory that temperatures around the world are increasing as a result of damage to the ozone layer which surrounds the earth coupled with the increasing carbon dioxide levels reported in the atmosphere. Data in Table* 21.2.1 *comprise annual means for daily minimum temperatures for the* 106 *years from* 1883 *to* 1988 *at the Hobart regional office of the Australian Bureau of Meterology. A plot of the data is provided in Figure* 21.2.1.

The analysis of physical data of which these are an example, may be examined for several reasons. One purpose may be to seek to understand the underlying driving force. In particular, is there a trend in the data—if so what is the form of the trend? Are there cyclical effects present? What is the nature and extent of correlations between successive years? As noted above, the particular reason for collecting these data is to seek evidence in support of or against the 'greenhouse' effect. Alternatively, or additionally, there may be interest in predicting temperatures over the next decade or in fifty years time.

The analysis may be extended to compare the temperature pattern with patterns in rainfall and pressure, or to compare temperature fluctuations in Hobart with those in other parts of the world over the same time period.

Table 21.2.1. Mean Annual Minimum Temperature for Hobart Regional Office (°C) for the Period 1883 to 1988 (*Source:* Australian Bureau of Meteorology records)

Yr	Temp	Yr	Temp	Yr	Temp	Yr	Temp	Yr	Temp	Yr	Temp
		1900	7.1	1920	8.3	1940	7.5	1960	8.0	1980	9.3
		1901	8.3	1921	8.7	1941	7.4	1961	8.8	1981	9.3
		1902	7.5	1922	8.4	1942	7.9	1962	8.1	1982	8.7
1883	8.9	1903	8.3	1923	8.1	1943	7.1	1963	7.7	1983	8.7
1884	7.6	1904	7.5	1924	8.0	1944	7.5	1964	8.1	1984	8.7
1885	7.6	1905	7.4	1925	7.9	1945	7.5	1965	8.1	1985	8.6
1886	7.4	1906	7.7	1926	8.5	1946	7.7	1966	8.3	1986	8.7
1887	7.7	1907	7.9	1927	8.0	1947	8.3	1967	8.3	1987	8.4
1888	7.2	1908	8.4	1928	8.7	1948	7.8	1968	8.3	1988	9.3
1889	8.1	1909	7.7	1929	8.1	1949	7.5	1969	8.6		
1890	8.9	1910	8.6	1930	8.5	1950	7.9	1970	8.5		
1891	7.8	1911	8.5	1931	8.0	1951	8.1	1971	8.7		
1892	7.8	1912	8.4	1932	8.3	1952	7.7	1972	8.9		
1893	8.5	1913	8.0	1933	7.9	1953	7.9	1973	8.9		
1894	8.8	1914	8.4	1934	8.6	1954	8.2	1974	9.1		
1895	8.2	1915	8.2	1935	8.5	1955	8.0	1975	8.9		
1896	8.0	1916	8.4	1936	8.2	1956	8.3	1976	8.6		
1897	7.8	1917	8.4	1937	8.4	1957	7.5	1977	8.6		
1898	8.4	1918	8.4	1938	8.2	1958	7.5	1978	8.7		
1899	8.2	1919	8.7	1939	8.2	1959	8.1	1979	8.8		

Example 21.2 (Estimated Population in Tasmania: 1900–1971). *Data provided in Table* 21.2.2 *record the estimated population in the state of Tasmania between the years* 1900 *to* 1971. *The data were obtained for the purpose of making forecasts of future population growth in the state. A plot of the time series is given in Figure* 21.2.3.

Example 21.3 (Monthly Figures for Production of Television Sets in Australia: 1962–1976). *Table* 21.2.3 *contains data from a series which has a strong seasonal component as is revealed in Figure* 21.2.4. *A common statistical task in respect of such data is to adjust the data for seasonal variations, i.e. to produce "seasonally adjusted" figures in order to allow comparisons to be made between results for different seasons in a year.*

Table 21.2.2. Estimated Population for the State of Tasmania: 1900–1971
(*Source:* Australian Bureau of Statistics records)

Year	Pop.*	Year	Pop.*	Year	Pop.*	Year	Pop.*
1900	172,631	1920	210,350	1940	241,134	1960	346,913
1901	172,525	1921	213,404	1941	240,389	1961	353,623
1902	175,173	1922	215,379	1942	241,087	1962	355,682
1903	180,375	1923	216,420	1943	242,869	1963	360,590
1904	183,007	1924	216,274	1944	245,618	1964	364,554
1905	184,478	1925	215,552	1945	248,596	1965	367,970
1906	184,272	1926	213,800	1946	252,192	1966	371,483
1907	184,791	1927	213,051	1947	257,636	1967	375,397
1908	187,485	1928	215,471	1948	263,445	1968	379,916
1909	190,227	1929	217,752	1949	270,327	1969	385,079
1910	191,005	1930	220,933	1950	278,785	1970	388,180
1911	190,120	1931	224,811	1951	288,294	1971	391,242
1912	190,796	1932	227,084	1952	298,361		
1913	194,361	1933	228,434	1953	306,318		
1914	196,041	1934	229,161	1954	311,055		
1915	196,230	1935	229,616	1955	315,565		
1916	194,265	1936	231,046	1956	321,039		
1917	194,210	1937	233,951	1957	328,435		
1918	198,193	1938	235,678	1958	335,382		
1919	204,959	1939	238,000	1959	341,423		

*Population figures are obtained from census data every five years and are estimated in intervening years.

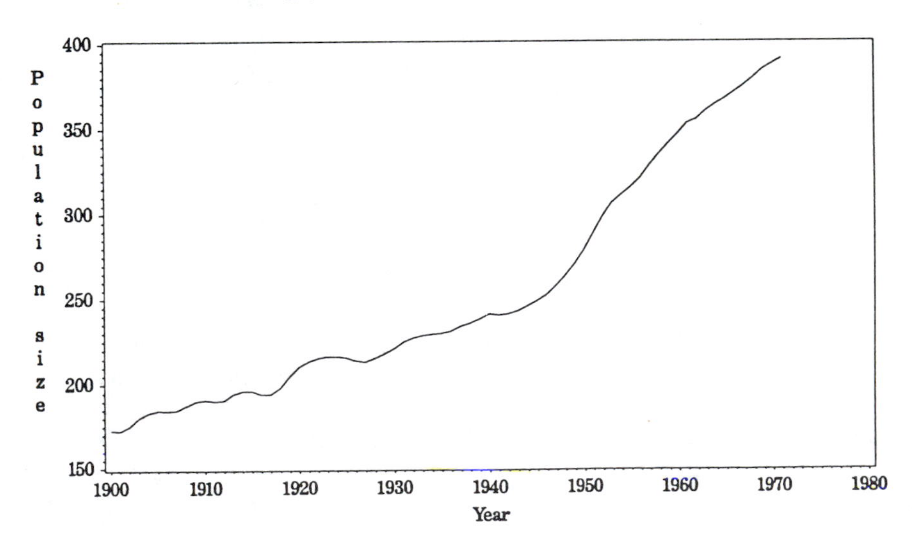

Figure 21.2.3. Change in population size in Tasmania: 1900–1971 (Data from Example 21.2).

Table 21.2.3. Television Set Production in Australia: 1962–1976

Number of sets produced by month

Year	J	F	M	A	M	J
1962	12972	21150	27588	27144	41860	39057
1963	15714	20653	24652	22361	29148	22924
1964	9153	23043	23170	27410	30725	27925
1965	10088	19917	25132	27094	29454	29060
1966	7453	19384	22790	22986	29739	27614
1967	7795	17453	20117	19656	26587	16066
1968	9164	21273	22696	21288	28006	22294
1969	11455	20024	21974	21273	29422	26790
1970	11490	26082	26923	29962	29843	27563
1971	12581	27451	32995	21756	34879	32941
1972	11298	23024	29471	23152	30357	31499
1973	14214	31376	38289	29056	43691	31765
1974	17161	39083	40034	34995	47623	39385
1975	16847	36978	37939	41542	48459	43763
1976	18069	43425	46687	42530	34508	43730

Year	J	A	S	O	N	D
1962	37510	31121	21237	23577	25521	18145
1963	25260	25135	24275	23640	20159	13893
1964	34639	30414	29999	28896	24797	21609
1965	29886	27237	28294	23783	19850	18275
1966	26057	25106	23934	20100	21854	16583
1967	23342	24390	23040	20690	21905	16184
1968	27862	26428	26189	27176	25658	16497
1969	30189	28439	30449	32137	26742	21921
1970	30337	25700	27449	28273	29297	24469
1971	36548	35527	35398	34328	35314	27700
1972	30082	33939	32951	33146	36357	29897
1973	42357	42903	36376	41142	43517	32355
1974	39356	38603	38549	44370	42223	29434
1975	54417	51495	52790	58321	49806	36828
1976	44652	47158	53235	50226	54884	42734

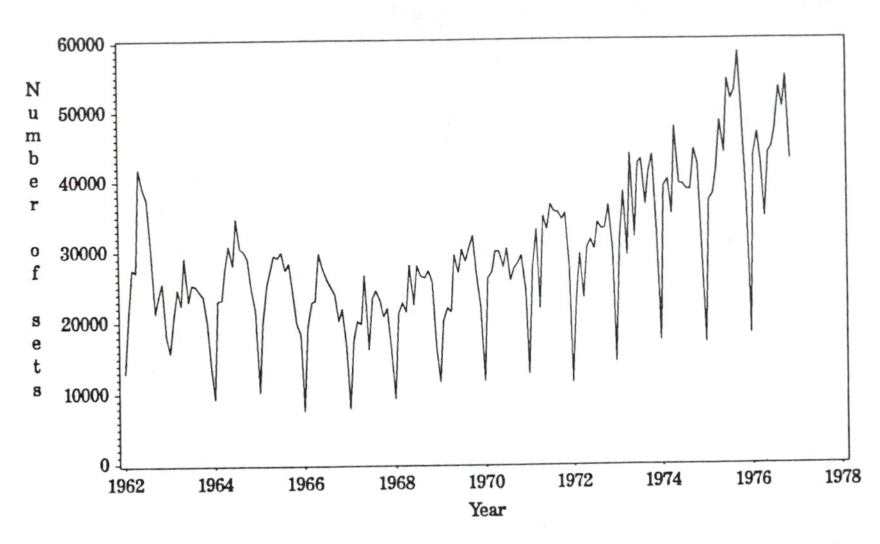

Figure 21.2.4. A plot of monthly television production figures based on data in Table 21.2.3.

21.2.3 The Anatomy of Time Series

Time series are composed from the following components:

- one or more *seasonal effects* which are effects that repeat in units of days, weeks, months or years;
- other *cyclical effects;*
- a *trend* which represents a non-periodic change in mean response over the time interval in which the time series applies; and
- a *random component* which is the combined effect of factors which are not identified in the description of the times series model.

The trend, seasonal and cyclical effects are *deterministic components* of the model and the random component is termed the *stochastic component* because of the unpredictability of its values.

Attempting to unravel the individual components by eye can be difficult. There is widespread application of statistical methods to determine possible structures for time series.

21.2.4 Autocorrelation

For most time series data there is correlation between the random components. Generally, the correlation is greatest between neighboring components. It is possible to measure the correlation between successive pairs of responses, or more generally, between pairs of responses which are k times units apart. Since the correlations are computed from sets of values on the same response variable, they are termed *autocorrelations.* If $y_1, y_2, \ldots, y_n$ are n statistics which provide the responses $y_{10}, y_{20}, \ldots, y_{n0}$ in a time series of length n for which y_i is recorded at time t_i $(i = 1, 2, \ldots, n)$, the (*sample*) *autocorrelation at lag* k is defined by (21.2.1). The formula derives from that for the product-moment correlation between two variables (17.2.1).

$$r_k = \frac{\sum_{i=1}^{n-k}(y_i - \overline{y}_{(1)})(y_{i+k} - \overline{y}_{(2)})}{\sqrt{[\sum_{i=1}^{n-k}(y_i - \overline{y}_{(1)})^2][\sum_{i=1}^{n-k}(y_{i+k} - \overline{y}_{(2)})^2]}} \quad \text{for } k = 1, 2, \ldots \tag{21.2.1}$$

where

$$\overline{y}_{(1)} = \sum_{i=1}^{n-k} y_i/(n-k) \quad \text{and} \quad \overline{y}_{(2)} = \sum_{i=k+1}^{n} y_i/(n-k).$$

In practice, autocorrelations are only of value when k is very much smaller than n, in which case (21.2.1) is well approximated by (21.2.2).

$$r_k \cong \sum_{i=1}^{n-k}(y_i - \overline{y})(y_{i+k} - \overline{y}) / \sum_{i=1}^{n}(y_i - \overline{y})^2 \quad \text{for } k = 1, 2, \ldots . \tag{21.2.2}$$

The autocorrelation coefficient of lag 1 is commonly referred to as the *serial correlation coefficient.*

The set of autocorrelation coefficients for $k = 1, 2, \ldots$ defines the (sample) *autocorrelation function* and the magnitudes and patterns of the values are important tools in the identification of the structure of time series and in the examination of proposed models.

21.2.5 Stationary Time Series

From a statistical viewpoint, a time series of length n comprises a set of values from response variables $y_1, y_2, \ldots, y_n$. If there is no trend, seasonal effect or other cyclical effect in the time series and all variables have the same variance, the time series is said to be *stationary.* If $y_t, y_{t+1}, \ldots, y_{t+k}$ is a subsequence of the sequence $\{y_i\}$, the joint distribution of $y_t, y_{t+1}, \ldots, y_{t+k}$ is identical for $t = 1, 2, \ldots, n-k$. Hence, it is immaterial from which part of a longer stationary times series the n responses are obtained.

The fact that there is no deterministic component in the variation in the observed set of responses in a stationary series means the statistical analysis can be employed to establish the nature of the autocorrelation structure in the set. As a general rule, the first stage in the analysis of a time series is the transformation of the series to a stationary series by (i) removal of deterministic components and/or (ii) transformation of the responses to a new scale of measurement to stabilize variance.

21.3 Time Series: Statistical Models and Methods

21.3.1 Models for Time Series

A statistical model for a time series must provide (i) a component equation which describes the construction of the variables from the component parts—both deterministic and stochastic; and (ii) a correlation structure for the set of variables. These are considered in turn.

Component equations. A component equation links the responses $\{y_i\}$ to the effect of the trend $\{T_i\}$, the seasonal effect $\{S_i\}$, other cyclical effects $\{C_i\}$ and the random component $\{r_i\}$. The specification of a component equation is conveniently broken into three parts.

1. *The primary equation* which specifies the manner in which the deterministic component (M_i) and stochastic component (r_i) are combined. There are two commonly used options.

— additive equations $y_i = M_i + r_i$; and
— multiplicative equations $y_i = M_i \cdot r_i$.

2. *The secondary equation* which describes the decomposition of the deterministic component into its component parts. The common options are

— additive equations $M_i = T_i + S_i + C_i$; and
— multiplicative equations $M_i = T_i \cdot S_i \cdot C_i$.

3. *The tertiary equations* which define the trend, seasonal effects and other cyclical effects as functions of time.

Practical considerations. There are many decisions to be made when selecting a component equation for a time series model. Making good decisions rests heavily on knowledge of the experimental process which generated the time series and experience in fitting time series models to data. While general comments are offered below which may be of assistance, expert guidance and/or good reference material specific to the area of application are essential if serious blunders are to be avoided.

1. *Primary equation.* The distinction between additive and multiplicative equations has been extensively canvassed elsewhere in the book (see Section 10.3.8 for an introduction). The need for a multiplicative model arises when the magnitude of random variability rises with increasing mean response. This commonly occurs when there is a large seasonal component which results in peak seasonal response exceeding minimum levels by a factor of ten or more.

2. *Secondary equation.* If there is an increasing trend in the time series and the size of the seasonal effect and/or the cyclical effect appear to be increasing with increasing values of the trend, a multiplicative equation

may be required. Commonly, the form of the secondary equation matches the form chosen for the primary equation.

3. *Tertiary equations.*

(a) *Trend lines.* As in regression analysis, there are an infinite variety of trend lines which may have application with time series. A variety of forms are in common use. Linear equations and other polynomials offer a simple but flexible choice for use with an additive component equation in cases where there is no independent basis for selection of a trend line. Where a multiplicative equation is required, note that the analog to the linear equation is the exponential trend (which is a linear equation on a logarithmic scale). The exponential trend has the property that there is a constant rate of change, e.g. mean response doubles every three years. This trend is likely to apply when the time series has been collected in a period of rapid growth. Both the Gompertz and Logistic trend lines are widely applied in growth situations where there is an establishment period, a period of rapid growth and finally a slowing down as a maximum level is reached.

(b) *Seasonal effects.* There is no inherent reason for seasonal variation to be described by a simple mathematical expression. A seasonal effect which is a period of p units in the time series may be identified by $p-1$ parameters which present the contribution of the seasonal effect at each point in a seasonal cycle.

(c) *Cyclical effects.* Particularly in physical systems, there may be included one or more cyclical components which can adequately be expressed by *sine* or *cosine* functions of time, e.g. $C_i = A\cos(\omega t_i + \theta)$.

Distributional assumptions. Corresponding to the set of variables $y_1, y_2, \ldots, y_n$ which produce the values of the time series is the set of random components $r_1, r_2, \ldots, r_n$ which are the stochastic components in the component equation. The assumption is made that $r_1, r_2, \ldots, r_n$ defines a stationary time series (see Section 21.2.5) with r_i having mean zero and variance σ^2. In general, there is no assumption imposed for the distributional form since the practical application of time series demands large values for n which permits the Central limit theorem (Section 7.3.5) to be applied in estimation and test procedures.

In theory, the correlation structure of $r_1, r_2, \ldots, r_n$ is defined by the set of pairwise autocorrelations, $\{\rho_{ij}\}$ for $i, j = 1, 2, \ldots, n$. Even for small sample sizes, the number of distinct parameters in the set exceeds the number of sample members, n. Hence, there is a need to express the the correlation coefficients as functions of a smaller number of parameters and/or set some of the correlations to zero if the models are to have practical application. In fact, this direct approach is rarely used for time series models.

Two alternatives are employed. One approach indirectly identifies the correlation structure by expressing the relationship between the random components. The manner in which this is done is explained below. The alternative approach, known as *spectral analysis,* relies on the representation of the time series model as a sum of cyclical functions in time and identifies the correlation structure by the relative importance of cycles of different frequency. The explanation of this approach requires a level of mathematical understanding above that assumed in this book. Interested readers may obtain introductions in Jenkins and Watts [1969] (from an engineering, physical science viewpoint), Gottman [1981] (from a social science viewpoint) or Chatfield [1975] (from a general viewpoint).

ARIMA models. At time t_i in the time series, there is presumed to be a variable e_i $(i = 1, 2, \ldots, n)$ which makes a random or chance contribution to the response and $e_1, e_2, \ldots, e_n$ are presumed to be independent random variables. The random components $r_1, r_2, \ldots, r_n$ which appear in the component equation for the times series model are generally presumed to arise from these independent random contributions according to either (21.3.1) or (21.3.2).

$$r_i = e_i + B_1 e_{i-1} + \cdots + B_k e_{i-k} \tag{21.3.1}$$

$$r_i = e_i + B_1 r_{i-1} + \cdots + B_k r_{i-k}. \tag{21.3.2}$$

Where (21.3.1) is employed, the random process underlying the time series is termed a *moving average* (MA) *process,* whereas models which employ (21.3.2) are said to include an *autoregressive* (AR) *process.* In both cases, the value of k determines the *order* of the process—if k equals one it is a *first order process,* if k equals two it is a *second order process,* etc.

There is scope for including both processes in the model and this had led to the abbreviation *ARMA* to describe such models. A further extension is introduced by Box and Jenkins [1970] which relies on the fact that a time series with a linear trend plus an autoregressive and/or moving average process can be reduced to a stationary time series by taking differences between successive responses. Where differencing is employed, the model is referred to as an *ARIMA* model.

21.3.2 Model Fitting — Statistical Tools

The process of attempting to identify one or more models which might reasonably fit the data is generally not easy and relies on graphical and other statistical information which derive from the time series data. Some of the more important tools used in the examination are described below.

Time series plot. A plot of the time series is an obvious first step in analysis. It gives an overall impression of the deterministic structure of the series, points to possible trends, seasonal effects and cyclical effects. If the random component is relatively small, the plot may also point to any obvious errors in the data. The fact that the trend, seasonal effects and cyclical effects are superimposed on one another and the deterministic contribution is masked by the correlated random components, reduces the value of visual examination of the time series plot as an aid in identifying the deterministic component.

Filters. A major obstacle in seeking to establish the deterministic component in a time series is the fluctuations introduced by the random components. These tend to camouflage the trend and other deterministic contributions. By locally averaging responses, the impact of random fluctuations is reduced. A comparison of Figure 21.2.1 and Figure 21.3.1 illustrates the smoothing which was produced by replacing y_i with $(y_{i-2} + y_{i-1} + y_i + y_{i+1} + y_{i+2})/5$. This process of averaging may be seen as the transformation of the original time series into a new series by use of a *linear filter.* In general, a linear filter applied to a time series $\{y_i\}$ to produce a time series $\{z_i\}$ is defined by (21.3.3). In the case where the set of weights $\{c_i\}$ are chosen to sum to one, the elements in the set $\{z_i\}$ are called *moving averages* of the time series $\{y_i\}$. The simplest moving averages are those defined by (21.3.4). The plot in Figure 21.3.1 is based on a time series using (21.3.4) with $m = 2$.

$$z_i = \sum_{t=-r}^{s} c_t y_{i+t} \tag{12.3.3}$$

$$z_i = \sum_{t=-m}^{m} y_{i+t}/(2m+1). \tag{21.3.4}$$

The choice of a suitable moving average transformation requires considerable skill. Kendall [1973] provides a detailed discussion.

Linear filters may also be used to remove trends from time series. If $y_i = A + Bt_i + r_i$ is the component equation, then $z_i = y_i - y_{i+1}$ is a stationary series if the points on the time scale are equally spaced and $y_1, y_2, \ldots, y_n$ have identical variances. This approach may be extended to remove quadratic and higher order trends.

Correlograms. The set of autocorrelations for lags $1, 2, \ldots$ when represented in the graphical form illustrated in Figure 21.3.2 is termed a *correlogram.* It is a valuable tool in the process of model identification. There are two stages in which it is employed. At the first stage it may be based on autocorrelations from the time series data which include both the deterministic and stochastic components. An upward or downward trend in the series is characterized by autocorrelations tending slowly to zero with increasing lag (see Figure 21.3.2). If there is a seasonal effect or other cyclical effect in the series, the autocorrelations will show a cyclical pattern.

The correlogram also plays a role in model identification when it is based on the set of estimated random components. In this situation, it is used for identification of the orders of the autoregressive and/or moving average

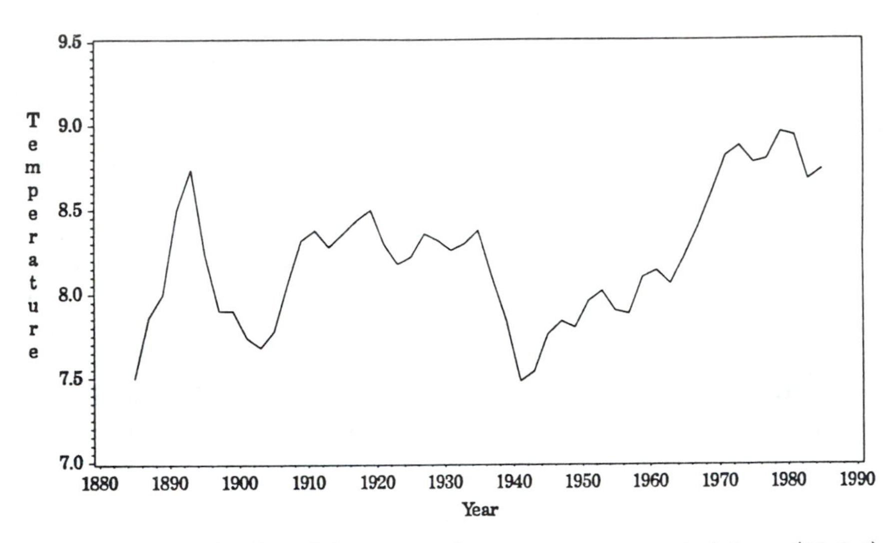

Figure 21.3.1. A plot of 5 year moving averages computed from (21.3.4) with $m = 2$ based on data in Table 21.2.1.

Lag	Corr.
1	0.500
2	0.467
3	0.391
4	0.398
5	0.360
6	0.224
7	0.304
8	0.253
9	0.249
10	0.098

−1.0 −0.8 −0.6 −0.4 −0.2 0.0 0.2 0.4 0.6 0.8 1.0

Figure 21.3.2. Correlogram based on data in Table 21.2.1.

processes of the time series models. Box and Jenkins [1970] give a comprehensive coverage of this role.

Residual analysis. Once a model has been selected and fitted to the data, it is possible to examine the goodness-of-fit of the model to the data. Fitted values can be constructed by including estimates of parameters in the deterministic components and in the expressions for random components (i.e. in (21.3.1) or (21.3.2)). The differences between the observed responses and the fitted values are termed *residuals*. There have been a number of residual-based procedures suggested for examining the fit of time series models. The worth of some tests has been questioned. Chatfield ([1975], Section 4.7) offers some pertinent comments on various procedures. Such research as has been done suggests that tests are generally of low power.

This is scarcely surprising since for any time series there are likely to be quite diverse models which could possibly fit the data.

21.3.3 Practical Considerations

The selection or identification of time series models which fit a given data set is, in general, a difficult task unless there is prior evidence to suggest a likely form. The choice of the deterministic components—trend, seasonal effect, other cyclical effects—is not necessarily separable from the choice of correlation structure presumed in the model. Scientists should clearly establish the purpose of analyzing the time series data before considering the method of analysis. Sophisticated methods do not necessarily equate with a high probability of determining the correct model or accurate forecasting. Perhaps simple descriptive tools, such as those presented in Neter et al. ([1988], Chapters 24 and 25) may suffice.

Unlike the selection of trend lines for regression models where there are independent random components, the deterministic component of time series models are not necessarily clearly visible. Furthermore, the autocorrelation which typically occurs in time series can itself cause patterns which may be mistaken for trends or cyclical effects. It is not inconceivable that two quite different models will give similar fits to the one data set. This fact should be kept in mind when using time series for projecting future responses—if a model with an upward trend fits the data, future projections will follow the upward trend line; if the apparent upward trend reflects autocorrelation, an upward movement in the series will reach a peak and, in the longer term, will decline towards the overall mean.

Establishing the nature of the stationary process in a time series is an essential part of constructing a mechanism for forecasting using the time series. Time series with fewer than one hundred readings are unlikely to provide adequate information for accurate forecasting unless the random component is making only a small contribution. Where there is strong autocorrelation, there is a high probability of confusing random and deterministic effects in short runs of a time series.

Over and above the length of the time series, there is need to consider the circumstances in which the series was produced. Particularly in demographic and economic areas, there may have been social, economic or political changes which affect the mechanism which is driving the time series. In the population study in Example 21.2 for instance, a distinct point of change can be identified—in the post Second World War period, there was an extensive period of immigration which is reflected in the change in trend. Less visible changes may affect the nature of the random component.

There is much experience required in model selection for time series. No attempt is made to provide guidelines in this book since, to do justice to the process of fitting and applying time series, a whole book is required. Where the aim is to use the time series for prediction, Nelson [1973] provides step-by-step instructions for identifying models based largely on correlograms and most comprehensive statistical packages provide a means of generating correlograms and of estimating parameters in the selected model.

If the task is to understand the structure of the time series or to compare time series from different sites or under different conditions, spectral analysis may be required. Practical guidelines to model fitting based on the study of the frequency domain information are not widely available. In part, this would seem to arise from the difficulties of clear and simple interpretation of the application of statistical tools for that purpose. Books such as Jenkins and Watts [1969], Gottman [1981] and Chatfield [1975] discuss the statistical tools used in model fitting, but not from the viewpoint of providing step-by-step instructions as to how to proceed down the path of model fitting.

22

Computers and Statistics

22.1 The Role of Computers in Statistics

There are five major areas of application of computers in Statistics. They are

- data storage and manipulation;
- summarizing features of the data;
- exploratory data analysis;
- calculations in the application of inferential procedures; and
- simulation studies.

These are considered in turn.

1. *Data storage and manipulation.* Whenever there is a requirement for lengthy calculations or repeated access to data for selected information, it is wise to consider the storage of data in a computer. The manner of storage may be critical to ease and speed of accessibility of the data and careful thought should be given both to the information to be stored and the manner of storage. If data are in the form of records collected over a long period of time, or if large quantities of data are involved, a wise course of action is to seek advice from a person with experience in data storage and retrieval to ensure (i) the input and storage mechanism is sufficiently flexible to allow modification if that should become necessary; and (ii) the program controlling data input and manipulation is transportable to another computer if there is an upgrade or replacement of the computer.

There is need to carefully consider the nature and quantity of information being stored. Two common faults are

- seeking to store and examine more data than can comfortably be handled by the available resources; and
- failing to include information which later proves to be essential and perhaps costly to input at a later data.

The first problem can be usually prevented by the simple action of counting how many pieces of information are to be stored. For example, if there are

100 units with 200 pieces of information per unit, there are a total of 20,000 pieces of information to be processed and checked. Do resources exist which can effectively handle this amount of information? The second problem normally arises when too little thought has been given to the reasons for collecting the data. A wiser researcher avoids the trap of collecting data first and thinking of question later.

2. *Summarizing features of the data.* There is an increasing use of graphs, tables and summary statistics for describing features of data sets and a rapidly growing set of statistical packages to perform the necessary operations. The only precautionary note is to choose carefully the method or methods of summarizing data. With the ease of application of the simpler statistical packages, it is tempting to produce a mass of summary information. Too much information can hide valuable features of the data as readily as too little information.

3. *Exploratory data analysis.* Exploratory data analysis is a branch of Statistics (or a complementary discipline, depending on your point of view) which has developed along with computers. If used sensibly, it has the capacity to assist in answering experimental questions and to suggest new questions which might lead to advances not foreseen by the researchers. The danger of the process lies in the difficulty of distinguishing real effects from artifacts of the sample. Virtually any set of data, if examined from many perspectives, will display a pattern which might have scientific significance. Experience suggests that plausible explanations are usually easy to discover after the patterns are reported, and, in time, the data find is thought to have supported the scientific explanation, thereby giving the idea a credence it does not warrant. Approach the exploration of a data set from a considered position. If new ideas do arise from the properties of the data, treat them as tentative until they have been reproduced from an independent data set.

From a practical viewpoint, there has been only slow development of standard packages which offer formalized methods of exploratory data analysis. In part, this reflects the lack of cohesion in the methods of exploratory data analysis and the demand for greater understanding and experience on the part of analysts.

A higher level of data analysis arises when inferential statistical models are proposed and the analysis is used to examine the acceptability of assumptions in the model which are crucial for valid application of statistical methods. Increasingly, this form of data analysis is being employed in an area which is referred to as *model fitting.* In regression analysis and the study of relationships between variables, a standard part of the analysis is concerned with selecting a suitable model from within a defined set of possible models.

4. *Calculations in the application of inferential methods.* With a few exceptions, commonly used statistical techniques are most easily applied by

using computers for the calculations. Indeed, some of the methods would be impractical to apply without the aid of computers.

5. *Simulation studies.* Knowledge of properties of statistical methods comes basically from mathematical analysis. However, for some important statistical methods, the mathematical development proves intractable. In such cases, computers can play an important role in providing solutions to mathematical problems by repeatedly simulating data from a model and examining the properties of the data sets which are produced. There are also circumstances in which data collected in an experiment are used to provide values of parameters in a model, and simulation studies based on that assumed model are used to study properties of the experimental set-up which generated the data.

22.2 Using Computers—Practical Considerations

There are four primary conditions to be met before a computer is useful in the analysis of data which are held in a computer. They are

- the means to perform the necessary calculations, e.g. a suitable computing program or package and human resources to operate the computer;
- expertise to convert the statistical problem into a computing problem, i.e. the knowledge to correctly use the program or package;
- expertise to correctly interpret the output arising from application of the program or package; and
- the resources to perform the analysis, e.g. access to a suitable computer, sufficient funds to pay for the computer analysis and sufficient time on the computer to complete the analyses.

While the necessity to meet these conditions should be self-evident, regrettably, many scientists are heavily committed to a project before considering the analysis of the data. Indeed, it is not uncommon for data to be collected before any consideration is given to the analysis. In such cases, the necessary resources to analyze the data may not be available and there is an (avoidable) waste of scientific resources.

Scientists who are not experienced in the use of computers for data storage and analysis are well advised to seek expert guidance at the time of designing an experiment. In most large organizations, a computer center or department exists which either has the staff to provide assistance or can direct an inquirer to the appropriate person. Alternatively, a consultant statistician or a 'local' computing expert should be sought for advice.

Statistical package or specialized program? An important consideration is the choice of the computing program. Broadly speaking, there are two groups of programs used for statistical analysis—statistical packages which contain general forms of data input, analysis and output, and specialized programs which are tailor-made for a particular application. Specialized programs are required

- when a method of analysis is required which is not readily available in statistical packages;
- there are savings in time or ease of usage which warrant the outlay of the resources required to produce and test the program.

The potential disadvantages of specialized programs should be clearly understood and considered before the decision is taken to use them. They are

- the cost of writing and the time and cost in checking the program to ensure its correctness;
- the likelihood that, when the data is transferred to another computer, the program will not be transportable;
- the possibility of a modification or addition being required at a later stage—will a programmer be available who can make the changes?

The decision to use a specialized program should never be taken lightly, especially when the development is outside the control of the user.

Statistical packages are not without their disadvantages. With any package, there is a time cost in becoming familiar with the operation of the package—in general, the more comprehensive the package, the more time which is required to understand its basic requirements and the format of input of data and output of results. Depending on the package, there may be severe restrictions on the format in which the data may be input into the package, and/or limited control of the output which is provided. With comprehensive packages, there is usually need for on-going expert assistance.

22.3 Statistical Packages

Increasingly, the use of statistical packages for analyses is becoming a part of scientific experimentation. With the availability of many of the major packages on microcomputers, scientists are in a position to make their own selection of a package rather than depending on the choice of a central computing department. In this environment it is essential the characteristics of a chosen package should match the needs and skills of the user. In this section, criteria are presented which may help in the choice of a statistical package.

General comments. Statistical packages can be classified according to two criteria—ease of operation and generality of statistical operations which can be performed. As a basic rule, increasing generality implies decreasing simplicity of operation. The task facing a scientist who is seeking a suitable statistical package is to strike the right balance between these conflicting requirements. Except for persons who are performing a wide range of analyses on a frequent basis, it is not usually efficient to attempt to become familiar with more than one or two packages. The reason lies not only in the time spent learning how to use packages but also in the amount of memorization required in respect of code words, forms of data input, etc.

Language. To a greater or lesser extent, all statistical packages introduce their own language. This involves words with special meanings and a syntax or grammar. At the primary level are single words, referred to as 'key words.' Thus, the word READ may represent an instruction to provide data. Many key words require qualifications. For example, the word READ will typically require additional information about what is to be read and perhaps information on the place where the data are held and the format in which the data are held. As packages become more general, so the number of options for each key word increase in number and become more cryptic in expression. When examining a package, spend some time looking at the key words and their options (i) in the area of data input; (ii) in applications of special interest; and (iii) in control of output. Except for high level packages which permit programming instructions to augment the standard options, users are bound absolutely by the limits of the options in the package.

Menu-driven versus command-line packages. There is a basic division of packages into *menu-* or *icon-driven* packages and *command-line* packages. In the menu- or icon-driven packages, a set of key words or symbols is provided. From these the user makes a selection. The selection of a key word either prompts a set of options from which a selection is made or operates on the selection if there is no choice of options. Command-line packages require users to type in key words with options typed in on the same or following lines.

Interactive versus batch operations. A package which operates on each instruction as it is given is said to be operating in an *interactive* mode. Menu- and icon-driven packages operate in interactive mode. If the package requires a specific instruction to apply the commands it is operating in *batch* mode. All major command-line packages have the option to operate in batch mode. Most also have the option to operate in interactive mode. A valuable property of many command-line packages is the option to operate in interactive mode but with the set of instructions being stored with the possibility of retrieval for modification and later operation in batch mode.

Comparing menu- or icon-driven and command-line packages. The menu- or icon-driven approach makes few demands on computing experience provided it possesses the properties that

(i) errors made by users (e.g. seeking an option which requires division by zero) do not cause the package to 'crash' thereby leaving the user in an unfamiliar environment; and

(ii) users (a) can always ascertain their position in the program (e.g. if there are several levels of options there may be confusion as to what the current options on the screen relate to), and (b) always has guidance as to possible forward and backward steps.

Additionally, the presence on the screen of all options available at a given point in the application, relieves users of searching through a manual for possible options. Furthermore, the interactive nature of the package means the application of the package to the statistical analysis in an automatic process. Command-line packages may possess an interactive mode of operation, but are more cumbersome in the provision of screen support. Many offer information when cued to do so, e.g. by the command HELP. However, experience suggests that inexperienced computer users are more comfortable with menu- or icon-driven packages.

There are two areas in which command-line packages are superior, and these tend to be of such importance that, at all but the most elementary level, command-line packages are the most widely used packages. The first relates to the recovery of previous steps in a sequence of operations and the second relates to the simplicity and speed of issuing instructions. These are considered in turn.

1. Recovering and editing instructions. If (i) an error is detected during data checking; (ii) repeated analyses are to be performed which require basically the same structure but with minor modifications; (iii) wrong instructions have been issued; or (iv) in model fitting, various combinations are being explored, the most efficient approach may be to recover earlier instructions and edit those instructions before reapplying them. If the list of command-line instructions which were used can be recovered, these may be modified and rerun under batch operation. By the nature of the menu- and icon-driven approaches, there is no simple equivalent.

2. Simplicity and speed of issuing instructions. In the learning phase, the prompting of users which occurs with the menu- and icon-driven packages has a great advantage. However, as users become more familiar with the operation, the more cryptic instructions of command-line packages offers a much faster means of issuing instructions. For example, the instruction READ/P [PROB1.DAT] X1..X10 $2,3X,8 can provide the instructions that the data should be read for variables X1 to X10 from the file PROB1.DAT using parallel read in which X1 and X2 are to be read from the first two columns, the next three columns are to be ignored and the remaining variables are to be read from the next 8 columns. To a person who is familiar

with the options and the syntax of the language, this contraction of information is a much faster way of inputting instructions than can be provided by a menu-driven approach.

In summary, for elementary packages which have few options and for users who use a package infrequently, the menu- or icon-driven approach is easier and faster to apply than the command-line approach. With more general packages and for persons who use packages frequently, the command-line approach is likely to be superior. Where there is likely to be a need to reproduce a sequence of instructions, the command-line approach is recommended.

Constructing and editing instructions. Command-line programs which operate in batch mode require a means of constructing a sequence of instructions and for modifying these instructions when a change is required. An *editor* is required for this purpose. The editor has a language of its own to allow for insertion, deletion and modification of instructions. This language and the method of applying the editor must be learnt by users. Some packages provide their own editor, while others assume an external editor is available and accessible to a user. It is important to determine the characteristics of editing for any package at the time it is being considered for purchase. In general, an editor which is external to the package will require some computing experience on the part of the user.

Flexibility of the package. Of considerable importance is the ability of the package to both receive and supply information in forms which suit a user. There is much value for a user in carefully considering the range of requirements and checking that they can be met by the package. Universal requirements involve

- restrictions on the form of input and output of information; and
- the range of statistical methods which are covered and the options which may be implemented with those methods.

As the level of statistical analysis increases, there is a need to examine

- the range of model and data checking facilities;
- the possibility of using results from one application of the package as input for a later application;
- facilities for storing sets of instructions (macros) which are repeatedly used or which provide a means of expanding the statistical options; and
- the integration of the package with other packages and with computing peripherals, e.g. data management packages, graphics packages, printers.

Support for users of a package. Support for the use of a package comes in four forms. They are

(a) reference and instruction manuals which come in book form or are available through HELP commands on the computer;

(b) worked examples of standard methodological applications;

(c) persons experienced in the use of the package; and

(d) courses on the operation of the package.

Of the written information, the following are usually required if a package is to be usefully employed;

- a reference manual which describes the syntax of the package and the limits of its application;
- a users' guide or instruction manual which provides basic information on the steps required to operate the package, describes the different applications of the package and contains sample applications;
- a booklet describing the features of the operation of the package on a specific computer (if the users' guide and/or reference manual are not machine specific);
- a quick-reference guide to the key words and options employed by the package (which may be contained as a tear-out addition to the users' guide).

In a large organization, the reference manual may be held in a central repository, e.g. a computer center. It is not generally required by persons who are making low level use of the package. However, the other documents are essential. Before purchasing a package, establish the availability and cost of manuals. If it is not possible to obtain a demonstration copy of the package, at least insist on access to the users' guide and check for simplicity of information, scope and availability of sample programs.

All levels of users require assistance in the use of statistical packages. The most effective help comes from (i) colleagues who are experienced in the use of the package; and (ii) sample outputs which may shed light on the correct method of inputting instructions. Help at this level should carry enormous weight when the purchase of a statistical package is being considered.

General purpose versus special purpose packages. There are a growing number of packages which have been developed specifically to meet the needs of scientists in a single discipline, e.g. in biology, psychology, medicine. The advantages of such packages are that (i) they tend to include specialized methods of interest in that discipline; (ii) they use jargon which is familiar to scientists working in that discipline; (iii) they tend to involve less learning than more general packages; and (iv) they may require less computer time to perform a specific task. The disadvantage lies in a degree of inflexibility. The statistical content relies heavily on the ideas of the authors as to what is important and this may result in absence of methods which some users regard as essential. There may be less support for the package and a limited range of computers on which the package can be applied. This can be serious in cases where an organization upgrades computers and loses access to a package as a result.

22.4 Choosing a Computer

This section relates to the selection of a computer for the purpose of statistical analysis where there is a choice. The basic decision is between a micro-computer under the user's sole control and a larger computer to which the user has access. Micro-computers are appealing on the grounds that they generally have the capacity for simpler operation, greater accessibility, more advanced graphical capabilities and more reliable operation. At a low level of statistical operation, given the ready availability of elementary statistical packages for micro-computers, it is difficult to find advantages for the use of larger computing systems. The reasons for maintaining statistical packages on larger computers are (i) the need for the greater power of the larger computers which is displayed in faster operation and capacity to handle larger sized problems; (ii) limitations on the flexibility of statistical packages for micro-computers; (iii) possible cost savings in that only one copy of the package must be bought for use by many scientists; and (iv) advantages in a centralized location for knowledge and support for the application of the package. To a greater or lesser extent, all of these reasons are losing strength as the power of micro-computers rapidly increases, the packages become more readily available and at more affordable prices, and there is a broader level of support for the packages. At least for the immediate future, there remains a need to consider carefully the choice of computing facility when both micro-computers and larger computers are available.

At the micro-computer level, be aware that a given statistical package may not be available for all brands of computers or there may be a minimal configuration required in order to run a package. Where possible, seek the opportunity to have an experienced user of the package try the package on your computer to establish if it runs efficiently.

Appendix A. Tables for Some Common Probability Distributions*

A.1. The Normal Distribution

Using the table

1. *To find the value of p which satisfies* $Pr(z \leq z_0) = p$

   (a) z_0 *in the interval* $[-2.99, +2.99]$:

Round off the value of z_0 to two decimal places and read the value of p directly from either Table A.1.1 or Table A.1.2.

Example: if $z_0 = -2.972$ round the value off to -2.97. Go to the row labelled -2.9 in Table A.1.1 and the column .07 and read off the value of p as 0.0015.

   (b) z_0 *in the interval* $[3.0, 3.9]$:

Round off the value of z_0 to one decimal place and read the value of p directly from Table A.1.3.

Example: if $z_0 = 3.28$ round the value off to 3.3. Go to the column 0.3 and read off the value of p as 0.99952.

   (c) z_0 *in the interval* $[-3.0, -3.9]$:

Use Table A.1.3 to find a value p' satisfying $\Pr(z \leq -z_0) = p'$ and compute $p = 1 - p'$.

Example: if $z_0 = 03.28$ round the value off to -3.3.. Go to the column 0.3 and read off the value of p' as 0.99952. Hence, $p = 1 - 0.99952 = 0.00048$.

2. *To find a value of p satisfying* $\Pr(|z| \geq |z_0|) = p$

---

*The tables are provided for applications with methods introduced in this book. For parameter combinations which are not included, simple linear interpolation may be used to compute approximate p-values or critical values for statistics. The values presented in these tables were computed using the MINITAB and SAS statistical package.

If z_0 is negative, remove the minus sign. Round off the (unsigned) value of z_0 to two decimal places and read the value of p directly from Table A.1.4.

Example: for $z_0 = -2.972$, remove the negative sign, round the value off to 2.97, go to the row labelled 2.9 and the column .07 in Table A.1.4 and read off the value of p as 0.0030.

3. *To find a value of* z_0 *satisfying* $\Pr(z \leq z_0) = p_0$

Find the value of p from Table A.1.1, A.1.2 or A.1.3 which is nearest to p_0 and read of the row and column values for z_0.

Example: if $p_0 = 0.6180$, establish that in Table A.1.2 there are values of p equal to 0.6179 and 0.6217. Since 0.6180 is closer to 0.6179, choose the value for p of 0.6179 and read off the value of z_0 as $.3 + .00 = 0.30$.

4. *To find a value of* z_0 *satisfying* $\Pr(|z| \geq |z_0|) = p_0$

If $P = 100p_0$ is one of the percentiles in Table A.1.5, read the corresponding value of z_0 from the table. Otherwise, use the relationship $\Pr(z \leq z_0) = 1 - \frac{1}{2}p_0$ and employ method 3 above.

Variations. See Section 7.3.4 and, in particular, Table 7.3.1 for variations based on the cumulative Normal probabilities.

Table A.1.1. Cumulative Probabilities of the Standard Normal Distribution for z from −2.99 to −0.00 in Increments of 0.01

z	.00	.01	.02	.03	.04	.05	.06	.07	.08	.09
−2.9	.0019	.0018	.0018	.0017	.0016	.0016	.0015	.0015	.0014	.0014
−2.8	.0026	.0025	.0024	.0023	.0023	.0022	.0021	.0021	.0020	.0019
−2.7	.0035	.0034	.0033	.0032	.0031	.0030	.0029	.0028	.0027	.0026
−2.6	.0047	.0045	.0044	.0043	.0041	.0040	.0039	.0038	.0037	.0036
−2.5	.0062	.0060	.0059	.0057	.0055	.0054	.0052	.0051	.0049	.0048
−2.4	.0082	.0080	.0078	.0075	.0073	.0071	.0069	.0068	.0066	.0064
−2.3	.0107	.0104	.0102	.0099	.0096	.0094	.0091	.0089	.0087	.0084
−2.2	.0139	.0136	.0132	.0129	.0125	.0122	.0119	.0116	.0113	.0110
−2.1	.0179	.0174	.0170	.0166	.0162	.0158	.0154	.0150	.0146	.0143
−2.0	.0228	.0222	.0217	.0212	.0207	.0202	.0197	.0192	.0188	.0183
−1.9	.0287	.0281	.0274	.0268	.0262	.0256	.0250	.0244	.0239	.0233
−1.8	.0359	.0351	.0344	.0336	.0329	.0322	.0314	.0307	.0301	.0294
−1.7	.0446	.0436	.0427	.0418	.0409	.0401	.0392	.0384	.0375	.0367
−1.6	.0548	.0537	.0526	.0516	.0505	.0495	.0485	.0475	.0465	.0455
−1.5	.0668	.0655	.0643	.0630	.0618	.0606	.0594	.0582	.0571	.0559
−1.4	.0808	.0793	.0778	.0764	.0749	.0735	.0721	.0708	.0694	.0681
−1.3	.0968	.0951	.0934	.0918	.0901	.0885	.0869	.0853	.0838	.0823
−1.2	.1151	.1131	.1112	.1093	.1075	.1056	.1038	.1020	.1003	.0985
−1.1	.1357	.1335	.1314	.1292	.1271	.1251	.1230	.1210	.1190	.1170
−1.0	.1587	.1562	.1539	.1515	.1492	.1469	.1446	.1423	.1401	.1379
−0.9	.1841	.1814	.1788	.1762	.1736	.1711	.1685	.1660	.1635	.1611
−0.8	.2119	.2090	.2061	.2033	.2005	.1977	.1949	.1922	.1894	.1867
−0.7	.2420	.2389	.2358	.2327	.2296	.2266	.2236	.2206	.2177	.2148
−0.6	.2743	.2709	.2676	.2643	.2611	.2578	.2546	.2514	.2483	.2451
−0.5	.3085	.3050	.3015	.2981	.2946	.2912	.2877	.2843	.2810	.2776
−0.4	.3446	.3409	.3372	.3336	.3300	.3264	.3228	.3192	.3156	.3121
−0.3	.3821	.3783	.3745	.3707	.3669	.3632	.3594	.3557	.3520	.3483
−0.2	.4207	.4168	.4129	.4090	.4052	.4013	.3974	.3936	.3897	.3859
−0.1	.4602	.4562	.4522	.4483	.4443	.4404	.4364	.4325	.4286	.4247
−0.0	.5000	.4960	.4920	.4880	.4840	.4801	.4761	.4721	.4681	.4641

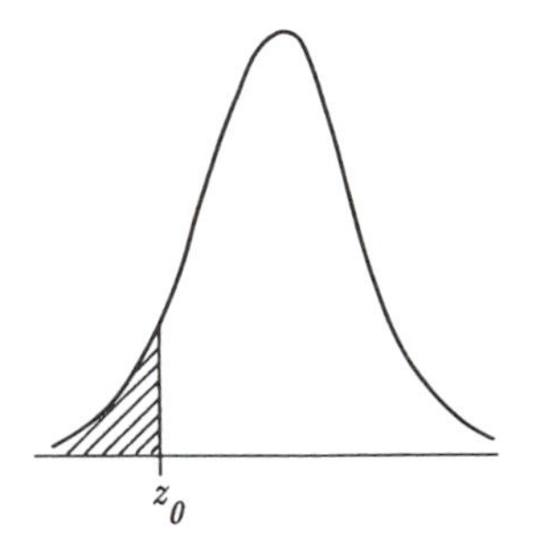

Table A.1.2. Cumulative Probabilities of the Standard Normal Distribution for z from 0.00 to 2.99 in Increments of 0.01

z	.00	.01	.02	.03	.04	.05	.06	.07	.08	.09
.0	.5000	.5040	.5080	.5120	.5160	.5199	.5239	.5279	.5319	.5359
.1	.5398	.5438	.5478	.5517	.5557	.5596	.5636	.5675	.5714	.5753
.2	.5793	.5832	.5871	.5910	.5948	.5987	.6026	.6064	.6103	.6141
.3	.6179	.6217	.6255	.6293	.6331	.6368	.6406	.6443	.6480	.6517
.4	.6554	.6591	.6628	.6664	.6700	.6736	.6772	.6808	.6844	.6879
.5	.6915	.6950	.6985	.7019	.7054	.7088	.7123	.7157	.7190	.7224
.6	.7257	.7291	.7324	.7357	.7389	.7422	.7454	.7486	.7517	.7549
.7	.7580	.7611	.7642	.7673	.7704	.7734	.7764	.7794	.7823	.7852
.8	.7881	.7910	.7939	.7967	.7995	.8023	.8051	.8078	.8106	.8133
.9	.8159	.8186	.8212	.8238	.8264	.8289	.8315	.8340	.8365	.8389
1.0	.8413	.8438	.8461	.8485	.8508	.8531	.8554	.8577	.8599	.8621
1.1	.8643	.8665	.8686	.8708	.8729	.8749	.8770	.8790	.8810	.8830
1.2	.8849	.8869	.8888	.8907	.8925	.8944	.8962	.8980	.8997	.9015
1.3	.9032	.9049	.9066	.9082	.9099	.9115	.9131	.9147	.9162	.9177
1.4	.9192	.9207	.9222	.9236	.9251	.9265	.9279	.9292	.9306	.9319
1.5	.9332	.9345	.9357	.9370	.9382	.9394	.9406	.9418	.9429	.9441
1.6	.9452	.9463	.9474	.9484	.9495	.9505	.9515	.9525	.9535	.9545
1.7	.9554	.9564	.9573	.9582	.9591	.9599	.9608	.9616	.9625	.9633
1.8	.9641	.9649	.9656	.9664	.9671	.9678	.9686	.9693	.9699	.9706
1.9	.9713	.9719	.9726	.9732	.9738	.9744	.9750	.9756	.9761	.9767
2.0	.9772	.9778	.9783	.9788	.9793	.9798	.9803	.9808	.9812	.9817
2.1	.9821	.9826	.9830	.9834	.9838	.9842	.9846	.9850	.9854	.9857
2.2	.9861	.9864	.9868	.9871	.9875	.9878	.9881	.9884	.9887	.9890
2.3	.9893	.9896	.9898	.9901	.9904	.9906	.9909	.9911	.9913	.9916
2.4	.9918	.9920	.9922	.9925	.9927	.9929	.9931	.9932	.9934	.9936
2.5	.9938	.9940	.9941	.9943	.9945	.9946	.9948	.9949	.9951	.9952
2.6	.9953	.9955	.9956	.9957	.9959	.9960	.9961	.9962	.9963	.9964
2.7	.9965	.9966	.9967	.9968	.9969	.9970	.9971	.9972	.9973	.9974
2.8	.9974	.9975	.9976	.9977	.9977	.9978	.9979	.9979	.9980	.9981
2.9	.9981	.9982	.9982	.9983	.9984	.9984	.9985	.9985	.9986	.9986

Table A.1.3. Cumulative Probabilities of the Standard Normal Distribution for z from 3.00 to 3.99 in Increments of 0.1

z	.0	.1	.2	.3	.4	.5	.6	.7	.8	.9
3.0	.99865	.99903	.99931	.99952	.99966	.99977	.99984	.99989	.99993	.99995

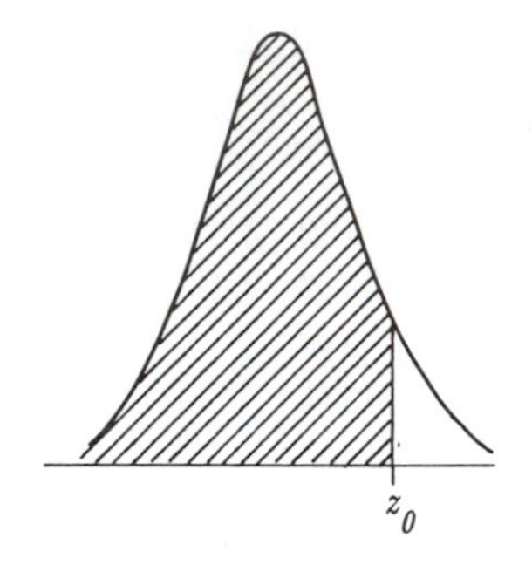

Table A.1.4. Probabilities of the Standard Normal Distribution
Probabilities Satisfying $\Pr\{|z| \geq |z_{\frac{1}{2}\alpha}|\} = \alpha$

z	.0	.1	.2	.3	.4	.5	.6	.7	.8	.9
0.0	1.0000	.9920	.9840	.9761	.9681	.9601	.9522	.9442	.9362	.9283
0.1	.9203	.9124	.9045	.8966	.8887	.8808	.8729	.8650	.8572	.8493
0.2	.8415	.8337	.8259	.8181	.8103	.8026	.7949	.7872	.7795	.7718
0.3	.7642	.7566	.7490	.7414	.7339	.7263	.7188	.7114	.7039	.6965
0.4	.6892	.6818	.6745	.6672	.6599	.6527	.6455	.6384	.6312	.6241
0.5	.6171	.6101	.6031	.5961	.5892	.5823	.5755	.5687	.5619	.5552
0.6	5485	.5419	.5353	.5287	.5222	.5157	.5093	.5029	.4965	.4902
0.7	.4839	.4777	.4715	.4654	.4593	.4533	.4473	.4413	4354	.4295
0.8	.4237	.4179	.4122	.4065	.4009	.3953	.3898	.3843	.3789	.3735
0.9	.3681	.3628	.3576	.3524	.3472	.3421	.3371	.3320	.3271	.3222
1.0	.3173	.3125	.3077	.3030	.2983	.2937	.2891	.2846	.2801	.2757
1.1	.2713	.2670	.2627	.2585	.2543	.2501	.2460	.2420	.2380	.2340
1.2	.2301	.2263	.2225	.2187	.2150	.2113	.2077	.2041	.2005	.1971
1.3	.1936	.1902	.1868	.1835	.1802	.1770	.1738	.1707	.1676	.1645
1.4	.1615	.1585	.1556	.1527	.1499	.1471	.1443	.1416	.1389	.1362
1.5	.1336	.1310	.1285	.1260	.1236	.1211	.1188	.1164	.1141	.1118
1.6	.1096	.1074	.1052	.1031	.1010	.0989	.0969	.0949	.0930	.0910
1.7	.0891	.0873	.0854	.0836	.0819	.0801	.0784	.0767	.0751	.0735
1.8	.0719	.0703	.0688	.0672	.0658	.0643	.0629	.0615	.0601	.0588
1.9	.0574	.0561	.0549	.0536	.0524	.0512	.0500	.0488	.0477	.0466
2.0	.0455	.0444	.0434	.0424	.0414	.0404	.0394	.0385	.0375	.0366
2.1	.0357	.0349	.0340	.0332	.0324	.0316	.0308	.0300	.0293	.0285
2.2	.0278	.0271	.0264	.0257	.0251	.0244	.0238	.0232	.0226	.0220
2.3	.0214	.0209	.0203	.0198	.0193	.0188	.0183	.0178	.0173	.0168
2.4	.0164	.0160	.0155	.0151	.0147	.0143	.0139	.0135	.0131	.0128
2.5	.0124	.0121	.0117	.0114	.0111	.0108	.0105	.0102	.0099	.0096

Table A.1.4. (*cont.*)

z	.0	.1	.2	.3	.4	.5	.6	.7	.8	.9
2.6	.0093	.0091	.0088	.0085	.0083	.0080	.0078	.0076	.0074	.0071
2.7	.0069	.0067	.0065	.0063	.0061	.0060	.0058	.0056	.0054	.0053
2.8	.0051	.0050	.0048	.0047	.0045	.0044	.0042	.0041	.0040	.0039
2.9	.0037	.0036	.0035	.0034	.0033	.0032	.0031	.0030	.0029	.0028

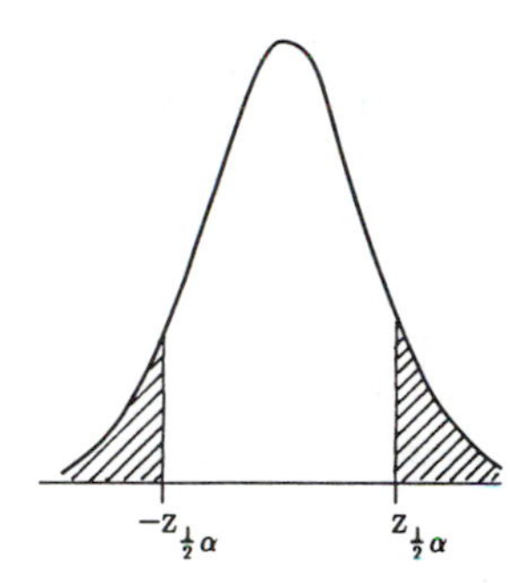

Table A.1.5. Selected Percentiles of the Standard Normal Distribution

P	0.1	1	2.5	5	10	25
Percentile	−3.090	−2.326	−1.960	−1.645	−1.282	−0.675

P	75	90	95	97.5	99	99.9
Percentile	0.675	1.282	1.645	1.960	2.326	3.090

A.2. The t-Distribution

Using the tables

1. *To find the value of* p *which satisfies* $\Pr(|t(\nu_0)| \geq |t_0|) = p$

For application in computing and interpreting a p-value, the steps are as follows.

Step 1. Determine the degrees of freedom, ν_0, to identify the appropriate t-distribution.

Step 2. From the row in Table A.2 for which $\nu = \nu_0$, establish whether

- t_0 is less than $t_{0.05}(\nu_0)$ in which case p exceeds 0.05;
- t_0 exceeds $t_{0.01}(\nu_0)$ in which case p is less than 0.01; or

– t_0 lies between $t_{0.05}(\nu_0)$ and $t_{0.01}(\nu_0)$ in which case p lies between 0.05 and 0.01.

Example. if $t_0 = 1.95$ and $\nu = 20$, it can be established from Table A.2 that $\Pr\{|t(20)| \geq |2.09|\} = 0.05$. Hence, p has a value in excess of 0.05.

2. *To find a value of* $t_\alpha(\nu_0)$ *satisfying* $\Pr\{|t(\nu_0)| \geq |t_\alpha(\nu_0)|\} = \alpha_0$

Find the row in Table A.2 for which $\nu = \nu_0$ and the column for which $\alpha = \alpha_0$. Read the value of $t_\alpha(\nu_0)$ from the body of the table.

Example. if $\nu = 10$ and $\alpha = 0.05$, then $t_{0.05}(10) = 2.23$

3. *To find cumulative probabilities for the t-distribution*

The tables have limited application in determining cumulative probabilities by utilizing the property of symmetry for the t-distribution which establishes that

$$\Pr\{t(\nu) \leq t_0\} = 1 - \frac{1}{2}\Pr\{|t(\nu)| \geq |t_0|\} \qquad \text{for } t_0 \geq 0$$
$$= \frac{1}{2}\Pr\{|t(\nu)| \geq |t_0|\} \qquad \text{for } t_0 < 0$$

Table A.2.1. Values of $t_\alpha(\nu)$ Satisfying $\Pr\{|t(\nu)| \geq |t_\alpha(\nu)|\} = \alpha$

d.f.	α				
ν	0.20	0.10	0.05	0.02	0.01
1	3.08	6.31	12.7	31.8	63.7
2	1.89	2.92	4.30	6.96	9.92
3	1.64	2.35	3.18	4.54	5.84
4	1.53	2.13	2.78	3.75	4.60
5	1.48	2.02	2.57	3.36	4.03
6	1.44	1.94	2.45	3.14	3.71
7	1.41	1.89	2.36	3.00	3.50
8	1.40	1.86	2.31	2.90	3.36
9	1.38	1.83	2.26	2.82	3.25
10	1.37	1.81	2.23	2.76	3.17
11	1.36	1.80	2.20	2.72	3.11
12	1.36	1.78	2.18	2.68	3.05
13	1.35	1.77	2.16	2.65	3.01
14	1.35	1.76	2.14	2.62	2.98
15	1.34	1.75	2.13	2.60	2.95
16	1.34	1.75	2.12	2.58	2.92
17	1.33	1.74	2.11	2.57	2.90

Table A.2.1. (*cont.*)

d.f. ν	α 0.20	0.10	0.05	0.02	0.01
18	1.33	1.73	2.10	2.55	2.88
19	1.33	1.73	2.09	2.54	2.86
20	1.33	1.72	2.09	2.53	2.85
21	1.32	1.72	2.08	2.52	2.83
22	1.32	1.72	2.07	2.51	2.82
23	1.32	1.71	2.07	2.50	2.81
24	1.32	1.71	2.06	2.49	2.80
25	1.32	1.71	2.06	2.49	2.79
26	1.31	1.71	2.06	2.48	2.78
27	1.31	1.70	2.05	2.47	2.77
28	1.31	1.70	2.05	2.47	2.76
29	1.31	1.70	2.05	2.46	2.76
30	1.31	1.70	2.04	2.46	2.75
31	1.31	1.70	2.04	2.45	2.74
32	1.31	1.69	2.04	2.45	2.74
33	1.31	1.69	2.03	2.44	2.73
34	1.31	1.69	2.03	2.44	2.73
35	1.31	1.69	2.03	2.44	2.72
36	1.31	1.69	2.03	2.43	2.72
37	1.30	1.69	2.03	2.43	2.72
38	1.30	1.69	2.02	2.43	2.71
39	1.30	1.68	2.02	2.43	2.71
40	1.30	1.68	2.02	2.42	2.70
45	1.30	1.68	2.01	2.41	2.69
50	1.30	1.68	2.01	2.40	2.68
60	1.30	1.67	2.00	2.39	2.66
∞	1.28	1.65	1.96	2.33	2.58

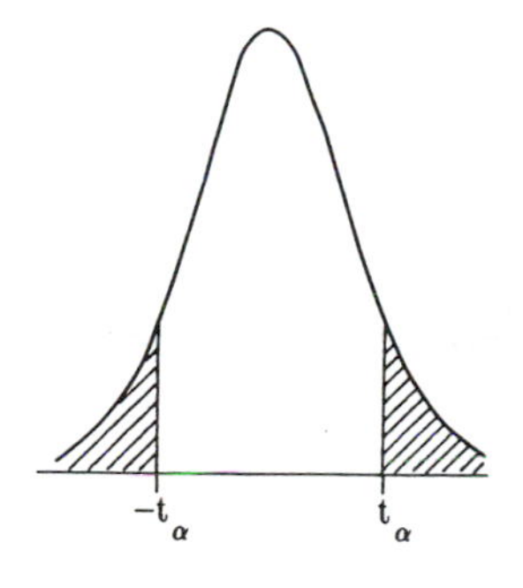

A.3. The Chi-Squared Distribution

Using the tables

1. *To find the value of p which satisfies* $\Pr\{\chi^2(\nu_0) \geq X_0^2\} = p$

For application in computing and interpreting a p-value, the steps to be followed are listed below.

Step 1. Determine the degrees of freedom, ν_0, to identify the appropriate Chi-squared distribution. If

- ν_0 is less than or equal to 100, proceed to step 2(a);
- v_0 exceeds 100, proceed to step 2(b).

Step 2(a). From the row in Table A.3.1 for which $\nu = \nu_0$, establish whether

- X_0^2 is less than $X_{0.05}^2(\nu_0)$ in which case p exceeds 0.05;
- X_0^2 exceeds $X_{0.01}^2(\nu_0)$ in which case p is less than 0.01; or
- X_0^2 lies between $X_{0.05}^2(\nu_0)$ and $X_{0.01}^2(\nu_0)$ in which case p lies between 0.05 and 0.01.

Example. if $\nu = 20$, it can be established from Table A.3.1 that $\Pr\{\chi^2(20) \geq 31.41\} = 0.05$ and $\Pr\{\chi^2(20) \geq 37.57\} = 0.01$. Hence,

– a value of X_0^2 which is less than 31.41 has an associated p-value which exceeds 0.05;

– a value of X_0^2 which exceeds 37.57 has an associated p-value which is less than 0.01; and

– a value of X_0^2 which lies between 31.41 and 37.57 has an associated p-value which lies between 0.05 and 0.01.

Example. if ν_0 is 54 it is necessary to use information for ν_1 equals 50 and ν_2 equals 60. If

X_0^2 is less than $X_{0.05}^2(50) = 67.50$, the associated p-value exceeds 0.05;

X_0^2 exceeds $X_{0.01}^2(60) = 88.38$, the associated p-value is less than 0.01.

Otherwise, it is necessary to use linear interpolation to compute

$$X_{0.05}^2(54) \cong (1-0.4)(67.50) + 0.4(79.08) = 72.1; \quad \text{and}$$
$$X_{0.01}^2(54) \cong (1-0.4)(76.15) + 0.4(88.38) = 81.0.$$

Step 2(b). If ν_0 exceeds 100, a large sample approximation can be invoked which states that $\sqrt{\{2\chi^2\}}$ is approximately $N(\sqrt{\{2\nu_0 - 1\}}, 1)$. Hence, Table A.1 may be used in conjunction with the formula

$$p \cong 1 - \Pr\{z \leq z_0\} \qquad \text{where } z_0 = [\sqrt{\{2X_0^2\}} - \sqrt{\{2\nu_0 - 1\}}].$$

Example. If $\nu_0 = 150$ and $X_0^2 = 168.3$, then $z_0 = [18.346 - 17.292] = 1.055$. From Table A.1.2, $\Pr(z \leq 1.06) = 0.8554$. Hence, $p \cong 0.14$.

2. *To find a value of* $X_\alpha^2(\nu_0)$ *satisfying* $\Pr\{\chi^2(\nu_0) \geq X_\alpha^2(\nu_0)\} = \alpha$

Step 1. Determine the degrees of freedom, ν_0, to identify the appropriate Chi-squared distribution. If

- ν_0 is less than or equal to 100, proceed to step 2(a);
- v_0 exceeds 100, proceed to step 2(b).

Step 2(a). From the row identified by ν_0 and the column identified by α, read the value of $X_\alpha^2(\nu_0)$ from the body of Table A.3.1.

Example. For $\nu = 10$ and $\alpha = 0.05$, $X_\alpha^2(10) = 18.31$.

Example. If $\nu = 43$ and $\alpha = 0.05$, it is necessary to use information for $\nu_1 = 40$ and $\nu_2 = 45$. From Table A.3.1, $X_{0.05}^2(40) = 55.76$ and $X_{0.05}^2(45) = 61.66$. Hence, using linear interpolation,

$$X_{0.05}^2(43) \cong (1 - 0.6)(55.76) + 0.6(61.66) = 59.3$$

Step 2(b). If ν_0 exceeds 100, a large sample approximation can be invoked which states that $\sqrt{\{2\chi^2\}}$ is approximately $N(\sqrt{\{2\nu_0 - 1\}}, 1)$. Hence,

$$X_\alpha^2(\nu_0) \cong \frac{1}{2}\{z_\alpha + \sqrt{2\nu_0 - 1}\}^2.$$

Example. For $\alpha = 0.05$, Table A.1.5 provides a value of 1.645 for z_α.

$$\text{If } \nu_0 = 150, \quad X_{0.05}^2(150) \cong \frac{1}{2}\{1.645 + \sqrt{2(150) - 1}\}^2 = 179.3.$$

Table A.3.1. Values of X_α^2 satisfying $\Pr\{\chi^2(\nu) \geq X_\alpha^2\} = \alpha$

d.f.	α									
ν	0.995	0.990	0.975	0.95	0.90	0.10	0.05	0.025	0.010	0.005
1	$0.0^4 39$	$0.0^3 16$	$0.0^3 98$	$0.0^2 39$	0.0158	2.706	3.841	5.024	6.635	7.879
2	0.0100	0.0201	0.0506	0.103	0.211	4.605	5.991	7.378	9.210	10.60
3	0.0717	0.115	0.216	0.352	0.584	6.251	7.815	9.348	11.34	12.84
4	0.207	0.297	0.484	0.711	1.064	7.779	9.488	11.14	13.28	14.86
5	0.412	0.554	0.831	1.145	1.610	9.236	11.07	12.83	15.09	16.75
6	0.676	0.872	1.237	1.635	2.204	10.64	12.59	14.45	16.81	18.55
7	0.989	1.239	1.690	2.167	2.833	12.02	14.07	16.01	18.48	20.28
8	1.344	1.646	2.180	2.733	3.490	13.36	15.51	17.53	20.09	21.95
9	1.735	2.088	2.700	3.325	4.168	14.68	16.92	19.02	21.67	23.59

Table A.3.1. (*cont.*)

d.f. ν	α 0.995	0.990	0.975	0.95	0.90	0.10	0.05	0.025	0.010	0.005
10	2.156	2.558	3.247	3.940	4.865	15.99	18.31	20.48	23.21	25.19
11	2.603	3.053	3.816	4.575	5.578	17.28	19.68	21.92	24.72	26.76
12	3.074	3.571	4.404	5.226	6.304	18.55	21.03	23.34	26.22	28.30
13	3.565	4.107	5.009	5.892	7.042	19.81	22.36	24.74	27.69	29.82
14	4.075	4.660	5.629	6.571	7.790	21.06	23.68	26.12	29.14	31.32
15	4.601	5.229	6.262	7.261	8.547	22.31	25.00	27.49	30.58	32.80
16	5.142	5.812	6.908	7.962	9.312	23.54	26.30	28.85	32.00	34.27
17	5.697	6.408	7.564	8.672	10.09	24.77	27.59	30.19	33.41	35.72
18	6.265	7.015	8.231	9.390	10.86	25.99	28.87	31.53	34.81	37.16
19	6.844	7.633	8.907	10.12	11.65	27.20	30.14	32.85	36.19	38.58
20	7.434	8.260	9.591	10.85	12.44	28.41	31.41	34.17	37.57	40.00
21	8.034	8.897	10.28	11.59	13.24	29.62	32.67	35.48	38.93	41.40
22	8.643	9.542	10.98	12.34	14.04	30.81	33.92	36.78	40.29	42.80
23	9.260	10.20	11.69	13.09	14.85	32.01	35.17	38.08	41.64	44.18
24	9.886	10.86	12.40	13.85	15.66	33.20	36.42	39.36	42.98	45.56
25	10.52	11.52	13.12	14.61	16.47	34.38	37.65	40.65	44.31	46.93
26	11.16	12.20	13.84	15.38	17.29	35.56	38.89	41.92	45.64	48.29
27	11.81	12.88	14.57	16.15	18.11	36.74	40.11	43.19	46.96	49.65
28	12.46	13.56	15.31	16.93	18.94	37.92	41.34	44.46	48.28	50.99
29	13.12	14.26	16.05	17.71	19.77	39.09	42.56	45.72	49.59	52.34
30	13.79	14.95	16.79	18.49	20.60	40.26	43.77	46.98	50.89	53.67
31	14.46	15.66	17.54	19.28	21.43	41.42	44.99	48.23	52.19	55.00
32	15.13	16.36	18.29	20.07	22.27	42.58	46.19	49.48	53.49	56.33
33	15.82	17.07	19.05	20.87	23.11	43.75	47.40	50.72	54.78	57.65
34	16.50	17.79	19.81	21.66	23.95	44.90	48.60	51.97	56.06	58.96
35	17.19	18.51	20.57	22.47	24.80	46.06	49.80	53.20	57.34	60.28
36	17.89	19.23	21.34	23.27	25.64	47.21	51.00	54.44	58.62	61.58
37	18.59	19.96	22.11	24.07	26.49	48.36	52.19	55.67	59.89	62.88
38	19.29	20.69	22.88	24.88	27.34	49.51	53.38	56.90	61.16	64.18
39	20.00	21.43	23.65	25.70	28.20	50.66	54.57	58.12	62.43	65.48
40	20.71	22.16	24.43	26.51	29.05	51.81	55.76	59.34	63.69	66.77
45	24.31	25.90	28.37	30.61	33.35	57.51	61.66	65.41	69.96	73.17
50	27.99	29.71	32.36	34.76	37.69	63.17	67.50	71.42	76.15	79.49
60	35.53	37.48	40.48	43.19	46.46	74.40	79.08	83.30	88.38	91.95
70	43.28	45.44	48.76	51.74	55.35	85.53	90.53	95.02	100.4	104.2
80	51.17	53.54	57.15	60.39	64.28	96.58	101.9	106.6	112.3	116.3
90	59.20	61.75	65.65	69.13	73.29	107.6	113.1	118.1	124.1	128.3
100	67.33	70.06	74.22	77.93	82.36	118.5	124.3	129.6	135.8	140.2

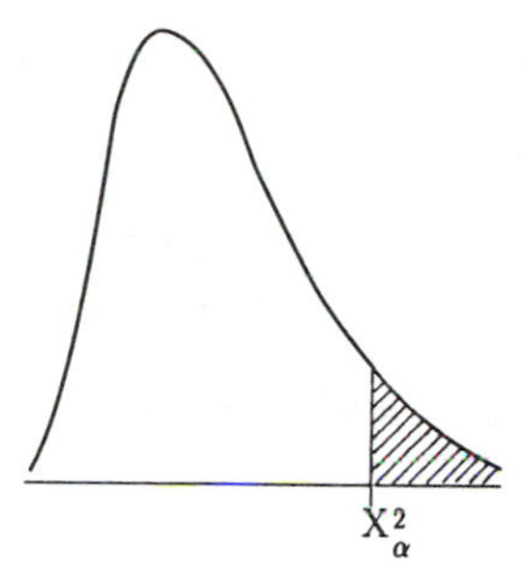

A.4. F-Distribution

Using the tables

1. *To find the value of* p_0 *which satisfies* $\Pr\{F(\nu_1, \nu_2) \geq F_0\} = p_0$

For application in obtaining and interpreting a p-value, the steps are as follows.

Step 1. Determine the degrees of freedom, ν_1 and ν_2, to identify the appropriate F-distribution.

Step 2. Find the entries in Table A.4.1 for the $F(\nu_1, \nu_2)$ distribution, and establish whether

- F_0 is less than $\mathrm{F}_{0.05}(\nu_1, \nu_2)$ in which case p exceeds 0.05;
- F_0 exceeds $\mathrm{F}_{0.01}(\nu_1, \nu_2)$ in which case p is less than 0.01;
- F_0 lies between $\mathrm{F}_{0.05}(\nu_1, \nu_2)$ and $\mathrm{F}_{0.01}(\nu_1, \nu_2)$ in which case p lies between 0.05 and 0.01.

Example. If $\nu_1 = 2$ and $\nu_2 = 20$, $\mathrm{F}_{0.05}(2, 20) = 3.49$ and $\mathrm{F}_{0.01}(2, 20) = 5.85$. Hence,

- a value of F_0 which is less than 3.49 has an associated p-value which exceeds 0.05;
- a value of F_0 which exceeds 5.85 has an associated p-value which is less than 0.01; and
- a value of F_0 which lies between 3.49 and 5.85 has an associated p-value which lies between 0.05 and 0.01.

Example. If $\nu_1 = 2$ and $\nu_2 = 45$ it is necessary to employ F-values from the $F(2, 40)$ and $F(2, 50)$ distributions. Check if

(i) F_0 is less than $\mathrm{F}_{0.05}(2,50)$ in which case p_0 exceeds 0.05; or

(ii) F_0 exceeds $\mathrm{F}_{0.01}(2,40)$ in which case p_0 is less than 0.01.

Otherwise, linear interpolation is required to obtain approximate 5% and 1% values for the $F(2,45)$ distribution, i.e.

$\mathrm{F}_{0.05}(2,45) \cong (1-a)\mathrm{F}_{0.05}(2,40)+a\mathrm{F}_{0.05}(2,50)$ where $a = (45-40)/(50-40)$.

Example. If $\nu_1 = 13$ and $\nu_2 = 45$ it is necessary to employ F-values from the $F(12,40)$, $F(12,50)$, $F(15,40)$ and $F(15,50)$ distributions. Check if

(i) F_0 is less than $\mathrm{F}_{0.05}(15,50)$ in which case p_0 exceeds 0.05; or

(ii) F_0 exceeds $\mathrm{F}_{0.01}(12,40)$ in which case p_0 is less than 0.01.

Otherwise, linear interpolation is required to obtain approximate 5% and 1% values for the $F(13,45)$ distribution. The linear interpolation can proceed in two stages. Obtain approximate 5% and 1% values for $F(12,45)$ and $F(15,45)$ distributions. Using these values, obtain approximate 5% and 1% values for the $F(13,45)$ distribution by linear interpolation.

2. *To find a value of* $\mathrm{F}_\alpha(\nu_1,\nu_2)$ *which satisfies* $\Pr\{F(\nu_1,\nu_2) \geq \mathrm{F}_\alpha(\nu_1,\nu_2)\} = \alpha$

(a) *for* $\alpha = 0.005, 0.01, 0.025, 0.05, 0.10$.

The value of $\mathrm{F}_\alpha(\nu_1,\nu_2)$ may be read directly from Table A.4.1. Linear interpolation is required for values of ν_1 and/or ν_2 which are not provided in Table A.4.1.

Example. If $\nu_1 = 2$ and $\nu_2 = 20$ then for $\alpha = 0.025$, $\mathrm{F}_{0.025}(2,20) = 4.46$.

Example. If $\nu_1 = 2$ and $\nu_2 = 45$, then for $\alpha = 0.025$, it is necessary to use the values for $\mathrm{F}_{0.025}(2,40)$ and $\mathrm{F}_{0.025}(2,50)$. Hence

$$\begin{aligned}\mathrm{F}_{0.025}(2,45) &\cong (1-a)\mathrm{F}_{0.025}(2,40) + a\mathrm{F}_{0.025}(2,50)\\ &\qquad \text{where } a = (45-40)/(50-40)\\ &= (1-0.5)(4.05) + 0.5(3.97)\\ &= 4.01.\end{aligned}$$

(b) *for* $\alpha = 0.90, 0.95, 0.975, 0.99, 0.995$.

There is a relationship for the F distribution which states that

$$\mathrm{F}_\alpha(\nu_1,\nu_2) = 1/\mathrm{F}_{1-\alpha}(\nu_2,\nu_1).$$

By employing this relationship, Table A.4.1 may be used to provide values for the above-listed values of α.

Example. $\mathrm{F}_{0.975}(20,2) = 1/\mathrm{F}_{0.025}(2,20) = 1/4.46 = 0.224$.

Table A.4.1. Values of $F_\alpha(\nu_1, \nu_2)$ satisfying
$\Pr\{F(\nu_1,\nu_2) \geq F_\alpha(\nu_1,\nu_2)\} = \alpha$

ν_2	$\nu_1=1$					ν_2	$\nu_1=2$				
	α						α				
	0.10	0.05	0.025	0.010	0.005		0.10	0.05	0.025	0.010	0.005
1	39.9	161	648	4050	16200	1	49.5	199	799	5000	19999
2	8.53	18.5	38.5	98.5	198	2	9.00	19.0	39.0	99.0	199
3	5.54	10.1	17.4	34.1	55.6	3	5.46	9.55	16.0	30.8	49.8
4	4.54	7.71	12.2	21.2	31.3	4	4.32	6.94	10.6	18.0	26.3
5	4.06	6.61	10.0	16.3	22.8	5	3.78	5.79	8.43	13.3	18.3
6	3.78	5.99	8.81	13.7	18.6	6	3.46	5.14	7.26	10.9	14.5
7	3.59	5.59	8.07	12.2	16.2	7	3.26	4.74	6.54	9.55	12.4
8	3.46	5.32	7.57	11.3	14.7	8	3.11	4.46	6.06	8.65	11.0
9	3.36	5.12	7.21	10.6	13.6	9	3.01	4.26	5.71	8.02	10.1
10	3.28	4.96	6.94	10.0	12.8	10	2.92	4.10	5.46	7.56	9.43
11	3.23	4.84	6.72	9.65	12.2	11	2.86	3.98	5.26	7.21	8.91
12	3.18	4.75	6.55	9.33	11.8	12	2.81	3.89	5.10	6.93	8.51
13	3.14	4.67	6.41	9.07	11.4	13	2.76	3.81	4.97	6.70	8.19
14	3.10	4.60	6.30	8.86	11.1	14	2.73	3.74	4.86	6.51	7.92
15	3.07	4.54	6.20	8.68	10.8	15	2.70	3.68	4.77	6.36	7.70
16	3.05	4.49	6.12	8.53	10.6	16	2.67	3.63	4.69	6.23	7.51
17	3.03	4.45	6.04	8.40	10.4	17	2.64	3.59	4.62	6.11	7.35
18	3.01	4.41	5.98	8.29	10.2	18	2.62	3.55	4.56	6.01	7.21
19	2.99	4.38	5.92	8.19	10.1	19	2.61	3.52	4.51	5.93	7.09
20	2.97	4.35	5.87	8.10	9.94	20	2.59	3.49	4.46	5.85	6.99
21	2.96	4.32	5.83	8.02	9.83	21	2.57	3.47	4.42	5.78	6.89
22	2.95	4.30	5.79	7.95	9.73	22	2.56	3.44	4.38	5.72	6.81
23	2.94	4.28	5.75	7.88	9.63	23	2.55	3.42	4.35	5.66	6.73
24	2.93	4.26	5.72	7.82	9.55	24	2.54	3.40	4.32	5.61	6.66
25	2.92	4.24	5.69	7.77	9.48	25	2.53	3.39	4.29	5.57	6.60
30	2.88	4.17	5.57	7.56	9.18	30	2.49	3.32	4.18	5.39	6.35
35	2.85	4.12	5.48	7.42	8.98	35	2.46	3.27	4.11	5.27	6.19
40	2.84	4.08	5.42	7.31	8.83	40	2.44	3.23	4.05	5.18	6.07
50	2.81	4.03	5.34	7.17	8.63	50	2.41	3.18	3.97	5.06	5.90
75	2.77	3.97	5.23	6.99	8.37	75	2.37	3.12	3.88	4.90	5.69
100	2.76	3.94	5.18	6.90	8.24	100	2.36	3.09	3.83	4.82	5.59
∞	2.71	3.84	5.02	6.64	7.88	∞	2.30	3.00	3.69	4.61	5.30

Table A.4.1 (*cont.*)

ν_2	$\nu_1=3$					ν_2	$\nu_1=4$				
	α						α				
	0.10	0.05	0.025	0.010	0.005		0.10	0.05	0.025	0.010	0.005
1	53.6	216	864	5403	21613	1	55.8	225	900	5625	22501
2	9.16	19.2	39.2	99.2	199	2	9.24	19.2	39.2	99.2	199
3	5.39	9.28	15.4	29.5	47.5	3	5.34	9.12	15.1	28.7	46.2
4	4.19	6.59	9.98	16.7	24.3	4	4.11	6.39	9.60	16.0	23.2
5	3.62	5.41	7.76	12.1	16.5	5	3.52	5.19	7.39	11.4	15.6
6	3.29	4.76	6.60	9.78	12.9	6	3.18	4.53	6.23	9.15	12.0
7	3.07	4.35	5.89	8.45	10.9	7	2.96	4.12	5.52	7.85	10.1
8	2.92	4.07	5.42	7.59	9.60	8	2.81	3.84	5.05	7.01	8.81
9	2.81	3.86	5.08	6.99	8.72	9	2.69	3.63	4.72	6.42	7.96
10	2.73	3.71	4.83	6.55	8.08	10	2.61	3.48	4.47	5.99	7.34
11	2.66	3.59	4.63	6.22	7.60	11	2.54	3.36	4.28	5.67	6.88
12	2.61	3.49	4.47	5.95	7.23	12	2.48	3.26	4.12	5.41	6.52
13	2.56	3.41	4.35	5.74	6.93	13	2.43	3.18	4.00	5.21	6.23
14	2.52	3.34	4.24	5.56	6.68	14	2.39	3.11	3.89	5.04	6.00
15	2.49	3.29	4.15	5.42	6.48	15	2.36	3.06	3.80	4.89	5.80
16	2.46	3.24	4.08	5.29	6.30	16	2.33	3.01	3.73	4.77	5.64
17	2.44	3.20	4.01	5.18	6.16	17	2.31	2.96	3.66	4.67	5.50
18	2.42	3.16	3.95	5.09	6.03	18	2.29	2.93	3.61	4.58	5.37
19	2.40	3.13	3.90	5.01	5.92	19	2.27	2.90	3.56	4.50	5.27
20	2.38	3.10	3.86	4.94	5.82	20	2.25	2.87	3.51	4.43	5.17
21	2.36	3.07	3.82	4.87	5.73	21	2.23	2.84	3.48	4.37	5.09
22	2.35	3.05	3.78	4.82	5.65	22	2.22	2.82	3.44	4.31	5.02
23	2.34	3.03	3.75	4.76	5.58	23	2.21	2.80	3.41	4.26	4.95
24	2.33	3.01	3.72	4.72	5.52	24	2.19	2.78	3.38	4.22	4.89
25	2.32	2.99	3.69	4.68	5.46	25	2.18	2.76	3.35	4.18	4.84
30	2.28	2.92	3.59	4.51	5.24	30	2.14	2.69	3.25	4.02	4.62
35	2.25	2.87	3.52	4.40	5.09	35	2.11	2.64	3.18	3.91	4.48
40	2.23	2.84	3.46	4.31	4.98	40	2.09	2.61	3.13	3.83	4.37
50	2.20	2.79	3.39	4.20	4.83	50	2.06	2.56	3.05	3.72	4.23
75	2.16	2.73	3.30	4.05	4.63	75	2.02	2.49	2.96	3.58	4.05
100	2.14	2.70	3.25	3.98	4.54	100	2.00	2.46	2.92	3.51	3.96
∞	2.08	2.61	3.12	3.78	4.28	∞	1.94	2.37	2.79	3.32	3.72

Table A.4.1 (*cont.*)

ν_2	$\nu_1=5$					ν_2	$\nu_1=6$				
	α						α				
	0.10	0.05	0.025	0.010	0.005		0.10	0.05	0.025	0.010	0.005
1	57.2	230	922	5764	23056	1	58.2	234	937	5859	23436
2	9.29	19.3	39.3	99.3	199	2	9.33	19.3	39.3	99.3	199
3	5.31	9.01	14.9	28.2	45.4	3	5.28	8.94	14.7	27.9	44.8
4	4.05	6.26	9.36	15.5	22.5	4	4.01	6.16	9.20	15.2	22.0
5	3.45	5.05	7.15	11.0	14.9	5	3.40	4.95	6.98	10.7	14.5
6	3.11	4.39	5.99	8.75	11.5	6	3.05	4.28	5.82	8.47	11.1
7	2.88	3.97	5.29	7.46	9.52	7	2.83	3.87	5.12	7.19	9.16
8	2.73	3.69	4.82	6.63	8.30	8	2.67	3.58	4.65	6.37	7.95
9	2.61	3.48	4.48	6.06	7.47	9	2.55	3.37	4.32	5.80	7.13
10	2.52	3.33	4.24	5.64	6.87	10	2.46	3.22	4.07	5.39	6.54
11	2.45	3.20	4.04	5.32	6.42	11	2.39	3.09	3.88	5.07	6.10
12	2.39	3.11	3.89	5.06	6.07	12	2.33	3.00	3.73	4.82	5.76
13	2.35	3.03	3.77	4.86	5.79	13	2.28	2.92	3.60	4.62	5.48
14	2.31	2.96	3.66	4.69	5.56	14	2.24	2.85	3.50	4.46	5.26
15	2.27	2.90	3.58	4.56	5.37	15	2.21	2.79	3.41	4.32	5.07
16	2.24	2.85	3.50	4.44	5.21	16	2.18	2.74	3.34	4.20	4.91
17	2.22	2.81	3.44	4.34	5.07	17	2.15	2.70	3.28	4.10	4.78
18	2.20	2.77	3.38	4.25	4.96	18	2.13	2.66	3.22	4.01	4.66
19	2.18	2.74	3.33	4.17	4.85	19	2.11	2.63	3.17	3.94	4.56
20	2.16	2.71	3.29	4.10	4.76	20	2.09	2.60	3.13	3.87	4.47
21	2.14	2.68	3.25	4.04	4.68	21	2.08	2.57	3.09	3.81	4.39
22	2.13	2.66	3.22	3.99	4.61	22	2.06	2.55	3.05	3.76	4.32
23	2.11	2.64	3.18	3.94	4.54	23	2.05	2.53	3.02	3.71	4.26
24	2.10	2.62	3.15	3.90	4.49	24	2.04	2.51	2.99	3.67	4.20
25	2.09	2.60	3.13	3.85	4.43	25	2.02	2.49	2.97	3.63	4.15
30	2.05	2.53	3.03	3.70	4.23	30	1.98	2.42	2.87	3.47	3.95
35	2.02	2.49	2.96	3.59	4.09	35	1.95	2.37	2.80	3.37	3.81
40	2.00	2.45	2.90	3.51	3.99	40	1.93	2.34	2.74	3.29	3.71
50	1.97	2.40	2.83	3.41	3.85	50	1.90	2.29	2.67	3.19	3.58
75	1.93	2.34	2.74	3.27	3.67	75	1.85	2.22	2.58	3.05	3.41
100	1.91	2.31	2.70	3.21	3.59	100	1.83	2.19	2.54	2.99	3.33
∞	1.85	2.21	2.57	3.02	3.35	∞	1.77	2.10	2.41	2.80	3.09

Table A.4.1 (*cont.*)

ν_2	$\nu_1=7$					ν_2	$\nu_1=8$				
	α						α				
	0.10	0.05	0.025	0.010	0.005		0.10	0.05	0.025	0.010	0.005
1	58.9	237	948	5929	23715	1	59.4	239	957	5981	23925
2	9.35	19.4	39.4	99.4	199	2	9.37	19.4	39.4	99.4	199
3	5.27	8.89	14.6	27.7	44.4	3	5.25	8.85	14.5	27.5	44.1
4	3.98	6.09	9.07	15.0	21.6	4	3.95	6.04	8.98	14.8	21.4
5	3.37	4.88	6.85	10.5	14.2	5	3.34	4.82	6.76	10.3	14.0
6	3.01	4.21	5.70	8.26	10.8	6	2.98	4.15	5.60	8.10	10.6
7	2.78	3.79	4.99	6.99	8.89	7	2.75	3.73	4.90	6.84	8.68
8	2.62	3.50	4.53	6.18	7.69	8	2.59	3.44	4.43	6.03	7.50
9	2.51	3.29	4.20	5.61	6.89	9	2.47	3.23	4.10	5.47	6.69
10	2.41	3.14	3.95	5.20	6.30	10	2.38	3.07	3.85	5.06	6.12
11	2.34	3.01	3.76	4.89	5.86	11	2.30	2.95	3.66	4.74	5.68
12	2.28	2.91	3.61	4.64	5.52	12	2.24	2.85	3.51	4.50	5.35
13	2.23	2.83	3.48	4.44	5.25	13	2.20	2.77	3.39	4.30	5.08
14	2.19	2.76	3.38	4.28	5.03	14	2.15	2.70	3.29	4.14	4.86
15	2.16	2.71	3.29	4.14	4.85	15	2.12	2.64	3.20	4.00	4.67
16	2.13	2.66	3.22	4.03	4.69	16	2.09	2.59	3.12	3.89	4.52
17	2.10	2.61	3.16	3.93	4.56	17	2.06	2.55	3.06	3.79	4.39
18	2.08	2.58	3.10	3.84	4.44	18	2.04	2.51	3.01	3.71	4.28
19	2.06	2.54	3.05	3.77	4.34	19	2.02	2.48	2.96	3.63	4.18
20	2.04	2.51	3.01	3.70	4.26	20	2.00	2.45	2.91	3.56	4.09
21	2.02	2.49	2.97	3.64	4.18	21	1.98	2.42	2.87	3.51	4.01
22	2.01	2.46	2.93	3.59	4.11	22	1.97	2.40	2.84	3.45	3.94
23	1.99	2.44	2.90	3.54	4.05	23	1.95	2.37	2.81	3.41	3.88
24	1.98	2.42	2.87	3.50	3.99	24	1.94	2.36	2.78	3.36	3.83
25	1.97	2.40	2.85	3.46	3.94	25	1.93	2.34	2.75	3.32	3.78
30	1.93	2.33	2.75	3.30	3.74	30	1.88	2.27	2.65	3.17	3.58
35	1.90	2.29	2.68	3.20	3.61	35	1.85	2.22	2.58	3.07	3.45
40	1.87	2.25	2.62	3.12	3.51	40	1.83	2.18	2.53	2.99	3.35
50	1.84	2.20	2.55	3.02	3.38	50	1.80	2.13	2.46	2.89	3.22
75	1.80	2.13	2.46	2.89	3.21	75	1.75	2.06	2.37	2.76	3.05
100	1.78	2.10	2.42	2.82	3.13	100	1.73	2.03	2.32	2.69	2.97
∞	1.71	2.01	2.29	2.64	2.90	∞	1.67	1.94	2.19	2.51	2.74

Table A.4.1 (*cont.*)

ν_2	$\nu_1=9$					ν_2	$\nu_1=10$				
	α						α				
	0.10	0.05	0.025	0.010	0.005		0.10	0.05	0.025	0.010	0.005
1	59.9	241	963	6023	24092	1	60.2	242	969	6055	24227
2	9.38	19.4	39.4	99.4	199	2	9.39	19.4	39.4	99.4	199
3	5.24	8.81	14.5	27.3	43.9	3	5.23	8.79	14.4	27.2	43.7
4	3.94	6.00	8.90	14.7	21.1	4	3.92	5.96	8.84	14.5	21.0
5	3.32	4.77	6.68	10.2	13.8	5	3.30	4.74	6.62	10.1	13.6
6	2.96	4.10	5.52	7.98	10.4	6	2.94	4.06	5.46	7.87	10.2
7	2.72	3.68	4.82	6.72	8.51	7	2.70	3.64	4.76	6.62	8.38
8	2.56	3.39	4.36	5.91	7.34	8	2.54	3.35	4.30	5.81	7.21
9	2.44	3.18	4.03	5.35	6.54	9	2.42	3.14	3.96	5.26	6.42
10	2.35	3.02	3.78	4.94	5.97	10	2.32	2.98	3.72	4.85	5.85
11	2.27	2.90	3.59	4.63	5.54	11	2.25	2.85	3.53	4.54	5.42
12	2.21	2.80	3.44	4.39	5.20	12	2.19	2.75	3.37	4.30	5.09
13	2.16	2.71	3.31	4.19	4.94	13	2.14	2.67	3.25	4.10	4.82
14	2.12	2.65	3.21	4.03	4.72	14	2.10	2.60	3.15	3.94	4.60
15	2.09	2.59	3.12	3.89	4.54	15	2.06	2.54	3.06	3.80	4.42
16	2.06	2.54	3.05	3.78	4.38	16	2.03	2.49	2.99	3.69	4.27
17	2.03	2.49	2.98	3.68	4.25	17	2.00	2.45	2.92	3.59	4.14
18	2.00	2.46	2.93	3.60	4.14	18	1.98	2.41	2.87	3.51	4.03
19	1.98	2.42	2.88	3.52	4.04	19	1.96	2.38	2.82	3.43	3.93
20	1.96	2.39	2.84	3.46	3.96	20	1.94	2.35	2.77	3.37	3.85
21	1.95	2.37	2.80	3.40	3.88	21	1.92	2.32	2.73	3.31	3.77
22	1.93	2.34	2.76	3.35	3.81	22	1.90	2.30	2.70	3.26	3.70
23	1.92	2.32	2.73	3.30	3.75	23	1.89	2.27	2.67	3.21	3.64
24	1.91	2.30	2.70	3.26	3.69	24	1.88	2.25	2.64	3.17	3.59
25	1.89	2.28	2.68	3.22	3.64	25	1.87	2.24	2.61	3.13	3.54
30	1.85	2.21	2.57	3.07	3.45	30	1.82	2.16	2.51	2.98	3.34
35	1.82	2.16	2.50	2.96	3.32	35	1.79	2.11	2.44	2.88	3.21
40	1.79	2.12	2.45	2.89	3.22	40	1.76	2.08	2.39	2.80	3.12
50	1.76	2.07	2.38	2.78	3.09	50	1.73	2.03	2.32	2.70	2.99
75	1.72	2.01	2.29	2.65	2.93	75	1.69	1.96	2.22	2.57	2.82
100	1.69	1.97	2.24	2.59	2.85	100	1.66	1.93	2.18	2.50	2.74
∞	1.63	1.88	2.11	2.41	2.62	∞	1.60	1.83	2.05	2.32	2.52

Table A.4.1 (*cont.*)

ν_2	$\nu_1=12$					ν_2	$\nu_1=15$				
	α						α				
	0.10	0.05	0.025	0.010	0.005		0.10	0.05	0.025	0.010	0.005
1	60.7	244	977	6107	24428	1	61.2	246	985	6157	24633
2	9.41	19.4	39.4	99.4	199	2	9.42	19.4	39.4	99.4	199
3	5.22	8.74	14.3	27.1	43.4	3	5.20	8.70	14.3	26.9	43.1
4	3.90	5.91	8.75	14.4	20.7	4	3.87	5.86	8.66	14.2	20.4
5	3.27	4.68	6.52	9.89	13.4	5	3.24	4.62	6.43	9.72	13.1
6	2.90	4.00	5.37	7.72	10.0	6	2.87	3.94	5.27	7.56	9.81
7	2.67	3.57	4.67	6.47	8.18	7	2.63	3.51	4.57	6.31	7.97
8	2.50	3.28	4.20	5.67	7.01	8	2.46	3.22	4.10	5.52	6.81
9	2.38	3.07	3.87	5.11	6.23	9	2.34	3.01	3.77	4.96	6.03
10	2.28	2.91	3.62	4.71	5.66	10	2.24	2.85	3.52	4.56	5.47
11	2.21	2.79	3.43	4.40	5.24	11	2.17	2.72	3.33	4.25	5.05
12	2.15	2.69	3.28	4.16	4.91	12	2.10	2.62	3.18	4.01	4.72
13	2.10	2.60	3.15	3.96	4.64	13	2.05	2.53	3.05	3.82	4.46
14	2.05	2.53	3.05	3.80	4.43	14	2.01	2.46	2.95	3.66	4.25
15	2.02	2.48	2.96	3.67	4.25	15	1.97	2.40	2.86	3.52	4.07
16	1.99	2.42	2.89	3.55	4.10	16	1.94	2.35	2.79	3.41	3.92
17	1.96	2.38	2.82	3.46	3.97	17	1.91	2.31	2.72	3.31	3.79
18	1.93	2.34	2.77	3.37	3.86	18	1.89	2.27	2.67	3.23	3.68
19	1.91	2.31	2.72	3.30	3.76	19	1.86	2.23	2.62	3.15	3.59
20	1.89	2.28	2.68	3.23	3.68	20	1.84	2.20	2.57	3.09	3.50
21	1.87	2.25	2.64	3.17	3.60	21	1.83	2.18	2.53	3.03	3.43
22	1.86	2.23	2.60	3.12	3.54	22	1.81	2.15	2.50	2.98	3.36
23	1.84	2.20	2.57	3.07	3.47	23	1.80	2.13	2.47	2.93	3.30
24	1.83	2.18	2.54	3.03	3.42	24	1.78	2.11	2.44	2.89	3.25
25	1.82	2.16	2.51	2.99	3.37	25	1.77	2.09	2.41	2.85	3.20
30	1.77	2.09	2.41	2.84	3.18	30	1.72	2.01	2.31	2.70	3.01
35	1.74	2.04	2.34	2.74	3.05	35	1.69	1.96	2.23	2.60	2.88
40	1.71	2.00	2.29	2.66	2.95	40	1.66	1.92	2.18	2.52	2.78
50	1.68	1.95	2.22	2.56	2.82	50	1.63	1.87	2.11	2.42	2.65
75	1.63	1.88	2.12	2.43	2.66	75	1.58	1.80	2.01	2.29	2.49
100	1.61	1.85	2.08	2.37	2.58	100	1.56	1.77	1.97	2.22	2.41
∞	1.55	1.75	1.95	2.19	2.36	∞	1.49	1.67	1.83	2.04	2.19

Table A.4.1 (*cont.*)

ν_2	$\nu_1=20$					ν_2	$\nu_1=24$				
	α						α				
	0.10	0.05	0.025	0.010	0.005		0.10	0.05	0.025	0.010	0.005
1	61.7	248	993	6209	24838	1	62.0	249	997	6235	24938
2	9.44	19.4	39.4	99.4	199	2	9.45	19.5	39.5	99.5	199
3	5.18	8.66	14.2	26.7	42.8	3	5.18	8.64	14.1	26.6	42.6
4	3.84	5.80	8.56	14.0	20.2	4	3.83	5.77	8.51	13.9	20.0
5	3.21	4.56	6.33	9.55	12.9	5	3.19	4.53	6.28	9.47	12.8
6	2.84	3.87	5.17	7.40	9.59	6	2.82	3.84	5.12	7.31	9.47
7	2.59	3.44	4.47	6.16	7.75	7	2.58	3.41	4.41	6.07	7.65
8	2.42	3.15	4.00	5.36	6.61	8	2.40	3.12	3.95	5.28	6.50
9	2.30	2.94	3.67	4.81	5.83	9	2.28	2.90	3.61	4.73	5.73
10	2.20	2.77	3.42	4.41	5.27	10	2.18	2.74	3.37	4.33	5.17
11	2.12	2.65	3.23	4.10	4.86	11	2.10	2.61	3.17	4.02	4.76
12	2.06	2.54	3.07	3.86	4.53	12	2.04	2.51	3.02	3.78	4.43
13	2.01	2.46	2.95	3.66	4.27	13	1.98	2.42	2.89	3.59	4.17
14	1.96	2.39	2.84	3.51	4.06	14	1.94	2.35	2.79	3.43	3.96
15	1.92	2.33	2.76	3.37	3.88	15	1.90	2.29	2.70	3.29	3.79
16	1.89	2.28	2.68	3.26	3.73	16	1.87	2.24	2.63	3.18	3.64
17	1.86	2.23	2.62	3.16	3.61	17	1.84	2.19	2.56	3.08	3.51
18	1.84	2.19	2.56	3.08	3.50	18	1.81	2.15	2.50	3.00	3.40
19	1.81	2.16	2.51	3.00	3.40	19	1.79	2.11	2.45	2.92	3.31
20	1.79	2.12	2.46	2.94	3.32	20	1.77	2.08	2.41	2.86	3.22
21	1.78	2.10	2.42	2.88	3.24	21	1.75	2.05	2.37	2.80	3.15
22	1.76	2.07	2.39	2.83	3.18	22	1.73	2.03	2.33	2.75	3.08
23	1.74	2.05	2.36	2.78	3.12	23	1.72	2.01	2.30	2.70	3.02
24	1.73	2.03	2.33	2.74	3.06	24	1.70	1.98	2.27	2.66	2.97
25	1.72	2.01	2.30	2.70	3.01	25	1.69	1.96	2.24	2.62	2.92
30	1.67	1.93	2.20	2.55	2.82	30	1.64	1.89	2.14	2.47	2.73
35	1.63	1.88	2.12	2.44	2.69	35	1.60	1.83	2.06	2.36	2.60
40	1.61	1.84	2.07	2.37	2.60	40	1.57	1.79	2.01	2.29	2.50
50	1.57	1.78	1.99	2.27	2.47	50	1.54	1.74	1.93	2.18	2.37
75	1.52	1.71	1.90	2.13	2.31	75	1.49	1.66	1.83	2.05	2.21
100	1.49	1.68	1.85	2.07	2.23	100	1.46	1.63	1.78	1.98	2.13
∞	1.42	1.57	1.71	1.88	2.00	∞	1.38	1.52	1.64	1.79	1.90

Table A.4.1 (*cont.*)

ν_2	$\nu_1=30$					ν_2	$\nu_1=40$				
	α						α				
	0.10	0.05	0.025	0.010	0.005		0.10	0.05	0.025	0.010	0.005
1	62.3	250	1002	6260	25042	1	62.5	251	1006	6287	25146
2	9.46	19.5	39.5	99.5	199	2	9.47	19.5	39.5	99.5	199
3	5.17	8.62	14.1	26.5	42.5	3	5.16	8.59	14.0	26.4	42.3
4	3.82	5.75	8.46	13.8	19.9	4	3.80	5.72	8.41	13.7	19.8
5	3.17	4.50	6.23	9.38	12.7	5	3.16	4.46	6.18	9.29	12.5
6	2.80	3.81	5.07	7.23	9.36	6	2.78	3.77	5.01	7.14	9.24
7	2.56	3.38	4.36	5.99	7.53	7	2.54	3.34	4.31	5.91	7.42
8	2.38	3.08	3.89	5.20	6.40	8	2.36	3.04	3.84	5.12	6.29
9	2.25	2.86	3.56	4.65	5.62	9	2.23	2.83	3.51	4.57	5.52
10	2.16	2.70	3.31	4.25	5.07	10	2.13	2.66	3.26	4.17	4.97
11	2.08	2.57	3.12	3.94	4.65	11	2.05	2.53	3.06	3.86	4.55
12	2.01	2.47	2.96	3.70	4.33	12	1.99	2.43	2.91	3.62	4.23
13	1.96	2.38	2.84	3.51	4.07	13	1.93	2.34	2.78	3.43	3.97
14	1.91	2.31	2.73	3.35	3.86	14	1.89	2.27	2.67	3.27	3.76
15	1.87	2.25	2.64	3.21	3.69	15	1.85	2.20	2.58	3.13	3.58
16	1.84	2.19	2.57	3.10	3.54	16	1.81	2.15	2.51	3.02	3.44
17	1.81	2.15	2.50	3.00	3.41	17	1.78	2.10	2.44	2.92	3.31
18	1.78	2.11	2.44	2.92	3.30	18	1.75	2.06	2.38	2.84	3.20
19	1.76	2.07	2.39	2.84	3.21	19	1.73	2.03	2.33	2.76	3.11
20	1.74	2.04	2.35	2.78	3.12	20	1.71	1.99	2.29	2.69	3.02
21	1.72	2.01	2.31	2.72	3.05	21	1.69	1.96	2.25	2.64	2.95
22	1.70	1.98	2.27	2.67	2.98	22	1.67	1.94	2.21	2.58	2.88
23	1.69	1.96	2.24	2.62	2.92	23	1.66	1.91	2.18	2.54	2.82
24	1.67	1.94	2.21	2.58	2.87	24	1.64	1.89	2.15	2.49	2.77
25	1.66	1.92	2.18	2.54	2.82	25	1.63	1.87	2.12	2.45	2.72
30	1.61	1.84	2.07	2.39	2.63	30	1.57	1.79	2.01	2.30	2.52
35	1.57	1.79	2.00	2.28	2.50	35	1.53	1.74	1.93	2.19	2.39
40	1.54	1.74	1.94	2.20	2.40	40	1.51	1.69	1.88	2.11	2.30
50	1.50	1.69	1.87	2.10	2.27	50	1.46	1.63	1.80	2.01	2.16
75	1.45	1.61	1.76	1.96	2.10	75	1.41	1.55	1.69	1.87	1.99
100	1.42	1.57	1.71	1.89	2.02	100	1.38	1.52	1.64	1.80	1.91
∞	1.34	1.46	1.57	1.70	1.79	∞	1.30	1.39	1.49	1.59	1.67

Table A.4.1 (*cont.*)

ν_2	$\nu_1=60$					ν_2	$\nu_1=120$				
	α						α				
	0.10	0.05	0.025	0.010	0.005		0.10	0.05	0.025	0.010	0.005
1	62.8	252	1010	6312	25254	1	63.1	253	1014	6339	25357
2	9.47	19.5	39.5	99.5	199	2	9.48	19.5	39.5	99.5	199
3	5.15	8.57	14.0	26.3	42.2	3	5.14	8.55	13.9	26.2	42.0
4	3.79	5.69	8.36	13.7	19.6	4	3.78	5.66	8.31	13.6	19.5
5	3.14	4.43	6.12	9.20	12.4	5	3.12	4.40	6.07	9.11	12.3
6	2.76	3.74	4.96	7.06	9.12	6	2.74	3.70	4.90	6.97	9.00
7	2.51	3.30	4.25	5.82	7.31	7	2.49	3.27	4.20	5.74	7.19
8	2.34	3.01	3.78	5.03	6.18	8	2.32	2.97	3.73	4.95	6.06
9	2.21	2.79	3.45	4.48	5.41	9	2.18	2.75	3.39	4.40	5.30
10	2.11	2.62	3.20	4.08	4.86	10	2.08	2.58	3.14	4.00	4.75
11	2.03	2.49	3.00	3.78	4.45	11	2.00	2.45	2.94	3.69	4.34
12	1.96	2.38	2.85	3.54	4.12	12	1.93	2.34	2.79	3.45	4.01
13	1.90	2.30	2.72	3.34	3.87	13	1.88	2.25	2.66	3.25	3.76
14	1.86	2.22	2.61	3.18	3.66	14	1.83	2.18	2.55	3.09	3.55
15	1.82	2.16	2.52	3.05	3.48	15	1.79	2.11	2.46	2.96	3.37
16	1.78	2.11	2.45	2.93	3.33	16	1.75	2.06	2.38	2.84	3.22
17	1.75	2.06	2.38	2.83	3.21	17	1.72	2.01	2.32	2.75	3.10
18	1.72	2.02	2.32	2.75	3.10	18	1.69	1.97	2.26	2.66	2.99
19	1.70	1.98	2.27	2.67	3.00	19	1.67	1.93	2.20	2.58	2.89
20	1.68	1.95	2.22	2.61	2.92	20	1.64	1.90	2.16	2.52	2.81
21	1.66	1.92	2.18	2.55	2.84	21	1.62	1.87	2.11	2.46	2.73
22	1.64	1.89	2.14	2.50	2.77	22	1.60	1.84	2.08	2.40	2.66
23	1.62	1.86	2.11	2.45	2.71	23	1.59	1.81	2.04	2.35	2.60
24	1.61	1.84	2.08	2.40	2.66	24	1.57	1.79	2.01	2.31	2.55
25	1.59	1.82	2.05	2.36	2.61	25	1.56	1.77	1.98	2.27	2.50
30	1.54	1.74	1.94	2.21	2.42	30	1.50	1.68	1.87	2.11	2.30
35	1.50	1.68	1.86	2.10	2.28	35	1.46	1.62	1.79	2.00	2.16
40	1.47	1.64	1.80	2.02	2.18	40	1.42	1.58	1.72	1.92	2.06
50	1.42	1.58	1.72	1.91	2.05	50	1.38	1.51	1.64	1.80	1.93
75	1.37	1.49	1.61	1.76	1.88	75	1.32	1.42	1.52	1.65	1.74
100	1.34	1.45	1.56	1.69	1.79	100	1.28	1.38	1.46	1.57	1.65
∞	1.24	1.32	1.39	1.47	1.53	∞	1.17	1.22	1.27	1.32	1.36

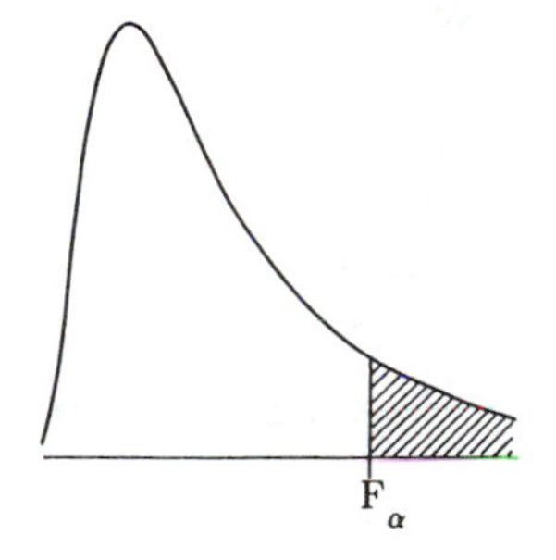

References

Bates, D.M. and Watts, D.G. [1988] *Non-linear Regression Analysis and its Applications.* Wiley, N.Y.

Bishop, Y.M.M., Fienberg, S.E. and Holland, P.W. [1975] *Discrete Multivariate Analysis: Theory and Practice.* MIT Press, Cambridge, Mass.

Box, G.E.P. and Draper, N.R. [1987] *Empirical Model Building and Response Surfaces.* Wiley, N.Y.

Box, G.E.P. and Tiao, G.C. [1973] *Bayesian Inference in Statistical Analysis.* Addison Wesley, Reading, Mass.

Box, G.E.P. and Jenkins, G.M. [1970] *Time Series Analysis, Forecasting and Practice.* Holden Day, San Francisco.

Chatfield, C. and Collins, A.J. [1980] *Introduction to Multivariate Analysis.* Chapman and Hall, London.

Chatfield, C. [1975] *The Analysis of Time Series: Theory and Practice.* Chapman and Hall, London.

Clarke, R.D. [1946] An application of the Poisson distribution. *J. Instit. Actuaries,* **72**, 48.

Cleveland, W.S. [1985] *The Elements of Graphing Data.* Wadsworth, Calif.

Cochran, W.G. [1977] *Sampling Techniques.* 2nd ed. Wiley, N.Y.

Cochran, W.G. and Cox, G.M. [1957] *Experimental Designs.* 2nd ed. Wiley, N.Y.

Conover, W.J. [1980] *Practical Nonparametric Statistics.* 2nd ed. Wiley, New York.

Darroch, J.N. [1962] Interactions in multi-factor contingency tables. *J. R. Statist. Soc. B,* **24**, 251–263.

Draper, N.R. and Smith, H. [1981] *Applied Regression Analysis.* 2nd ed. Wiley, New York.

Duncan, O.D. [1975] *Introduction to Structural Equation Models.* Academic Press, N.Y.

Fienberg, S.E. [1978] *The Analysis of Cross-Classified Categorical Data.* MIT Press, Cambridge, Mass.

Fleiss, J.L. [1981] *Statistical Methods for Rates and Proportions.* 2nd ed. Wiley, N.Y.

Fox, J. [1984] *Linear Statistical Models and Related Methods.* Wiley, N.Y.

Freedman, D.A. [1983] A note on screening regression equations. *Amer. Statist.*, **37**, 147–151.

Goodman, L.A. and Kruskal, W.H. [1979] *Measures of Association for Cross Classification.* Springer-Verlag, N.Y.

Gordon, H.D. [1981] *Classification: Methods for Exploratory Data Analysis.* Chapman and Hall, London.

Gottman, J.M. [1981] *Time Series Analysis: A Comprehensive Introduction for Social Scientists.* Cambridge University Press, Cambridge.

Harman, H.H. [1976] *Modern Factor Analysis.* 3rd ed. Univ. of Chicago Press, Chicago.

Hocking, R.R. [1976] The analysis and selection of variables in linear regression. *Biometrics,* **32**, 1–49.

Hollander, M. and Wolfe, D.A. [1973] *Nonparametric Statistical Methods.* Wiley, N.Y.

Jenkins, G.M. and Watts, D.G. [1968] *Spectral Analysis and its Applications.* Holden Day, San Francisco.

Jensen, A.R. [1980] *Bias in Mental Testing.* Free Press, N.Y.

Jolliffe, I.T. [1986] *Principal Component Analysis.* Springer-Verlag, N.Y.

Kang, K.M. and Seneta, E. [1980] Path analysis: an exposition. *Developments in Statistics,* (P.R. Krishnaiagh, ed.) **3**, 217–246.

Kendall, M.G. [1973] *Time Series.* Griffin, London.

Kendall, M.G. [1955] *Rank Correlation Methods.* Hafner, N.Y.

McCullagh, P. and Nelder, J.A. [1983] *Generalized Linear Models.* Chapman and Hall, London.

McNeil, D.R. [1977] *Interactive Data Analysis: A Practical Primer.* Wiley, N.Y.

Mood, A.M. [1971] Partitioning variance in multiple regression analysis as a tool for developing learning models. *Amer. Educational Research J.,* **8**, 191–202.

Nelson, C.R. [1973] *Applied Time Series for Managerial Forecasting.* Holden Day, San Francisco.

Neter, J., Wasserman, N. and Whitmore, G.A. [1988] *Applied Statistics.* 3rd ed. Allyn and Bacon, Mass.

Plackett, R.L. [1981] *The Analysis of Categorical Data.* 2nd ed. Griffin, London.

Pratt, J.W. [1964] Robustness of some procedures for the two sample location problem, *J. Amer. Statist. Assoc.,* **59**, 665–680.

Ratkowsky, D.A. [1983] *Non-linear Regression Modeling.* Marcel Dekker, N.Y.

Schiffman, S.S., Reynolds, M.L. and Young, F.W. [1981] *Introduction to Multivariate Scaling.* Academic Press, N.Y.

Searle, S.S. [1971] *Linear Models.* Wiley, N.Y.

Siegel, S. [1956] *Non-Parametric Statistics.* McGraw-Hill, Tokyo.

Snedecor, G.W. and Cochran, W.G. [1980] *Statistical Methods.* 7th ed. Iowa State Univ. Press, Ames, Iowa.

Tasmanian Year Book [1984] Commonwealth Bureau of Statistics, Hobart Office.

Theil, H. [1971] *Principles of Econometrics.* Wiley, N.Y.

Tukey, J.W. [1977] *Exploratory Data Analysis.* Addison Wesley, Reading, Mass.

Tukey, J.W. [1954] Causation, Regression and Path Analysis, in Kempthorne, O., Bancroft, T.A., Gowers, J. and Lush, J.T. (eds.) *Statistics and Mathematics in Biology.* Chap. 3, Iowa State Univ. Press, Ames, Iowa.

Williams, E.J. [1959] *Regression Analysis.* Wiley, N.Y.

Winer, B.J. [1971] *Statistical Principles in Experimental Design.* McGraw-Hill, N.Y.

Wright, S. [1923] The theory of path coefficients: a reply to Nile's criticism. *Genetics,* **8**, 239–255.

Figure, Table, and Example Index

Chapter 6

Chapter 7

Chapter 9

Chapter 10

Chapter 11

Chapter 12

Chapter 13

Chapter 14

Chapter 15

Chapter 16

Chapter 16 (*cont.*)

Chapter 17

Chapter 18

Chapter 19

Chapter 20

Chapter 21

Index

Springer Texts in Statistics *(continued from p. ii)*

Pfeiffer Probability for Applications

Santner and Duffy The Statistical Analysis of Discrete Data